SOIL MECHANICS

Principles and Applications

WILLIAM H. PERLOFF, Ph.D.

FORMERLY PROFESSOR OF SOIL MECHANICS
PURDUE UNIVERSITY

WILLIAM BARON, Ph.D.

ASSOCIATE PROFESSOR OF CIVIL ENGINEERING
CLEMSON UNIVERSITY

John Wiley & Sons
NEW YORK • CHICHESTER • BRISBANE • TORONTO

ISBN 0 471 06671-0

Library of Congress Catalog Card Number: 74–22543

PRINTED IN THE UNITED STATES OF AMERICA

10 9 8 7 6 5

Preface

This text developed out of our concern to enliven the interest with which our undergraduate civil engineering students addressed the study of principles and applications of soil mechanics. What we had in mind was to introduce a series of case histories, and in turn introduce and discuss the subject matter as it applies in the solution of the problems these pose. From the start students have before them the relation of the course material to real-world concerns and an enhanced reason to learn growing from their immediate view of the course as more than a series of assignments. This approach leads to a sequence of coverage which, though it departs from the more traditional, has proven to be quite satisfactory.

The case studies were selected with two main criteria in mind: (1) a case should involve one or more principles that can be investigated by the students, and (2) where possible the case should be presented as a reprint of a published article written at the technical level appropriate for third-year and fourth-year courses. While the articles provide a firm foundation for the subject material, it is fully expected that an instructor will want to amplify his presentation with examples from his own experience.

Rather than overburden students with an exhaustive coverage of soil mechanics and foundation engineering, we have selected certain subject areas to cover thoroughly, concentrating on the areas for which the design process offers a rational basis. We believe that this approach adds coherence to the treatment, and has certain pedagogical merits beyond the specific topics discussed.

Our experience with the more than three hundred students who have used the text in note form in our classes at Purdue University has led us to the view that the case study orientation, combined with the use of numerous additional examples, has produced an increased interest and desire to learn as well as improved understanding.

We are grateful to the many colleagues and students who have contributed to this text. We especially appreciate the thoughtful critical comments offered by Professors G. A. Leonards and M. E. Harr of Purdue University, and H. E. Wahls of North Carolina State University. The illustrations were prepared by Emile Y. Baladi and Gilbert Y. Baladi, and the text was typed by Gayle Pribie and Janice Wait; without their help, the book would never have been completed.

WILLIAM H. PERLOFF
WILLIAM BARON

January, 1976

Contents

1   Introduction   **1**

   1–1   Use of Case Studies   1
   1–2   Introductory Case Studies   1
   1–3   Elementary Definitions   17

2   Mechanical Behavior of Cohesionless Soils   **26**

   2–1   Introduction   26
   2–2   Failure of Cohesionless Soil   27
   2–3   Phase Relations   39
   2–4   Grain Size and Grain-Size Distribution   44
   2–5   Stress, Strain, and Strength   51
   2–6   Fort Peck Slide Reconsidered   80
   2–7   Summary of Key Points   84

3   Elements of Stability Analysis for Cohesionless Soils   **92**

   3–1   Introduction   92
   3–2   Rankine Earth Pressure Theory   100
   3–3   Earth Pressure at Rest   124
   3–4   Generalized Limiting Equilibrium Analysis   127
   3–5   Wedge Analysis—Coulomb Theory   128
   3–6   Reconsideration of Crib Wall Failure   149
   3–7   Summary of Key Points   154

4   Stresses Within an Earth Mass   **160**

   4–1   Introduction   160
   4–2   Stress   167
   4–3   Displacement and Strain   168

4–4 Constitutive Relations   173
4–5 Solution to Linear-Elastic Problems   178
4–6 Application to Building 301   191
4–7 Summary of Key Points   191

5 Compressibility and Settlement of Cohesive Soils   196

5–1 Introduction   196
5–2 Immediate Settlement   198
5–3 Consolidation Settlement—One-Dimensional Case   212
5–4 Application to Building 301   245
5–5 Summary of Key Points   252

6 Fundamentals of Fluid Flow Through Porous Media   258

6–1 Introduction   258
6–2 Conservation of Mass   259
6–3 Free Energy   262
6–4 Capillarity   270
6–5 Effect of Capillarity on Soil Behavior   277
6–6 Darcy's Law—Conservation of Linear Momentum   283
6–7 The General Diffusion Equation   300
6–8 Summary of Key Points   301

7 Consolidation and Time Rate of Settlement   305

7–1 Introduction   305
7–2 One-Dimensional Consolidation   306
7–3 Numerical Solution   309
7–4 Degree of Consolidation   319
7–5 Results from Analytical Solution   328
7–6 Laboratory Determination of the Coefficient of
     Consolidation   339
7–7 Application to Building 301   347
7–8 Summary of Key Points   350

8 Shearing Resistance of Cohesive Soils   354

8–1 Introduction   354
8–2 Hvorslev Hypothesis   355
8–3 Pore Water Pressures   359

8-4   Normally Consolidated Clays   363
8-5   Over-Consolidated Clays   369
8-6   Unconsolidated-Undrained Test   373
8-7   Relationship Between Stresses Imposed in the Field
        and Those Applied During Laboratory Testing   380
8-8   Summary of Key Points   387

9   Structure of Cohesive Soils   **392**

9-1   Introduction   392
9-2   Clay Particles—Preliminary Considerations   396
9-3   Clay Particles as Colloids   398
9-4   Ion Exchange   403
9-5   Structural Arrangement of Clay Particles   405
9-6   Observed Features of Clay Behavior   411
9-7   Effect of Water on Clays   416
9-8   Soil Classification   426
9-9   Summary of Key Points   427

10   Shallow Foundations, Bearing Capacity   **436**

10-1   Overview of Problems Facing the Foundation
         Engineer   436
10-2   Method of Approach of Foundation Designer   438
10-3   Types of Foundations   442
10-4   Bearing Capacity of Shallow Foundations   444
10-5   Reconsideration of Transcona Grain Elevator
         Failure   470
10-6   Factor of Safety   473
10-7   Importance of Proper Subsurface Exploration   473
10-8   Summary of Key Points   474

11   Shallow Foundations; Settlements   **479**

11-1   Introduction   479
11-2   Effect of Size on Settlements of a Loaded Area   482
11-3   Distortion Settlement of Footings on Sand   489
11-4   Three-Dimensional Effects on Compression
         Settlements of Cohesive Soils   504
11-5   Three-Dimensional Drainage Effects   507

11–6   Settlements Due to Secondary Compression   513
11–7   Effect of Quasi-Preconsolidation   514
11–8   Effect of Settlement on Structures   516
11–9   Reduction of Detrimental Settlements   517
11–10 Evaluation of Preloading at Building 301   521
11–11 Summary of Key Points   522

12   Stability of Slopes   528

12–1   Introduction   528
12–2   Varieties of Slope Failures   528
12–3   Processes Responsible for Slope Failures   538
12–4   Assumptions of Stability Analysis   538
12–5   Stability of Infinite Slope of Cohesionless Soil 542
12–6   Stability of Slopes in Purely Cohesive Soils   543
12–7   Stability of Slopes in Homogeneous Soils
         Possessing Friction   548
12–8   Application of Methods of Slices to Stability
         Analysis   556
12–9   Composite Slip Surfaces   563
12–10 Maximum Shear Stress Method   565
12–11 Selection of Strength Parameters for Stability
         Analysis   567
12–12 Stability of Tip No. 7 at Aberfan, Wales   579
12–13 Summary of Key Points   581

13   Elements of Design of Rigid Retaining Structures   587

13–1   Introduction   587
13–2   Types of Retaining Structures   588
13–3   Magnitude and Distribution of Earth Pressure
         Forces on Rigid Retaining Structures   590
13–4   Stability of Retaining Walls   610
13–5   Other Design Considerations   615
13–6   Summary of Key Points   616

14   Steady-State Seepage and Its Effects   619

14–1   Introduction   619
14–2   Review of Assumptions and Concepts   620
14–3   LaPlace's Equation   622
14–4   Boundary Conditions   627

14–5   Solution Procedures   629
14–6   Construction of the Flow Net   637
14–7   Application of Results   660
14–8   Summary of Key Points   665

ᴀppendix   Influence Diagrams for Stresses
           in Elastic Media   671

A.1   Stresses Due to Surface Loads   671
A.2   Stresses Due to Self-Weight   681

References Cited   718

Index   735

SOIL
MECHANICS
Principles and Applications

1

Introduction

1.1 USE OF CASE STUDIES

Throughout this book we describe and discuss a variety of actual geo-technical engineering projects and events related to their construction. These cases studies are presented in the form of reprints of articles written about the projects near the time the events occurred. Some of the articles pay tribute to the skill and ingenuity of engineers responsible for a successful innovative design. Others record catastrophic failures, whose only virtues are the lessons to be learned from their study. Common to all these projects, however, is the fact that we cannot understand the salient features of the situation without first learning certain pertinent concepts. Thus the case histories challenge us to understand the principles underlying these phe-nomena, principles which are the major subject matter of this text.

1.2 INTRODUCTORY CASE STUDIES

The three articles reprinted in this chapter introduce certain problems that geotechnical engineers have faced, not always with success. Although the articles do little more than suggest the scope and range of such problems, they do illustrate important aspects of one of the civil engineer's key func-tions: the design of structures on, of, and in the earth. The three projects are the subjects of which these articles provide points of departure for dis-cussing certain of these aspects.

Golden Gateway Development

During the early 1960's, 45 acres of San Francisco's old produce market area was converted into a handsome residential and business development. Almost adjacent to San Francisco Bay, the area originally contained narrow crowded streets, rotting buildings (and produce), rats—and little promise for the future. Through the vision of the San Francisco Redevelopment Agency, this has been replaced by an attractive and functional combination

1

of high- and low-rise structures; the result is an innovative urban development. The article on pages 4 to 9, reprinted from *Civil Engineering* of
January 1964, describes the design of the novel foundation used to support
these structures.

Building such foundations in the San Francisco Bay area is expensive
because, to a depth of 70 to 140 feet below the ground surface, the earth
materials are too weak and compressible to support structures of the size
involved in this project. Thus it was necessary to bypass these unsuitable
materials and transmit the weight of the structures to the bedrock underlying them. Bypassing unsuitable subsoils by means of a deep foundation,
in order to gain support from stronger materials at greater depths, is hardly
new: Julius Caesar had wood piling driven to support a bridge across the
Rhine. However, the manner in which the type of caisson described in the
article is designed to transmit loads to the underlying rock is a recent
development.

This example illustrates several important points:

1. In contrast to many other types of engineers who can specify the
   characteristics of the materials used in their designs, the foundation
   engineer usually needs to adjust his design to accommodate the
   natural materials at the site. Thus investigation of the nature and
   distribution of subsurface soils at a site is an important part of the
   design process.
2. Strength and compressibility of the subsurface materials are major
   factors in determining the foundation type that represents the best
   combination of technical suitability and economy.
3. Innovative design that exploits the peculiarities of the site conditions
   can effect significant economies. The type of innovation described
   often warrants, and requires, full-scale field testing to verify its
   efficacy.

Electroosmotic Stabilization of Clay Soils

The West Branch Dam is an 80-foot-high compacted earthfill dam on the
west branch of the Mahoning River in Ohio. Its construction was distinguished by a partial failure within one of the soil strata underlying the dam,
and the method by which a complete failure was arrested and the offending
material strengthened to the extent that the dam could be completed safely.
This is described in the article on pages 10 to 16, originally published in
Engineering News-Record of June 23, 1966.

The development of large water pressures within the clay soil underlying
the dam, which led to the failure, had not been anticipated. It would have
taken as much as ten years to create, through dissipation of water pressure
by normal flow processes, an increase in the strength of the clay sufficient
to permit completion of the dam.

An understanding of the electrochemical forces that play a major role in the behavior of earth materials like the clay described in the article permitted a dramatic hastening of this process by application of *electroosmosis*. In Chapter 9 we shall consider these basic electrochemical forces, and the way in which electroosmosis can increase strength of clay soils.

Several other features of the article are noteworthy:

1. Earth materials are used by civil engineers in at least two ways: first, as the foundation material from which structures must derive their support; and second, as a construction material. In fact, for many purposes a *soil* structure (dam, embankment, levee, etc.) is less expensive than any other type yet devised.

2. Important characteristics of soils, including strength and compressibility, change with time. Such changes can be influenced by works of man, either beneficially or, as described below, adversely.

3. A useful tool for safe and economical construction in circumstances where the consequences of the construction cannot be predicted with the necessary accuracy is the observational method. This involves monitoring important parameters in the field during construction to provide information required to adjust the construction rate or otherwise modify a preliminary design.

Vaiont Reservoir Disaster

At 875 feet, Vaiont Dam is the world's second-highest dam. Its construction in 1960 eventuated in one of the most catastrophic failures of all times. The site of the dam and the reservoir behind it is underlain by a limestone layer containing many seams of soft soils, and by open solution channels formed by water flowing through the rocks. This formation had been marginally stable for some time, as evidenced by the continuous creep of the rock slopes into the Vaiont Canyon and the frequent rock and earth slides in the area. The increased groundwater elevation resulting from the reservoir created by the dam appears to have upset a very delicate balance which existed between the strength of the rock mass and the internal stresses.

After a significant increase in rate of creep of the rock slopes toward the reservoir, one side of the canyon failed completely, causing more than 300 million cubic yards of rock and soil to slide into the reservoir within a period of 15 to 30 seconds. This completely filled the reservoir for a distance of over $1\frac{1}{2}$ miles upstream from the dam and generated an enormous wave of water that swept over the dam to a height of some 328 feet above the crest. The wave raced down the valley destroying everything in its path, and was still over 230 feet high one mile downstream. More than 3000 lives were lost. Remarkably, the dam itself survived the unexpected forces created by

Foundation Design

WILLIAM W. MOORE, F. ASCE
Senior Consulting Partner
Dames & Moore, San Francisco, Calif.

The site of the Golden Gateway Project was reclaimed from San Francisco Bay by man-made filling operations. These operations were started in Gold Rush times, about 1849, and in the next 20 years the shoreline was rapidly extended into the mud flats. In some instances, old ships, piers, and dolphins were abandoned and covered over. The fill loads have caused a subsidence of several feet over the past one hundred years, principally due to the consolidation of the soft underlying soils. The rate of subsidence in the Gateway area has gradually decreased in recent years to a maximum of about 0.1 in. per year.

The site is generally underlain by about 20 ft of sand and rubble fill below street grades. These fills are in turn underlain by about 20 to 70 ft of soft plastic marine clays, locally known as "bay mud." In parts of the area, the bay mud is underlain by alternating layers of stiffer clays, sandy clays, and occasional layers of sand. In other parts of the site, the deposits of bay mud extend almost to the top of the bedrock formation.

Usually there is a thin layer of residual soil composed of sandy clay with rock fragments or weathered rock immediately above bedrock. The depth to rock in the area of current construction varies from 70 to 140 ft below street grade. The bedrock formation at this location consists of highly fractured and unevenly weathered sandstones and shales of the Franciscan series. See Fig. 1 for a typical soil profile.

In some areas, firm sands or sandy clay soils suitable for foundation support are found, but they are often underlain by more compressible clays above the rock. Thus the bedrock is the only material underlying the entire site that can support high loads with small settlements.

The foundation problems encountered were due to a number of factors. The soft and compressible bay mud requires deep foundations for major structures. Unfortunately, the bedrock is both deep and extremely variable in strength because it contains crushed and unevenly weathered zones. Of course the presence of a high water table—within 10 ft of street grade—and the soft plastic soils complicate the installation of foundations to rock.

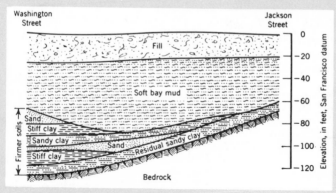

FIG. 1. Typical soil conditions across the Golden Gateway Project. "Fill" is mostly sand with broken concrete and rubble. "Soft bay mud" is silty clay with a moisture content of 45 percent and a dry weight of 75 pcf. "Bedrock" is fractured and weathered shale and sandstone.

4

Types of foundations considered

During the preliminary foundation investigations, the idea of using shallow spread or mat foundations for the high-rise buildings was quickly discarded because of the low strength and high compressibility of the bay mud. For the garage structures, some consideration was given to the use of mat foundations, which would impose loads only slightly greater than the weight of the soil to be excavated. However, mat foundations were discarded even for the garage structures because of the probable differences in settlement between these structures and the adjacent high-rise structures, and also because fairly wide column spacings were planned for the garages. Such spacing would result in column loads large enough to make a mat foundation relatively expensive.

To support the high-rise buildings, the most obvious and conventional type of support would be driven piling. In most of the current construction area, such piles would have to be driven to rock because of the limited thickness of firm soil below the bay mud. Studies of several types of concrete and steel bearing piles indicated that piles of fairly high capacity, in the range of 90 to 150 tons per pile, would offer the most desirable combination of economy and satisfactory behavior of the supported structures. The probability of differential settlement between piles supported on bedrock and others supported by skin friction in the soils above bedrock led to the decision that all piling should gain support from bedrock.

Consideration was given to the use of high-capacity caissons bearing on the rock formation to support column loads of 500 to 2,000 kips in the high-rise structures. Previous experience with caisson foundations in this area indicated that considerable difficulty might be encountered because of weak zones in the bedrock formation. When caissons are designed for end-bearing on this variable bedrock formation, it is necessary either to limit the bearing pressures to low values, on the order of 10,000 psf, or to be sure each caisson is founded on an adequate thickness of hard rock, which requires a test hole beneath each caisson to a depth of one or two base diameters. Such holes have disclosed crushed clayey or gouge-type material in the rock below the planned caisson base, which then requires that the caisson be deepened. On the Gateway site, it was concluded that the problems of field inspection and construction would be extremely difficult since the shafts could probably not be dewatered for inspection or for a satisfactory cleanout of the bottom bearing area without installing expensive air locks to complete the work as pneumatic caissons.

The solution developed was to use high-capacity friction caissons gaining their support from skin friction or bond between the poured concrete shaft and the rock formation. The advantages of such friction caissons are that it is not necessary to completely clean out the bottom bearing area, and that the effects of weaker zones in the rock formation can be compensated for by simply extending the depth of the straight shaft into the rock formation. Concrete can be placed by tremie methods. The feasibility and practicability of high-capacity friction caissons is largely the result of the development of high-powered equipment capable of excavating large-diameter holes into rather hard materials.

Final plans for the foundations provided for alternate bids on pile foundations and on high-capacity Drilled-In caissons. The piles could be 12-in. metal-cased concrete, 14-in. prestressed concrete, or 14-in. H-piles generally driven into the rock formation. Pile design loads were 120 tons for the high-rise buildings and 100 tons for the garages. Other combinations of bearing piles or caissons that would provide equivalent support for the structures on the underlying bedrock formation were invited and several were received. After study, high-capacity skin-friction caissons were selected. See Fig. 2.

Initial planning for the caissons was based on a design skin-friction stress

Test caisson was loaded to demonstrate its feasibility and verify capacity that could be developed by skin friction or bond between the tremie-placed concrete and the bedrock formation.

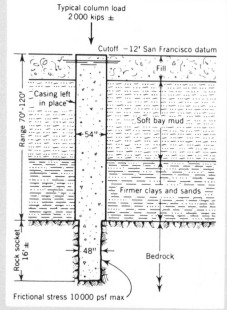

Typical column load
2 000 kips ±

Cutoff −12' San Francisco datum

Fill

Casing left in place

Range 70'-120'

Soft bay mud

54"

Firmer clays and sands

Rock socket 16' ±

48"

Bedrock

Frictional stress 10 000 psf max

▲ FIG. 2. Typical caisson (not to scale)

or shear-bond of 5,000 psf between the concrete and the rock formation, arrived at by an evaluation of laboratory data from rock samples that had been tested. For the highly variable and fractured rock formation, the laboratory-determined skin-friction value would be subject to many indeterminate factors. There was reason to believe from the test data that skin-friction values as high as 10,000 psf might possibly be used if confirmed by a full-size test on an actual caisson installation.

The test caisson consisted of a shaft of 24-in. diameter from the ground surface to bedrock, at a depth of about 85 ft. Above bedrock, an outer casing of 30-in. diameter was added in order to isolate any frictional support from the upper soils. Below the 85-ft depth, the shaft was 20 in. in diameter and penetrated 10 ft into the bedrock formation. A specially constructed plug was placed in the bottom 18 in. of the shaft to prevent the development of end-bearing during the test. See Fig. 3.

The isolation plug was so designed that it would support the wet weight of the concrete placed in the caisson but would collapse if loads appreciably in excess of that weight were applied. Three Carlson stress meters were installed on top of the plug to measure any end-bearing that might develop. (Measurements taken during the test indicated that the end-bearing never amounted to more than a total

6

Caissons were drilled by augers without the use of a casing. Full-length casing was installed when the shaft reached bedrock.

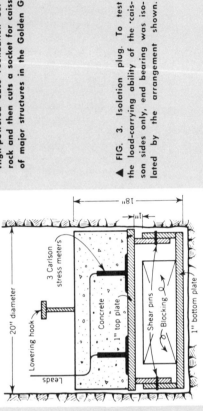

High-powered Case Foundation Co. auger drills to rock and then cuts a socket for caissons for support of major structures in the Golden Gateway Project.

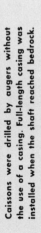

▲ FIG. 3. Isolation plug. To test the load-carrying ability of the caisson sides only, end bearing was isolated by the arrangement shown.

20' diameter

3 Carlson stress meters

Lowering hook

Leads

Concrete

1" top plate

18"

1"

Shear pins

Blocking

1" bottom plate

of about 30 tons.) End-bearing that might develop in the actual caissons was not considered in the design because of uncertainties as to cleaning and bearing.

For the test caisson, skin-frictional support was available to the caisson on a cylindrical surface 20 in. in diameter and 8.5 ft long, which gave a side area for test purposes of 44 sq ft. Concrete of 5,000 psi, with a slump of 6 in., was placed in the caisson by the tremie method. Test loads were applied by hydraulic jacks pushing up against a reaction frame held down by four tension caissons.

Loads were applied in increments of 25 or 50 tons, and deflections were measured by micrometer dial gages attached to independent reference beams. Loads were maintained until the rate of deflection had dropped to about 0.01 in. per hour. At intervals, the load was removed and reapplied to evaluate the behavior of the caisson under repetitive loading.

The total deflection under a 500-ton load was 1¼ in. (Fig. 4). About ½ in. was due to elastic compression in the caisson itself. After rebound, the total permanent deflection was about ⅝ in. Under this 500-ton test load, the skin-friction stress imposed on the test section was slightly more than 20,000 psf. No failure developed within the capacity of the loading system. In view of these results, the use of a design skin-frictional stress up to

10,000 psf was recommended and was approved by the City of San Francisco Building Department. The use of this stress was contingent upon a very careful inspection of the character of the rock penetrated in the project caissons, to make sure that the rock was equivalent to that at the location of the test caisson. Where variations in the character and strength of the rock were noted by inspection, the penetration into the rock of the actual caissons was modified as required.

Construction inspection

During the installation of the caissons, continuous inspection was provided to verify the elevation of the top of the rock formation, to classify the rock according to strength, and to confirm or modify the planned penetration into the rock. The rock types penetrated were classified from cuttings brought to the surface during the drilling operation and from the behavior of the drilling tools and rate of drill penetration.

Caisson drilling was done by high-powered augers operated by the Case Foundation Company. These augers drilled the hole from the surface to the rock formation without the introduction of any casing, or the removal of cuttings except those that stuck to the auger. Thus the entire hole was filled with a slurry made up of the soil

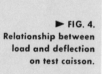

► FIG. 4.
Relationship between load and deflection on test caisson.

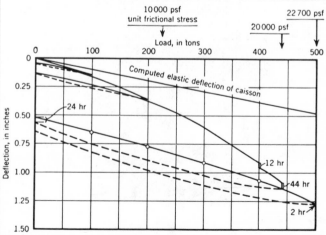

cuttings and ground water. The clayey nature of the bay muds produced an effective driller's mud which kept the holes from caving.

After the shaft had reached bedrock, a full-length casing was lowered and seated on the weathered rock formation. Then the augers drilled the rock socket to the depth required by the type of rock and the caisson loads. When this drilling had been completed, the hole would often be filled, to depths of as much as 100 ft, with remolded plastic solids mixed with rock fragments, sand, and other materials. It was necessary to remove most of these solid materials from the caisson and to clean the walls of the rock socket before placing concrete.

During this cleaning operation, it was particularly important to be sure that plastic material was not smeared on the walls of the rock socket in such a way as to reduce the bond or skin friction. To clean the rock socket, scraper teeth were extended from the outer sides of the cleanout buckets to scrape the walls. In some instances a water jet was directed against the walls of the socket. Experimentally, some rock sockets were drilled a few inches smaller than the planned final diameter and reamed after all the cuttings had been cleaned out, just ahead of concreting.

In addition to removing the cuttings from the caisson, it was necessary that the fluid in the caisson be thin enough not to interfere with the proper placement of the tremie concrete. Fluid at the bottom of the caisson was obtained by a "thief" sampler for visual verification of the cleaning and thinning. At first the fluid was thinned to the consistency of a heavy cream or buttermilk. Later the criterion was established that the density of the fluid in the rock socket would not be greater than 85 pcf.

The difficulties encountered in installing the caissons at the Gateway site was due partly to the presence of old wood piling, which hampered drilling in some places. But the main problems were not as great as originally related to the difficult soil, rock, and water conditions and the need to develop new techniques for satisfactory caisson installation at relatively great depths. Despite these problems, the caissons have been satisfactorily completed for all the highrise buildings in Phase I. It is interesting to note that similar types of high-capacity friction caissons have been satisfactorily installed on two other projects in San Francisco since the Gateway Project was started.

It was intended to use caisson foundations also for the garage columns and some of these were installed. However, because the time and cost advantages of such caissons for the garages were not as great as originally expected, cast-in-place concrete piling was driven for most of the garage area. The 100-ton piles were driven to the rock formation primarily to restrict differential settlement between the caisson and the pile foundations.

The use of high-capacity friction caissons on this project constitutes a very interesting development in methods of foundation construction. Such a design is not known to have been previously utilized elsewhere. Several aspects of the drilled-caisson design are attractive from an engineering and construction standpoint. It is believed that the development of high-powered, high-capacity drilling equipment and improved installation procedures will make similar caisson foundations more feasible and more attractive economically in the future.

POWER to develop electric potential between hundreds of anodes and cathodes, piercing dam (background), comes from 12 diesel-powered generators with total capacity of 2,600 kw.

For the past 10 months the Pittsburgh District of the U. S. Army Corps of Engineers has been doing something shocking to a relatively small, nearly complete earthfill dam in Ohio. But West Branch Dam deserved it.

In the fall of 1964, the 80-ft-high embankment on the West Branch of the Mahoning River gave the Corps quite a shock. With nearly 9,000 ft of embankment all topped out and the last 1,000-ft section near the center almost completed, part of the dam started to move (ENR 8/19/65 p. 14).

The Corps successfully stopped the movement, diagnosed a foundation problem, and solved it by consolidating the clay under the dam through the application of electro-osmosis to the soil.

Early this month, earthmoving equipment moved back onto the dam to top it out by this October. And the Pittsburgh district engineer, Col. James E. Hammer, is so happy about how things worked out that the district is holding a seminar this week to present a West Branch case history to interested Army Engineers from around the country.

• **The first ugly clue**—Like any fresh embankment, West Branch settled some. But when resident engineer Rudy Croft saw evidence of lateral movement in the construction joints of the con-

Electro-Osmosis Stabilizes Earth Dam's Tricky Foundation Clay

crete outlet conduits running through the dam, he sounded the alarm.

As soon as survey crews confirmed that things were in fact moving, the Corps had the contractor strip 12 ft of fill off 700 ft of the embankment. Settlement went back to normal and lateral movement stopped.

The dam's foundation in the central 2,000 ft, which also is where it reaches its maximum height, consists in descending order of 20 to 30 ft of alluvial silty sands; 50 to 60 ft of grey silty clay; 10 ft of irregularly layered silt, sands and plastic glacial till; and other compact till layers above rock.

The Corps established the founda-

tion clay's characteristics during design of the structure, and construction plans called for wide berms both up and downstream in the central section (see plan, p. 40). The upstream berm is 150 ft wide, downstream it's 180 ft wide.

● **Clay the culprit**—Once movement and abnormal settlement of the embankment was detected, the Corps immediately began an elaborate program of sampling and piezometer installation.

The sampling program pretty well confirmed design data, but the piezometers in the clay showed high pore water pressures, which had a very slow rate of decline. Some readings showed

ELECTRODES, 20 ft c-c, line surface of upstream berm. Water pipes service eductors, buried in sand columns, that pump as cathode nearby attracts water trapped in clay foundation.

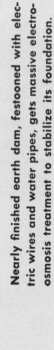

Nearly finished earth dam, festooned with electric wires and water pipes, gets massive electro-osmosis treatment to stabilize its foundation.

pressure heads that were as high as 30 to 40 ft above the top of the dam and the berm.

(Pore pressure in soil, clay in this case, is the pressure exerted by water trapped in the interstices between minute particles of the soil. Under normal conditions, there is little migration of water from the soil, but under extreme pressure, such as the load exerted by an earthfill dam, the water is forced out causing settlement.)

The problem that can develop—and did at West Branch—is that these high pore pressures do not dissipate quickly enough, hence normal settlement of the fill is delayed. Top soils men at the Chief of Engineers' office in Washington, estimate that the pore pressures developed at West Branch might have taken 10 years to dissipate if no remedial measures had been taken.

When such high pressures develop, clay exhibits a strength modified by its failure to consolidate under load.

Since this "strength" is not developed through contact between soil particles, the condition is actually one of weakness. The soil tends to lose some of its shear strength. Under load, the weakened soil moves, like a sandwich with too much mustard. At West Branch, the foundation clay layer was on the verge of moving.

The embankment settled as much as 2 ft within the 1,000-ft, second-stage construction area. Some cracks developed parallel to the axis of the dam, the longest running nearly 700 ft.

The extent of horizontal movement was most pronounced near the axis of the dam, where one construction joint in the intake conduit opened up 9 in. Fortunately, design of the conduit called for a 4-ft concrete collar to be cast around each construction joint, so despite the movement, the conduit remained intact.

The few piezometers installed in the embankment itself, mainly to check the structure after the reservoir is filled, didn't provide a clue to the trouble, but once engineers put in about 80 instruments at various levels in the foundation clay, the nature of the trouble became all too clear.

● **Experts brought in**—With a serious foundation failure on its hands, the Corps called in Leo Casagrande, professor of foundation engineering at Harvard, as a special consultant.

Mr. Casagrande recommended that they eliminate the excessive pore pressures and consolidate the foundation clay under West Branch Dam's ailing midsection with electro-osmosis.

Electro-osmotic consolidation of soil involves a remarkable, but not new, process in which a direct current electric potential set up in the soil by means of electrodes induces the flow of water in the pores of even fine-grained clay toward the negative electrode, or cathode.

Professor Casagrande pioneered in the application of electro-osmosis, a phenomenon discovered in 1807, when he used electricity to stabilize a German railroad excavation in 1939. Since then, electro-osmosis has been used occasionally for stabilization of excavation slopes, and building and bridge foundations. But West Branch not only is the largest installation ever (slightly more than 1,000 electrodes—cathodes and anodes), it is the first electro-osmotic stabilization of the foundation of a substantial earth embankment after placing the fill.

● **And then, the stabilization**—Under an agreement with the Corps, the prime contractor for the dam, Lane Construction Co., of Meriden, Conn., subcontracted the soil treatment project to Wellpoint Dewatering Corp., a subsidiary of Griffin Wellpoint Corp., New York City. Wellpoint Dewatering moved on the job late last summer and began installing a maze of piping and electrical wires.

The cathode well, which attracts the water then brings it to the surface, is a complex installation. First, 14-in.-dia casings are jetted anywhere from 100 to 140 ft below the top of the embankment. Next, two 2-in.-dia steel pipes are inserted inside each casing to serve as the cathode and the eductor well.

Next the casing is filled with sand, then extracted from the ground leaving a sand column surrounding the two small units (see the drawing).

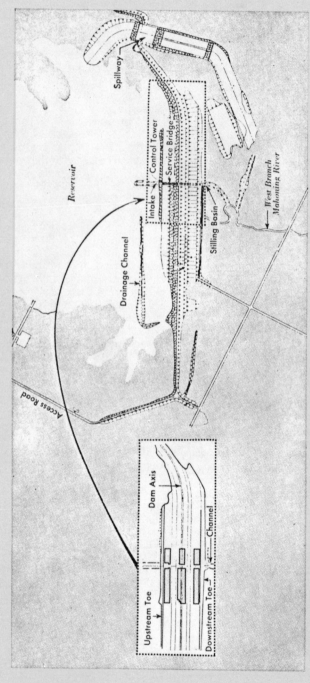

WEST BRANCH DAM is 9,900 ft long. Only a short section of its foundation (shaded rectangles, inset) had to be treated with electro-osmosis. This reduced the high pore pressures that were causing the embankment to move laterally.

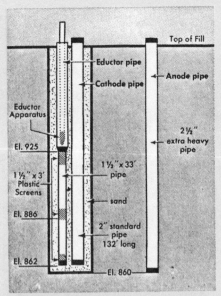

ELECTRODES are steel pipes, inserted through dam into clay. Eductor, installed with cathode in sand, withdraws water.

The eductor apparatus is a kind of jet pump in which water forced down the outer pipe, passes through a venturi at the bottom and back up the inner pipe at a much higher velocity. The resulting pressure differential at the venturi draws groundwater up the inner pipe along with the water injected from the surface.

Groundwater enters the eductor through a check valve that is located at the top of a 1½-in.-dia pipe which extends into the clay formation below the eductor. The eductor unit is kept just above the natural watertable, El 930.

The anodes are merely 2½-in.-dia pipes inserted 20 ft away from the cathodes. Extra-heavy pipe is used for the anodes because of the deterioration caused by electrolysis. The anodes and the cathodes are both attached to the power source through wires laid on the ground.

• A pattern, but not really—Mr. Casagrande had the contractor install rows of anodes and cathodes, 20-ft c-c, in three rectangular strips, one on top of the embankment, and one on the outside edge of each berm (drawing).

Of the 1,000 electrodes, roughly two-thirds are anodes. There is a pattern in which several anodes surround a cathode, but the pattern varies, depending on Mr. Casagrande's assessment of the soil conditions.

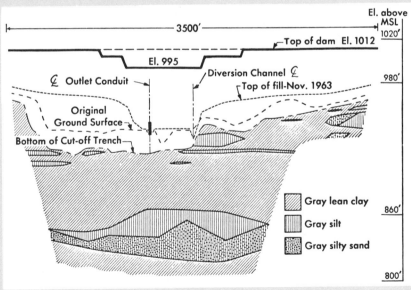

GEOLOGY is similar under 2,000-ft section of dam, but clay failed only under central 1,000 ft. Embankment on either side was placed in two stages (dotted line); winter shutdown may have allowed pressures to dissipate before upper portion was placed.

15

In areas of high electrolyctic action close to the outlet conduit, old railroad rails serve as anodes to avoid having to replace too many.

Wellpoint Dewatering set up its power generation and pumping station on the berm. The station contained a 3,000-gpm eductor pump and 10 diesel-powered d-c generators with a capacity of 2,600 kw.

The water system with its double headers appeared relatively simple, but once in place, the electrical hookup looked like a splicer's nightmare. Cables ran like giant spaghetti in all directions uphill and down, over and under pipes.

Other than the initial failure, nothing startled the Corps more than when Wellpoint began jetting the first pipes into the embankment. Pore pressure readings, already alarmingly high, soared when the additional water pressure of the jets hit the foundation clay.

The dam's stability was so uncertain at this point that just the additional pressure of the jetting operation caused the ground beyond the toe of the embankment to move.

"For a minute there, it looked like we might lose it trying to fix it," says Garth Fuquay, chief soils engineer for the Pittsburgh District. But by decentralizing the electrode installation, instead of jetting one hole after another in a limited area, and immediately activating the installed electrodes and eductor wells, this problem was eliminated.

• **After the rise, the fall**—With these initial scares over and all electrodes and eductors in place, Wellpoint's crews applied 100 to 150 volts to the clay early last fall. "Then," Mr. Fuquay says, "we waited to see the pore pressures go down."

And down they went. It took time—until early spring, in fact—before construction division chief Jack Minnotte could say, "We're pretty sure we've got

CONTROL TOWER tilted slightly when dam moved, but returned to vertical after electro-osmosis stabilized dam foundation.

it licked. If Dr. Casagrande agrees, we'll be ready to resume fill placement in early June."

Late in May, Wellpoint Dewatering began pulling the system on the main embankment. Early this month, earthmoving equipment was back at work on the dam. The electro-osmosis plant was moved downstream from the dam so treatment of both berms can continue while the dam is completed.

Fortunately, Lane Construction was well ahead of schedule when the trouble developed, so if all goes well, and there's no reason to think it won't, West Branch Dam will be topped out this October and ready to go into service on schedule for the 1966-67 flood season

The Corps estimates that the West Branch's foundation troubles will boost its $16-million price tag by about $2 million.

the slide. The section of an article reprinted on pages 18 to 24, from *Civil Engineering* of March 1964, describes many important details of the events that led to this disaster.

In retrospect, it is clear that there was considerable evidence warning of potential trouble of this sort. The geologic evidence concerning the character of the various strata, the slide history of the area, the measured creep rates of the valley wall—all pointed to trouble. The potential for failure was recognized by some investigators, but the validity of their concern was not accepted until too late.

This terrible event also illustrates a number of important points with which the civil engineer engaged in the design of structures on the earth must concern himself. Among these are:

1. The geologic setting of the project is always important. It may often be of crucial importance, and should be thoroughly understood.
2. Almost any work of man introduces a change in the natural environment. The effects of this change must be considered with regard to surrounding areas in addition to those on the materials with which the structure is in direct contact.
3. The influence of water on the strength and deformation of earth materials subjected to applied loads is always of major importance, and may never be overlooked.
4. The properties of earth materials generally change with time, even if no man-made change is introduced into the environment. The design of a structure must take such changes into account.
5. There are always uncertainties in the prediction of the behavior of earth and rock materials. The consequences of any risks associated with these uncertainties must be kept within limits acceptable to the designer, his client, and the public at large. This requires that the designer take full advantage of all available knowledge, and exercise good judgment in applying it to his design.

Between 1959 and 1975, eight major dams, and many smaller ones, at various places around the world failed. Most of these failures have occurred because one or another "small" point was overlooked. It is evident that a thorough appreciation of the principles underlying the development and behavior of earth materials is essential to the design of such structures.

1.3 ELEMENTARY DEFINITIONS

The three cases described above are a sampling of others that will be discussed in subsequent chapters. The subject matter common to these problems is soil mechanics.

Soil mechanics is the name given to the scientific approach to understanding and predicting the behavior of soil in an engineering context—as a foundation or construction material, for instance. To the engineer, *soil* is

VAIONT RESERVOIR DISASTER

*Geologic causes of tremendous
landslide accompanied
by destructive flood wave*

GEORGE A. KIERSCH, F. ASCE
Professor of Engineering Geology, Cornell University, Ithaca, N.Y.

Findings and views expressed here are those of the author, or as cited, and are not those of any board or organization or other person or persons. As a Senior Postdoctoral Fellow of the National Science Foundation, Professor Kiersch is on leave from his teaching post at Cornell University for special research in Europe on stresses in rock masses and how they affect conditions at the sites of high dams and open cuts. He is at the Technical University, Vienna, and was priviledged to study the Vaiont slide. His were among the first observations made outside of those by the authorities in immediate charge. CIVIL ENGINEERING is pleased to present this report for study as a means of reducing the danger of such disasters.

The worst dam disaster in history occurred on October 9, 1963, at the Vaiont Dam, Italy, when almost 3,000 lives were lost. The greatest loss of life in any similar disaster was 2,209 in the Johnstown Flood in Pennsylvania in 1899. The Vaiont tragedy is unique in many respects because:

• It involved the world's second highest dam, of 265.5 meters (875 ft).

• The dam, the world's highest thin arch, sustained no damage to the main shell or abutments, even though it was subjected to a force estimated at 4 million tons from the combined slide and overtopping wave, far in excess of design pressures.

• The catastrophe was caused by subsurface forces, set up wholly within the area of the slide, 2.0 kilometers long and 1.6 km wide.

• The slide volume exceeded 240 million cu m (312 million cu yd), mostly rock.

• The reservoir was completely filled with slide material for 2.0 km and up to heights of 175 m (574 ft) above reservoir level, all within a period of 15 to 30 sec. (A point in the mass moved at a speed of 50 to 100 ft per sec.)

• The slide created strong earth tremors, recorded as far away as Vienna and Brussels.

The quick sliding of the tremendous rock mass created an updraft of air accompanied by rocks and water that climbed up the right canyon wall a distance of 260 m (850 ft) above reservoir level. (References to right and left assume that the observer is looking downstream.) Subsequent waves of water swept over both abutments to a height of some 100 m (328 ft) above the crest of the dam. It was over 70 m (230 ft) high at the confluence with the Piave valley, one mile away. Everything in the path of the flood for miles downstream was destroyed (Fig. 1).

Reprinted from *Civil Engineering*, March 1964.

18

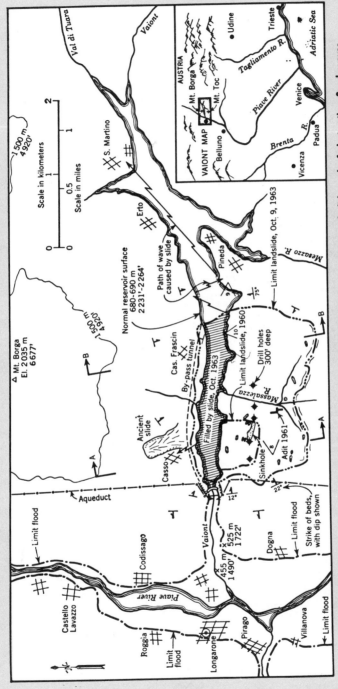

FIG. 1. Map of Vaiont Dam area and Piave River valley shows geographic features, limits of slide and of destructive flood waves.

19

FIG. 2. Completed Vaiont Dam is seen with reservoir at about El. 680 m. On the right of the photograph, the white slope near the dam marks the 1960 slide. All the steep slope beyond was involved in the 1963 slide. (Wide World photo)

A terrific, compressive air blast preceded the main volume of water. The overtopping jet of water penetrated all the galleries and interior works of the dam and abutments. Air currents then acted in decompression; this tensional phase opened the chamber-locked safety doors of all the galleries and works and completed destruction of the dam installations, from crest to canyon floor.

This catastrophe, from the slide to complete destruction downstream, occurred within the brief span of some 7 min. It was caused by a combination of: (1) adverse geologic features in the reservoir area; (2) man-made conditions imposed by impounded water with bank storage, affecting the otherwise delicately balanced stability of a steep rock slope; and (3) the progressive weakening of the rock mass with time, accelerated by excessive ground-water recharge.

On the day after the catastrophe, October 10, 1963, the Italian Government appointed a 15-man technical board to review the circumstances leading up to the slide and establish responsibility for preslide policies and actions. The report of this board was released on January 16, 1964, and cited a lack of coordination between the technical and governmental officials. Findings of another government board are scheduled for release some time in 1964. Understandably, this article omits reference to such aspects of the disaster, and is confined to an assessment of the geologic setting and the influence of engineering works on these conditions.

Design and construction

Vaiont Dam is a double-curved, thin-arch, concrete structure completed in the fall of 1960 (Fig. 2). The dam is 3.4 m (11.2 ft) wide at the top and 22.7 m (74.5 ft) wide at the plug in the bottom of the canyon. It has an overflow spillway, carried a two-lane highway on a deck over the crest, and had an underground powerhouse in the left abutment. Reservoir capacity was 150 million cu m (196 million cu yd, or 316,000 acre-ft).

The way in which the dam resisted the unexpected forces created by the slide is indeed a tribute to designer Carlo Semanza and the thoroughness of construction engineer Mario Pancini. The anchor tie-rods which strengthened the abutments were devised and supervised by the engineers, L. Muller and F. Pacher, of Salzburg, Austria.

Design and construction had to overcome some disadvantages both of the site and of the proposed structure. The foundation was wholly within limestone beds, and a number of unusual geologic conditions were noted during the abutment excavation and construction. A strong set of rebound (relief) joints parallel to the canyon walls facilitated extensive scaling within the destressed, external rock "layer." Excessive stress relief within the disturbed outer zone caused rock bursts and slabbing in excavations and tunnels of the lower canyon. Strain energy released within the external, unstable "skin" of the abutment walls was recorded by seismograph as vibrations of the medium. This active strain phenomenon in the abutments was stabilized with a grout curtain—to 150 m (500 ft) outward at the base—and the effects were verified by a seismograph record. Grouting was controlled through variations of the elastic modulus.

The potential for landslides was considered a major objection to the site by some early investigators ; others believed that "the slide potential can be treated with modern technical methods."

Vaiont Dam was constructed by SADE (Societa Adriatica Di Elettricita, Venezia) as part of its extensive hydroelectric system in northeastern Italy. In 1962, the Italian national electric monopoly (ENEL) began to take over all of SADE's power facilities; during 1963 the operation of Vaiont Dam was under ENEL direction.

The geologic setting

The Vaiont area is characterized by a thick section of sedimentary rocks, dominantly limestone with frequent clayey interbeds and a series of alternating limey and marl layers. The general subsurface distribution is shown in the geologic cross sections, Fig. 5. A brief description, progressing from the oldest to the youngest formations follows:

Lias formation, Lower Jurassic. Thin beds of gray limestone alternating with thin beds of reddish, sandy marl (shalysandy limestone). The soft beds aided fault movement in the overlying rock units (Fig. 5). The Lias formation does not crop out in the Vaiont Reservoir, but underlies the region and is near canyon level at the dam site.

Dogger formation, Middle Jurassic. Medium to thickly bedded gray, dense limestone; massive series over 300 m (1,000 ft) thick. Parting seams of clay are common. Upper part is thin bedded. Dissolved by solution action to produce some openings. The dam foundation is wholly within Dogger beds and the series is well exposed in the walls of the canyon and on the slopes of Mt. Toc and Mt. Borga (Figs. 1 and 5).

Malm formation (Titonico), Upper Jurassic. White to reddish, platy to very thin-bedded limestone with some siliceous beds. Clay seams are common along bedding planes and some claystone interbeds. Dissolved by solution action so that sinkholes, tubes, and openings are present. This formation crops out in the walls of the reservoir (Fig. 3), mainly within 1 km (3,280 ft) of the dam; it was involved in the slides of 1960 and 1963.

Lower Cretaceous formation. White, very thin- to medium-bedded limestone with some interbeds of siliceous limestone and claystone. Solutioning of the limestone has taken place and openings are common. This formation crops out in the walls of the reservoir, mainly on the left bank and upper sector of the slide (Fig. 5, Section B-B); it was involved in the slides of 1960 and 1963.

Upper Cretaceous formation (Senon). Red, thin beds of marl alternating with light red, thin beds of limestone. There is one zone of grayish marl with red to gray clayey sandstone. This formation crops out in the upper part of reservoir on the left bank and channel; it had a strong influence on the slide plane in 1963 (Fig. 5, Section B-B).

Glacial debris, Pleistocene. Irregular boulders and gravels, largely limestone with sand and silt. morainal remnants deposited on the floor of the glacial valley. This material occurs as a thin mantle overlying bedrock on the sides of the outer valley (Fig. 5); it was involved in the slides of 1960 and 1963.

Slide debris, Recent. Irregular blocks of talus, slope wash and old landslide material. This material occurs as a thin to thick mantle overlying the bedrock of both the outer and the inner valleys (Fig. 5); it was involved in the slide of 1963.

Retained stress

The young folded mountains of the Vaiont region retain a part of the active tectonic stresses that deformed the rock sequence. Faulting and local folding accompanied the regional tilting along with abundant tectonic fracturing. This deformation, further aided by bedding planes and relief joints, created blocky rock masses.

The development of rebound joints beneath the floor and walls of the outer valley is shown in Fig. 4. This destressing effect creates a weak zone of highly fractured and "layered" rock, accentuated by the natural dip of the rock units. This weak zone is normally 100 to 150 m (330 to 500 ft) thick. Below this a stress balance is reached and the undisturbed rock has the natural stresses of mass.

Rapid carving of the inner valley resulted in the formation of a second set of rebound joints—in this case parallel to the walls of the present Vaiont canyon. The active, unstable "skin" of the inner canyon was fully confirmed during the construction of the dam.

The two sets of rebound joints, younger and older, intersect and coalesce within the upper part of the inner valley (Fig. 4). This sector of the canyon walls, weakened by overlapping rebound joints, along with abundant tectonic fractures and inclined bedding planes, is a very unstable rock mass and prone to creep until it attains the proper slope.

Causes of slide

Several adverse geologic features of the reservoir area contributed to the landslide on October 9:

Rock units that occur in a semicircular outcrop on the north slopes of Mt. Toc are steeply tilted. When deformed, some slipping and fault movement between the beds weakened frictional bond.

Steep dip of beds changes northward to Vaiont canyon, where rock units flatten along the synclinal axis; in three dimensions the area is bowl-shaped (Fig. 5). The down-dip toe of the steep slopes is an escarpment offering no resistance to gravity sliding.

Rock units involved are inherently weak and possess low shearing resistance; they are of limestone with seams and clay partings alternating with thin beds of limestone and marl, and frequent interbeds of claystone (Fig. 5).

Steep profile of the inner canyon walls offers a strong gravity force to produce visco-elastic, gravitational creep and sliding (Fig. 4).

Semicircular dip pattern confined the tendency for gravitational deformation to the bowl-shaped area (Fig. 1).

Active solutioning of limestone by ground-water circulation has occurred at intervals since early Tertiary time.

The result has been subsurface development of extensive tubes, openings, cavities and widening of joints and bedding planes. Sinkholes formed in the floor of the outer valley (Fig. 1), particularly along the strike of the Malm formation on the upper slopes (Fig. 5); these served as catchment basins for runoff for recharge of the ground-water reservoir. This interconnected ground-water system weakened the physical bonding of the rocks and also increased the hydrostatic uplift. The buoyant flow reduced gravitational friction, thereby facilitating sliding in the rocks.

Two sets of strong rebound joints, combined with inclined bedding planes and tectonic and natural fracture planes, created a very unstable rock mass throughout the upper part of the inner canyon (Fig. 4).

Heavy rains for two weeks before October 9th produced an excessive inflow of ground-water from the drainage area on the north slopes of Mt. Toc. This recharge raised the natural ground-water level through a critical section of the slide plane (headward part) and subsequently raised the level of the induced water table in the vicinity of their junction (critical area of tensional action). The approximate position of both water levels at the time of the slide is shown in Fig. 5.

Excessive ground-water inflow in early October increased the bulk density of the rocks occurring above the initial water table; this added weight contributed to a reduction in the gross

Top of slide

Crest of dam

FIG. 3. Vaiont canyon is seen looking toward the reservoir site during early construction and stripping of abutments. Terrain of 1960 and 1963 slides is shown at right, with the steep cliff of Malm formation. The overtopping flood wave scoured the abutments above the dam, destroying the aqueduct and bridge (black) and stripped away the highway, lower right. Photo courtesy of *Water Power.*

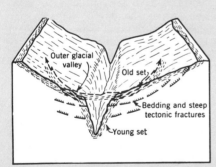

Outer glacial valley

Old set

Bedding and steep tectonic fractures

Young set

FIG. 4. On sketch of inner Vaiont canyon and remnants of the outer glacial valley, are shown rebound joints—old and young set—from stress relief within the walls of the valley to depths of 100 to 150 m (330 to 500 ft).

shear strength. Swelling of some clay minerals in the seams, partings and beds created additional uplift and contributed to sliding. The upstream sector (Fig. 5, Section B-B) is composed largely of marl and thin beds of lime-stone with clay partings—a rock sequence that is inherently less stable than the downstream sector (Fig. 5, Section A-A).

The bowl-shaped configuration of the beds in the slide area increased the confinement of ground water within the mass; steeply inclined clay partings aided this containment on the east, south, and west.

The exploratory adit driven in 1961 reportedly exposed clay seams and small-scale slide planes. Drill holes bored near the head of the 1960 slide (Figs. 1 and 5, Section A-A) were slowly closed and sheared off. This confirmed the view that a slow gravitational creep was in progress following the 1960 slide and probably even before that—caused by a combination of geologic causes. Creep and the accompanying vibrations due to stress relief were later described by Muller.

23

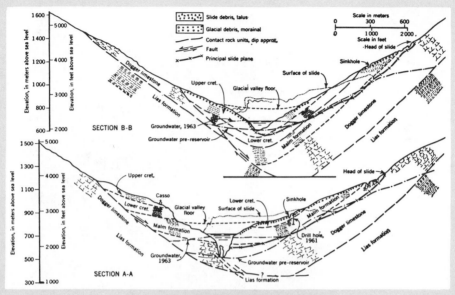

FIG. 5. On geologic cross-sections of slide and reservoir canyon, running from north to south, principal features of the slide plane, rock units and water levels are shown. For location of Sections A-A and B-B, see Fig. 1.

Effects of man's activities

Construction of Vaiont Reservoir created an induced ground-water level which increased the hydrostatic uplift pressure throughout a triangular subsurface mass (Fig. 5) aided by fractures and the interconnected system of solution openings in the limestone.

Before April 1963, the reservoir was maintained at El. 680 m. In September, five months after the induced water table was raised 10 m (33 ft), the slide area increased its rate of creep. This action has three possible explanations: (1) a very delicate balance existed between the strength of the rock mass and the internal stresses (shear and tensile), which was destroyed by the 10-m rise of bank storage and accompanying increase in hydrostatic pressure; (2) the same reaction resulted from the large subsurface inflow in early October due to rains; or (3) the induced ground-water level from the reservoir at El. 680 m during 1961-1962 did not attain maximum lateral infiltration until September 1963, when creep accelerated. In any case, the rate of ground-water migration into bank storage is believed to have been critical. To ascertain which condition actually prevailed, observation wells would have been needed for measurement of the transmissibility factor.

Evidence indicates that the immediate cause of the slide was an increase in the internal stresses and a gross reduction in the strength of the rock mass, particularly the upstream sector where this mass consists largely of marl and alternating thin beds of limestone and marl. Actual collapse was triggered by an excess of ground water, which created a change in the mass density and increased the hydrostatic uplift and swelling pressures along planes of inherent weakness, combined with the numerous geologic features that enhanced and facilitated gravitational sliding.

The final movement was sudden— no causes from "outside" the affected area are thought to have been responsible.

24

all unconsolidated accumulations of solid particles produced by mechanical or chemical disintegration of rocks.

When a knowledge of soil mechanics and other pertinent engineering disciplines is combined with a knowledge of geology, and is applied with judgment to the creation of a design, the process is called *foundation engineering*, *soil engineering*, or *highway engineering*, for example, depending upon the specific problem considered. Thus, design of a foundation for a building would fall under foundation engineering. Design of an earth dam would be an example of soil engineering. Design of a highway pavement and the embankment upon which it rests would be highway engineering. For purposes of discussion, we may refer to all of these functions as *geotechnical engineering*.

The case histories focus on important facets of geotechnical engineering, and we shall examine and evaluate many of the principles of soil mechanics required to understand them.

2

Mechanical Behavior of Cohesionless Soils

2.1 INTRODUCTION

The article on pages 28 to 33, which originally appeared in *Engineering News-Record* of May 11, 1939, describes the failure of the east end of the upstream face of the Fort Peck Dam. At the time it was constructed, this dam of 108 million cubic yards was the largest earthfill in the world. Figure 2.1 shows a simplified cross-section through the dam before and after the slide failure. The dam itself consists of two basic parts: the inner *silt core*, which has the primary function of reducing the amount of water seeping through the dam; and the outer *sand shells* which provide the strength to hold up the core.

The dam was constructed by the *hydraulic fill method*. This economical method for moving large quantities of earthfill involves *hydraulic dredging* of the borrow material, and transporting it in suspension through pipes to the top of the dam, where it is expelled. The outlet pipes are located at the outer edges of the upper surface of the dam and directed in toward the center. As the suspension flows toward the center and the flow velocity decreases, the coarser materials drop from suspension to form the outer *shells*. The finer particles flow toward the center, where they form the *core pool*. The fine-grained soils settle relatively slowly from suspension so that the core material is usually of low density and weak compared to the coarser-grained solids in the outer portions of the dam.

When the slide occurred, approximately $5\frac{1}{2}$ million cubic yards moved as much as a thousand feet upstream. This quantity of material is sufficient to fill a typical classroom more than 20,000 times. An estimate of the surface along which sliding took place is shown in Figure 2.1. The consulting board engaged to investigate the slide, and others who studied their results, agreed that it was triggered by failure of the badly weathered shale rock underneath the dam (Middlebrooks, 1942).* However, there is some reason to believe that the severity of the slide was significantly affected by the character of the materials within the dam and the manner in which they were placed. It has been suggested that the hydraulic fill process leaves the granular material in the vicinity of the clay core in a potentially unstable condition. As a consequence, no major hydraulic fill dam has been built in this country since the Fort Peck Dam.

We shall consider in the following section the factors involved in the development of an unstable condition such as appeared to occur in the Fort Peck slide, and the degree to which it would be possible today to predict a failure of this sort.

2.2 FAILURE OF COHESIONLESS SOIL

We begin by asking ourselves the question: How can we create failure in an incoherent mass of soil such as sand? The answer to this question is not obvious. (In fact, generalized to engineering materials as a whole, this question has been the subject of intensive investigation for more than two hundred years.) We may postulate a variety of mechanisms by which failure occurs, but must resort to experiment for any authoritative answer.

One might intuitively suppose that it would be possible to crush a mass of soil under a large hydrostatic (fluid) pressure. Such *is* the case for pressure magnitudes which are large compared to those usually encountered in civil engineering applications. At moderate pressure levels, however, a surprising result occurs: if we fill a rubber bag with dry sand it is quite flexible and can be squeezed in the hand; if we then evacuate the bag, which has the effect of applying one atmosphere of hydrostatic pressure to the sand, the mass becomes rigid, and can no longer be kneaded. It has been generally observed that at the pressure levels customary in civil engineering applications, failure of a granular mass cannot be produced by a pure hydrostatic compressive pressure.†

Inquiring further, we might postulate that failure occurs as the result of tensile forces; in fact, we recognize immediately that sand and similar *coarse-grained* materials can sustain no tensile force. We know, however, that failure occurs in many situations where tensile forces are not present. Thus we must look still further.

If we examine the results of many experiments, we would conclude that failure of granular soils is brought about by shear forces. To understand this

---

* These bibliographical references are given in full at the end of the book.

† Recent developments in design and construction methods have led to the construction of some very high earth- and rockfill dams (Oroville dam in California is more than 700 ft high). The crushing of individual particles as a result of the large pressures produced by such structures is considered to be an important problem in their design (ASCE Report, 1967). This problem is not, however, significant in the majority of soil engineering designs.

Large Slide in Fort Peck Dam Caused by Foundation Failure

Summary—The technical inquiry board that investigated the Fort Peck Dam slide of Sept. 22, 1938, reports that weakness of the shale strata underlying the valley fill caused the failure, and recommends reconstruction with flatter upstream slope. The new part of the core, built of rolled fill, is to be bonded to the old hydraulic fill core by steel sheeting. One member of the board favors more extensive reconstruction.

REPORTING on the great slide of Sept. 22, 1938, in the upstream bank of Fort Peck Dam, an inquiry board of engineers and geologists attributes the accident to shearing failure of the shale which underlies the valley fill on which the dam rests. As briefly noted in our news pages of April 27, p. 44, the board recommends rebuilding the destroyed section with flatter upstream slope and adding a wider and higher berm on the upstream side of the standing part.

The report, dated March 2 but released only recently, is signed by seven of the nine members of the board; it is abstracted below, together with a separate report by Thaddeus Merriman. The latter agrees with the general finding of cause but emphasizes the part played by hydrostatic uplift and calls for removal of more material from slide and abutment.

Prof. W. J. Mead, ninth member of the board, did not join in the report, though expressing agreement with the views of both the main report and Mr. Merriman in a letter. He holds that the usefulness of the dam is too small to warrant incurring the risk of failure that any engineering structure involves.

Our abstract is preceded by a description of the advance warnings that preceded the slide, and of the slide itself, from data furnished from official records by Maj. Clark Kittrell, U. S. District Engineer in charge of the project.

Settlement preceded the slide

Fort Peck Dam is an earth embankment 220 ft. high by 9,000 ft. long, with slopes of about 4 to 1 upstream and 8 to 1 downstream, across the valley of the Missouri River at the former site of Fort Peck, 20 mi. southeast of Glasgow, in northeastern Montana. It will form a reservoir of 19,500,000 acre-ft. capacity for improvement of navigation from Sioux City to the mouth by increasing the low-water flow, for flood protection, for irrigation and for generation of power. (For fuller data on the project see *ENR*, Aug. 29, 1935, p. 279.) When the slide occurred the dam proper was nearing completion, having been under construction since 1933 by force account under the Engineer Corps of the Army, with Col. T. B. Larkin and later Maj. Clark Kittrell in charge.

At the site the river valley is a wide flat of glacial and river fill mainly sand and clay, on a substratum of "shale" or hard clay 80 to 100 ft. down, between bluffs rising 300 to 350 ft. high above the river. The ma-

Reprinted from *Engineering News-Record*, May 11, 1939.

terial for building the dam by the hydraulic-fill method was pumped out of the valley bottom upstream and downstream of the dam site. The easterly and westerly parts of the dam were built first, to a height of about 150 ft., and later the intervening embankment across the river gap was built to the same height after diverting the flow through four tunnels driven through the east hillside. Finally, beginning late in 1937, the full length of the dam was carried up from this initial level.

On Sept. 22 work had proceeded to within 30 ft. of the crest, and the core pool and its confining banks were correspondingly narrow. Construction was going on at full capacity, to complete the hydraulic fill work by mid-November so that the dredging fleet could be dismantled. Four dredge units were pumping about 200,000 cu.yd. per day of solid material into the dam. Dam and core pool had been raised about 30 ft. in two months. A depth of about 65 ft. of water was impounded in the reservoir, though two months earlier the water had stood 19 ft. higher.

On the morning of Sept. 22 inspectors reported insufficient freeboard of the upstream bank above the core pool; the dredge pipe which rested on this bank appeared to be about 2 ft. lower than it should be. An immediate survey confirmed the observation; at Stations 15, 16 and 17 the pipe was only 3 ft. above pool level instead of at its proper height of $4\frac{1}{2}$ ft. above pool, and Station 17 was the lowest point. When this result was reported (11:45 a.m.) a conference was called to meet at 1:15 in the afternoon at the point of greatest settlement, for discussion of the situation. The slide occurred just as the conference party was assembling.

At this time the core pool was at El. 2252, the reservoir water at 2117.50. Two lines of railroad track for quarry-rock placement and a roadway extended along the upstream berm at El. 2212, and quarry rock revetment was in place up to this berm. A pump boat, tender and a jetting barge were in the core pool near the east abutment, two draglines were working on the upstream bank crest near Station 17+50, and about 180 men were working in or near the area of the slide.

No movement of tracks other than normal settlement caused by passage of the heavy rock trains had been observed prior to the day of the slide. The structure of the dam, both core and banks, near the east abutment was normal. The core pool here had been held close to minimum width, and no sliding of shale from the abutment into the pool had been observed other than some surface raveling of the slope due to scour from the dredge discharge and a minor slide of the shale fill on the ridge west of the first gate shaft.

Eyewitness reports gave the following composite account of the slide:

The core pool began to settle, slowly at first and then more rapidly, and about the same time cracks were observed on the upstream face 30 ft. below the crest. Then portions of the upstream shell nearest the pool began to slide into the sinking core pool, and some cracking and slumping took place also on the downstream beach. Simultaneously the main mass of the upstream shell, almost intact, was moving out into the reservoir, in a swing similar to that of a gate hinged at the east abutment, breaking away from the west or main part of the dam at about Station 27; the waters of the core pool poured out at the break.

A man at the pump barge, which was in the core pool near the east abutment, said that looking westward it appeared as if the fill to the west was falling into a big hole and that the hole progressed rapidly toward

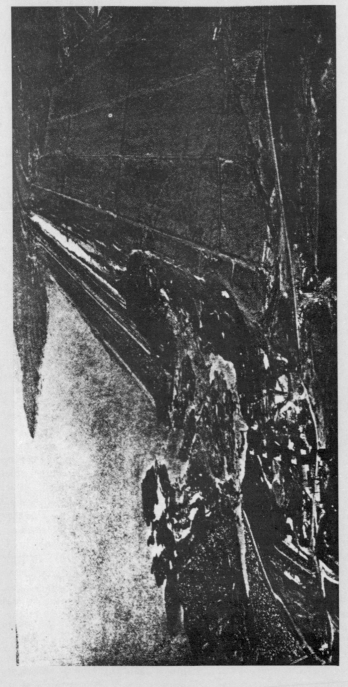

Fig. 1. Airview of Fort Peck Dam immediately after the slide. As water stood about 65 ft. deep in the reservoir, the lower part of the slide material cannot be seen. The larger blocks of material showing above water farthest from the face of the break are sections of the upstream face, still covered with the heavy stone riprap that mantled the dam nearly up to final water line.

30

the pump boat; that as it reached the boat its forward end dipped down sharply and disappeared.

Estimates of the total time of movement of the slide indicate that it occupied about ten minutes. A total of 5,217,000 cu. yd. moved out into the reservoir pool; the maximum distance of movement was 1,200 ft. The upstream shell was partially destroyed for a length of 1,700 ft. (Station 10 to Station 27). Of the men in the slide area, 34 were carried along and 8 lost their lives.

The upstream shell of the section that slid out moved out almost intact, and after movement stopped the debris lay in a fan-shaped area roughly centering on the axis at the east abutment (see air view Fig. 1 and sketch plan Fig. 2). Large parts of the upstream face were exposed in almost horizontal position, still evenly covered with their quarry-rock revetment almost undisturbed.

An inquiry board of nine called to report on the slide and its reconstruction comprised William Gerig, W. J. Mead and Thaddeus Merriman of the original consulting board, and A. Casagrande, Irving B. Crosby, Glennon Gilboy, J. D. Justin, W. H. McAlpine and C. W. Sturtevant. Extensive investigations were made in the course of the inquiry, including soil sampling and testing, 30,000 ft. of core drilling (200 drill holes 2 in. to 6 in. in diameter, seven holes 12 in. to 36 in. in diameter drilled in prefrozen ground to show the undisturbed position and character of the material), two exploration tunnels about 150 ft. long in the hillside, and five shafts averaging 60 ft. deep in the dam outside the slide area.

As to the cause of the slide the board says:

"After a careful consideration of all the pertinent data the board has concluded that the slide in the upstream portion of the dam near the right abutment was due to the fact that the shearing resistance of the weathered shale and bentonite seams in the foundation was insufficient to withstand the shearing force to which the foundation was subjected. The extent to which the slide progressed upstream may have been due, in some degree, to a partial liquefaction of the material in the slide."

For reconstruction in the slide area the board recommends building a narrow core of rolled impervious glacial till, 50 ft. wide at the bottom and 15 ft. at the top, surrounded by banks of either hydraulic fill or rolled fill (see section at Station 20 in Fig. 3) with upstream slope ranging from 3.5:1 at the top to 23:1 at the base.

The board agrees that the shell of the reconstructed section may be built by hydraulic fill or rolled-fill methods, and recommends that the core be constructed of rolled impervious glacial till from the left abutment, to minimum widths of 15 ft. at El. 2255 and 50 ft. at El. 2120. Mr. Gerig dissents, believing that either a rolled-fill or a hydraulic-fill core would be satisfactory.

The board recommends that the plan of bonding the old core to the new core by means of a single row of steel sheetpiling, as proposed by the district engineer, be adopted.

The board approves the exploratory program outlined by the district engineer covering installation of piezometer pipes in the shale at both abutments for the purpose of measuring hydrostatic pressures. The board further approves the plan of the district engineer to install drainage wells at such points as the piezometer measurements indicate drainage to be desirable. The board recommends that the existing exploratory holes should be backfilled with pervious material.

The board has considered in detail the question of stripping disintegrated shale from the east abutment and has examined the results of shear tests and stability analyses made by the district engineer. As a result of this examination the board believes that the flat slopes of the reconstructed section will result in shearing stresses in the disintegrated shale of such low intensities as to provide an ample factor of safety, and that, therefore, re-

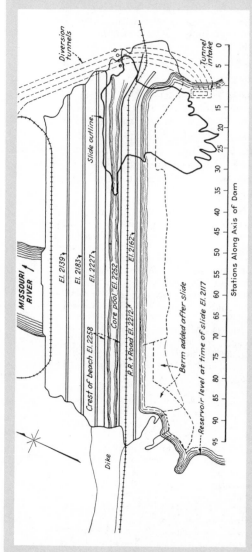

Fig. 2. Fort Peck Dam, whose right end was damaged by a 5,000,000-yd. slide last September, is to be repaired by filling the break with new fill of much flatter slope. The map indicates the condition of the dam before the slide and the approximate outline of break and slide debris. The proposed reconstruction is shown in Fig. 3. The map was prepared from official data and views and is not part of the board's report.

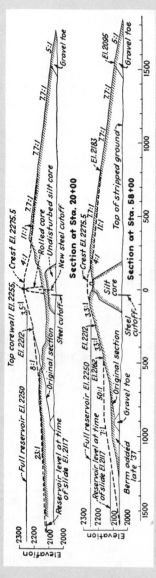

Fig. 3. Old as well as new parts of the dam are to be strengthened along the upstream face by giving the embankment a flatter slope or by adding a berm. Reconstruction outline is shown by heavy dash lines.

32

moval of the disintegrated shale is not necessary.

The board has considered the question of grouting the fault zones in the shale in the east abutment, and approves the plan of the district engineer for further study of this problem.

As to those parts of the dam not affected by the slide, the board recommends that they be strengthened by adding an upstream berm (see cross-section at Station 58 in Fig. 3), built of material the same as or coarser than the material now forming the shell of the dam, and that this material be compacted.

The board recommends that, in the construction of the berms on the upstream slope of the dam by hydraulic means, tractors be utilized on the beach to secure as great a degree of compaction as may be practicable without interference with hydraulic placing of the material. The board recommends that the use of these tractors be continued to such an extent as may be necessary to demonstrate whether or not a worthwhile additional compaction is obtained.

Above the present top of the dam (El. 2250) the embankment is to be completed by rolled fill. The dike which extends 11,000 ft. beyond the west end of the dam is considered satisfactory as built. Some hillside grading and minor structural changes are proposed in the vicinity of the entrance to the diversion tunnels to safeguard against hillside slides and to close the front entrance to the intake, blocked by the new slope.

Merriman—lower reservoir level

Thaddeus Merriman, in a separate report, lays special stress on the part played by the weakness of the weathered shale on top of the hard blue shale, the slipperiness of the clay layer formed by the top of the weathered shale, and the hydrostatic pressure in the cracks and seams of the weathered layer. He recommends more extensive reconstruction than the majority report. He urges that all

weathered shale within reach be removed from the abutments and the slide area, and that because of the impossibility of removing the objectionable material below about El. 2100 the spillway of the dam be cut down 50 ft. in height so as to lower the reservoir by this amount.

During the period of sixty days prior to the slide, the difference in level between the core pool and the water in the reservoir had increased from 83 ft. to 134 ft. Because of the open joints in the disintegrated and sub-firm shale, a substantial part of this pressure head was transmitted through them by the drainage water from the core pool as it passed through both the core and the shell. This pressure acting as an uplift under the upstream portion of the dam reduced the effective weight of the toe so that finally it was unable to restrain the slope above it.

Failure was thus initiated. The toe in the vicinity of Sta. 18 and range U 8 moved outward on the lubricated and disintegrated shale . . . The failure was distinctly in the disintegrated shale at depth under the dam . . . first, because it transmitted water pressures and, second, because it was slippery. All of the other materials involved behaved admirably. The core gave an excellent account of itself. . . . The shell material everywhere performed remarkably well. It was strong and exhibited a remarkable degree of solidity under most difficult circumstances. The frozen drill cores fully confirm the stability of this material. No blue unweathered shale showed any sign of motion. The valley fill materials above the weathered shale were badly deformed and disturbed, but the more sandy portions gave the best account of themselves.

The plans referred to constitute a reasonable solution of the problem, but, as a condition precedent to their execution, all wholly or partly disintegrated shale should be removed down to at least El. 2100. This removal must be so complete that the fill sections will everywhere rest on and abut against the blue shale from El. 2100 to the crest level of the dam.

As it will be impossible to remove the disintegrated shale from below El. 2100 some compensating measure must be adopted. I recommend that the pool level be dropped from El. 2250 to El. 2200 by lowering the spillway and removing the gates now installed. It will thus not be necessary to raise the main dam above its present elevation.

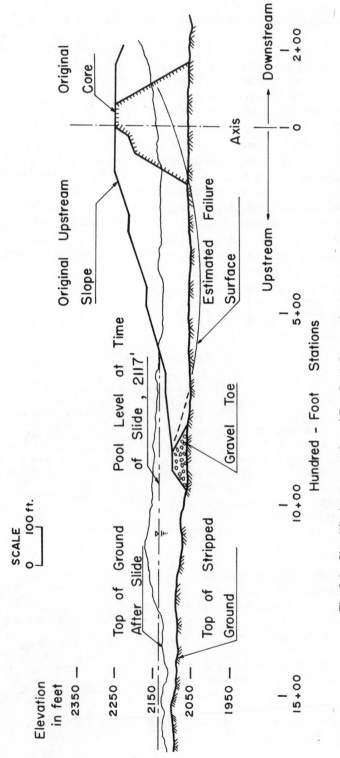

Fig. 2.1—Simplified cross-section of Fort Peck Dam at Station 20 + 00, before and after slide.

34

concept better, let us visualize the following experiment, illustrated schematically in Figure 2.2a. A small metal box filled with dry sand is split
along a horizontal plane through the middle. Let us apply a downward
force N to the top of the box, normal to the plane of the split, and a force
T into and parallel with the plane of the split. Let us suppose that the force
N is held constant, while the force T is increased in small increments. If we
measure the displacement of the upper part of the box relative to the bottom
as T is increased, we will find that an increment in T produces a corresponding
increment in the relative displacement of the two halves of the box. The
magnitude of this relative displacement will depend upon the magnitude of
T, but will be relatively small until some limiting value of T is reached.
When this limiting value, T_{max}, is reached, the upper half of the box will
slide with respect to the lower half of the box. This is considered to constitute
failure of the sand mass. Because the relative movement of the material on
either side of the surface along which failure has occurred is parallel to the
failure surface, the failure is considered to be one of *shear*. Thus the effect
of the force T which produces failure is not that of the normal force on the
side of the box, but rather that of the shearing force in the plane along which
failure occurs.

Fig. 2.2—Direct shear test of a granular soil.

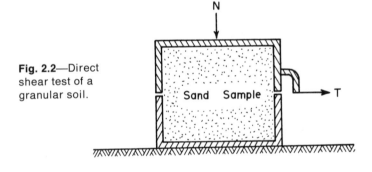

a) - Direct Shear Test

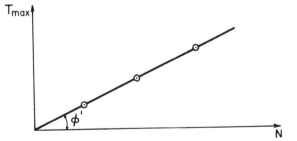

b) - Plot Obtained from Direct Shear Test

If a similar experiment is performed on an identical sand sample, with a magnitude of N other than that of the previous experiment, a different T_{max} will result. The relationship between T_{max} and N for a series of such experiments is shown in Figure 2.2b. This result is representative of granular materials in general. It indicates that *at failure* there is a fixed relation between the normal and shearing forces on the failure plane, for specimens which are identical before testing. The equation of this relationship may be written

$$T_{max} = N \cdot f \tag{2-1}$$

where $f = \tan \phi'$. The angle ϕ' is called the *angle of shearing resistance*. Equation 2-1 is one form of the relation first reported by the French engineer Charles A. Coulomb (1776). Because it is also a special case of the failure theory for materials proposed by Otto Mohr (1900), Equation 2-1 is often called the Mohr-Coulomb failure theory for granular soils.

This result is analogous to that observed for a weightless block resting on a surface where friction exists between the block and the surface. This is illustrated in Figure 2.3. The relationship between T_{max} and N for the friction block is the same as it was for the sand mass in the direct shear test, except that f is usually called the coefficient of friction. For this reason, granular soils are commonly referred to as *frictional* or, more commonly, *cohesionless* materials.

The kind of shear surface that develops in the direct shear test, or already exists in the case of the friction block, is similar to the surface of sliding shown in Figure 2.1. In the case of the more complicated earth slide, the sliding surface may not be plane, because the forces change from point to point in the mass. However, the principle discussed still applies.

Factors Affecting Failure

The failure of soil differs from the failure of many other civil engineering materials because of the particulate nature of soil. Failure usually occurs by relative movement between the grains, rather than by breaking up of the grains themselves (by contrast, when steel fails, the failure surface passes through individual crystals). The relative movement of the grains occurs as rolling or sliding; thus the resistance to shear developed by the soil mass depends upon the forces *between* the grains, rather than forces existing *within*

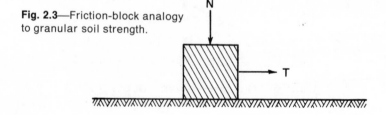

Fig. 2.3—Friction-block analogy to granular soil strength.

the grains. For cohesionless materials, these forces occur primarily as a result of frictional resistance between the grains, and interference of one grain with another. Thus, factors which affect this grain-to-grain interaction will affect the shearing resistance of the soil. These factors include:

1. *Mineral type(s) of the individual grains.* Some typical coefficients of friction for common minerals are given in Table 2.1. Obviously, the slope of the "failure line" in Figure 2.2b for a given mineral type will not be the same as that for the coefficients of friction in Table 2.1, because of the many other factors affecting this slope. However, all other things being equal, as the coefficient of friction between minerals increases, the slope of the failure line may be expected to increase.*

TABLE 2.1

Coefficient of Friction for Selected Dry Minerals

Mineral	Static Coefficient of Friction
Clear quartz	0.11
Milky quartz	0.14
Rose quartz	0.13
Microcline feldspar	0.12
Calcite	0.14
Muscovite	0.43
Chlorite	0.53
Talc	0.36

SOURCE: Horn and Deere (1962).

2. *Grain shape.* The grain shape of cohesionless soils is generally bulky; that is, the dimensions are all of the same order of magnitude. Even so, the angularity of the grains is important, as may be seen in Figure 2.4a. It seems intuitively reasonable that the very angular grains shown at the right of the figure will offer more shearing resistance than the rounded grains shown at the left, all other things being equal. The degree of angularity is described qualitatively by descriptive terms: rounded, subrounded, subangular, angular. The roughness of the individual grains is also important.

3. *Density of packing of grains.* From Figure 2.4b, it is easy to see how much more difficult it would be to shear the densely packed grains

---

* The relationship between the coefficient of friction of the mineral and the slope of the failure line has been studied by a number of investigators (Thurston and Deresiewicz, 1959; Duffy and Mindlin, 1957; Rowe, 1962; Horne, 1965) for idealized arrangements of spherical particles. Their results are difficult to apply to real heterogeneous soils, but they do suggest that a large change in the particle-to-particle coefficient of friction would be required to produce a modest change in ϕ' (if other parameters remain constant).

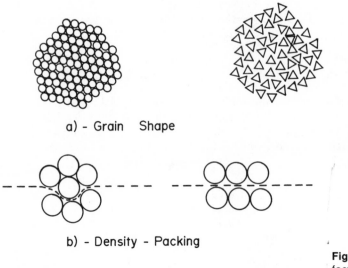

a) - Grain   Shape

b) - Density - Packing

Fig. 2.4—Some factors affecting failure of granular materials.

c) - Distribution of Grain Sizes

than the loosely packed grains along some failure surface such as that of the dashed line. This factor is influenced not only by the geometric arrangement of the grains but also by the distribution of sizes. Figure 2.4c shows two different grain-size distributions for an idealized mass of spheres. The largest grains are arranged in identical fashion on both sides. However, the material with the range of grain sizes is likely to be packed in a fashion that offers more resistance to shear than the uniform material. The density of packing is the most important factor in determining the angle of shearing resistance of a cohesionless soil.

4. *Unit weight of material.* The weight of the soil is frequently the factor providing the normal force, N, on some potential failure surface (along which $T_{max} = N \tan \phi'$). Therefore, it is significant to the problem. The weight is a function of the weights of the individual grains as well as their arrangement and packing.

To understand the effect of these various factors, and their combinations, on pertinent soil properties, it is desirable to quantify them. Toward this end, our next step is to consider some fundamental relations between the constituent parts of a soil mass.

2.3 PHASE RELATIONS

Phase Diagram

In general, soil is a three-phase system consisting of solid (mineral) particles, liquid (such as water or sometimes oil), and gas (usually air). Quantifying the relationships among these three phases permits us to describe such things as relative density of packing, unit weight of soil, and the like. A convenient tool for doing this is the *phase diagram* shown in Figure 2.5. If we could take a mass of soil in a container and combine all the grains into a completely solid mass with no space between them, with all the liquid floating on top of this and the gas above the liquid, we would have a phase

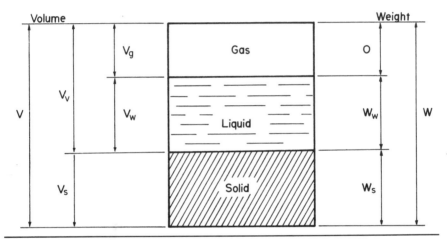

V	Total volume of a mass of soil;
V_s	Volume occupied by solid matter in mass of soil (i.e., the soil particles);
V_v	Volume occupied by voids (i.e., total volume not occupied by solid matter);
V_w	Volume occupied by water;
V_g	Volume occupied by gas (usually air);
W	Total weight of mass of soil (usually called wet or moist weight);
W_s	Weight of solid matter (i.e., dry weight. This quantity may depend upon the drying temperature);
W_w	Weight of water contained in mass soil.

Fig. 2.5—Phase diagram and definitions of quantities.

diagram. The phase diagram is simply a conceptual device to assist us in visualizing various relations among the three phases.

In Figure 2.5 the weights and volumes of the components are symbolized alongside the phase diagram, and defined below the sketch. Some relationships among these parameters are especially useful in quantifying the factors discussed above. These defined relations are given below:

$\gamma_w = W_w/V_w$ Unit weight of water (FL^{-3}). (For most purposes this may be assumed to be the unit weight of water at $4°C$: 1.0 gm/cc or 62.4 lb/ft^3.)

$\gamma_s = W_s/V_s$ Unit weight of solid matter (FL^{-3}).

$G = W_s/V_s\gamma_w$ Specific gravity of solid matter (*dimensionless*). The

$= \gamma_s/\gamma_w$ specific gravity of solids (or γ_s) is a function of mineral type, as shown in Table 2.2. For gravel grains G varies typically from 2.55–2.7, for sand from 2.65–2.75, for inorganic clays from 2.7–2.85.

$\gamma_m = W/V$ Unit weight of mass of soil (FL^{-3}) (often called wet, moist, bulk or mass unit weight). Typical values are given in Table 2.3.

$G_m = W/V\gamma_w$ Specific mass gravity of soil (*dimensionless*) (also called

$= \gamma_m/\gamma_w$ *bulk specific gravity*).

$\gamma_d = W_s/V$ Dry unit weight of soil mass (FL^{-3}). Typical values are given in Table 2.3.

$n = V_v/V$ Porosity (*dimensionless*).

$e = V_v/V_s$ Void ratio (*dimensionless*). Void ratio (and porosity) are indicators of the density of packing of soil, and therefore significantly affect strength. Typical values of void ratio and porosity are shown in Table 2.3.

$w = W_w/W_s$ Water content (*dimensionless*); usually expressed as a percentage. Water content may be used in combination with S_r to compute the void ratio. Water content is easy to measure. Typical values are given in Table 2.3.

$S_r = V_w/V_v$ Degree of saturation (*dimensionless*); usually expressed as a percentage.

These fundamental relations serve as definitions for quantities which are useful descriptors of the state of a mass of soil.

TABLE 2.2
Specific Gravity of Selected Minerals

Mineral	Specific Gravity G	Mineral	Specific Gravity G
Quartz	2.67	Calcite	2.72
Feldspar		Magnetite	5.17–5.18
Orthoclase	2.57	Illite	2.64–2.69
Plagioclase	2.62–2.76	Kaolinite	2.60–2.68
Mica		Montmorillonite	2.20–2.74
Muscovite	2.76–3.00	Hornblende	3.00–3.47
Biotite	2.95–3.00		

TABLE 2.3
Unit Weight, Porosity, and Void Ratio of Typical Soils in Natural State

Description	Unit Weight (pcf) γ_d	Unit Weight (pcf) γ	Porosity n (%)	Void Ratio e	Water Content Fully Saturated w (%)
Uniform sand					
Loose	90	118	46	0.85	32
Dense	109	130	34	0.51	19
Mixed-grained sand					
Loose	99	124	40	0.67	25
Dense	116	135	30	0.43	16
Glacial till, very mixed-grained	132	145	20	0.25	9
Glacial clay					
Soft	—	110	55	1.20	45
Stiff	—	129	37	0.60	22
Soft organic clay					
Slight organic content	—	98	66	1.90	70
High organic content	—	89	75	3.00	110
Soft bentonite	—	80	84	5.20	194

SOURCE: Terzaghi and Peck (1967).

Note that the unit weight provides the only link between the two sides of the phase diagram (G is the ratio of two unit weights). Unit weights are frequently called "densities." This is a misnomer; however, a careful distinction between unit weight and density is necessary only in a few cases.

The phase diagram is generally useful in two kinds of problems. The first type is that in which it is desired to derive a general relationship between certain given or measured quantities and some other quantity of interest. Examples 2.1 and 2.2 illustrate this type of problem.

Example 2.1

The task here is, knowing the void ratio e, to determine the porosity n. To do this, we may imagine that the phase diagram in Figure 2.6 represents an arbitrary, convenient quantity of soil. Suppose that we choose an amount such that the volume of solids is

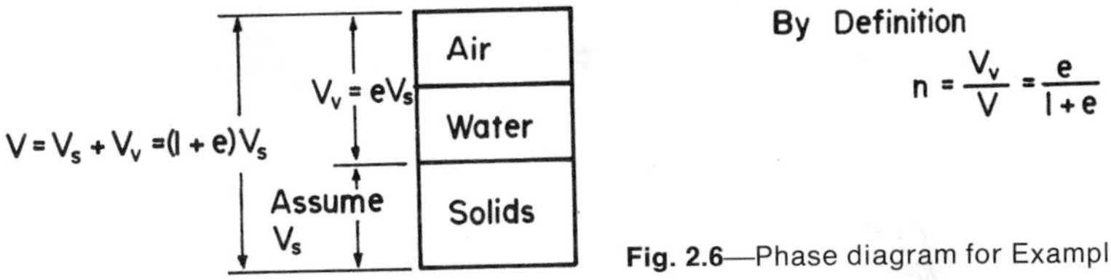

By Definition

$$n = \frac{V_v}{V} = \frac{e}{1+e}$$

Fig. 2.6—Phase diagram for Example 2.1.

V_s; the quantities shown in the diagram then follow directly from definitions. When the choice of quantity of material is arbitrary, it is often convenient to choose the denominator of one of the given items.

Example 2.2

In this case we wish to find e in terms of the specific gravity of solids G, and the water content w, for a *saturated* soil. The term "saturated" implies that $S_r = 100\%$, that is, the voids are completely filled with water. Therefore $V_w = V_v$, and the other quantities on the phase diagram in Figure 2.7 follow from definitions.

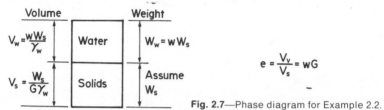

Fig. 2.7—Phase diagram for Example 2.2.

The second type of problem in which the phase diagram is useful is one where numerical values are available for certain quantities, and it is desired to compute the numerical value of other parameters.*

Example 2.3

Knowing that $G = 2.70$, $W_s = 10$ gm, $w = 20\%$, we wish to find γ and e for a fully saturated soil. When numerical values are given, the choice of amount of material is no longer arbitrary. One may either solve the problem in general terms and then substitute the numerical values, or just substitute numerical values into the phase diagram, as shown in Figure 2.8.

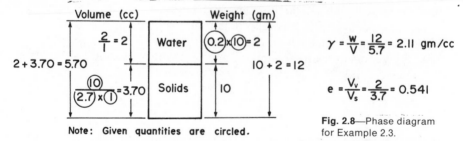

Note: Given quantities are circled.

Fig. 2.8—Phase diagram for Example 2.3.

Relative Density

If sufficient information is given, the phase diagram can be used to quantify two of the important factors affecting failure (Section 2.2). Example 2.3 above illustrates a case in which the unit weight is found. Of much more significance, however, is the density of packing of the grains, in the sense that it affects the relationship between normal and shearing forces on the failure surface at failure (i.e., the angle ϕ'). This is often referred to as the "state of

* Variations on this type of problem include cases in which the quantity sought is not apparent in the problem statement. These are illustrated in the exercises provided at the end of the chapter.

compaction" of the material. Experience has shown that it is not the absolute value of the density of packing that is important, but rather the state of compaction relative to the loosest and densest possible conditions into which the particular cohesionless soil can be placed. This relation is called the *relative density*, D_r, and is ordinarily defined in terms of the void ratio of the material:

$$D_r = \frac{e_{max} - e}{e_{max} - e_{min}} \tag{2-2}$$

where e is the void ratio for which the relative density is being computed, e_{max} is the maximum value of the void ratio, that is, the loosest state in which the material can be placed in a dry or completely submerged condition, and e_{min} is the minimum void ratio, the densest condition in which the material can be placed in a dry or completely submerged condition. The range of D_r is from 0 to 1 for the loosest and densest states, respectively.

Figure 2.9 illustrates schematically the loosest and densest possible conditions for a cohesionless mass consisting of perfect spheres with each grain in contact with all adjacent ones. The void ratio varies from $e_{max} = 0.91$ to $e_{min} = 0.35$. The corresponding values of the porosity are 48 and 26 per cent respectively. For real soils, the grains are not generally spherical and vary in size. Hence, e_{max} and e_{min} cannot be determined theoretically, but must be found by experiment.

Because the experimental values of e_{max} and e_{min} are functions of the method of placing the granular material, and no standard method exists at this time, it is difficult to define precisely the maximum and minimum void ratios for a real soil. However, natural cohesionless soils rarely exhibit e_{max} greater than 1 or e_{min} less than about 0.2.

When e_{max} and e_{min} are determined for a particular soil, the relative density at any other e can then be determined. This parameter is the most important

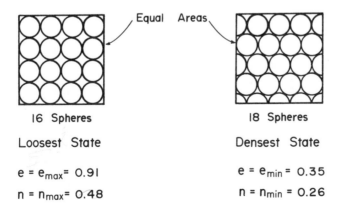

Equal Areas

16 Spheres   18 Spheres

Loosest State   Densest State

$e = e_{max} = 0.91$ $e = e_{min} = 0.35$

$n = n_{max} = 0.48$ $n = n_{min} = 0.26$

Fig. 2.9—Limiting void ratios for spheres.

single factor affecting the angle of shearing resistance of a given cohesionless soil.

2.4 GRAIN SIZE AND GRAIN-SIZE DISTRIBUTION

Grain Size

The size of the individual grains in a mass of cohesionless soil is important primarily in a quantitative rather than a qualitative sense. That is, mechanical behavior of the soil with large grain sizes is not basically different from that of a material with smaller grain size, although the numerical values of ϕ' or other relevant parameters may be different. A mass of pills or dry corn exhibits similar mechanical response to that of dry sand, as long as the individual grains do not crush. Nonetheless, it has been found convenient to subdivide cohesionless soils on the basis of grain size, and assign descriptive names to various grain-size ranges. A common system of classification is given in Table 2.4.

TABLE 2.4
Classification by Grain Size

Descriptive Name	Grain "Size" D
Coarse-grained	
Rock	> 1 cu yd
Boulders	1 cu yd–12 in.
Cobbles	12 in.–3 in.
Gravel	
Coarse	3 in.–3/4 in.
Fine	3/4 in.– #4 sieve (4.76 mm)
Sand	
Coarse	#4– #10 (2.00 mm)
Medium	#10– #40 (0.42 mm)
Fine	#40– #200 (0.074 mm)[a]
Fine-grained[b]	
Silt	#200–0.002 mm (2 μ)
Clay	$< 2 \mu$[c]

[a] The smallest size which can be detected with the unaided eye is approximately 0.074 mm.
[b] Grain size is not a significant indicator of properties for $D < $ #200 sieve.
[c] Colloidal size $\leqslant 0.2 \mu$.

We must recognize that the grain sizes described do not represent the actual "size" of the grains. For instance, the gravel and sand sizes are determined by sieving the soil with a set of standard sieves. Thus these grain "sizes" are in fact the sizes of the sieve opening through which the grains will just pass, or upon which the grains are retained. Consequently the grain sizes described probably indicate a measure of the intermediate dimension of the particle and are properly described as *equivalent* grain diameters. (The "fine-grained" soils listed in Table 2.4 will be discussed later.)

Grain-Size Distribution

Real soils almost invariably consist of a variety of grain sizes. For coarse-grained soils the distribution of these sizes is customarily determined by sieving a sample of the soil through a set of nested sieves with openings which decrease in size from top to bottom. Thus the weight of soil retained on each sieve represents that proportion of the total sample which has an equivalent grain diameter smaller than that of the sieve above (through which it has just passed) and larger than that of the sieve on which it is retained. We can plot the results of our sieve analysis as a *histogram* showing the proportion of the total sample corresponding to each sieve size. Three such histograms are shown in Figure 2.10. If a smooth curve were drawn through the values on the histogram it would represent a *frequency-distribution curve* rather than discrete values determined by experiment.

Note that the equivalent grain diameter in these three figures is plotted to a logarithmic scale. It is customary to plot grain-size distributions in this way because of the wide range of grain sizes usually encountered.

The grain-size distribution can yield information about the geologic origin of the particular soil being investigated. For example, in Figure 2.10a the lognormal distribution of grain sizes suggests deposition by a relatively nonselective geologic agent. Thus the soil represented by this figure could have been deposited by glacial ice, or might be a residual soil of moderate maturity. A soil exhibiting a wide range of grain sizes approximately log-normally distributed, is considered to be *well graded*.

By contrast, the soil described by Figure 2.10b was evidently deposited by a selective agent such as flowing water or wind. The agent was capable of carrying few, if any, grains much coarser than the dominant size, and carried most of the finer grains beyond the point from which the sample was taken. Such a uniform distribution is *poorly graded*.

The material represented by Figure 2.10c is also poorly graded, because certain grain sizes are almost absent. It is called *gap-graded*. Such gap-grading occurs among sand-gravel mixtures deposited by rapidly flowing rivers which also carried large amounts of fine-grained sediments. Alternatively, gap-grading may indicate the presence of two different depositional agents, such as river flow into a glacial lake combined with melting from glacial ice floating on the surface of the lake.

It is more convenient to describe the grain-size distribution quantitatively by plotting it in another form. By summing the ordinates of the histogram for all grain sizes less than those of a particular grain diameter, we can plot a *cumulative frequency-distribution curve*. Such curves are shown in Figure 2.11 for the three materials described in Figure 2.10. The abscissa is the

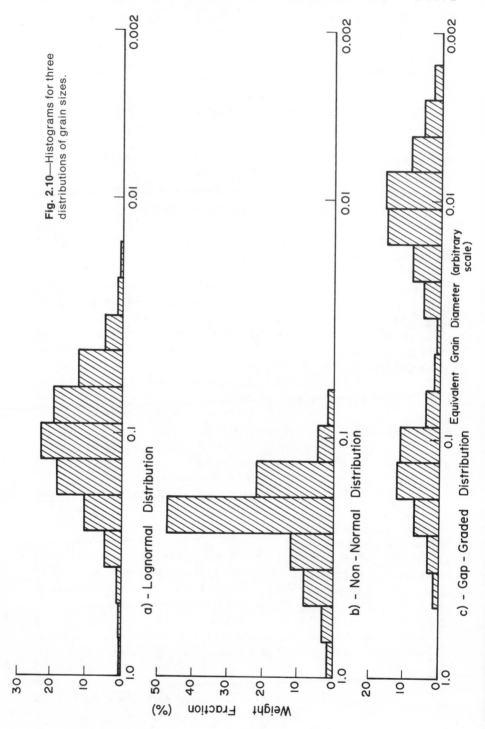

Fig. 2.10—Histograms for three distributions of grain sizes.

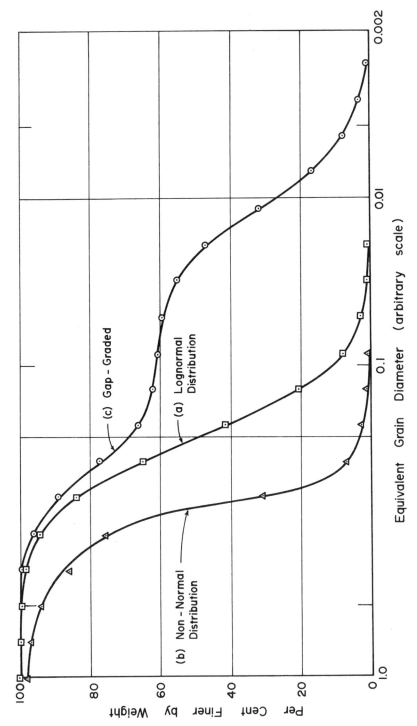

Fig. 2.11—Cumulative grain-size distribution curves for histograms shown in Figure 2.10.

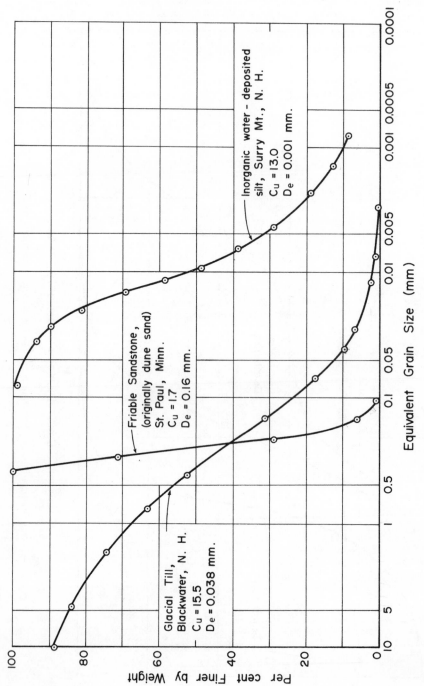

Fig. 2.12—Grain-size distribution curves for several soils.

equivalent grain diameter, shown to a logarithmic scale. The ordinate is the proportion of the total sample finer than the specific grain size.

These cumulative frequency-distribution curves, generally called *grain-size-distribution curves*, are commonly characterized by two parameters, to which numerical values may be assigned. By doing this it becomes possible to compare grain-size-distribution curves for different soils without plotting them on the same diagram. The first of these parameters is a characteristic grain size, which serves to locate the grain-size-distribution curve along the abscissa. Any arbitrary point would serve for this purpose. However, a useful value was determined by Hazen (1911), who found that the ease with which water could flow through a soil was a function of the *effective grain size*, D_{10}. D_{10} is the grain size for which 10 per cent of the soil by weight is finer.

The range of grain sizes is indicated by the uniformity coefficient, C_u:

$$C_u = \frac{D_{60}}{D_{10}} \tag{2–3}$$

where D_{60} is the equivalent grain diameter for which 60 per cent of the soil by weight is finer, and D_{10} is the effective size. Thus, for a perfectly uniform (very poorly graded) material, $C_u = 1$. C_u indicates the *shape* of the grain-size-distribution curve whereas D_{10} *locates* the curve. A natural soil is considered very uniform if C_u is less than 5, and very well graded if C_u equals 15 to 20.*

Grain-size-distribution curves for three natural soils are shown in Figure 2.12. The differences in these distribution curves are clearly functions of their geologic origin.

The grain-size distribution of a soil has a marked influence on the possible density of packing of the grains, and therefore on the shear strength. This is illustrated in Figure 2.13, which shows the effect of grain-size distribution on e_{max} and e_{min}. These results were obtained by mixing prescribed quantities of sand and gravel together to obtain different grain-size-distribution curves. Three of these curves are shown in the upper part of the figure. For each grain-size distribution, the values of e_{max} and e_{min} are shown in the lower part of the figure as a function of the gravel content. There is clearly a particular distribution of grain sizes which will produce the lowest (i.e., densest) void ratios for both e_{max} and e_{min}. It is intuitively reasonable that this should be the case.

The combined effects of the factors discussed above on the angle of shearing resistance (ϕ') are shown in Table 2.5.

Retrospect

The Fort Peck slide and the general concern about hydraulic fill structures have led us to consideration of failure of cohesionless soils. We have

---

* Gap-grading cannot be identified by knowing only D_{10} and C_u for a soil.

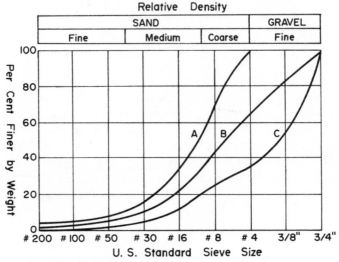

a) - Grain - Size Distribution Curves

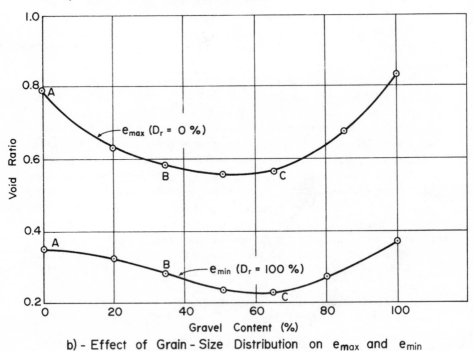

b) - Effect of Grain - Size Distribution on e_{max} and e_{min}

Fig. 2.13—Maximum and minimum void ratios for sand-gravel mixtures. (Data from *Earth Manual*, 1960.)

TABLE 2.5

Approximate Values of ϕ' for Granular Soils as Affected by State of Compaction, Size, Gradation, and Angularity of Grains

		Values of ϕ' (deg)	
Size of Grain	State of Compaction	Rounded Grains, Uniform Gradation	Angular Grains, Well Graded
Medium sand	Very loose	28–30	32–34
	Moderately dense	32–34	36–40
	Very dense	35–38	44–46
Gravel–sand ratio (%)			
65–35	Loose	—	39
65–35	Moderately dense	37	41
80–20	Dense	—	45
80–20	Loose	34	—
Blasted rock fragments	—	45–55	

SOURCE: Leonards (1962).

examined briefly the shear mechanism causing failure in such materials by considering the results of a direct shear test conducted on a dry sand. We then investigated certain physical features of a soil which affect its strength. And in the process, we have learned to quantify certain elementary but useful properties of the soil.

However, in order to extrapolate from the particular test described to the field conditions in which we are really interested, there are two other basic problems which must be considered: First, we must examine how forces are transmitted through the soil so that we can relate the results of our small direct shear test to the large field condition; and second, we must consider the pronounced effects which result when water is introduced into the system. These features are discussed below.

2.5 STRESS, STRAIN, AND STRENGTH

Stress at a Point

In order to relate what we have learned about failure of cohesionless soils in a direct shear test to conditions in the field, we shall investigate the way in which forces are distributed throughout the soil mass. We begin by considering a body of material subjected on its surface to a variety of concentrated and distributed forces, as shown in Figure 2.14. Let us pass a plane through this body (Figure 2.14a) and examine the interaction between the material on either side of the plane. Figure 2.14b shows the lower half of the body cut by the plane, exposing area A. A small area ΔA, with outward normal n on the cut surface is also indicated. The force acting on this small area is $\Delta \mathbf{F}$. The distribution of the forces is then described by a *defined*

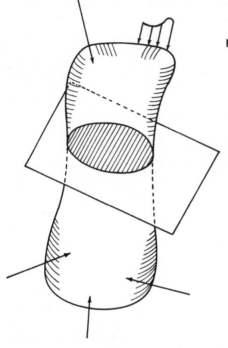

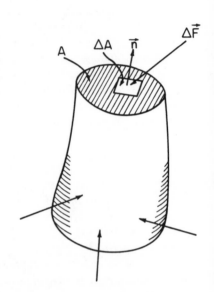

Fig. 2.14—Body with forces.

a) - Whole Body   b) - Split Body

quantity called the *stress*:

$$\text{Stress} = \lim_{\Delta A \to 0} \frac{\Delta \mathbf{F}}{\Delta A} \tag{2–4a}$$

We shall find it convenient to consider components of stress normal and tangential to the area:

$$\text{Normal stress} = \sigma = \lim_{\Delta A \to 0} \frac{\Delta F_n}{\Delta A} \tag{2–4b}$$

$$\text{Shear stress} = \tau = \lim_{\Delta A \to 0} \frac{\Delta F_t}{\Delta A} \tag{2–4c}$$

where ΔF_n is the magnitude of the component of $\Delta \mathbf{F}$ normal to the area ΔA, and ΔF_t is the magnitude of the component tangent to the area.

Physically, ΔA approaching zero implies that the area becomes a point, that is, the notion of stress refers to a condition *at a point*. Thus, the whole concept of stress is based upon faith in the fact that the ratio of the force to the area *does* in fact approach a finite limit as the area shrinks to a point. This is a basic assumption of continuum mechanics.

Although we usually accept stress as a physical entity, it seems unlikely that the stress defined in Equations 2–4 actually ever exists in a real material. This is a consequence of the fact that, as we shrink the area further and further, it will eventually become small compared to an individual grain or crystal or even molecule. Figure 2.15 illustrates this point. An apparently "homogeneous" soil mass is shown in Figure 2.15a to the scale indicated below the figure. If we now enlarge the scale by 20 times so that we are viewing an area only 1/400 as large, we see in Figure 2.15b discrete particles. Nonetheless the force distributed over such an area might still be approximately representative of the force per unit area corresponding to Figure 2.15a. However, if we increase the scale by another factor of 20 (Figure 2.15c), we observe a large void space in the middle of the area with several grains of different sizes presumably supporting the force, ΔF, applied to the area. It would be coincidence indeed if the ratio $\Delta F/\Delta A$ is the same for the two areas shown in Figures 2.15a and 2.15c. In soils which are multiphase systems, this effect may be more pronounced than in some other materials.*

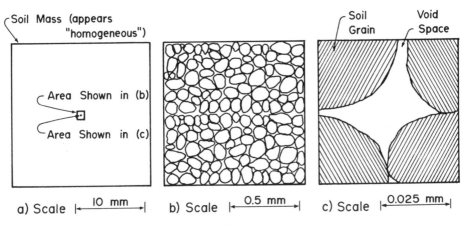

a) Scale |⟵ 10 mm ⟶| b) Scale |⟵ 0.5 mm ⟶| c) Scale |⟵ 0.025 mm ⟶|

Fig. 2.15—Effect of scale on view of soil mass.

In spite of this evident inconsistency, we shall discover that stress is an exceedingly useful quantity, and we shall make much use of it from this point on. However, it behooves us to remember that it is only a defined parameter, and does not actually exist in a soil mass. We shall later define other equally artificial quantities whose virtue, as for stress, is only that they are useful.

* The consequences of this assumption for granular materials are explored in detail by Skempton (1961), and are shown to be significant at very high pressures.

Mohr Circle of Stress

Accepting the concept of stress at a point, we must now consider how the stress on a plane passing through a particular point varies as the orientation of that plane varies. Figure 2.16a is a two-dimensional representation of a

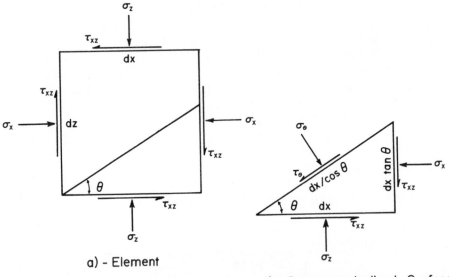

a) – Element

b) – Stress on Inclined Surface

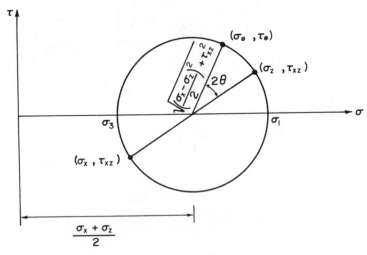

c) – Mohr Circle of Stress

Fig. 2.16—Rotation of stress at a point.

"small" element, for which the stresses on two mutually perpendicular planes are shown. We shall assume that the element is sufficiently small so that the stresses can be considered uniform on each of the faces. (There are, of course, stresses which act on the faces parallel to the plane of the page. However, for purposes of the present discussion we shall ignore them.)

The question which we seek to answer is then: What will be the stresses on a plane inclined at some angle θ to one face of the element, as shown in Figure 2.16b? This figure is a freebody diagram of that portion of the element in Figure 2.16a which lies below the inclined plane.* Equilibrium requires that the sum of the *forces* (*not* stresses) in any direction must be zero. Thus we must multiply each stress by the area over which it acts. Satisfying this requirement first in a direction normal to the θ plane, and then in a direction parallel to the θ plane, leads to

$\sum F_\theta = 0$:

$$\sigma_\theta \left(\frac{dx}{\cos \theta} \right) - \sigma_x(dx \tan \theta) \sin \theta + \tau_{xz}(dx \tan \theta) \cos \theta$$
$$- \sigma_z(dx) \cos \theta + \tau_{xz}(dx) \sin \theta = 0$$

or

$$\sigma_\theta = \sigma_x \sin^2 \theta + \sigma_z \cos^2 \theta - 2\tau_{xz} \sin \theta \cos \theta \qquad (2\text{-}5)$$

$\sum F_{\theta+90°} = 0$:

$$\tau_\theta \left(\frac{dx}{\cos \theta} \right) + \sigma_x(dx \tan \theta) \cos \theta + \tau_{xz}(dx \tan \theta) \sin \theta$$
$$- \sigma_z(dx) \sin \theta - \tau_{xz}^*(dx) \cos \theta = 0$$

or

$$\tau_\theta = (\sigma_z - \sigma_x) \sin \theta \cos \theta - \tau_{xz}(\sin^2 \theta - \cos^2 \theta) \qquad (2\text{-}6)$$

where the quantities are as shown in Figure 2.16a and b. Rearranging and applying some trigonometric identities enables us to express Equations 2–5 and 2–6 in terms of the double angle 2θ:

$$\sigma_\theta = \frac{\sigma_z + \sigma_x}{2} + \frac{\sigma_z - \sigma_x}{2} \cos 2\theta - \tau_{xz} \sin 2\theta$$

$$\tau_\theta = \frac{\sigma_z - \sigma_x}{2} \sin 2\theta + \tau_{xz} \cos 2\theta$$

$$(2\text{-}7)$$

* Once we have selected a sign convention for normal stress based upon convenience, we are not free to choose an arbitrary sign convention for shear stress in the development of the equation for the Mohr circle. In most soil engineering problems, the normal stresses are compressive. Thus we usually choose compressive stress as positive. However we must now choose positive shear stress as that stress which creates a *clockwise* moment about a point just *outside* the element at the plane considered. Had the opposite convention been selected the resulting equation would have been a hyperbola rather than a circle.

Squaring and adding these equations leads to

$$\left(\sigma_\theta - \frac{\sigma_z + \sigma_x}{2}\right)^2 + \tau_\theta^2 = \left(\frac{\sigma_z - \sigma_x}{2}\right)^2 + \tau_{xz}^2 \qquad (2\text{–}8)$$

This is the equation of a circle of radius $\sqrt{[(\sigma_z - \sigma_x)/2]^2 + \tau_{xz}^2}$ with its center at $[(\sigma_z + \sigma_x)/2, 0]$. The circle is shown in Figure 2.16c.

The parameter 2θ is the counterclockwise angle between the radius vector to the point (σ_z, τ_{xz}) and the radius vector to the point $(\sigma_\theta, \tau_\theta)$. Thus, the angle between the radius vectors to two points on the Mohr circle is *twice the space angle* between the planes on which the stresses represented by those points act.

The angle between σ_z and the principal stresses may be found by setting the second of Equations 2–7 equal to zero. This leads to

$$\tan 2\theta = \frac{-2\tau_{xz}}{\sigma_z - \sigma_x} \qquad (2\text{–}9)$$

Alternatively, if the x and z directions are themselves principal directions, such that $\sigma_z = \sigma_1$ (the major principal stress), $\sigma_x = \sigma_3$ (the minor principal stress), and $\tau_{xz} = 0$, then Equations 2–7 become

$$\sigma_\theta = \frac{\sigma_1 + \sigma_3}{2} + \frac{\sigma_1 - \sigma_3}{2} \cos 2\theta$$

$$\tau_\theta = \frac{\sigma_1 - \sigma_3}{2} \sin 2\theta \qquad (2\text{–}10)$$

where 2θ is the angle between the radius vector to the major principal stress $(\sigma_1, 0)$ and the radius vector to the point $(\sigma_\theta, \tau_\theta)$. This may be seen in Figure 2.16c if we visualize the point (σ_z, τ_{xz}) as rotated clockwise so that it coincides with the major principal stress. Note also from this figure that the planes on which the maximum shear stress acts are inclined at $45°$ ($2\theta = 90°$) to the principal planes.

Knowing the state of stress on any two planes and their orientation permits construction of the circle shown in Figure 2.16c and, thereby, graphical determination of stresses on planes passing through the same point with any other orientation. Because this idea was first suggested in 1882 by Mohr the circle is called the Mohr circle of stress.

The Pole

There is an especially useful point on the Mohr circle called the *pole*: Any straight line drawn through it will intersect the circle at a point representing the stress on a plane inclined at the *same* space orientation as the line. If the stresses on any one plane are known, the pole can be determined by drawing a line parallel to the plane and through the point on the Mohr circle corresponding to the stresses on that plane. The other intersection of that line with the circle is the pole. This is demonstrated in Figure 2.17. In the upper part of the figure two planes inclined at angles α and β, respectively, to the horizontal are shown, as well as the stresses on these planes. In the lower part of the figure is the Mohr circle describing the state of stress at the point through which these planes pass. Point A is the point representing the

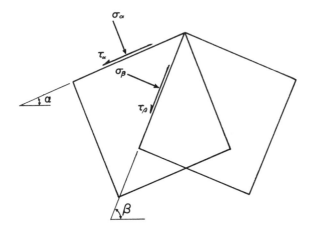

a) - Stresses on Two Planes

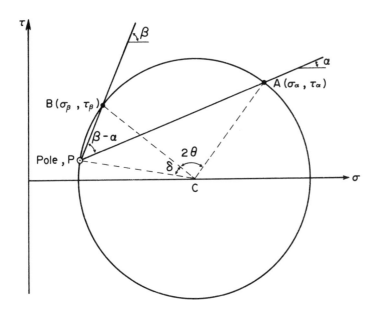

b) - Mohr Circle Showing Pole

Fig. 2.17—Illustration of pole method.

stresses on the plane inclined at the angle α. The pole has been determined by passing a line through point A inclined at the angle α. Point B, representing stress on the β plane, has been determined by passing a line through the pole inclined at the angle β. The angle 2θ is, of course, twice the space angle between the two planes, that is, $\theta = \beta - \alpha$.

The validity of the pole construction may be demonstrated by showing that angle APB is in fact $\beta - \alpha$. From Figure 2.17b,

$$\angle APB = \angle CPB - \angle CPA$$
$$= \tfrac{1}{2}(\pi - \delta) - \tfrac{1}{2}[\pi - (2\theta + \delta)]$$
$$= \theta$$

but the angle θ is simply $\beta - \alpha$, thus demonstrating the point.

Mohr-Coulomb Failure Theory

Having defined stress at a point, it is now possible to consider the Mohr-Coulomb failure theory for cohesionless soils in stress fields more general than that imposed by the direct shear test. The Mohr-Coulomb failure theory states that failure will occur along that plane for which the ratio of shear stress to normal stress reaches a critical limiting value. The equation, analogous to Equation 2–1, that expresses this in terms of stress is

$$\tau_{ff} = \sigma'_{ff} \tan \phi' \qquad (2\text{--}11)$$

where τ_{ff} is the shear stress on the failure plane *at failure*, σ'_{ff} is the normal stress on the failure plane *at failure*, and ϕ' is the angle of shearing resistance (i.e., the slope of the failure line). The shear stress on the failure plane at failure, τ_{ff}, is called the *shear strength*. Thus failure is said to occur when the shear stress on the potential failure plane equals the shear strength.

To see how this condition develops, let us visualize various stages during the conduct of a direct shear test. We shall assume that a normal force which corresponds to a particular normal stress, σ', is applied to the top of the specimen, as shown in Figure 2.2. A small shear force is applied, producing a shear stress τ. The Mohr circle corresponding to this condition is shown in Figure 2.18 as the smallest circle.* Only the upper half of the circle is shown. Under these circumstances the soil is in a stable condition. If we imagine that the shear force T (and the shear stress τ) is increased, the Mohr circle will become larger. This is shown as the second circle representing a stable condition. If the shear force T is increased sufficiently, the Mohr circle will just touch the failure line. The state of stress at this point of tangency represents the limiting condition that is, the stress on the failure plane equals the strength. *At that point, failure occurs.* The dashed Mohr

---

* In a conventional direct shear test, the stresses are not known on any plane other than that on which failure will be produced. Thus, a special measurement would be required in order to draw the Mohr circle. We shall assume that such a measurement has been made, and that we can construct the Mohr circle for each stage of the test.

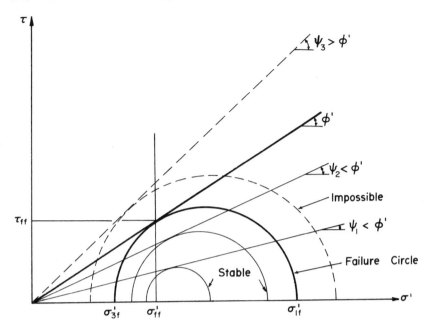

Fig. 2.18—Relation between Mohr circles and Mohr-Coulomb failure criterion.

circle which crosses the failure envelope is impossible. This state of stress can never be achieved because failure will occur first.

This can also be viewed from the point of view that the slope of the line from the origin, tangent to any Mohr circle, represents the maximum ratio of shear stress to normal stress for that circle. This ratio is termed the *obliquity* (Taylor, 1948). Thus we see that the obliquity of the tangents to the two stable circles is less than the angle ϕ', and that when the obliquity equals ϕ', failure occurs. Having selected the maximum obliquity as the criterion for failure, it is obviously impossible that there be a circle for which the obliquity is greater.*

Because the Mohr-Coulomb failure line is a limiting envelope to all of the possible Mohr failure circles, it is often called the Mohr-Coulomb failure envelope.

Some useful relationships that can be derived from the Mohr failure diagram are given below:

$$\sigma'_{ff} = \sigma'_{3f}(1 + \sin \phi') \tag{2-12}$$

$$\sigma'_{ff} = \sigma'_{1f}(1 - \sin \phi') \tag{2-13}$$

---

* This statement is true because we have defined it to be so. However, we shall find subsequently that "failure" is not quite so clearcut a phenomenon, and we may wish to define failure differently for different applications.

$$\tau_{ff} = \frac{\sigma'_{1f} - \sigma'_{3f}}{2} \cos \phi' \qquad (2\text{-}14)$$

$$\frac{\sigma'_{1f}}{\sigma'_{3f}} = \frac{1 + \sin \phi'}{1 - \sin \phi'} = \tan^2 (45° + \phi'/2) \qquad (2\text{-}15)$$

$$\frac{\sigma'_{3f}}{\sigma'_{1f}} = \frac{1 - \sin \phi'}{1 + \sin \phi'} = \tan^2 (45° - \phi'/2) \qquad (2\text{-}16)$$

Note also that, from the geometry of the failure diagram, there will be two failure surfaces which are oriented $\pm(45° + \phi'/2)$ to the major principal plane. An example of the use of the Mohr diagram is given below.

Example 2.4

A specimen of sand tested with a direct shear device failed owing to a shear stress of 2.40 kg/cm^2 when the normal stress on the failure surface was 4.16 kg/cm^2.

Determine the magnitude and direction of the principal stresses in the zone of failure.

Solution.   The solution is illustrated in Figure 2.19.

$$\phi' = \tan^{-1} \left(\frac{\tau_{ff}}{\sigma_{ff}}\right) = \tan^{-1} \left(\frac{2.40}{4.16}\right) = 30°$$

$$\sigma'_{3f} = \frac{\sigma'_{ff}}{1 + \sin \phi'} = \frac{4.16}{1 + 0.5} = \underline{2.77} \text{ kg/cm}^2$$

$$\sigma'_{1f} = \frac{\sigma'_{ff}}{1 - \sin \phi'} = \frac{4.16}{1 - 0.5} = \underline{8.32} \text{ kg/cm}^2$$

$$\theta = 45° + \frac{\phi'}{2} = 45° + 30°/2 = \underline{60°}$$

Directions of the principal stresses are shown in Figure 2.19b. The same result may be found, perhaps more conveniently, by constructing the Mohr circle to scale and determining the solution graphically as in Figure 2.19a, or by using the pole method illustrated in Figure 2.19c.

Triaxial Compression Tests

The direct shear test is simple in principle, but offers at least three major drawbacks, which invalidate its use for some purposes:

1. Stress and deformation conditions are markedly nonuniform along the failure surface, which leads to progressive failure from the edges of the specimen toward the center. Thus failure occurs at different points on the failure surface at different stages of the test. This effect is illustrated graphically by Roscoe (1953), and Morgenstern and Tchalenko (1967).

2. Shearing deformations occur over a limited, and unpredictable, depth of the specimen. We cannot, therefore, determine the *strains* corresponding to the developed average shear stress.

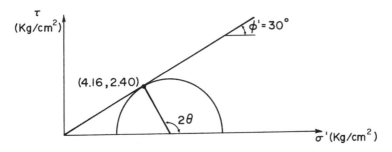

a) - Illustration of Mohr Diagram at Failure

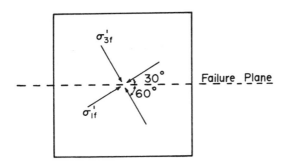

b) - Principal Direction

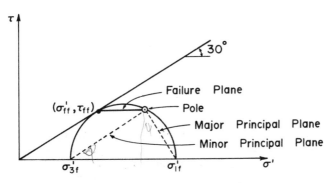

c) - Solution by Pole Method

Fig. 2.19—Sketch of solution for Example 2.4.

3. Drainage of fluid into or out of the specimen is difficult to control. We shall see later that this has important undesirable implications for the interpretation of test results.

A more commonly used strength test, which offers a number of advantages with regard to experimental details as well as interpretation of results, is the *triaxial compression test*. It is illustrated schematically in Figure 2.20a. The test specimen is a right circular cylinder of soil usually 2 or $2\frac{1}{2}$ times as high as it is wide. It rests upon a rigid pedestal, is covered at the top with a rigid loading cap, and is surrounded by a flexible membrane. The specimen is placed within a chamber filled with a convenient fluid. Pressure is then applied to this fluid to provide confinement. An additional axial load is applied to the specimen through a piston at the top.

Assumed stress conditions at a typical point within the triaxial compression specimen are shown in Figure 2.20b. Three important assumptions are customarily made with respect to the distribution of these stresses and corresponding strains to simplify interpretation of triaxial compression test results:

1. Stress conditions are assumed *homogeneous* throughout the test specimen. This means that the stresses at each point in the specimen are the same as at every other point. Thus the state of stress in the specimen can be considered a state of stress at a point, which permits description of the two-dimensional state of stress everywhere by a single Mohr circle. The result of this assumption is that the axial stress σ_a is equal to the total axial load applied to the specimen divided by the specimen area, the radial stress σ_r is equal to the fluid confining pressure, and these are principal stresses.
2. σ_θ is assumed to be the intermediate principal stress. This assumption is especially important because, according to the Mohr-Coulomb failure theory, the intermediate principal stress has no effect upon failure. Thus it is not essential to know the *magnitude* of σ_θ as long as it is the intermediate principal stress. However, in experimental investigations of the validity of the Mohr-Coulomb theory it is conventionally assumed that $\sigma_\theta = \sigma_r$.
3. Deformations of the specimen are also assumed homogeneous. Hence the sample is presumed to retain its cylindrical shape as it deforms. This permits computation of the specimen area normal to the longitudinal axis simply by knowing the axial displacements and the volume change. The average axial stress is determined using the area obtained from this computation.

The degree of validity of these assumptions is of considerable interest to us, because our understanding and interpretation of soil strength behavior have evolved, to a major extent, from the results of triaxial compression tests. A number of theoretical and experimental efforts have been made to

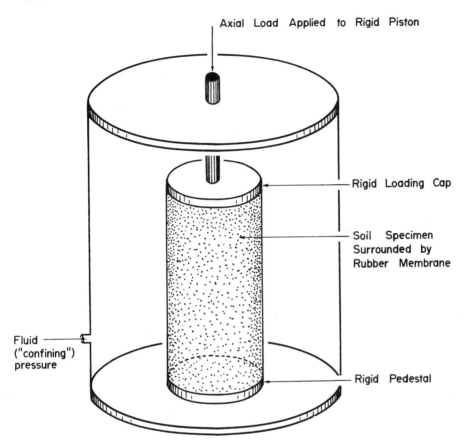

a) - Schematic Diagram of Triaxial Compression Test

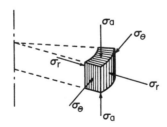

b) - Assumed Stress Conditions Within Test Specimen

Fig. 2.20—Description of the triaxial compression test.

elucidate this point.* On the basis of these studies, the following conclusions may be drawn:

1. Because of the frictional restraint between the end platens and the specimen, stresses are not homogeneous throughout the specimen. This is illustrated by the results shown in Figure 2.21. In Figure 2.21a, a nonlinear elastic cylinder is shown loaded by two end platens for which the frictional restraint between the specimen and the platens is sufficient to prevent any relative movement (Perloff and Pombo, 1969). The specimen is subjected to an average axial strain of 1 per cent, which produces an average axial stress of 24.3 pounds per square inch. Contours of equal stress are shown on a schematic representation of the specimen in Figure 2.21b. Because of symmetry, only one-quarter of the specimen need be shown. Hence the stress contours for σ_z, σ_r, σ_θ, and τ_{rz} are all shown in the figure.

2. The magnitude of σ_θ differs from that of σ_r at various points within the specimen. Throughout major portions of the specimen, σ_θ is the smallest normal stress. This places the conventional interpretation of the triaxial compression test in question.

3. Deformation conditions are nonhomogeneous in most cases. Recent efforts to reduce this effect have improved the quality of test results (Rowe and Barden, 1964; Blight, 1965, Kirkpatrick and Belshaw, 1968). However, the actual stress and strain conditions in the failure plane at failure are still not generally known with certainty.†

4. A number of other experimental artifacts may influence the reliability of triaxial compression test results, including piston friction, membrane leakage, and testing-strain-rate effects. Nonetheless, the triaxial compression test offers significant advantages over other current methods of strength testing. This point is discussed further in Chapter 8.

In spite of these manifest difficulties, we shall make the conventional assumptions in the discussion which follows.

Using the assumptions discussed above, we may calculate failure-plane conditions from a knowledge of the principal stresses at failure.

---

* Analyses of linear elastic solid cylinders loaded axially and subjected to varying types of end restraint have been conducted by Filon (1902), Pickett (1944), D'Appolonia and Newmark (1951), and Balla (1964). A recent numerical study of restrained elastic-plastic cylinders by Perloff and Pombo (1969) has indicated significant inhomogeneity of stress and strain. Experimental studies of nonuniform deformations, including Crawford (1959), Shockley and Ahlvin (1960), Roscoe et al., (1963), Rowe and Barden (1964), Bishop and Green (1965), and Kraft (1965) have shown that axial strains in the central portion of triaxial compression specimens differ significantly from the "average" axial strains. Recent developments in testing devices (Bjerrum and Landva, 1966; Ko and Scott, 1967) circumvent some of these difficulties.

† In general, the ϕ' determined from direct shear tests and triaxial compression tests would differ even if account were taken of the experimental artifacts. This is so because the Mohr-Coulomb failure law is a simplified representation of conditions obtained from fundamental consideration of particle-to-particle interaction (Rowe, 1962, 1969; Horne, 1965). Nonetheless it is still useful at present to view strength phenomena in terms of the Mohr-Coulomb criterion.

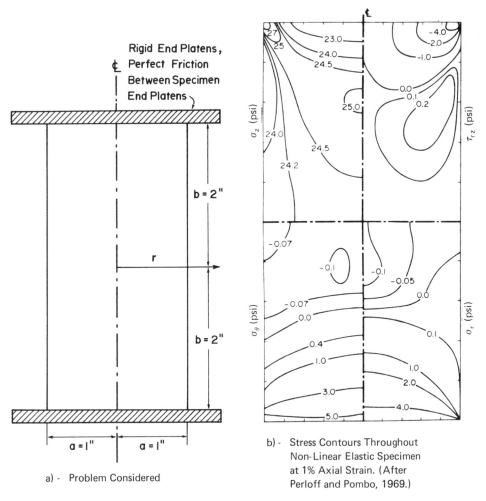

Rigid End Platens,
₵ Perfect Friction
Between Specimen
End Platens

b = 2"

r

b = 2"

a = 1"   a = 1"

a) - Problem Considered

b) - Stress Contours Throughout
Non-Linear Elastic Specimen
at 1% Axial Strain. (After
Perloff and Pombo, 1969.)

Fig. 2.21—Stresses in elastic cylindrical specimen with end restraint.

Example 2.5

A cylindrical specimen of dry sand failed in a triaxial compression test when the nominal stresses were $\sigma'_{af} = \sigma'_{1f} = 5.88$ tsf (tons/square foot) and $\sigma'_{rf} = \sigma'_{3f} = 1.50$ tsf. Determine the stresses on, and the orientation of the failure plane at failure. What is ϕ'?

Solution. Using Equation 2–15:

$$\tan^2(45° + \phi'/2) = \frac{\sigma'_{1f}}{\sigma'_{3f}} = 3.92$$

$$\phi' = 2(63.2 - 45) = \underline{36.4°}$$

The solution by use of the pole method is illustrated in Figure 2.22. Note that, in this case, the pole is at σ'_{3f}.

Alternatively, from Equation 2–12:

$$\sigma'_{ff} = \sigma'_{3f}(1 + \sin \phi') = (1.5)(1 + .594) = \underline{2.39} \text{ tsf*}$$

From Equation 2–14:

$$\tau_{ff} = \frac{\sigma'_1 - \sigma'_3}{2} \cos \phi' = (1/2)(5.88) - (1.50)(.805) = \underline{1.76} \text{ tsf}$$

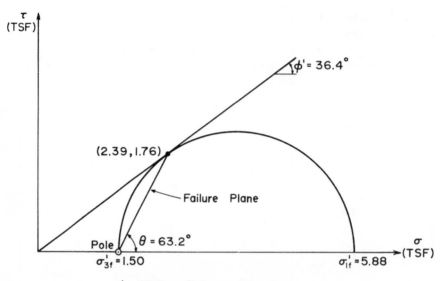

a) - Mohr Failure Diagram

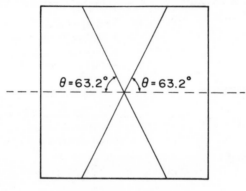

b) - Failure Planes

Fig. 2.22—Sketch of solution for Example 2.5.

* We can check our answer with Equation 2–13:

$$\sigma'_{ff} = \sigma'_{1f}(1 - \sin \phi') = (5.88)(1 - .594) = \underline{2.39} \text{ tsf}$$

The failure surfaces are oriented with respect to the major principal plane (a horizontal plane in this case) at the angle

$$\theta = \pm(45° + \phi'/2) = \pm63.2°$$

as shown in Figure 2.22b.

Effect of Water on Shearing Resistance

The effect of water on the observed strength of soils can be very important. To illustrate this, let us look at the results of several experiments. Consider a triaxial compression test performed on a dry, well-graded, medium-to-fine sand in a moderately loose condition, say $D_r = 0.5$. The Mohr circle at failure for this test is shown in Figure 2.23a as a dashed line. Suppose now that we conduct a similar test under the *same confining pressure* on an identical specimen, except for the fact that the specimen is saturated with water. If we conduct the test sufficiently slowly so that water is free to move into or

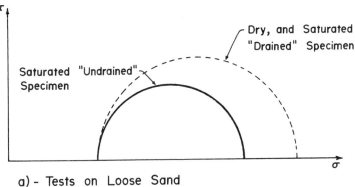
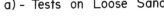

a) - Tests on Loose Sand

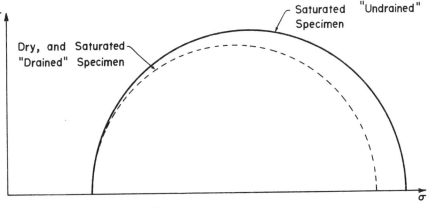

b) - Tests on Dense Sand

Fig. 2.23—Mohr failure circles for triaxial compression tests on sand.

out of the specimen (say, for this soil, a test that lasts approximately one-half-hour), the result will be essentially identical to that for the dry sand. This is shown in Figure 2.23a. Although the presence of water has a pronounced effect on the coefficient of friction between solid polished minerals, it does not seem to be reflected in the Mohr failure .envelope for assemblages of mineral particles.

Let us now perform a similar experiment on an identical specimen of the saturated loose sand, *except* that after the application of the confining pressure, we shall prevent drainage of water into or out of the sample. Such an *"undrained"* test can be conducted either by closing all drainage lines leading to the saturated sample, or by carrying out the test at a sufficiently rapid rate that the water does not have time to drain out of or into the sample. The result of this test is shown as the solid curve in Figure 2.23a. This circle is very much smaller than that for the other two tests on the same loose sand! What is the difference between these tests? Why does the identical specimen appear so much weaker when tested in a saturated *"undrained"* condition than it does when tested in a dry, or saturated, *"drained"* condition?

In a similar fashion, consider another set of these experiments performed on the same well-graded, medium-to-fine sand, which has now been compacted to a dense condition, say $D_r = 0.9$. In this series of experiments, the *dry* and saturated *drained* specimens will fail at essentially the same state of stress, shown by the dashed circle in Figure 2.23b. However, the saturated *undrained* specimen will appear stronger than the other two, as indicated by the solid circle in Figure 2.23b. This is just the reverse of what we observed for the test on loose sand, and appears to confuse the picture still further.

In order to understand the basis for this apparently anomalous behavior, we must consider a property of particulate materials called *dilatancy*. Dilatancy is the tendency to change volume upon the application of a shear strain. This very important property of soils was first noticed by Osborne Reynolds (1885).

Suppose that we had measured the quantity of water flowing into or out of our saturated samples during the course of the *drained* tests. The results of such observations along with the corresponding "stress-strain" curves are shown in Figure 2.24. There are several important features to be noticed in these curves. The stress-strain relationships for the dense and loose sand (Figure 2.24a) are clearly quite different. The curve for the dense sand exhibits a distinct peak at relatively small strain; continued straining causes a reduction in the shear stress. The curve for the loose sand gradually increases to a maximum value at much larger strain. Note that, after very large strains, the "ultimate strength" appears to be approximately the same for the sand in both dense and loose states.

The volume change in response to strain (Figure 2.24b) is especially interesting. The loose sand contracts as it is sheared producing a net reduction in volume at failure. However, the rate of volume change with respect to

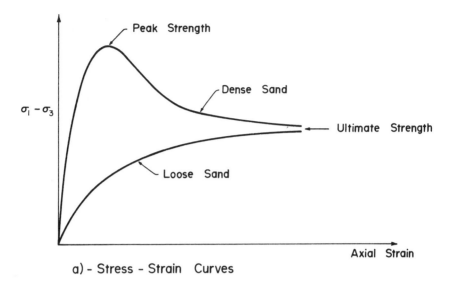

$\sigma_1 - \sigma_3$

Peak Strength

Dense Sand

Ultimate Strength

Loose Sand

Axial Strain

a) - Stress - Strain Curves

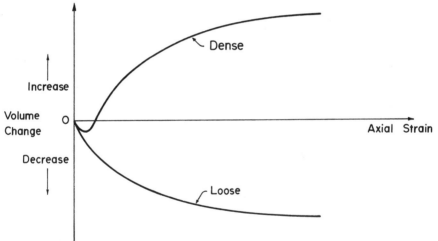

Increase

Volume
Change   O

Decrease

Dense

Axial Strain

Loose

b) - Volume Change - Strain Curves

Fig. 2.24—Volume change during triaxial compression test.

strain at the "ultimate strength" is approximately zero.* By contrast, the
dense sand expands, following a slight contraction upon initial straining.

---

* The stage of yielding at which continued straining occurs without a change of stress or
volume change has been termed the "*critical state*" (Roscoe, et al., 1958). The yielding process
by which soils approach this critical state has been viewed from the perspective of plasticity
theory by Roscoe and his co-workers. An exposition of this approach is given by Schofield and
Wroth (1968). We shall make use of certain aspects of these developments subsequently.

This occurs in spite of the fact that the applied increase in axial stress has a component which produces a decrease in volume. Expansion continues up to strains larger than those required to produce peak stress; at large strains the rate of volume change is also nearly zero.

The general tendency for dense soils to expand when sheared and for loose soils to contract when sheared is intuitively reasonable, as illustrated in Figure 2.25. If a sand is in a very dense condition, the only way that shearing can occur is for grains to move apart. Conversely, if the sand is in a very loose condition, the grains will tend to move together. Thus, a dense sand exhibits a tendency for volume increase (dilatation) when it is sheared to failure, and a loose sand exhibits a tendency for volume decrease when it is sheared to failure.

Pore Water Pressure

Now let us consider what happened to the *fully saturated* sand specimens, for which no drainage of water was allowed. If we assume that water is

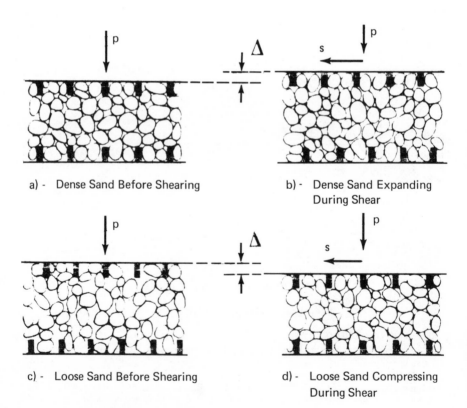

a) - Dense Sand Before Shearing

b) - Dense Sand Expanding During Shear

c) - Loose Sand Before Shearing

d) - Loose Sand Compressing During Shear

Fig. 2.25—Dilatancy of sand (From Leonards, 1962).

incompressible compared to the soil skeleton, and no flow of water into or out of the specimen will occur, we see that the volume of the specimen during such a test must remain constant. In order for this to happen, the water must sustain stresses sufficient to prevent volume change of the specimen, despite the tendency shown in Figure 2.24 for such volume change to occur. For example, in the case of the loose sand, which tends to contract, it seems reasonable to suppose that the water must be experiencing some compressive stress, the magnitude of which is related to the amount of volume change which would occur if it were permitted. If this is the case, then the water is carrying some of the normal stress applied to the specimen (of course, the water cannot sustain a static shear stress). If the water carries some of this normal stress, then the soil skeleton carries less normal load and therefore has less shear strength than it would in the drained test. The converse of this reasoning applies to the test on the dense sand.

This explains in qualitative terms the test results shown in Figure 2.23. It also illustrates a very important notion: The shear strength of a granular soil is related to the normal forces carried by the *soil skeleton*. If water in the soil voids carries some of the applied force, then the shear strength is reduced. Thus, the strength of a cohesionless soil is determined not by the *total* normal load applied to the failure plane, but rather to the *effective* normal load, that is, that portion of the total load which is actually carried by the skeleton.

In order to quantify this important concept, and express it in terms of applied stresses, consider a saturated soil mass, shown in Figure 2.26, subjected to an applied *average* normal stress σ. Imagine that the soil mass is cut along a surface so that a free-body diagram could be drawn. Let us suppose that this surface is approximately horizontal, but is wavy, so that it always passes *between* particles rather than *through* particles, as shown in the figure. Then the surface will pass through areas of solid-to-solid contact, and through void spaces filled with water. If we let:

A_{th} be the total horizontal projection of the cutting surface, for the soil mass considered

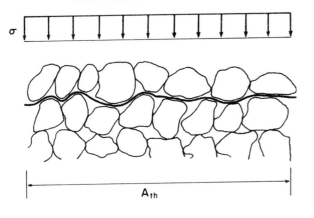

Fig. 2.26—Soil mass subjected to average stress.

A_{ch} be the horizontal projection of the contact area between the solids lying in the cutting surface

A_{wh} be the horizontal projection of the portion of the cutting surface which passes through water, then by the requirements of equilibrium in the vertical direction,

$$\sigma A_{th} = \sigma^* A_{ch} + u A_{wh} \qquad (2\text{-}17)$$

where σ^* is the actual intergranular stress at points of contact, and u is the pressure in the water. Because the water referred to occupies the pore spaces in the soil, this pressure is commonly called the *pore water pressure*.

Dividing by A_{th} gives:

$$\sigma = \sigma^* \frac{A_{ch}}{A_{th}} + u \frac{A_{wh}}{A_{th}} \qquad (2\text{-}18)$$

In this equation, we can measure the pore water pressure, u, the total area, A_{th}, and at least an average value of σ. Furthermore, if the soil is saturated, $A_{wh} = A_{th} - A_{ch}$. Therefore, if we were able to determine A_{ch}, we could calculate σ^*. Although this is not a simple matter, A_{ch} has been measured for some soils.* It appears that for coarse-grained materials, A_{ch} is very small, approaching zero (Bishop and Eldin, 1950). Therefore, A_{wh} approaches A_{th}, and σ^* must be very large. Equation 2–18 becomes

$$\sigma = \sigma^* \frac{A_{ch}}{A_{th}} + u \qquad (2\text{-}19)$$

From Equation 2–19, it is evident that the product $\sigma^* A_{ch}$ must approach a finite limit even though σ^* is very large and A_{ch} is very small. In fact, the first term on the right side of Equation 2–19 must be some measure of the average stress carried by the soil skeleton even though we cannot give it a precise physical interpretation. Because this quantity is used frequently, it is accorded a special name. It is called *effective stress*, σ', defined by

$$\sigma' \equiv \sigma^* \frac{A_{ch}}{A_{th}} \qquad (2\text{-}20)$$

Thus, Equation 2–19 may be rewritten as

$$\sigma = \sigma' + u \qquad (2\text{-}21)$$

which is the *principle of effective stress*, proposed by Terzaghi (1923).†

Although the effective stress is clearly a fictitious quantity (as is the total stress), experience has shown it to be extremely useful. The shearing resistance, and many other important properties of granular and fine-grained soils, depend to a marked extent upon the effective stress. In fact, *the relationship between shear strength and normal stress on the failure plane at failure is*

---

* Good discussions of the problems inherent in such measurements and their interpretation are given by Lambe and Whitman (1959) and Skempton (1961a).

† The development given describes a special case of a more general approach, which, for most practical purposes is sufficient (Skempton, 1961a).

really *a relationship between shear strength and* effective *normal stress.* That is why Equation 2–11 was written in terms of σ'_{ff}, the *effective stress* on the failure plane at failure, rather than σ_{ff}, the *total stress.*

We are now able to interpret the apparently anomalous behavior shown in Figure 2.23. The results from Figure 2.23 are repeated in Figure 2.27, along

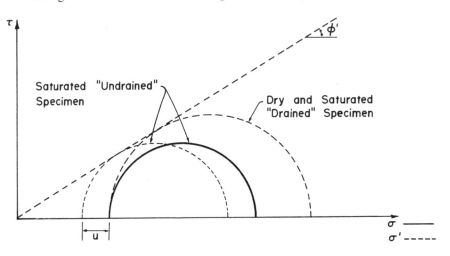

a) - Tests on Loose Sand

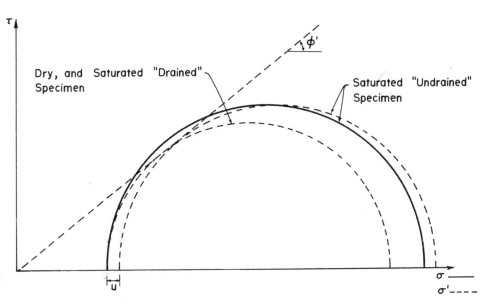

b) - Tests on Dense Sand

Fig. 2.27—Failure conditions for triaxial compression tests on sand.

with the Mohr failure envelopes for the two densities (recall that these are, of course, in terms of *effective* stresses). The total-stress circles are shown as solid lines, the effective-stress circles as dashed lines. From Figure 2.27, it is evident why the tests on the dry and saturated "drained" specimens produced the same result: there was no pore water pressure in these tests, and therefore the *total* and *effective* normal stresses were the same. Thus, the shear strength was the same.

In the case of the "undrained" test on the saturated specimen of loose sand, however, the tendency for volume reduction created a positive pore water pressure which caused a reduction in effective stress. Hence, the effective-stress Mohr circle was shifted to the left of the total-stress Mohr circle by an amount equal to the pore water pressure, as seen in Figure 2.27a. Because failure occurs when the *effective*-stress Mohr circle touches the failure envelope (which is in terms of effective stresses), the development of positive pore water pressure causes a *reduction* in the shear strength. Note that the pore water pressure does not affect existing shear stress directly, nor does it change ϕ'. Its only influence is to change the effective *normal* stress, which in turn modifies the shear strength. This can be seen from the fact that in Figure 2.27, the total and effective stress Mohr circles at failure *for the same test* are the same size; they are just displaced along the normal stress axis.

In the case of the "undrained" test on saturated dense sand, there is a tendency for volume dilatation (increase), which is counteracted by a negative pore water pressure (below atmospheric pressure). Thus, in the case of dense sand, the negative pore water pressure causes the effective stress to be *larger* than the total stress, and the Mohr circle is shifted to the right. This explains why the strength in the undrained test was larger than the Mohr circle at failure for the "drained" specimen.*

Example 2.6 —————————————————————————

A specimen of sand is tested in an undrained triaxial compression test in which the total stresses at failure are: $\sigma_{3f} = 1.00$ tsf, $\sigma_{1f} = 2.42$ tsf, and the pore water pressure, u_f, is measured to be 0.50 tsf.

---

* The angle ϕ' obtained in a "drained" test will not, in general, be the same as that obtained in an "undrained" test in which pore water pressures are measured. The reason for this is that in a constant-volume test, the work done on the soil acts entirely to overcome shearing resistance. When the rate of volume change at failure is not zero, work must be done against the normal stresses on the specimen. Consequently, a part of the measured maximum shear resistance in such a case comes from the work required to produce volume change. A detailed discussion of this problem is given by Skempton and Bishop (1954), who show that a major portion of the difference in ϕ' for loose and dense sands, when measured at peak stress, can be explained by the energy involved in changing the soil volume at failure.

For loose sands, in which the rate of volume change at peak stress is approximately zero (see Figure 2.24), the ϕ' determined in drained and undrained tests will be the same. For dense sands, however, the rate of volume change at peak strength is positive, and ϕ' measured in a drained test may be as much as several degrees larger than that measured in an undrained test.

The specific correction to be applied to the drained strength to bring the two values of ϕ' into correspondence has been a subject of some controversy (Rowe et al., 1964); for the present discussion we shall assume that ϕ' in the drained tests will be the same as that in the undrained tests.

(a) What are the major and minor effective principal stresses at failure?
(b) What is ϕ'?
(c) What are the shear and effective normal stresses on the plane of failure?

Solution.   The solution is shown diagrammatically in Figure 2.28.
(a)   From Equation 2–21,

$$\sigma'_{3f} = \sigma_{3f} - u_f = 1.00 - 0.50 = \underline{0.50} \text{ tsf}$$
$$\sigma'_{1f} = \sigma_{1f} - u_f = 2.42 - 0.50 = \underline{1.92} \text{ tsf}$$

(b)   From Equation 2–16,

$$\tan^2 (45° + \phi'/2) = \frac{\sigma'_{1f}}{\sigma'_{3f}} = \frac{1.92}{0.50} = 3.84, \phi' = \underline{36°}$$

(c)   From Equations 2–12 or 2–13,

$$\sigma'_{ff} = \sigma'_{3f}(1 + \sin \phi') = \sigma'_{1f}(1 - \sin \phi') = \underline{0.791} \text{ tsf}$$

(d)   From Equations 2–11 or 2–14,

$$\tau_{ff} = \sigma'_{ff} \tan \phi' = \frac{\sigma'_{1f} - \sigma'_{3f}}{2} \cos \phi' = \underline{0.574} \text{ tsf}$$

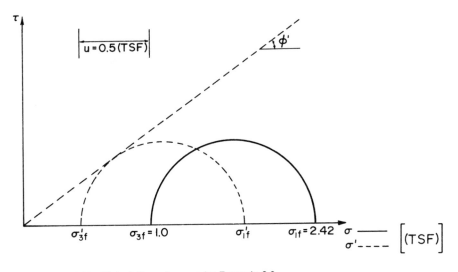

Fig. 2.28—Mohr failure diagram for Example 2.6.

From the point of view of the stability of structures made of cohesionless soils, it is desirable to have the soil in a dense condition. This is so, in part, because ϕ' will be greater for the dense soil than for the loose soil. Of more importance, however, is the fact that a rapidly applied shear stress which creates a tendency for expansion in the dense soil leads to a reduction in pore water pressure, accompanied by an increase in shear strength. Even if this increase is only temporary, the strength will be larger than or equal to that for the "drained" test. Conversely, a rapidly applied load will produce increased pore water pressures in a loose granular material and a temporary reduction in strength. This may be disastrous, as apparently was the case at the Fort Peck Dam.

The problem is especially pronounced if the load is cyclic in nature, as for example, in an earthquake (Seed, 1968). Seed and Lee (1966) have shown that cyclic loading produces increased pore water pressures, even in dense sands, which may lead to drastic reductions in shear strength. However, the resistance of dense soils to this effect is much greater than that of loose soils.

Critical Void Ratio

If dense sands expand when sheared to failure and loose sands contract, it seems reasonable to suppose that there is some intermediate state in which the volume change at failure will be zero. In fact, such a state does exist and may be observed in the following way: Imagine that a series of "drained" triaxial compression tests (or some other equivalent shear test) is performed, at a particular effective confining pressure, on a series of saturated specimens of the same sand with different initial void ratios. If stress, strain, and volume change are measured during the course of the tests, the results will be of the sort illustrated in Figure 2.24. If the net volume change *at failure* is plotted as a function of the void ratio at the beginning of shearing, a curve such as that shown in Figure 2.29a results. The initial void ratio for which there is no net volume change at failure is called the *critical void ratio, e_c*. If the void ratio is higher than the critical void ratio, then the sand behaves as a loose sand, and will contract when sheared to failure. If the void ratio is lower than the critical void ratio, then the sand will behave as a dense material and will exhibit a net expansion at failure. This permits us to define precisely the terms *loose* and *dense* for granular materials. A *loose* soil is one which exhibits a net reduction in volume when sheared to failure. A *dense* soil is one which exhibits a net increase in volume when sheared to failure.

Unfortunately, the critical void ratio is not a unique quantity for a granular soil. For if a similar series of experiments are performed on the same sand, but at different effective confining pressures, a family of curves, such as those shown in Figure 2.29b, will result. Thus, the greater the confining pressure, the lower is the critical void ratio. Alternatively, we can say that a higher pressure produces a stronger tendency for contraction in response to shear strain. Hence, our definition of loose and dense cohesionless soils depends upon the confining *pressure* as well as the void ratio or relative density. A large enough confining pressure will cause a sand with $D_r = 0.7$ to behave as if it were loose.

In the case of similar tests conducted "undrained," the pore water pressure at failure can be plotted as a function of the initial void ratio and effective

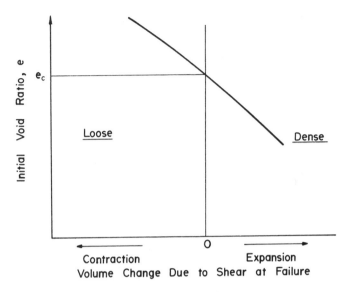

a) - Effect of Initial Void Ratio

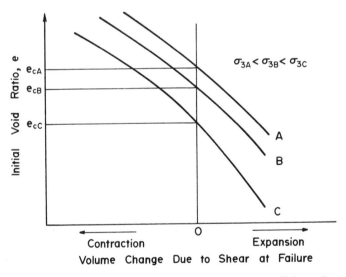

b) - Effect of Initial Void Ratio and Confining Pressure

Fig. 2.29—Volume change at failure for granular soils, as a function of initial void ratio and pressure in drained tests.

confining pressure, in a fashion analogous to that shown in Figure 2.29b. The result is shown in Figure 2.30. Thus, if a granular material is placed in a condition looser than the critical void ratio, a rapidly applied stress will create an increase in pore water pressure and a corresponding reduction in strength.

It appears that e_c is also a function of the type of test used to determine it. This results, at least in part, from the fact that test specimens do not generally possess the point-like attributes which we generally ascribe to them. Thus, one part of a sample may be expanding while another part is contracting. This is due both to the unavoidable nonuniformities within test specimens, and the inhomogeneous stress fields imposed by the testing apparatus as discussed above.

Strength in Terms of Total Stresses

The strength of cohesionless soils, when viewed from the point of view of total stresses, is a confusing phenomenon. In those cases where total and effective stresses are the same, no confusion results. However, when pore water pressures exist, total stress analysis of strength may be very over-conservative in some cases and unsafe in other cases. This is illustrated in Figure 2.31. The results of a series of shear tests performed on what we intuitively would consider to be a loose sand are shown in Figure 2.31a. The Mohr-Coulomb failure envelope is drawn as a dashed line. At most confining pressures, the pore water pressure is positive and the results, plotted

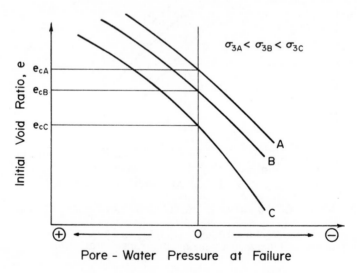

Fig. 2.30—Pore water pressure at failure for cohesionless soils as a function of initial void ratio and confining pressure in undrained tests.

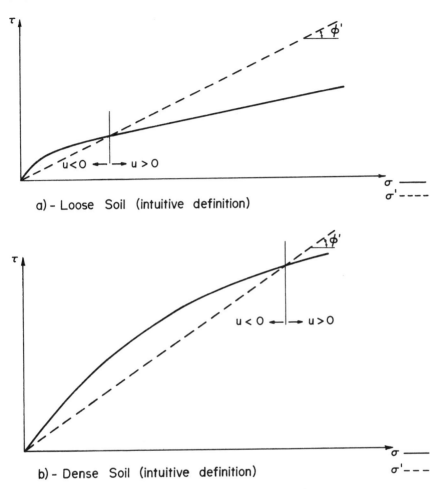

a) - Loose Soil (intuitive definition)

b) - Dense Soil (intuitive definition)

Fig. 2.31—Comparison of strength of cohesionless soils
analyzed in terms of total and effective stresses.

in terms of total rather than effective stresses, will lie to the right of the
failure envelope. At sufficiently small confining pressures, however, the soil
will tend to expand when sheared, that is, behave as a dense material, and
the total stress envelope will lie to the left of the Mohr-Coulomb failure
envelope. Although the total stress line is not actually a straight line, the
slope of that portion corresponding to larger stresses will be less than ϕ'.

Test results for a cohesionless soil in what we intuitively would consider
to be a dense condition are shown in Figure 2.31b. In this case, the pore
water pressures will be negative under low and moderate confining pressures,
and the failure envelope in terms of total stresses lies to the left of the Mohr-
Coulomb failure envelope. When the confining pressure becomes sufficiently

large, however, the dense sand behaves as if it were loose: the pore water pressures are positive and the failure envelope in terms of total stresses lies to the right of the Mohr-Coulomb failure envelope. Thus, we see that a "loose" sand, or a "dense" sand can behave in *either* a loose or dense fashion depending upon the combination of void ratio and confining pressure.

For granular soils which tend to contract when sheared, total stress analysis from undrained tests on saturated specimens underestimates the true strength. For soils which tend to expand when sheared, total stress analysis for undrained tests overestimates strength. On the other hand, an understanding of the strength of cohesionless soils in terms of *effective* stresses, combined with insight into the tendency for volume change, assists us in predicting the behavior of cohesionless soils under any combination of stresses and drainage conditions.

2.6 FORT PECK SLIDE RECONSIDERED

On the basis of our consideration of the shearing resistance of granular soils, we can now investigate probable conditions within the Fort Peck Dam before and during the slide. Figure 2.32a shows a portion of the estimated sliding surface, and stress conditions at a typical point on that surface before failure. This stress may be resolved into two components: the effective normal stress applied to the failure surface, σ', and the shear stress applied to the failure surface, $\tau_{applied}$. These stresses are shown in Figure 2.32b on a Mohr diagram.* Note that $\tau_{applied}$ is *not* the strength corresponding to σ'. To emphasize this, τ_{ff} corresponding to σ' is shown on the same figure. Thus, Figure 2.32 implies that the dam will be stable under the conditions shown.

When the failure was initiated in the underlying weathered shale, shown in Figure 2.1, two effects probably resulted. The shearing resistance of the shale reduced from its peak value to some lower value, thereby causing more shear stress to be transmitted to the potential failure surface. In addition, the shearing strains which propagated back into the dam as a result of this movement in the shale appear to have caused a positive pore water pressure to develop in the shell material. The probability of a pore water pressure response was enhanced by the fact that, at the time of the failure, the reservoir level was at elevation 2117.5 ft, to which it had been lowered from elevation 2137 ft. Thus the shell material was likely saturated. This combination of

---

* There are a number of ways that the magnitude of stresses smaller than those required to produce failure can be determined. One approach is to assume that the stress will be the same as in a linear elastic dam of the same shape, for which a solution has been obtained (Perloff et al., 1967). Alternatively, numerical solutions permit determination of stresses for non-linear materials case by case (Clough and Woodward, 1967). More commonly, the dam is divided into vertical slices, and the stresses estimated from equilibrium considerations of the slice. These methods and their relative merits are considered in greater detail in subsequent chapters.

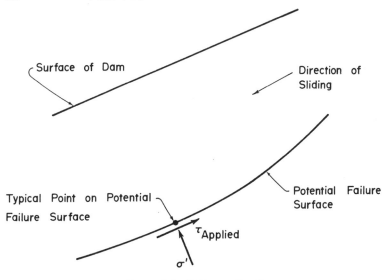

a) - Typical Point on Potential Failure Surface

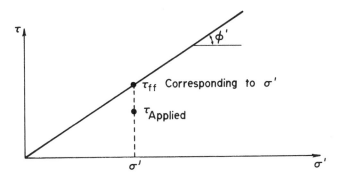

b) - Mohr Diagram at Typical Point Shown in (a) Before Slide

Fig. 2.32—Conditions along potential failure surface before slide

events would create the effect shown in Figure 2.33. The figure indicates that the conditions at failure at the typical point, illustrated in Figure 2.32, were brought about by a combination of a reduction in effective stress and an increase in applied shear stress. Thus the shearing *strength* was decreased even as the applied shear stress was probably increased somewhat.

The question then arises: If the pore water pressure created were larger than that required to just cause failure, what would happen? In the case of a strong imbalance of forces, we would expect relatively large accelerations in response to the disequilibrium. For a cohesionless soil, an apparent "flowing" or *liquefaction* of the material results. It has been suggested, and

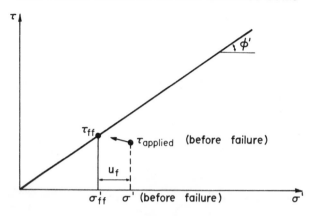

Fig. 2.33—Conditions at failure.

indeed seems reasonable, that such liquefaction along the failure surface was the mechanism which caused the large mass of slide material to be carried for distances up to 1000 ft in a few minutes, and to come to rest at such a flat slope.*

In view of this postulated explanation, the question which remains to be answered is: Was the pore water pressure created in the shell material by the initial slide movement positive or negative? For if the pore water pressure was negative, the temporary stability of the dam would have been enhanced rather than reduced. Hence, we are led to inquire whether the hydraulic fill shell material was at a void ratio above or below the critical void ratio.

Middlebrooks (1942) attempted to shed light on this question. Figure 2.34 shows results presented by him, of drained triaxial compression tests conducted to determine the critical void ratio of the shell material. The critical void ratio is shown as a function of the effective vertical pressure at failure, which in the triaxial compression test, is the major principal stress. Because the tests were drained, the major principal stress was proportional to the initial effective confining pressure (Equation 2–15). Thus this diagram may be interpreted as the relationship between critical void ratio and confining pressure. The points on the figure could be obtained from curves such as those shown in Figure 2.29 or 2.30. The two limiting lines in Figure 2.34 were given by Middlebrooks (1942) to indicate the range of values observed in the various tests.

Also plotted in this figure are measurements of the void ratio at various depths within the fill as a function of the estimated vertical effective stress. The *in situ* void ratios fall on, or just below, the lower limiting line, suggesting that the dam material was at or slightly below the critical void ratio.

---

* Seed (1968) describes numerous cases where landslides occurred during earthquakes, owing to liquefaction of saturated sands.

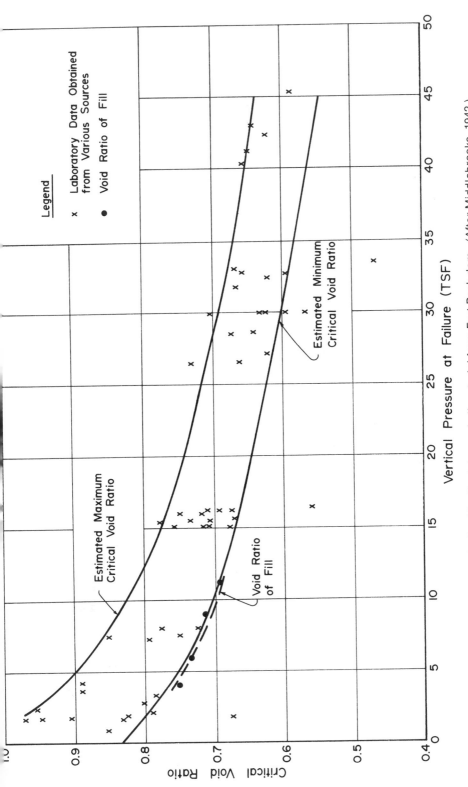

Fig. 2.34—Summary of critical void ratio data on shell material from Fort Peck dam. (After Middlebrooks, 1942.)

However, important points oppose this conclusion. First, we note that, at the time of the laboratory testing (*circa* 1938), a number of potential artifacts were not well recognized. In particular, piston friction probably led to *overestimation* of the vertical effective stress at failure. Thus it is likely that the "Estimated Minimum (and Maximum) Critical Void Ratio" lines in Figure 2.34 should be displaced to the left.

A second factor to be considered is one of the differences between the triaxial compression test and the field conditions. We saw above that the vertical stress is proportional to the effective confining pressure in a drained triaxial compression test on a granular soil. This is illustrated in Figure 2.35a. At a typical point in the field, however, the vertical stress will *not* be the major principal stress at failure, because failure is induced along some surface other than that inclined at $45° + \phi'/2$ to the horizontal. Consequently, the effective confining pressure corresponding to the drained triaxial test results will be greater in the field than in the laboratory, as shown in Figure 2.35b. Therefore, the field data (dashed curve) in Figure 2.34 should be displaced to the right in order to represent the field and laboratory data on an equivalent basis.

Several other discrepancies between assumed and actual conditions can be identified, but their effect is not known. These include: the difference between axisymmetric conditions in the triaxial compression test and approximately plane strain conditions in the field; and variations in grain-size distribution in the shell from the outside to the core.

Based on the discussion above, it appears likely that, in fact, the void ratio of the fill was above the critical void ratio, and positive pore water pressures did develop, which drastically reduced the strength. The excessive movement of the slide mass certainly supports such a view.

Although we are not yet in a position to quantify our conclusion, we have already developed a sufficient appreciation for the primary factors influencing the shearing resistance of granular materials that we can identify such a problem. Having done so, we can investigate solutions or perhaps alternatives.

If it were possible to insure that a hydraulic fill dam could be placed or compacted to a void ratio sufficiently below e_c, this method of construction would become most popular because of its economy. Since this has not been possible, unfortunately, the method has not been used for a major earthfill dam in this country since the Fort Peck Dam was constructed, although it has been used extensively in the Soviet Union.

2.7 SUMMARY OF KEY POINTS

Summarized below are the key points discussed in this chapter. They are arranged in a somewhat different sequence from that in which they were presented originally. The section in which they were considered is noted so

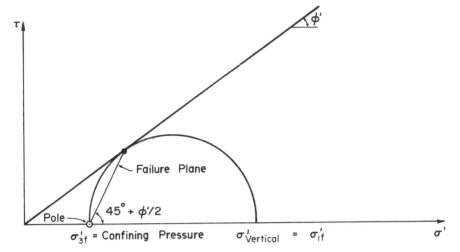

a) - Conditions in Drained Triaxial Compression Test

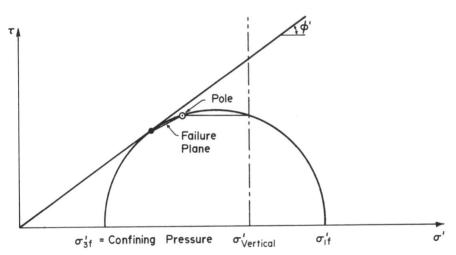

b) - Condition Corresponding to Drained Test for Typical Point on Failure Surface Within Dam with Same $\sigma'_{Vertical}$ as in (a) Above

Fig. 2.35—Comparison of failure conditions in laboratory and field.

that reference can be made to the original, more detailed discussion:

1. Useful descriptors of a coarse-grained soil include:
   a. Mineral type and shape of grains (Section 2.2).
   b. Unit weight of the soil mass (Sections 2.2–2.3).
   c. Density of packing of the grains. This is indicated by the relative

density, D_r, which is the most important factor affecting the strength of a particular coarse-grained soil (Sections 2.2–2.4).
   d.   Grain-size distribution. This is a useful qualitative indicator of mechanical properties and geologic origin of a deposit (Section 2.4).
2. a.   Stress is an artificial quantity, defined, at a point, as $\lim\limits_{\Delta A \to 0} \Delta F/\Delta A$. Despite the inconsistency of the assumption of a continuum, we find stress to be an exceedingly useful concept (Section 2.5).
   b.   Effective stress at a point is also an artificial quantity, defined as $\sigma' = \sigma - u$. It is useful, however, because we find that the mechanical behavior of coarse-grained soils is governed by the effective stress (Section 2.5).
3.   Failure in a granular mass is caused by shearing forces. The Mohr-Coulomb failure criterion states that failure occurs at a point in a cohesionless soil mass when the ratio of shearing stress to normal effective stress reaches a critical value (Section 2.5):

$$\tau_{ff} = \sigma'_{ff} \tan \phi'$$

The planes along which this critical value is reached, and failure will occur, are inclined at angles $\pm(45° + \phi'/2)$ to the major principal plane.
4.   The shear strength of soils is most commonly measured in the direct shear test or the triaxial compression test. Each of these test methods has a variety of limitations which must be considered in interpreting test results (Section 2.5).
5.   The critical void ratio, e_c, defined as the void ratio at which there is no tendency for volume change at failure, is an important parameter in the prediction of the response of coarse-grained soils to shearing stresses (Section 2.5):
   a.   The tendency for volume change at failure depends upon the magnitude of the void ratio relative to e_c; the volume increases if $e < e_c$ and the volume decreases when $e > e_c$.
   b.   The confining pressure affects the critical void ratio; e_c decreases as confinement increases.
   c.   The pore water pressure in saturated coarse-grained soils sheared under undrained conditions is a function of the tendency for volume change. A tendency for volume decrease produces a positive pore water pressure. A tendency for volume increase produces a negative pore water pressure. When $e > e_c$ for a granular soil the positive pore water pressure created by shearing effects leads to a reduction in strength, and may produce liquefaction.

PROBLEMS

2.1.   Use a "phase diagram" to find each of the following relationships in terms of the given quantities. (*Note*: The unit weight of water, γ_w, is always considered to be a known quantity.)

(a) Given n, find e.
(b) Given G, e, find w for a fully saturated soil.
(c) Given G, w, S_r, find γ.
(d) Given G, w, find γ', the unit weight of the soil when submerged in water, for a fully saturated soil.

2.2. A soil is at a void ratio of 0.9, with a specific gravity of the solid particles of 2.70.

(a) Can the water content be determined from the information given?
(b) If the water content cannot be determined, can upper and lower limits be determined? If so, what are they?

2.3. A sample of silty sand is compacted in a mold whose volume is $1/30$ ft^3. The moist weight of the soil is 4.00 lb. When dried, the soil weighs 3.50 lb. If the specific gravity of the solids is 2.70, compute—

(a) water content
(b) void ratio
(c) dry unit weight
(d) degree of saturation
(e) saturated unit weight

2.4. A proposed earth dam will contain 5,400,000 cubic yards of soil. The soil will be compacted to a void ratio of 0.80 when it is placed in the dam. There are three available borrow pits, designated as A, B, and C. The void ratio of the soil in each pit and the estimated cost of moving the soil to the dam are shown in the following table:

Pit	Void Ratio	Cost of Moving
A	0.9	$0.28/cubic yard
B	2.0	$0.20/cubic yard
C	1.6	$0.26/cubic yard

(a) If only one of the pits is used, what will be the least total earth-moving cost, and which pit will be the most economical from this point of view?
(b) In deciding which pit to use, are there other factors which you might wish to consider in addition to the earth moving costs? Discuss briefly.

2.5. In a specific gravity test, the following parameters are measured or known:

W_s —Weight of dry soil solids used in test
W_0 —Weight of pycnometer flask when filled with distilled water at temperature T
W_1 —Weight of pycnometer flask when filled with soil solids of weight W_s and water at temperature T
γ_{w_0} —Unit weight of water at 4°C
γ_{w_T} —Unit weight of water at temperature T

Determine the specific gravity of the soil grains: $G = \gamma_s/\gamma_{w_0}$

2.6. In a borrow pit, the natural soil ($G = 2.70$) was found to have a moist unit weight of 112 pcf and a water content of 12.0%.

It is decided to compact this soil in a highway embankment to a *dry* unit weight of 115 pcf at a water content of 15%. If 10,000 cubic yard of *compacted* fill are required:

(a) How many cubic yards of borrow material are required?

(b) How many *gallons* of water (total) must be added to the borrow before compaction ($1 \text{ ft}^3 = 7.48$ gal)?

(c) If the compacted fill later becomes saturated without change in the void ratio, what will be its water content and unit weight?

2.7. Laboratory tests on dry samples of cohesionless soil, for which $G = 2.70$, indicate that the maximum and minimum *dry* unit weights which can be obtained for this material are, respectively. $\gamma_{d_{max}} = 115$ pcf and $\gamma_{d_{min}} = 100$ pcf. Field density tests reveal that the *moist* unit weight *in situ* for this soil is $\gamma_m = 120$ pcf at a water content $w = 10\%$.

(a) What is the relative density, D_r, *in situ*?

(b) It has been decided to compact this material to improve its shearing resistance. The soil is to be compacted *without adding or removing water*, until its dry density $\gamma_d = 0.98\gamma_{d_{max}}$. What volume of compacted soil will be obtained from each cubic foot of the natural *in situ* material?

(c) What will be the degree of saturation, S_r, of the compacted material?

2.8. A well-graded granular fill is to be placed at a relative density of 90%. The total volume of the fill is 650 cubic yards. It is to be obtained from a borrow area whose water content is 12% and dry unit weight is 102 lb/ft³. Estimate the weight of soil which must be removed from the borrow area for the fill.

2.9. Two samples of dry soil, each weighing 500 g, are subjected to mechanical analysis. The test results are shown in the accompanying table.

Test Data for Prob. 2.9

Sieve No.	Diameter (mm)	Soil A Weight Retained (gm)	Soil B Weight Retained (gm)	% Finer by Weight of Material Passing 200 Sieve
3/8 in.	9.52		20.0	
4	4.76		12.5	
8	2.38		22.5	
16	1.19		35.0	
40	0.420	5.0	95.0	
60	0.250	40.0		
80	0.177	57.5	110.0	
120	0.125	112.5		
140	0.105	135.0		
200	0.074	125.0	92.5	
Hydrometer	0.050			75.6
	0.020			46.3
	0.010			34.1
	0.005			26.8
	0.002			17.1

(a) Plot the grain-size distribution curves for both soils on four-cycle semilogarithmic paper, showing equivalent grain diameter vs. per cent finer by weight.

(b) Determine D_e and C_u for each soil.

(c) Discuss the distinguishing characteristics of each curve and suggest a mode of deposition which would result in such a curve.

2.10. Consider the results of the following strength tests:

(a) A specimen of dry sand tested in a direct-shear device subjected to a normal stress of 3 tsf failed when the applied shear stress was 2.1 tsf. What is the value of ϕ'?

(b) If a cylindrical specimen of the same sand, at the same initial void ratio, were confined in a rubber membrane and subjected to an all-round external pressure, $\sigma'_1 = \sigma'_3 = 1.0$ kg/cm², what would be the maximum possible axial pressure if the radial pressure were held constant?

2.11. At a point in the interior of a mass of dry sand and gravel, the stresses acting on vertical and horizontal planes are found to be

$$\sigma_x = 1.5 \text{ tsf} \qquad \tau_{xz} = 0.5 \text{ tsf}$$
$$\sigma_z = 2.5 \text{ tsf} \qquad \tau_{xz} = -0.5 \text{ tsf}$$

(a) What are the magnitude and orientation of the principal stresses?

(b) Is the soil likely to be at a state of failure?

2.12. A cylindrical specimen of a dry cohesionless sand is enclosed in an air-tight membrane and *fully* evacuated. The evacuated specimen is then tested to failure in compression. Failure occurs under an *additional* vertical compression of 3.6 kg/cm². Assume atmospheric pressure $= 1.0$ kg/cm².

(a) What is the angle of shearing resistance of this sand?

(b) What angle will the failure planes make with the horizontal?

(c) What will be the normal and shearing stress on the plane of failure?

(d) What will be the *maximum* shear stress on any plane in the specimen at the instant of failure, and how will the plane in question be oriented with respect to the horizontal?

(e) If the specimen fails in shear, why doesn't it fail on one of the planes determined in part d?

2.13. A direct-shear test is conducted on a specimen of dry sand. The test results indicate that a failure occurs along the plane of sliding:

$$\sigma'_{ff} = 2.00 \text{ tsf} \qquad \tau_{ff} = 1.60 \text{ tsf}$$

(a) Is the sand loose or dense? Why?

(b) What is the magnitude of the maximum shear stress in the specimen? What are the orientations with respect to the horizontal of the planes on which the maximum shear stress acts?

(c) If the specimen were saturated, and the pore water pressure were related to the shear stress on the failure plane, τ_f, by the expression

$$u = -0.2\tau_f$$

determine τ_{ff} for the case where the *total* normal stress $\sigma_f = 2.0$ tsf, and remains constant throughout the test.

2.14. A cylindrical specimen of cohesionless silt failed in an unconfined compression test ($\sigma_3 = 0$) under a vertical stress of 10 kg/cm^2. The rate of testing was sufficiently high to prevent any drainage. A well-defined failure plane was formed at an angle of 60° with the horizontal:

    (a)   Draw Mohr diagrams for both total and effective stresses at failure.

    (b)   What was the pore water pressure at failure?

2.15. A budding soil mechanics major has devised a new "pure shear" test which he suggests will revolutionize the soil-testing industry. The device exerts stresses on a rectangular specimen as shown in the accompanying sketch. To verify the usefulness of the device, the proud inventor conducts

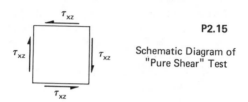

P2.15

Schematic Diagram of "Pure Shear" Test

two shear tests on identical specimens of *well-compacted* saturated sand: the first test lasts 15 sec. to failure; the second test lasts 15 min. The largest measured shear stress in the first test was 0.25 tsf. The measured shear stress throughout the entire second test was zero:

    (a)   Explain these apparently anomalous results.

    (b)   Was the sand loose or dense? How do you know?

    (c)   What is your opinion of the "pure shear" test?

2.16. Undaunted, the student described in Prob. 2.15 above has developed a "modified pure shear" test, Model Mark II. The test imposes a hydrostatic state of stress on a cubical sample, and then superimposes shear stress

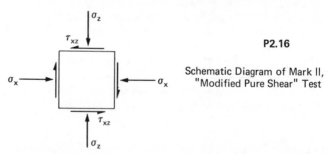

P2.16

Schematic Diagram of Mark II, "Modified Pure Shear" Test

on two mutually perpendicular planes, as shown schematically in the sketch. As a demonstration of the versatility of his new testing device, the inventor conducts shear tests to failure on three specimens of the same cohesionless soil at the same void ratio:

    (a)   The first test was conducted *drained* on a dry specimen with the following conditions at failure:

$$\sigma_{x_f} = \sigma_{z_f} = 3.5 \text{ kg/cm}^2 \qquad \tau_{xz_f} = \pm 1.86 \text{ kg/cm}^2$$

What were the orientations of the failure planes, and what were the stresses on them at failure?

(b) The second test was conducted on a saturated sample by applying the hydrostatic pressure under fully *drained* conditions, after which further drainage was prevented and the shear stresses were applied. Conditions at failure were

$$\sigma_{x_f} = \sigma_{z_f} = 3.0 \text{ kg/cm}^2 \qquad \tau_{xz_f} = \pm 1.25 \text{ kg/cm}^2$$

What was the pore water pressure at failure? Was the soil loose or dense? Why?

(c) The third test, designed to illustrate the portability of the device, was a *drained* test, conducted by a diver at the bottom of San Francisco Bay (depth of water = 15 m., $\gamma_w = 0.001 \text{ kg/cm}^3$). Results were

$$\sigma_{x_f} = \sigma_{z_f} = 5.5 \text{ kg/cm}^2 \qquad \tau_{xz_f} = \pm 2.11 \text{ kg/cm}^2$$

What was the pore water pressure at failure? Can you estimate from the results of this test whether the soil was loose or dense? Briefly justify your answer.

3

Elements of Stability Analysis for Cohesionless Soils

3.1 INTRODUCTION

In Chapter 2 we found that an appreciation of the factors producing failure *at a point* in a cohesionless soil was beneficial in understanding failure of a large mass of similar material. Soil engineers are often confronted with the necessity for predicting the degree of safety against failure of such large earth masses. Sometimes the concern is solely with an earth structure, such as a dam or embankment; at other times the question involves the interaction between a soil mass and another structure. Examples of this latter case are the ability of a highway bridge abutment to withstand the pressures imposed upon it by the approach earthfill, or the capacity of subsurface soils to sustain loads applied by a building foundation.

An appropriate margin of safety is provided by analyzing a trial design to determine the conditions required to produce failure of the soil mass, and then modifying the design until the actual imposed loads are sufficiently less than those calculated to produce a failure. Analysis of this sort is called *stability analysis.** This is in contrast to design methods in which the stresses or deformations of the structure in question are calculated and the design is adjusted until the calculated stresses or deformations are less than some appropriate limiting value.

Because the state of stress generally changes from point to point in an earth mass, we cannot always use conditions at a *single* point to predict the onset of failure throughout the mass. In this chapter, we shall extend the concept of failure at a point to failure of a mass of cohesionless soil.

---

* This design approach also is often used for concrete and steel structures.

Failure of a Retaining Wall

One class of soil engineering problems which is commonly investigated by stability analysis is the introduction of changes in elevation in an earth mass. Such changes are necessitated by a variety of causes, of which highway grade requirements are one of the most common. There are a number of ways by which such an elevation change may be accomplished. The simplest solution might be to create a stable slope in the soil mass as a transition between the two elevations, as illustrated in Figure 3.1a. However, there are occasions when such a solution is impractical because of the lateral extent of a stable slope. In such cases, where right-of-way or other space limitations dictate a sharper change in elevation, the sloping ground may require supplementary support, as shown in Figure 3.1b. A structure that provides such support to an earth mass is called a *retaining structure*. Because such structures are often long in one horizontal dimension they are frequently referred to as *retaining walls*.

The article "Analysis of a High Crib Wall Failure," (pages 94–99) reprinted from the *Proceedings of the Sixth International Conference on Soil Mechanics and Foundation Engineering*, describes the failure of a 30-ft-high retaining wall which had been constructed in a highway cut to support the cut slope. The structure described was a *crib wall** of greater than usual height. As a consequence, the wall consisted of two sections; the lower section was of double width to enhance stability of the wall-soil system.

Such structures are usually constructed free-standing after the cut has been over-excavated. When the wall is complete, the earth *backfill* is placed behind it. As the backfill is placed, it exerts forces on the wall. If equilibrium is maintained, the wall exerts equal and opposite forces on the soil. In the case of the crib wall described in the accompanying article, the pressure exerted by the backfill on the wall was sufficient to cause the upper part of the central portion of the wall to bulge and crack, necessitating costly reconstruction of some sections. The structural damage was accompanied by large movements of the backfill, indicating that it too had failed. Thus we

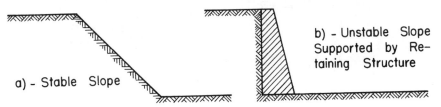

b) - Unstable Slope Supported by Retaining Structure

a) - Stable Slope

Fig. 3.1—Some methods of providing transition between elevation differences in surface of earth mass.

* A crib wall consists of a series of cells, or cribs, constructed of interlocking structural units, not unlike some versions of the common corn crib. The cells are filled with soil, which provides the weight and much of the strength required to make the wall stable.

Analysis of a High Crib Wall Failure

G. P. TSCHEBOTARIOFF, DR.ING., DR.H.C.

Associate of King & Gavaris, Consulting Engineers, New York, N.Y., U.S.A.

SUMMARY

A section of a crib wall, 34 ft high and concave in plan, experienced such severe deformations that it had to be demolished and rebuilt. A study showed that the central one of the three longitudinal footings which supported the skeleton of the crib had settled more than the outer ones. This reversed the usual direction of the wall friction along the rear face of the upper single cell part of the wall, reducing the factor of safety against sliding along its base to an unsafe value of 0.95, as compared to the 2.82 value computed for a conventionally directed wall friction force.

EARTH RETAINING CRIB WALLS built of precast reinforced concrete units are comparatively simple to analyse when their height does not require a width exceeding that of one cell (Tschebotarioff, 1951). Such crib walls easily adjust themselves to longitudinal differential settlements parallel to their face (Tschebotarioff, 1962). In recent years, attempts have been made to increase the height of crib walls to an extent requiring the use of two interlocking rows of cells in the lower portion of the wall in order to increase the width of its base. Such walls can be very sensitive to transverse settlements, however. This is illustrated by the following case of a concave 24-ft to 34-ft-high wall (Figs. 1 and 2).

Fig. 1. General view of 34-ft-high concave crib wall.

The upper part of the wall had noticeably bulged outwards (Fig. 3). As a result of this outward movement and of the concave curvature in plan of the wall, the expansion joints between double vertical rows of headers had widened (Fig. 4). This widening

Reprinted from the *Proceedings of the 6th International Conference on Soil Mechanics.* 1965.

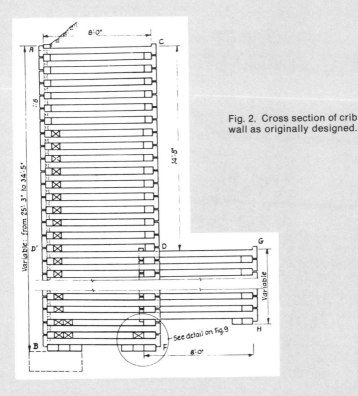

Fig. 2. Cross section of crib wall as originally designed.

occurred only in the upper part of the wall, thereby giving another indication that the outward sliding had taken place only where the wall was one cell in width.

It should be noted that a convex (Fig. 5a) crib wall of approximately the same height in the vicinity did not show any signs of distress. The severe cracking of the reinforced concrete headers and stretchers of the concave concrete wall (Fig. 3) had to be attributed, therefore, to tensile stresses induced in the headers (Figs. 5b and 5d) as a result of the outward sliding (Fig. 5c) of the upper one-cell wide portion of the wall. This sliding was, therefore, the primary cause of the trouble. The original designers of the wall had detailed it (Fig. 2) and had analysed its stability in a conventional manner (Fig. 6), computing a satisfactory factor of safety $F = 2.82$ against sliding along the surface where the sliding later actually occurred.

Field density determinations and laboratory tests were performed under the writer's direction indicating that the fill used within the cells and behind them was clean, well-packed sand with an angle of internal friction at least equal to the $\phi = 33°$ assumed by the original designers (Fig. 6). The soil beneath the crib wall was also compact sand.

An examination of the upper surface A–C of the crib wall (Fig. 2), however, indicated that the fill within the cells must have settled after it had been placed. Further, construction records showed that the placing of the fill within the upper part ACDD′ of the wall (Figs. 2, 6, and 7) was done in freezing weather when adequate compaction was not possible. However, the fill behind the wall was compacted by construction equipment passing over it.

The lowest value of the angle of internal friction determined in the laboratory for the loosest state of the fill was $\phi = 28°$. The actual slope of the fill surface immediately behind the wall was not as steep as was assumed in the original design (Figs. 6 and 7),

Fig. 3. Outward bulge of upper part of crib wall.

Fig. 4. Opening of upper part of expansion joint.

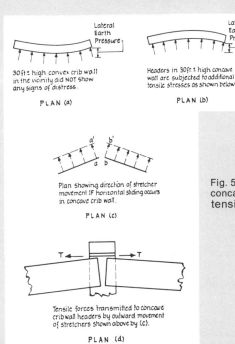

30 ft ± high convex crib wall in the vicinity did NOT show any signs of distress.

PLAN (a)

Headers in 30 ft ± high concave wall are subjected to additional tensile stresses as shown below

PLAN (b)

Plan showing direction of stretcher movement IF horizontal sliding occurs in concave crib wall.

PLAN (c)

Fig. 5. Outward lateral movement of concave crib wall induces transverse tensile stresses in headers.

Tensile forces transmitted to concave crib wall headers by outward movement of stretchers shown above by (c).

PLAN (d)

but was appreciably flatter ($\omega = 12°$). By repeating the analysis of Fig. 6 with $\phi = 28°$, $\delta = +18°, \beta = -9°$, and $\omega = +12°$, a satisfactory factor of safety against sliding equal to $F = 2.1$ was obtained. Thus the low original shearing strength of the fill within the cells could not explain the actual sliding. The following deductions, however, fitted all the observed facts.

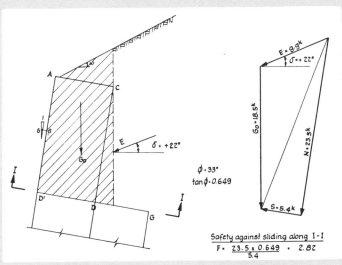

$E = 9.9^k$

$\delta = +22°$

$G_0 = 18.5^k$

$N = 23.5^k$

$S = 5.4^k$

$\delta = +22°$

$\phi = 33°$

$\tan\phi = 0.649$

Safety against sliding along I-I

$$F = \frac{23.5 \times 0.649}{5.4} = 2.82$$

Fig. 6. Conventional analysis of upper single-cell section of crib wall.

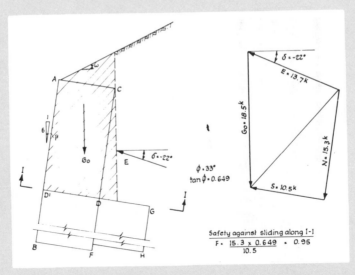

Fig. 7. Stability analysis of upper single-cell section
of crib wall if CDF settled more than AD'B and GH.

Settlement of the fill within the cells was bound to produce so-called "arching" or "bin effects" within the cells. This, as well as the 1:6 batter of the crib, loaded its inner longitudinal wall CDF to a greater extent than the outer walls AD'B and GH (Fig. 2).

The greater loading of the central longitudinal wall produced its greater settlement as a consequence of which the conventionally assumed direction of the angle of wall friction $(+\delta)$ was changed to $(-\delta)$. The stability analysis was therefore repeated for this changed condition (Fig. 7), all other assumptions of Fig. 6 remaining unchanged. This analysis, shown on Fig. 7, indicated that the reversal of the direction of wall friction decreased the safety against sliding to $F = 0.95$. Actually this factor must have been even smaller when sliding began since the angle of internal friction must have been smaller than $\phi = 33°$ before seasonal temperature and saturation variations, in combination with shearing deformations, compacted the originally loose frozen sand fill to its present satisfactory density.

The tendency towards greater settlement of the central longitudinal wall of the crib skeleton was accentuated by the details of its foundation. In accordance with the customary assumption that a crib wall acts like a massive gravity wall, a substantial continuous reinforced concrete footing was provided at its toe. It is shown by broken lines on Fig. 2. However, only three precast stretcher units—marked E, D, and C on Fig. 9—were laid down below the longitudinal centre wall. They were not structurally connected to each other in a transverse direction, except by the headers. A numerical estimation of the downward force P, transmitted by the longitudinal wall of the crib skeleton in the case of some soil "bin action" within its cells, indicated that the headers above and below the stretchers B and the block G could not have transmitted the resulting loads to the soil through the stretchers E, D, and C without cracking, as shown on Fig. 9.

To check this point, three cells were fully excavated and the anticipated cracks were actually found in the headers (see Fig. 8). The measured deflection of the headers indicated that the central longitudinal crib wall must have settled at least two inches more than the outer face of the crib.

Fig. 8. Photo of crack illustrated by Fig. 9 (taken from opposite direction).

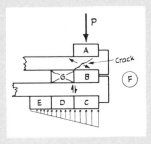

Fig. 9. Causes of shear cracks in headers at point F of crib wall (Fig. 2).

CONCLUSIONS

1. Concave crib walls are much more susceptible to damage by transverse deformations than are convex walls.

2. The reinforced concrete skeleton of crib walls over 20 feet high will not act as one massive unit with its earth fill unless the latter is compacted with special care.

3. Structural detailing should aim at preventing the greater settlement of the central longitudinal wall of the crib skeleton since such settlement may reverse the direction of friction between soil and the upper part of the wall, thereby strongly decreasing its resistance to lateral sliding.

REFERENCES

TSCHEBOTARIOFF, G. P. (1951). *Soil Mechanics, Foundations and Earth Structures,* p. 488. New York, McGraw-Hill.

————(1962). Chapter 5 on retaining structures in *Foundation Engineering* (ed. G. A. Leonards). New York, McGraw-Hill, p. 493.

see that the investigation of the stability of such a wall requires consideration of the interaction between the structure and the soil.

In order to predict correctly the failure of this wall we need to recognize that the *earth pressure* exerted on the wall by the backfill depends to a marked degree upon the magnitude and direction of the relative movement between the wall and the fill. In the accompanying article, the failure of this wall is attributed to an inadequate design resulting from an incorrect assumption about the relative movement, which in turn led to an erroneous conclusion about the magnitude and direction of the earth pressure applied to the wall. There were two major consequences of this error:

1. The component of the earth pressure tending to produce sliding of the top of the wall relative to the bottom was *underestimated.*
2. The component of the earth pressure normal to the potential sliding surface, and hence the frictional resistance to sliding of the top of the wall relative to the bottom, was *overestimated.*

Unfortunately, the actual resistance to sliding was insufficient to prevent failure.

In the following sections we shall investigate some of the principles which are applied to the analysis of retaining structures supporting an earth mass on the verge of failure, and shall see the basis for the erroneous assumption and its implications for the stability of the crib wall described in the article.

3.2   RANKINE EARTH PRESSURE THEORY

Equations of Equilibrium

In Chapter 2 we were concerned with the state of stress at a point. We considered failure at a point, and those planes through that point upon which failure would be predicted by the Mohr-Coulomb failure theory. In investigating the interaction between a structure and an earth mass on the verge of failure, it is again necessary to evaluate these parameters. We must recognize, however, that these quantities vary from point to point, and establish the rules governing their distribution. In doing this we shall restrict our attention to a two-dimensional framework. This simplification of the general case is useful for a great many problems, and can be generalized to three dimensions if necessary.

We shall consider the small plane element of unit thickness, illustrated in Figure 3.2 with the stresses shown applied to each face. The fact that we are using stresses implies the assumption of a continuum, but does not specify its nature. As shown, it has been assumed that the resultant stresses act at the center of each face. The position of each stress is described by the coordinates within the parenthesis following the stress symbol. Although

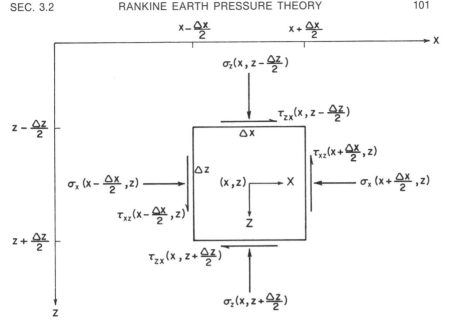

Fig. 3.2—Stresses and body forces acting
on small element of a continuum.

the element is assumed small, it is of finite size, and the stresses on opposite
faces are not necessarily equal.

The z axis is assumed positive in a downward direction. This convention
is a convenience because one of the principal forces in a soil mass results
from the unit weight of the soil itself; thus we can sum weights from the
ground surface down to a point of interest. Recall that we find it convenient
to assume compressive stresses positive, as shown.

The vectors X and Z are *body forces* per unit volume in the x and z directions, respectively. Body forces arise from force fields, which act on every
part of a material. Such fields include gravity (which leads to unit weight),
magnetic, and electrical fields. Only two body forces will be of interest to
us in the investigation of static problems. The first is the unit weight of the
material. The second is a *seepage force* resulting from the dissipation of
energy of a fluid flowing through the soil mass. Unlike the unit weight which
always acts vertically, a seepage force acts in the direction of the gradient of
the flow-producing energy, which in most cases changes from point to point
within a soil mass. In both cases these body forces are assumed to act at
the center of the element.

Restricting our attention to static problems, the summation of forces
acting on the element shown in Figure 3.2, in any direction, must equal zero.

$\sum F_x = 0$:

$$\tau_{zx}\left(x, z + \frac{\Delta z}{2}\right) \cdot \Delta x - \tau_{zx}\left(x, z - \frac{\Delta z}{2}\right) \cdot \Delta x + \sigma_x\left(x + \frac{\Delta x}{2}, z\right) \cdot \Delta z$$
$$- \sigma_x\left(x - \frac{\Delta x}{2}, z\right) \cdot \Delta z - X \cdot \Delta x \cdot \Delta z = 0 \qquad (3\text{--}1a)$$

and

$\sum F_z = 0$:

$$\sigma_z\left(x, z + \frac{\Delta z}{2}\right) \cdot \Delta x - \sigma_z\left(x, z - \frac{\Delta z}{2}\right) \cdot \Delta x + \tau_{xz}\left(x + \frac{\Delta x}{2}, z\right) \cdot \Delta z$$
$$- \tau_{xz}\left(x - \frac{\Delta x}{2}, z\right) \cdot \Delta z - Z \cdot \Delta x \cdot \Delta z = 0 \qquad (3\text{--}1b)$$

The requirements of equilibrium with respect to moments lead to the conclusion that $\tau_{xz} = \tau_{zx}$. So, dividing by the volume, $\Delta x \cdot \Delta z \cdot 1$, and taking the limit as $\Delta x \to 0$ and $\Delta z \to 0$, we have

$$\lim_{\Delta z \to 0}\left[\frac{\tau_{xz}\left(x, z + \frac{\Delta z}{2}\right) - \tau_{xz}\left(x, z - \frac{\Delta z}{2}\right)}{\Delta z}\right]$$
$$+ \lim_{\Delta x \to 0}\left[\frac{\sigma_x\left(x + \frac{\Delta x}{2}, z\right) - \sigma_x\left(x - \frac{\Delta x}{2}, z\right)}{\Delta x}\right] = X \qquad (3\text{--}2a)$$

$$\lim_{\Delta z \to 0}\left[\frac{\sigma_z\left(x, z + \frac{\Delta z}{2}\right) - \sigma_z\left(x, z - \frac{\Delta z}{2}\right)}{\Delta z}\right]$$
$$+ \lim_{\Delta x \to 0}\left[\frac{\tau_{xz}\left(x + \frac{\Delta x}{2}, z\right) - \tau_{xz}\left(x - \frac{\Delta x}{2}, z\right)}{\Delta x}\right] = Z \qquad (3\text{--}2b)$$

The terms on the left sides of Equations 3–2 are the definitions of the partial derivatives of the stresses. Thus,

$$\frac{\partial \sigma_x}{\partial x} + \frac{\partial \tau_{xz}}{\partial z} = X \qquad \frac{\partial \tau_{xz}}{\partial x} + \frac{\partial \sigma_z}{\partial z} = Z \qquad (3\text{--}3)$$

These relations are the static *equations of equilibrium*, which apply to any continuum in which only two dimensions are being considered.

For the discussion which follows, we shall assume that the only body force acting is the unit weight; so that $X = 0$, $Z = \gamma$.

Solution of Equilibrium Equations for Semi-infinite Mass

The two equations of equilibrium contain three unknowns, which prevents our obtaining a complete solution without introducing additional conditions. However for certain cases, a partial solution is possible. One such case is a *semi-infinite, homogeneous, isotropic* mass supporting a uniform loading over its entire surface.* Assuming that the boundary surface of the body is horizontal, there can be no variation of stresses in the x direction. Such a deduction is intuitive, that is, we cannot visualize a mechanism for creating changes in the x direction for the assumptions which have been made. As a result, the partial derivatives with respect to z in Equations 3–3 become ordinary derivatives:

$$\frac{d\sigma_z}{dz} = \gamma$$

$$\frac{d\tau_{xz}}{dz} = 0$$

(3–4)

Integrating, we have

$$\sigma_z = \gamma z + C_1$$

$$\tau_{xz} = C_2$$

(3–5)

in which C_1 and C_2 are constants that must be evaluated from the boundary conditions. If, as in Figure 3.3, the uniform load is applied at an angle ω from the horizontal, then at $z = 0$,

$$\sigma_z|_{z=0} = q \sin \omega$$

$$\tau_{xz}|_{z=0} = q \cos \omega$$

(3–6)

from which the constants C_1 and C_2 can be determined. Thus the stresses at any depth for this case are

$$\sigma_z = \gamma z + q \sin \omega \qquad\qquad \text{(3–7a)}$$

$$\tau_{xz} = q \cos \omega \qquad\qquad \text{(3–7b)}$$

These results apply irrespective of the nature of the material. That is, for a semi-infinite homogeneous continuum with a horizontal boundary surface, σ_z and τ_{xz} depend only upon the depth below the surface and the applied boundary load. We see further that *if the boundary loading is applied normal to the boundary, or is zero, the horizontal and vertical planes are principal planes.*

---

* A *semi-infinite* body is bounded by a plane surface of infinite extent; the body extends indefinitely in all directions below (or above) this plane. If the material characteristics are the same at every point of the body, it is said to be *homogeneous*. A material with point properties which are independent of direction is called *isotropic*.

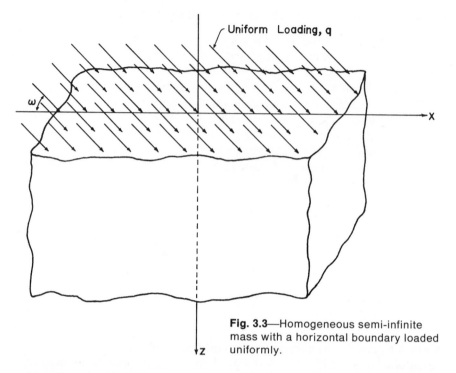

Fig. 3.3—Homogeneous semi-infinite mass with a horizontal boundary loaded uniformly.

Limiting Equilibrium

In order to determine σ_x, we must introduce an additional equation without introducing any more unknowns. It seems intuitively reasonable to suppose that such an expression should describe appropriate characteristics of the specific material being investigated. Such an expression is called a *constitutive law* because it describes a manifestation of the constitution of the material.

We have already established that we wish to investigate the conditions when the soil is at failure. Thus we shall assume that *every* point in the soil mass is just on the verge of failure. This is called a case of *limiting equilibrium.** The failure criterion which describes conditions at failure is the Mohr-Coulomb criterion, Equation 2–11. A more convenient form of the Mohr-Coulomb criterion can be determined from the Mohr circle at failure:

$$\left(\frac{\sigma_z' - \sigma_x'}{2}\right)^2 + \tau_{xz}^2 = \left(\frac{\sigma_z' + \sigma_x'}{2}\right)^2 \sin^2 \phi' \qquad (3\text{–}8)$$

With this additional relationship we now have a set of three equations and three unknowns which can be solved to yield the state of stress at

---

* There are alternatives to this assumption, the usefulness of which depend upon the application of the results obtained. We shall consider some of these alternatives in Chapters 4 and 5.

failure consistent with the assumed boundary conditions. Recall, however that some restrictive assumptions have been made in bringing us to this point. The Mohr-Coulomb criterion assumes that failure is independent of the intermediate principal stress. In the analysis of a semi-infinite soil body, we have used this, and two further assumptions:

1. The intermediate principal stress acts normal to the x–z plane. That is, $\sigma_2 = \sigma_y$.
2. The stresses σ_x, σ_z, τ_{xz} are independent of y.

Because of the quadratic nature of Equation 3–8, there are two possible states of stress that can produce failure. These can be obtained by solving Equation 3–8 for σ'_x. This leads to

$$\sigma'_x = \sigma'_z \left(\frac{1 + \sin^2 \phi'}{1 - \sin^2 \phi'}\right) - \frac{2 \sin \phi'}{1 - \sin^2 \phi'} \sqrt{\sigma'^2_z - \tau^2_{xz}\left(\frac{1 - \sin^2 \phi'}{\sin^2 \phi'}\right)} \qquad (3\text{–}9a)$$

or

$$\sigma'_x = \sigma'_z \left(\frac{1 + \sin^2 \phi'}{1 - \sin^2 \phi'}\right) + \frac{2 \sin \phi'}{1 - \sin^2 \phi'} \sqrt{\sigma'^2_z - \tau^2_{xz}\left(\frac{1 - \sin^2 \phi'}{\sin^2 \phi'}\right)} \qquad (3\text{–}9b)$$

The Mohr circles corresponding to these two states of limiting equilibrium are shown schematically in Figure 3.4. Both circles coincide at the point

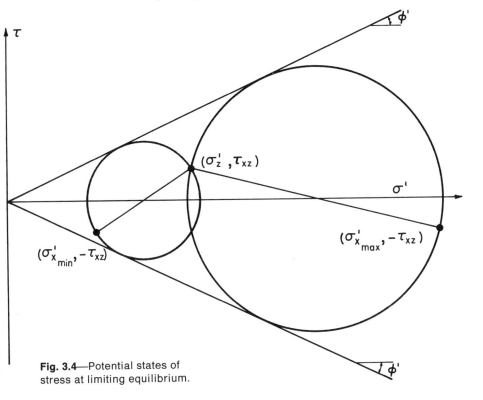

Fig. 3.4—Potential states of stress at limiting equilibrium.

(σ'_z, τ_{xz}). However, the left-side circle, which corresponds to Equation 3–9a, represents the case in which σ'_x is sufficiently *smaller* than σ'_z that the failure condition is reached at that point. This is the minimum possible magnitude of σ'_x for the particular (σ'_z, τ_{xz}). The right-side circle, corresponding to Equation 3–9b, illustrates the case in which σ'_x is so much *larger* than σ'_z that the point is at a state of failure. This produces the maximum possible σ'_x for the particular (σ'_z, τ_{xz}).

The two states of limiting equilibrium shown in Figure 3.4 are defined as conditions of *active* and *passive* earth pressure. The smaller circle depicts the active state; the larger circle represents the passive state.

It is generally simplest to solve Equations 3–9 by determining σ'_z and τ_{xz} from Equations 3–7 and then constructing the Mohr failure circles. From these circles, the orientation of the failure planes and the stresses acting on them, as well as the orientation of the principal planes and the magnitude of the principal stresses, may be determined.

Development of Limiting Equilibrium

In order to see how limiting equilibrium develops in an earth mass, we shall investigate an artificial example: a large mass of cohesionless soil containing a thin embedded rigid wall of infinite depth. We shall assume, as shown in Figure 3.5a, that the soil mass is initially in a state of equilibrium and that the wall extends indefinitely into and out of the figure. Two devices for measuring the state of stress are located at points A and B. We imagine further that the wall is so thin that it does not initially influence the state of stress in the earth mass, and that the idealized devices for measuring stress at the two points within the earth mass do not influence the state of stress at those points. We shall then consider the changes in the state of stress at points A and B as a consequence of movement of the wall.

The initial state of stress in the soil mass at any point, called the *at-rest condition*, is shown schematically in Figures 3.5b and c. The vertical effective stress σ'_{z_0} is determined by Equation 3–7a *from equilibrium conditions only*. The horizontal stress in the initial condition is denoted σ'_{x_0}. In most natural deposits of granular soil, this horizontal *earth pressure at rest* is less than the effective vertical stress. We shall discuss the factors influencing the relative magnitudes of these stresses in the next section.

Let us suppose now that the thin rigid wall shown in Figure 3.5a is displaced horizontally a small amount to the right in the diagram. We assume that the changes in stress which occur in response to this displacement are not influenced by the surface texture and composition of the wall. The movement of the wall causes strains to occur in the soil mass. The soil to the right side of the wall is compressed in a horizontal direction with horizontal compressive strains resulting. The soil to the left side of the wall is permitted to expand, with horizontal extensive strains occurring. These strains result in

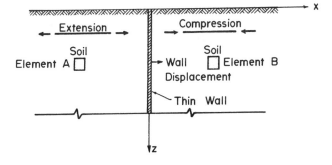

a) - Wall and Soil Mass

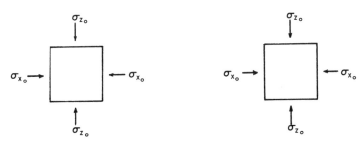

b) - Element A at Rest   c) - Element B at Rest

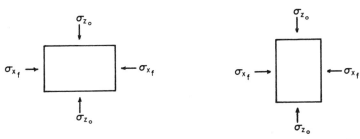

d) - Element A, Active Failure,   e) - Element B, Passive Failure,

$\sigma_{x_f} < \sigma_{x_o}$ $\sigma_{x_f} > \sigma_{x_o}$

Fig. 3.5—Displacements and stresses associated with conditions of limiting equilibrium

shape changes of the soil at elements A and B as indicated in Figures 3.5d and e, respectively.

The horizontal stresses measured at points A and B as this process occurs are shown in Figure 3.6 as a function of the wall displacement. The earth pressure at rest is shown in this diagram as point ①. This is the initial pressure in the ground before the wall is displaced. When the wall is displaced

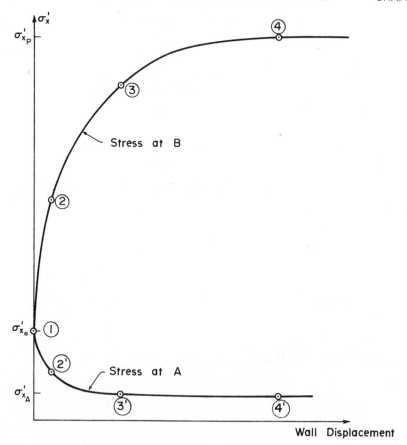

Fig. 3.6—Development of active and passive pressures in cohesionless earth mass

a small amount to the right, the horizontal stress at point B is increased as shown by point ② in Figure 3.6. At the same time, the horizontal stress at point A decreases as shown by point ②′. Further displacement of the wall increases the stress at point B to that shown by point ③ while the stress at point A is decreased to that shown by point ③′. Further increase of displacement increases the stress at point B to a maximum value indicated by point ④ while the stress at point A remains at its minimum value, point ④′. During this process the vertical stress has not changed.

We may view this change in horizontal stress as a development of the shearing resistance of the soil. Loads applied to the soil induce shear stresses. As the soil deforms in response to these stresses its resistance to shear increases until a condition of limiting equilibrium is reached at which point no increase in resistance is possible. The conditions of limiting equilibrium at points A and B are shown in Figure 3.6 by points ③ and ④ respectively. Thus we see that the curves in Figure 3.6 are analogous to stress-strain curves for the soil. An increase or decrease in the stresses from the at-rest condition *must be accompanied by strains* without which the limiting equilibrium condition cannot be reached.

We can see why there are two such limiting equilibrium conditions. Because the soil fails in response to shear effects, failure can be produced by increasing the horizontal stress until the induced shear stresses are sufficient to effect it, or by reducing the horizontal stress until the induced shear stresses lead to failure. The magnitude of the larger horizontal stress which brings about failure is the *passive earth pressure*, and denoted by σ'_{x_P}. The smaller horizontal stress producing failure is the *active earth pressure* and denoted by σ'_{x_A}.

To visualize how failure is brought about by this process, let us consider the various Mohr circles corresponding to the wall displacements indicated by the numbered points in Figure 3.6. These circles are shown in Figure 3.7. In Figure 3.7a we see circle ①, which is a typical at-rest condition. In Figure 3.7b Mohr circles corresponding to points ② and ②′ from Figure 3.6 are shown. The Mohr circle corresponding to the at-rest condition is shown as circle ① for reference. We note that, in the case of circle ②, the horizontal stress has increased while the vertical stress has remained constant so that the circle has actually shrunk in size and the shear stress on all nonprincipal planes has decreased. Conversely the horizontal stress for circle ②′ has decreased so that the Mohr circle for this condition is larger than for the at-rest case with a concomitant increase in shear stresses. Further wall displacements leads to the Mohr circles shown in Figure 3.7c. In the case of circle ③, the displacement has caused the horizontal stress to increase beyond the condition where the circle has shrunk to a single point; horizontal stress is now the major principal stress. In the case of circle ③′ the horizontal stress has decreased to its minimum possible value and the Mohr circle is just touching the failure envelope, that is, the soil is at the point of failure. Again the at-rest condition is shown as circle ① for reference. Finally, still further wall displacement leads to the passive failure shown by circle ④, Figure 3.7d, while the material at point A is still in the active condition, circle ④′.

Stress Distribution at Limiting Equilibrium

We shall now consider the stress distribution throughout a semi-infinite cohesionless mass for three cases which illustrate a number of important points.

Example 3.1 ————————————————————————

A semi-infinite mass of cohesionless soil with unit weight γ and angle of shearing resistance ϕ' is bounded by an unloaded horizontal surface. We wish to determine the

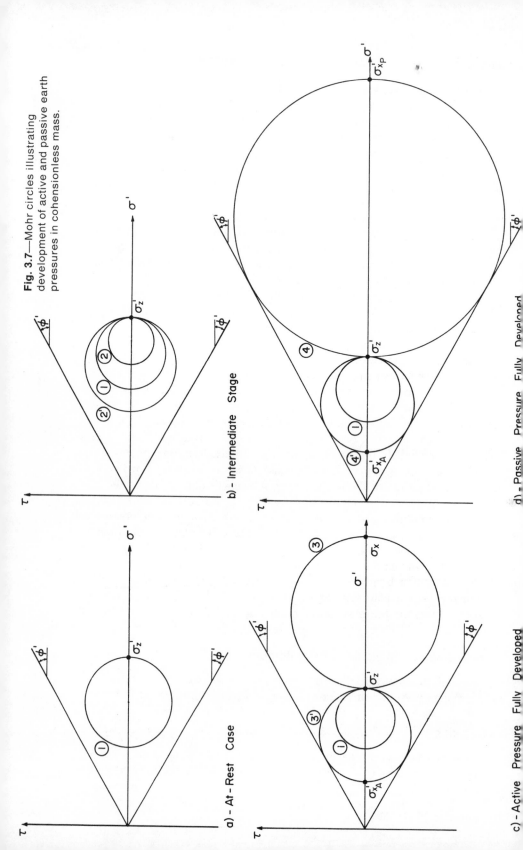

Fig. 3.7—Mohr circles illustrating development of active and passive earth pressures in cohensionless mass.

a) - At - Rest Case

b) - Intermediate Stage

c) - Active Pressure Fully Developed

d) - Passive Pressure Fully Developed

following for conditions of limiting equilibrium:

(a)   The magnitude of the resultant stresses on a vertical plane as a function of depth below the ground surface.

(b)   The shape of the failure surfaces.

Solution. (a) From Equations 3–7 we see that the horizontal and vertical planes are principal planes and that $\sigma_z = \gamma z$. Thus the *resultant* stress on a vertical plane is the *horizontal normal* stress which is determined by substituting for σ_z in Equations 3–9. This leads to the minimum horizontal stress σ'_{x_A},

$$\sigma'_{x_A} = \left(\frac{1 - \sin\phi'}{1 + \sin\phi'}\right)\gamma z = \tan^2\left(45° - \frac{\phi'}{2}\right)\gamma z \qquad (3\text{–}10a)$$

and the maximum horizontal normal stress σ'_{x_P},

$$\sigma'_{x_P} = \left(\frac{1 + \sin\phi'}{1 - \sin\phi'}\right)\gamma z = \tan^2\left(45° + \frac{\phi'}{2}\right)\gamma z \qquad (3\text{–}10b)$$

for the active and passive cases respectively.

(b) Because the horizontal and vertical stresses are both linear functions of depth in this case, the two Mohr circles illustrating failure conditions for active and passive cases shown in Figure 3.8 represent the limiting equilibrium states at *all* depths to some scale. Consequently the directions of the failure surfaces at limiting equilibriums are

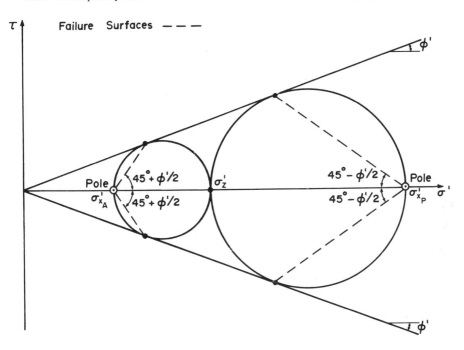

Fig. 3.8—Mohr circles for active and passive states of limiting equilibrium for Example 3.1.

also the same at all depths, inclined at $\pm(45° + \phi'/2)$ and $\pm(45° - \phi'/2)$ from the horizontal, for the active and passive cases respectively. The distribution of stresses on all vertical planes and the directions of the failure surfaces at limiting equilibrium for the active and passive cases are illustrated in Figure 3.9. We must note that the *linear* distribution of stresses with depth, and the *planar* surfaces occurred only because the horizontal surface was subjected solely to a normal stress (in this case, of zero magnitude).

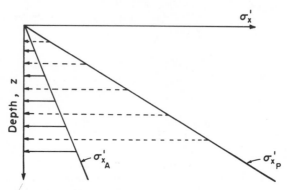

a) - Distribution of Horizontal Normal Stress at Limiting Equilibrium

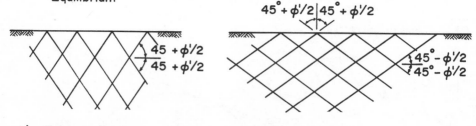

b) - Failure Surfaces for Active Case

c) - Failure Surfaces for Passive Case

Fig. 3.9—Stress distribution and failure surfaces for active and passive states of limiting equilibrium for Example 3.1.

The ratio of the horizontal active pressure to the vertical stress is designated K_A, the *coefficient of active earth pressure*,

$$\frac{\sigma'_{x_A}}{\sigma'_z} = K_A = \tan^2\left(45° - \frac{\phi'}{2}\right) \tag{3-11a}$$

The ratio of the horizontal passive pressure to the vertical stress is denoted by K_P, the *coefficient of passive earth pressure*,

$$\frac{\sigma'_{x_P}}{\sigma'_z} = K_P = \tan^2\left(45° + \frac{\phi'}{2}\right) \tag{3-11b}$$

For the special case of the horizontal boundary that is either unloaded or subjected to a uniform vertical normal stress, these coefficients are functions only of the angle of shearing resistance, as indicated by Equations 3–11. Typical values for these limiting earth pressure coefficients are given in Table 3.1.

In the next example we shall investigate the effect of a nonnormal boundary loading on the distribution of stresses and the direction of the failure surfaces.

TABLE 3.1
Values of Coefficients of Active and Passive
Earth Pressure

ϕ' (deg)	K_A	K_P
30	0.333	3.00
35	0.271	3.69
40	0.217	4.60
45	0.172	5.83

Example 3.2

A semi-infinite mass of cohesionless soil with a unit weight of 125 pcf and an angle of shearing resistance of $\phi' = 35°$ is bounded by a horizontal surface. The boundary is subjected to a uniformly distributed pressure of 1000 psf inclined at an angle from the horizontal, $\omega = 60°$. We wish to determine:

(a) The minimum and maximum normal pressure distribution on any vertical plane, to a depth of 12 ft.
(b) The shape of the failure surfaces, for the active case, to a depth of 12 ft.

Solution. (a) The stresses on any horizontal plane, are, from Equations 3–7,

$$\sigma'_z = \gamma z + q \sin \omega = (125z + 867) \text{ psf} \qquad (3\text{–}12a)$$
$$\tau_{xz} = q \cos \omega = 500 \text{ psf} \qquad (3\text{–}12b)$$

Knowing σ'_z and τ_{xz}, we could use Equations 3–9 to determine σ'_x for the active and passive cases. Recalling that the angles between the major principal plane and the failure plane are $\pm(45° + \phi'/2)$ we could solve Equations 2–9 to determine the angle between the horizontal and principal planes and consequently the slope of the failure surface at any depth. Thus the solution is obtainable by strictly analytical methods. However, for this example, we shall use instead a graphical construction of Mohr's circle at failure to illustrate the simplicity of the graphical procedure as well as to assist in visualizing the rotation of the principal planes as the depth increases.

Mohr circles at failure, for the active condition, corresponding to depths of 0, 4, 8 and 12 ft, are shown in Figure 3.10. Note that the position of the poles on the failure circles changes as a function of the depth. Thus we see that the orientations of the failure surfaces change with depth, and that the surfaces are *not* planar. Consequently the distribution of normal stresses on the vertical plane will also not be linear with depth. It is interesting to observe that the vertical plane considered is a failure surface at a depth of 4 ft (see Figure 3.10b).

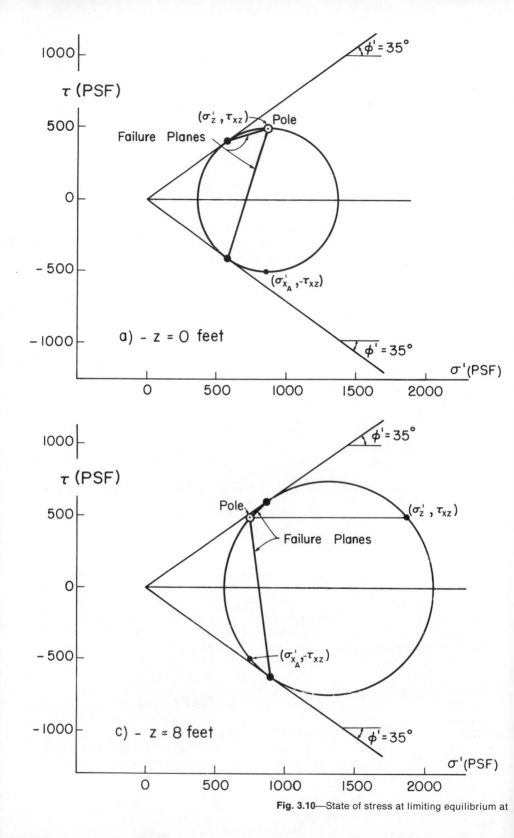

Fig. 3.10—State of stress at limiting equilibrium at

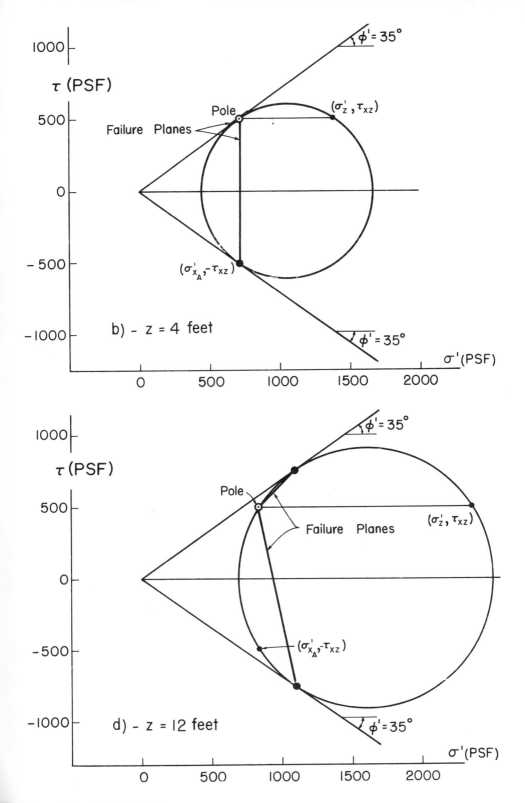

various depths for Example 3.2.

Similar Mohr circles for the passive case can be visualized. From these Mohr circles, the horizontal normal stresses, acting on a vertical plane, for the active and passive cases have been determined, and are shown in Figure 3.11a. The results are also tabulated in Table 3.2.

(b) The families of failure surfaces for the active cases, are sketched in Figure 3.11b. Such curved failure surfaces are typical when the surface loading is not normal to the boundary.

Example 3.3

A semi-infinite mass of cohesionless soil with unit weight γ and angle of shearing resistance ϕ' is bounded by an unloaded surface inclined to the horizontal at an angle

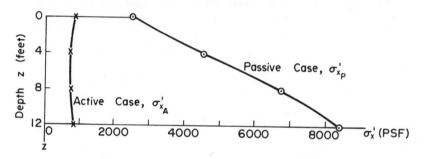

a) – Horizontal Normal Stress σ_x' at Limiting Equilibrium

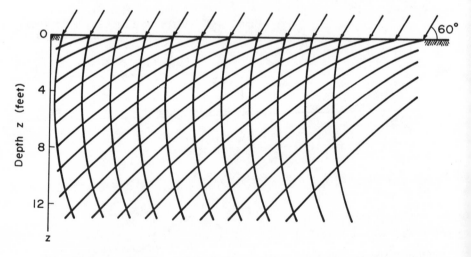

b) – Shape of The Failure Surfaces for Active Condition

Fig. 3.11—Distribution of horizontal normal stresses and failure surfaces at limiting equilibrium for Example 3.2.

TABLE 3.2
Stresses at Limiting Equilibrium for Example 3.2

Depth z (ft)	σ'_z (psf)	τ_{xz} (psf)	σ'_{x_A} (psf)	σ'_{x_P} (psf)
0	870	500	870	2530
4	1370	500	710	4550
8	1870	500	750	6750
12	2370	500	850	8410

$\beta = 20°$. We wish to determine, for conditions of limiting equilibrium:

(a) The resultant active earth pressure on a plane inclined at an angle $\alpha = 80°$ to the horizontal as a function of depth.

(b) The maximum angle of inclination β to which the surface may be inclined.

Solution. (a) The solution to this problem may be obtained most readily by establishing a new coordinate system, shown in Figure 3.12, with axes normal and parallel to the boundary surface. Rewriting the equations of equilibrium (Equations 3–3) in terms of the new coordinate system, and assuming that the unit weight is the only body force acting, we have

$$\frac{\partial \sigma'_u}{\partial u} + \frac{\partial \tau_{uv}}{\partial v} = \gamma \sin \beta$$

$$\frac{\partial \tau_{uv}}{\partial u} + \frac{\partial \sigma'_v}{\partial v} = \gamma \cos \beta$$

(3–13)

in which σ'_u and σ'_v are the effective normal stresses parallel to and normal to the boundary surface, respectively, and τ_{uv} is the shear stress on planes parallel and normal to the boundary surface.

Because of the semi-infinite nature of the earth mass, we again recognize that there can be no variation in stresses in the u direction. Hence the first terms in Equation

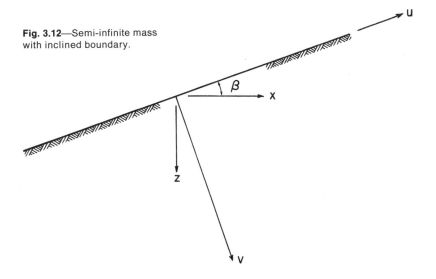

Fig. 3.12—Semi-infinite mass with inclined boundary.

3–13 must be zero, and the resulting expressions are ordinary differential equations:

$$\frac{d\tau_{uv}}{dv} = \gamma \sin \beta$$

$$\frac{d\sigma_v'}{dv} = \gamma \cos \beta \tag{3-14}$$

Integrating Equations 3–14 and expressing v in terms of the depth z, we find

$$\tau_{uv} = \gamma z \cos \beta \sin \beta$$

$$\sigma_v' = \gamma z \cos^2 \beta \tag{3-15}$$

These expressions describe the normal and shear stresses acting on planes within the earth mass inclined at an angle β to the horizontal. Thus we can also, using the notation of Chapter 2, denote τ_{uv} as τ_β and σ_v' as σ_β'. The obliquity on the β plane is the ratio of τ_β to σ_β',

$$\frac{\tau_\beta}{\sigma_\beta'} = \frac{\tau_{uv}}{\sigma_v'} = \tan \beta \tag{3-16}$$

which is independent of depth.

The Mohr circle corresponding to the active case of limiting equilibrium is shown in Figure 3.13a. The stress on the β plane for any depth is the intersection of the Mohr circle and a line through the origin inclined at an angle β to the σ' axis, as shown by Equation 3–16. (We must, of course, consider effective stresses when describing conditions at failure.) The pole is the other intersection of the β line with the Mohr circle, and leads to immediate determination of the stresses on a plane inclined at an angle $\alpha = 80°$ to the horizontal. These are found from the Mohr circle to be

$$\sigma_{\alpha_A}' = 0.37\gamma z$$

$$\tau_\alpha = -0.235\gamma z \tag{3-17}$$

And so, for this case, the stresses increase linearly with depth.

The resultant active earth pressure is

$$p_A' = (\sigma_{\alpha_A}'^2 + \tau_\alpha^2)^{1/2} = 0.44\gamma z \tag{3-18}$$

and is oriented as shown in Figure 3.13b.

(b) The maximum value of β is ϕ'! This may be determined by inspection of Figure 3.13a, in which we see that the β line cannot exceed the obliquity at failure. Thus we find the general rule that the limiting inclination of an unloaded infinite slope in a cohesionless soil is ϕ'.

Application to Analysis of Earth Pressure on Retaining Walls

In the nineteenth century, William J. M. Rankine applied the analysis of semi-infinite media in a state of limiting equilibrium to calculation of maximum and minimum earth pressures against retaining structures (1857). Consequently the approach is commonly termed *Rankine earth pressure analysis*. This approach assumes that the normal and shear stresses which act at the interface between the retaining wall and the soil are identical to those which would have existed on the same surface within a semi-infinite

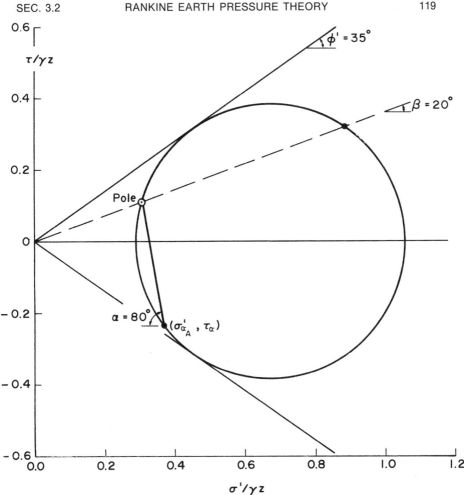

a) - Mohr Diagram for Active Case of Limiting Equilibrium

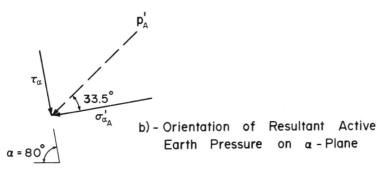

b) - Orientation of Resultant Active
Earth Pressure on α - Plane

Fig. 3.13—Active earth pressure on inclined plane in semi-infinite
cohesionless mass with sloping surface, Example 3.3.

mass in a state of limiting equilibrium. That is, we assume that the inter-actions between the soil and the wall is the same as would occur between soil and more soil.

This assumption will be valid if two events occur:

1. The wall must move laterally an amount sufficient for the develop-ment of strains in the backfill large enough to simultaneously mobilize the full strength of the soil, as illustrated in Figure 3.6.
2. The wall itself must deform as the soil it is replacing would have, so that every point along the wall-soil interface is on the verge of failure.

The first condition is often satisfied in the case of unbraced retaining structures in which lateral movements can be tolerated. In the case of cohesionless backfills, a lateral movement of the order of 0.005 times the height of the wall will produce the active earth pressure condition (Terzaghi, 1936). The second condition is rarely, if ever, the case. However, the dis-crepancies are often small enough that useful estimates of earth pressures on retaining walls can be made using the Rankine earth pressure theory.

Retaining walls which correspond to the earth pressure conditions calcu-lated in Examples 3.1 to 3.3 are shown in Figure 3.14. In Figure 3.14a, a wall with a vertical back is in contact with a backfill with a horizontal surface. Assuming that sufficient yielding occurs, the failure surfaces behind the wall would be as shown for the active and passive conditions. Although the theory requires the assumption that the entire semi-infinite mass is in a state of limiting equilibrium, only the portion shown behind the wall would be affected by the wall movement. Thus, in principle, the earth pressure against the wall is independent of whether the backfill to the right of the failure zones illustrated is yielding or not.

The vertical wall and failure surfaces corresponding to the analysis in Example 3.2 are shown in Figure 3.14b. In Figure 3.14c, a wall with a back face inclined at 80° to the horizontal, corresponding to the plane investigated in Example 3.3, and the inclined backfill are shown.

We must keep in mind that the Rankine earth pressure theory is an approximation to the real conditions at the interface between a retaining wall and the backfill it supports. Paramount among the assumptions which must be satisfied to make it a reasonable approximation is the condition that sufficient yielding has occurred to bring the backfill to the point of failure.

Effect of Submergence

In some cases retaining walls must be designed with a portion of the wall below the *groundwater table*,* that is, part of the backfill is submerged. In

---

* The *groundwater table*, also called the *free-water surface*, is defined as that surface within a soil mass on which the pore water pressure is equal to atmospheric pressure. Generally the soil below the groundwater table is fully saturated, and that above the groundwater table is partially saturated.

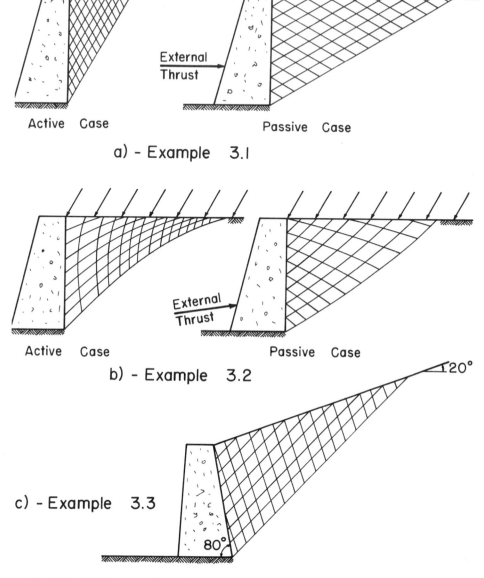

Active   Case   Passive   Case

External
Thrust

a) - Example   3.1

Active   Case   Passive   Case

External
Thrust

b) - Example   3.2

c) - Example   3.3

20°

80°

Fig. 3.14—Retaining walls analyzed by Rankine earth
pressure theory, results from Examples 3.1 to 3.3.

some instances, a groundwater table may not be present in the vicinity of
a retaining wall at the time of construction; but the wall itself intercepts
a local flow of groundwater, causing it to back up and create a temporary,
or even permanent, free-water surface behind the wall. Such local flows are
likely to occur as rainwater leaches through a soil mass. Owing to slow

soil drainage, there may be a significant amount of water retained behind a wall for considerable periods after a rainfall. As we shall see, the presence of this water may be of major importance to the design of the wall.

In Figure 3.15a, a retaining wall is shown supporting a horizontal backfill of considerable extent. The free-water surface is located at the top of the fill. At any depth within the soil mass:

Vertical normal stress: $\qquad$ $\sigma_z = \gamma_{sat} z$ $\hfill$ (3–19)

Pore water pressure: $\qquad$ $u = \gamma_w z$ $\hfill$ (3–20)

Effective vertical normal stress: $\qquad$ $\sigma'_z = (\gamma_{sat} - \gamma_w)z = \gamma' z$ $\hfill$ (3–21)

Because the horizontal and vertical planes are principal planes $\tau_{xz} = 0$, and from Equations 3–10 and 3–11,

$$\sigma'_{x_A} = K_A \gamma' z \tag{3–22a}$$

$$\sigma'_{x_P} = K_P \gamma' z \tag{3–22b}$$

In both cases the reduction in the vertical effective stress results in a reduced horizontal effective stress.

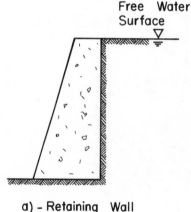

a) - Retaining Wall

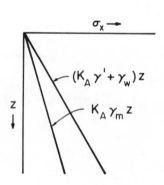

b) - Active Case

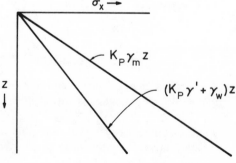

Fig. 3.15—Effect of Submergence of backfill on pressures on wall.

c) - Passive Case

But what does the wall feel? The wall must either resist in the active case. or overcome in the passive case, the total force applied to it. In addition to the earth pressure acting on the wall, there is, in a standing body of water, a fluid pressure equal to that on any horizontal section. Thus the total force on the wall is that due to the limiting equilibrium earth pressure plus the (hydrostatic) pore water pressure. This horizontal stress in the *active* case (Figure 3.15b) is

$$\sigma_{xA} = (K_A\gamma' + \gamma_w)z$$

or

$$\sigma_{xA} = [K_A\gamma_{sat} + (1 - K_A)\gamma_w]z \qquad (3-23a)$$

and in the *passive* case (Figure 3.15c) is

$$\sigma_{xP} = (K_P\gamma' + \gamma_w)z$$

or

$$\sigma_{xP} = [K_P\gamma_{sat} + (1 - K_P)\gamma_w]z \qquad (3-23b)$$

By contrast, the horizontal earth pressures in a partially saturated mass in which the pore water pressure is assumed zero are

$$\sigma'_{xA} = K_A\gamma_m z \qquad (3-24a)$$
$$\sigma'_{xP} = K_P\gamma_m z \qquad (3-24b)$$

Recognizing that the unit weights of a partially saturated soil and a saturated soil are usually quite close we see that submergence causes the total pressure on a retaining wall to be—

1. *increased* by approximately $(1 - K_A)\gamma_w z$ for the active case, and
2. *decreased* by approximately $(K_P - 1)\gamma_w z$ for the passive case.

We can visualize the significance of these differences by considering a numerical example.

Example 3.4

Let us assume the backfill shown in Figure 3.15a is initially a partially saturated medium loose sand with the following properties:

$$S_r = 60\% \qquad G = 2.72 \qquad e = 0.62 \qquad \phi' = 33°$$

(a) Calculate the pressure on the wall, using a Rankine analysis for both the active and passive cases.
(b) If the free-water surface is brought to the top of the fill, determine the total pressure on the wall for both cases.

Solution. Calculate unit weights of soil:

$$\gamma = \gamma_w \frac{(eS_r + G)}{(1 + e)} \qquad (3-25a)$$

Partially saturated soil: $\gamma_m = 62.4 \left[\dfrac{0.62(0.60) + 2.72}{1.62} \right] = 119 \text{ pcf} \qquad (3-25b)$

Saturated soil: $\gamma_{\text{sat}} = 62.4 \left[\dfrac{0.62 + 2.72}{1.62} \right] = 129 \text{ pcf}$ (3–25c)

Calculate coefficients of active and passive earth pressure:

$$K_A = \tan^2 (45 - \phi'/2) = 0.294 \tag{3–26a}$$
$$K_P = \tan^2 (45 + \phi'/2) = 3.39 \tag{3–26b}$$

(a) Partially saturated soil: From Equations 3–24,

$$\sigma'_{x_A} = 0.294(119)z = 35.0z \text{ psf} \tag{3–27a}$$
$$\sigma'_{x_P} = 3.39(119)z = 403z \text{ psf} \tag{3–27b}$$

(b) Saturated soil: From Equations 3–12 and 3–13,

$$\sigma_{x_A} = [0.294(129) + (1 - 0.294)62.4] = 81.9z \text{ psf} \tag{3–28a}$$
$$\sigma_{x_P} = [3.39(129) - (3.39 - 1)62.4] = 288z \text{ psf} \tag{3–28b}$$

Note that in the active case there was a 184 per cent increase in the pressure on the wall. In the passive case there was a decrease of 29 per cent in the pressure.

The influence of water can be accounted for in a Rankine analysis only if within the semi-infinite mass the pore pressures and seepage forces exerted by the water are independent of the coordinate parallel to the ground surface. Except in rare cases, the Rankine earth pressure analysis is applicable to submerged masses only in situations where there is no flow, and both the ground surface and free-water surface are horizontal.

3.3 EARTH PRESSURE AT REST

Many structures which support earth backfill cannot yield a sufficient distance to mobilize the full shearing resistance of the soil. Examples are basement walls of buildings and box culverts. When such a structure does not displace, the soil in contact with it experiences no lateral strain. As a consequence, the magnitude of the earth pressure upon the wall is intermediate between the active and passive values. The lateral pressure acting in this circumstance is termed the *earth pressure at rest*. The ratio between the horizontal and vertical effective normal stresses is called the *coefficient of earth pressure at rest*, K_0,

$$K_0 = \frac{\sigma'_x}{\sigma'_z} \tag{3–29}$$

Because no horizontal strain is permitted, the soil cannot mobilize its shearing resistance and approach either condition of limiting equilibrium. Thus a limiting equilibrium analysis is of no assistance in determining K_0. Experiments have shown that K_0 depends upon the angle of shearing resistance of the soil, ϕ', and its stress history (Kane et al., 1965). The stress history is described in quantitative terms by the *over-consolidation ratio*, OCR. The over-consolidation ratio is the ratio of the *past maximum* vertical effective stress to the *present* vertical effective stress at a point in the mass. If some depth of soil is eroded after deposition, then the material remaining would have experienced a vertical effective stress greater than that presently applied. Such a material is *over-consolidated*. If however the present vertical effective stress is the maximum to which the soil has ever been subjected, it is *normally consolidated*. Magnitudes of K_0 for normally consolidated cohesionless soils have been reported to range from about 0.35 to 0.47.* It is apparent from the test results available that K_0 for a normally consolidated granular soil is a function of the same properties which determine its angle of shearing resistance. Although no theoretical relationship between K_0 and ϕ' has been derived, a useful empirical relation has been presented (Jaky, 1948):†

$$K_0 = 1 - \sin \phi' \tag{3-30}$$

In the case of over-consolidated granular soil, K_0 may be much greater than for the normally consolidated material. This is demonstrated in Figure 3.16, which shows the relationship between the effective horizontal and effective vertical stresses during loading and unloading of a cylindrical specimen of uniform sand in which no lateral strains were permitted. Note that on the initial loading (normally consolidated material) K_0 is constant. Upon unloading (over-consolidated material) K_0 increases, and becomes larger as the over-consolidation ratio increases, that is, as more load is removed. The relationship between K_0 and the over-consolidation ratio obtained by Hendron (1963) for a medium-dense sand, and reported by Brooker and Ireland (1965), is illustrated in Figure 3.17.

At the present time there is no reliable method known to the authors for determining the *in situ* (in the natural soil deposit in the field) value of K_0 for cohesionless soils. A reasonable interpretation of its value can sometimes be made from an estimate of the stress history of a natural deposit based upon known recent loadings, or geologic evidence combined with laboratory tests.

Even if we do not have much information about K_0 in undisturbed deposits, an appreciation of the significance of the OCR on K_0 is important when placing a granular backfill against an unyielding retaining structure. If in the placement and compaction of the fill, heavy equipment is used close to the wall, the over-consolidation of the fill will be reflected in a high value of K_0. Thus it is usually desirable to reduce the compactive effort in the vicinity of an unyielding retaining structure.

---

* K_0 has been measured by several investigators using different testing procedures ranging from medium scale retaining wall tests (Terzaghi, 1936), to small-scale tests on cylindrical samples in which no lateral strains were permitted (Bishop and Henkel, 1962; Kane et al., 1965; to mention just a few). However, the magnitudes of K_0 reported in all of these investigations are quite similar.

† Although alternative expressions have been suggested (summaries of these are given by Brooker and Ireland, 1965, and Alpan, 1967), Equation 3–30 is sufficiently accurate for our state of knowledge concerning K_0.

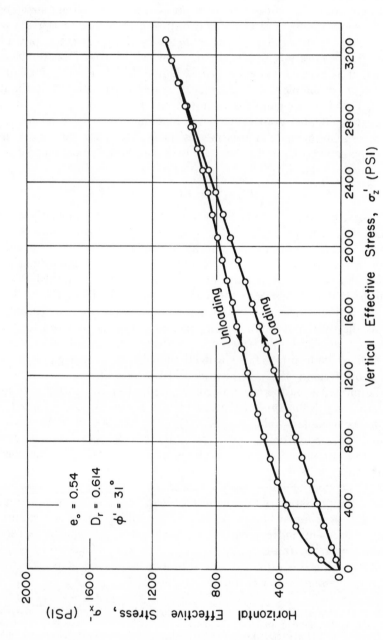

Fig. 3.16—Relation between horizontal and vertical stress for first cycle of loading of minnesota sand in one-dimensional compression. (From Kane et al, 1965)

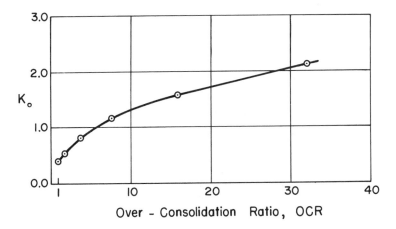

Fig. 3.17—Relationship between K_0 and OCR for medium-dense sand. (Data reported by Brooker and Ireland, 1965.)

3.4 GENERALIZED LIMITING EQUILIBRIUM ANALYSIS

Retaining structures of masonry, concrete, timber, or steel do not exhibit the same deformation and strength characteristics as the soils they support. Composed of material stiffer than soils, such walls generally experience negligible vertical strain as they move laterally to permit development of active (or passive) earth pressure. Thus, as the soil mass undergoes vertical strain, it slides with respect to the wall. In the case of cohesionless soils, the shear stress developed at the interface depends upon the effective normal stress acting on the wall and the friction developed between the wall and the soil. By contrast, the Rankine theory *assumes* the shear stress at the wall-soil interface to be the same as that which would exist in a semi-infinite mass of the soil on a plane with the same orientation as the wall. In the case of a horizontal unloaded backfill in contact with a vertical wall, the Rankine theory predicts that the vertical plane is a principal plane; that is, there is no shear stress between the wall and the soil!

Thus we see that a Rankine analysis would not have accounted for the shear stresses on the wall described in the article, even though the correct evaluation of the magnitude and direction of these stresses was clearly of decisive importance in the design of the wall. It is also evident that the Rankine theory is incapable of considering the effects of irregular boundaries of the backfill or loadings which do not extent uniformly over the surface of the backfill. Consequently, a more comprehensive analysis of earth pressures on retaining structures is often required.

One approach to this problem is to consider the equations of equilibrium in two dimensions in conjunction with the Mohr-Coulomb strength criterion, for more complex geometries and boundary conditions than the uniformly loaded half-space discussed in the preceding sections. The assumptions made in this *generalized limiting equilibrium analysis* are:

1. The medium in contact with the retaining structure consists of homogeneous, isotropic material which obeys the Mohr-Coulomb failure criterion (Equation 3–8).
2. A condition of plane strain exists in the direction of the long axis of the wall, so that the problem may be considered two-dimensional.
3. Limiting equilibrium is reached at all points within certain zones of failure associated with the geometry and boundary conditions of the specific problem considered. Boundary displacements, not calculated by the analysis, are assumed to be sufficient to mobilize fully the shearing resistance of the soil.
4. The locations of the slip surfaces of which the failure zones are composed, are determined by assuming the medium to be weightless. The actual slip surfaces for the soil mass with body weight will coincide with those calculated assuming weightlessness.

In general, such complex boundary conditions require that the simultaneous equations be solved numerically, not analytically. A numerical procedure for this purpose was first developed by Kotter (Brinch Hansen, 1953) and subsequently extended and applied to a variety of stability problems by Sokolovski (Sokolovski, 1960; Harr, 1966). We should note that the equations are exact, even though the solutions to them are obtained by approximate means.

The generalized limiting equilibrium approach has not found acceptance in practice because it is arduous to generate solutions manually, or requires the use of a computer. Furthermore approximate procedures which are simpler in application are available. This approach, called *wedge analysis*, is discussed in the following section, and in Chapter 13. It is considerably simpler to apply than generalized limiting equilibrium analysis, particularly in the case of retaining walls, and is still more flexible with regard to the boundary conditions which can be considered. The generalized limiting equilibrium analysis is especially useful, however, in providing a standard of comparison for the approximate wedge analysis method. Thus, results obtained by the general approach help us develop confidence in the approximate methods.

3.5 WEDGE ANALYSIS—COULOMB THEORY

In developing the wedge analysis, we take an approach different from that described above. Rather than describing the state of stress at every point of the backfill which is in a state of limiting equilibrium, we shall assume that, when the backfill is on the verge of failure, sliding will occur along a *single surface* through the soil, such as that shown in Figure 3.18a. The minimum

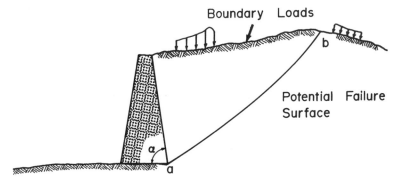

a) - Retaining Wall and Backfill

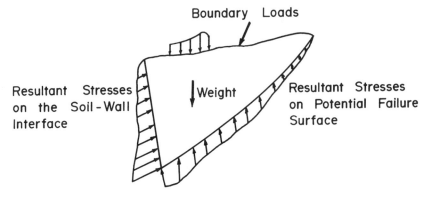

b) - Loads on Soil Wedge

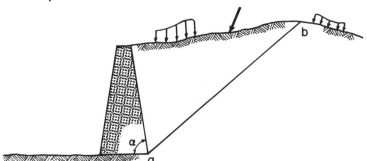

c) - Assumed Planar Failure Surface

Fig. 3.18—Active failure of retaining
wall supporting an irregular fill.

force (active case) or maximum force (passive case) between the wall and the soil *wedge* shown can be determined without considering the detailed description of the stresses within the wedge. If all the forces acting on the wedge in Figure 3.18b are known, except for those acting between the wall and the soil, these forces can be determined by considering the equilibrium of the wedge. In order to do this, we shall make the following assumptions:

1. A condition of plane strain exists in the direction of the long axis of the wall, so that the problem can be considered two-dimensional.
2. Along the sliding surface $\overline{ab}$, the cohesionless material is in a state of limiting equilibrium described by the Mohr-Coulomb relation, $\tau_{ff} = \sigma'_{ff} \tan \phi'$.
3. The sliding surface $\overline{ab}$ is *planar*, as shown in Figure 3.18c. That is, its intersection with the plane of the figure is a straight line.
4. As sliding takes place between the wall and the soil, shear stresses develop at the interface to resist the relative motion. The magnitude of the shear stress depends upon the normal effective stresses applied to the wall and an assumed constant coefficient of friction, $\tan \delta$. Thus, $\tau_\alpha = \sigma'_\alpha \tan \delta$, in which α describes the orientation of the back of the wall (Figure 3.18a). The direction of the shear stress is determined by the relative motion between the wall and the soil.

Implicit in these assumptions is the requirement that sufficient slip occur along a potential failure surface and at the interface between the soil and the wall to insure that the appropriate shearing resistance is fully mobilized.* We shall also limit our attention to those cases where the wall-soil interface is a plane; that is, there can be no break or curvature to the wall.

Active Case

In general, the forces acting on the soil wedge at an active earth pressure state include:

1. *Weight of soil within wedge.* $W = \gamma A$, where A is the area of the wedge, as shown in Figure 3.19a. Note the irregularity of the boundary. For the present, we shall assume the soil is not submerged and, therefore, that pore water pressure need not be considered. The weight acts vertically, so its magnitude and direction are both known.
2. *Body forces, such as seepage forces.* We shall not consider seepage forces at this time. Seepage effects are discussed in Chapter 14.
3. *Boundary loads.* The resultant magnitude and direction of the boundary loadings may be determined by either an analytical or graphical

---

* The approach of assuming a planar failure surface along which the normal and shear stresses are related by a coefficient of friction was first proposed by C. A. Coulomb, the city engineer of Paris, in a paper presented to the French Academy of Sciences in 1773 (Coulomb, 1776). He considered only the problem of a smooth vertical wall supporting a horizontal fill. However, the extension of this analysis to consideration of the potential friction developed between the wall and the soil as well as a planar surface is generally called the *Coulomb earth pressure analysis*. We discussed a form of Coulomb's hypothesis as the Mohr-Coulomb failure theory in Section 2.5.

summation of forces over the boundary. The assumption of plane strain requires that the loads be uniform in the direction of the long axis of the wall (normal to the page). However, they need not be uniform nor continuous along the boundary (see Figure 3.19a).

4. *Normal and shear force on failure surface.* The normal and shear forces on the failure plane are the summation of the normal and shear stresses. The magnitude and distribution of the stresses are not known, but because the failure surface is plane, the directions in which they act are known. It was for this reason that the failure surface was assumed to be planar. Because the stresses act over the same area and are related by the Mohr-Coulomb strength criteria, the forces are related by

$$\int_A \tau_{ff}\, dA = \int_A \sigma'_{ff} \tan \phi'\, dA \qquad (3\text{--}31a)$$

or

$$F_s = F'_n \tan \phi' \qquad (3\text{--}31b)$$

and may be replaced by a single force F' at an angle of ϕ' from the

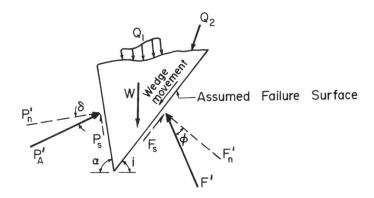

a) - Forces Acting on Assumed Failure Wedge

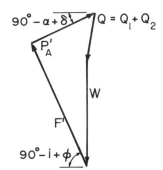

b) - Force Polygon

Fig. 3.19—Summation of forces acting on assumed failure wedge in active condition

normal to the failure surface (Figure 3.19a). Note that F_s is directed
to resist the movement of the sliding wedge. Therefore the direction
of F' is known; its magnitude is unknown, however.

5. *Normal and shear force on soil-wall interface.* The force applied by
the wall to the wedge is the summation of the normal and shear
stresses on the sliding surface between the wall and soil. Since the
wall is planar and the stresses act over the same area,

$$P_s = P'_n \tan \delta \qquad (3-32)$$

in which the forces are shown in Figure 3.19a. We see that, as in the
case of the failure surface, the shear force P_s is directed to oppose the
wedge motion with respect to the wall. The most common case for
the active condition is for the wedge to move down with respect to
the wall, producing a shear force which acts on the wedge in an upward sense, as indicated in the figure. We may substitute the resultant P'_A of the shear and normal forces acting on the wall for these
components. P'_A is called the active earth pressure force between the
wall and the soil wedge. As shown in Figure 3.19a, it acts at an angle
δ from the normal to the wall and is directed in opposition to the
wedge movement. The direction of P'_A is known; its magnitude is
unknown.

Knowing the direction of all the forces acting on the wedge and the magnitude of all but F' and P'_A, a force polygon may be drawn, as in Figure 3.19b,
from which the magnitude of the force on the wall, P'_A, may be determined.
The magnitude of force P'_A depends, however, on the surface of failure chosen.
Therefore, there are an *infinite* number of magnitudes for P'_A, corresponding
to an infinite number of plane surfaces through the heel of the wall that we
might consider. To decide upon the correct magnitude, we must recognize
that P'_A is the force which the wall must provide in order to prevent the soil
mass from sliding along a critical surface. If the wall is designed to provide
less than this force, the soil mass is unstable. Therefore the actual earth
pressure force P'_A is the *maximum* value calculated corresponding to the
various assumed failure surfaces, and the critical surface is the surface
corresponding to that force.

If all the forces on the wedge can be described as a function of i, the angle
between the assumed failure surface and the horizontal, then the magnitude
of the active earth pressure force may be determined analytically by maximizing P'_A with respect to i, by differentiation. If this cannot be easily accomplished, as is usually the case, then a trial procedure must be used, that is,
the potential value of P'_A is determined for a number of likely planar failure
surfaces. The active earth pressure force on the wall is the maximum value
of P'_A.

For those problems involving an irregular boundary of the backfill and
nonuniform loadings on the surface, a graphical solution procedure is most
practical. A number of potential planar failure surfaces are chosen, and for

each, the polygon of forces acting on the sliding wedge is constructed graphically. From the force polygon, the force P' is measured. With a sufficient number of force polygons drawn to define the change in P' as a function of i, the magnitude of P' may be plotted versus the angle i. P'_A, the maximum value of P', is then measured from the plot. A systematic procedure for doing this is described in the next section.

Passive Case

The Coulomb procedure may also be applied to determination of the passive earth pressure force on a retaining wall. The analysis will differ in two ways.

1. The relative motion of the sliding wedge and thus the direction of the forces F' and P'_P, which are directed in opposition to the movement of the wedge, are likely to be different. In the passive case, as the wall forces the soil to yield, the wedge will usually move upward relative to the wall and underlying soil, as shown in Figure 3.20a. A typical force polygon for the passive case is shown in Figure 3.20b. The force between the wall and the soil, P'_P, is clearly being required to overcome the soil resistance F'. This is in contrast to the active case where the force F' was acting to support the wedge. The angle that the failure surface makes with the horizontal is considerably less in the passive case than that observed in the active earth pressure case.
2. The passive pressure P'_P is the minimum value of P' obtained from the various assumed planar failure surfaces. No larger magnitude can be applied because a state of limiting equilibrium has already been reached on one surface within the fill and failure is then presumed to be imminent on this surface.

The validity of assuming a planar failure surface for both the active and passive cases is examined in Chapter 13.

Example 3.5 —————————————————————————————

A 15-ft-high rigid retaining wall, shown in Figure 3.21a, supports a cohesionless backfill. The back of the wall is inclined at an angle of 85° to the horizontal. The free-draining backfill has a unit weight of 120 pcf, and is estimated to exhibit an angle ϕ' of 35°. It is assumed that the soil will slide downward with respect to the wall with a friction angle between the wall and the soil of $\delta = 24°$. Determine the active earth pressure on the wall, using the Coulomb analysis.

Solution. The steps required for the Coulomb analysis of this problem are described below, and illustrated in Figures 3.21b and c:

1. A cross-section of the backfill and back of the retaining wall is drawn to scale (Figure 3.21b). The orientation and position assumed for the active earth pressure force P'_A is shown in the diagram for reference.
2. A trial failure plane is drawn from the *heel* of the wall to the ground surface, say, for example, line ① (Figure 3.21b).

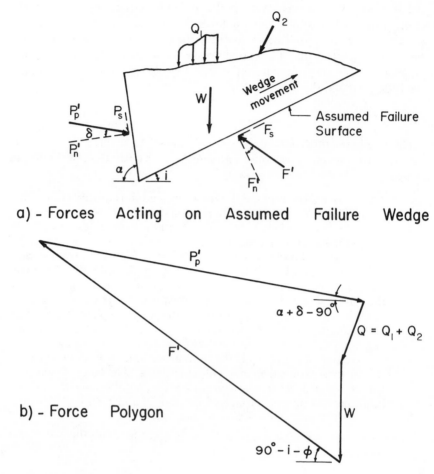

a) - Forces Acting on Assumed Failure Wedge

b) - Force Polygon

Fig. 3.20—Summation of forces acting on
assumed failure wedge in passive condition.

3. The weight of the soil within the trial failure wedge defined by the back of the wall, the ground surface, and the trial failure surface is calculated by computing the area *abc* (Figure 3.21b) and multiplying this area by the unit weight of the soil.

4. A vertical vector representing the weight of the trial failure wedge abc is constructed to a convenient scale, and is shown as weight vector ① in Figure 3.21c.

5. The magnitude of the resultant force F' on the trial failure plane is not known, but its direction is inclined at the angle ϕ' to the normal to the trial failure surface. The F' vector is shown for illustrative purposes in Figure 3.21b. Thus a line is constructed from the top of the weight vector in the direction of the resultant force F' (Figure 3.21c).

6. The direction of the active pressure force P'_A is also known, as shown in Figure 3.21b. From the base of the weight vector, a line is constructed parallel

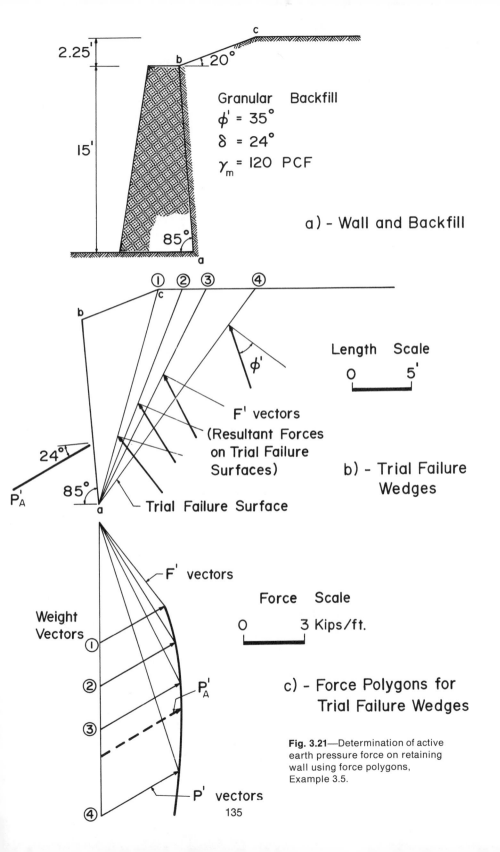

2.25'

b

c

20°

Granular Backfill
$\phi' = 35°$
$\delta = 24°$
$\gamma_m = 120$ PCF

15'

85°

a

a) - Wall and Backfill

① ② ③ ④

c

b

ϕ'

Length Scale

0   5'

F' vectors
(Resultant Forces
on Trial Failure
Surfaces)

24°

85°

P_A'

a

Trial Failure Surface

b) - Trial Failure
Wedges

F' vectors

Force Scale

0   3 Kips/ft.

Weight
Vectors

①

②

③

④

P_A'

c) - Force Polygons for
Trial Failure Wedges

Fig. 3.21—Determination of active
earth pressure force on retaining
wall using force polygons,
Example 3.5.

P' vectors

135

to this direction (Figure 3.21c). This line closes the force polygon and determines the magnitudes of the earth pressure force for the trial wedge P' and the resultant force on the trial failure surface F'.

7. Steps 2 through 6 are repeated for several additional trial failure surfaces through the heel of the wall shown as ②, ③ and ④ (Figure 3.21b). For convenience, each of the weight vectors originates from the same point (Figure 3.21c). Thus we see that the F' vectors all terminate at this same point.

8. A curved line connecting the ends of the P' vectors is drawn (Figure 3.21c). This line is the locus of the ends of all such P' vectors within the range of trial failure surface inclinations considered.

9. Because the P' vectors are all parallel, the distance from the weight vector where they originate to the curved line (Figure 3.21c) is proportional to their length. Thus a line parallel to the weight vector is tangent to the curved locus of the ends of the P' vectors at the point corresponding to the maximum such vector.

10. The active earth pressure P'_A is taken as the maximum value of P', determined in step 9. The P'_A vector is drawn (Figure 3.21c), and its magnitude is scaled as 4.8 kips per lineal foot along the long axis of the wall.

The orientation of the failure surface predicted by the Coulomb analysis can be determined from the direction of the F' vector corresponding to P'_A. We can see that the curve from which P'_A was determined is flat, and that the magnitude of P'_A is relatively insensitive to small changes in the location of the failure surface. Conversely, the precise determination of a failure surface predicted by the Coulomb analysis depends markedly upon the selection of the point of tangency to the curve. This fact encourages us to believe that the selection of an assumed plane failure surface may not have influenced the calculated magnitude of P'_A to a major degree.

The insensitivity of P'_A to the orientation of the failure surface also suggests that small inhomogeneities in the soil mass may alter the location of the failure surface. In fact, recent experimental observations of strains in backfills of model retaining walls indicate that there is no failure *surface* but rather a *zone* of material within which rupture is occurring at all points (Roscoe, 1970).

Culmann Construction

A convenient graphical procedure for constructing the force polygons corresponding to the various trial failure wedges is available. Called the *Culmann construction* after its originator (Culmann, 1866), it consists in rotating the force polygons so that they can be constructed directly on a scale drawing of the back of the retaining wall and backfill. This process is illustrated in Figure 3.22.

A retaining wall supporting a cohesionless fill is shown in Figure 3.22a. The back of the wall is inclined at an angle α to the horizontal. As in the previous example, the active earth pressure force is assumed to act on the wall at an angle δ from the normal to the wall. In Figure 3.22a one trial failure surface inclined at the angle i from the horizontal is shown. The force polygon corresponding to this trial failure surface is illustrated in Figure

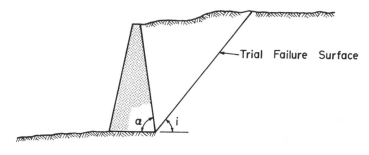

a) - Wall and Backfill

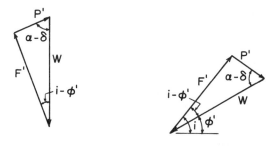

b) - Force Polygon
    Normal Orientation

c) - Force Polygon Rotated

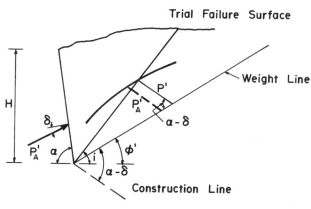

d) - Culmann Construction

Fig. 3.22—Illustration of Culmann graphical procedure
for determining active earth pressure.

3.22b. Note that the angle between the F' vector and the weight vector is $(i - \phi')$, and the angle between the P' vector and the weight vector is $(\alpha - \delta)$.

So far we have elected to construct the force polygon by drawing the various vectors with their orientation in the polygon the same as their orientation in space. Thus the weight vector is drawn vertical, and so on. It is not necessary, however, to do this, as long as the angular relationships between the vectors are preserved. In this case it is more convenient to construct the force polygon rotated from its original position in Figure 3.22b so that the weight vector is directed at an angle ϕ' from the horizontal, as shown in Figure 3.22c. All of the vectors in the force polygon are rotated by the same amount so that the angular relationships between them are the same, as we may see by comparing the two figures. The advantage of doing this arises from the fact that the F' vector for any failure surface which is oriented at an angle i from the horizontal will also lie at that angle i in the rotated force polygon. Thus in the rotated force polygon the F' vector assumes the orientation of the failure surface on which it acts. The P' vectors in the rotated force polygon (Figure 3.22c) are directed at an angle $(\alpha - \delta)$ from the rotated weight vector.

The Culmann procedure takes advantage of the properties of the rotated force polygon by using the trial failure surface on a scale drawing of the backfill to define the direction of the F' vectors in a series of force polygons for the various trial surfaces. The specific steps required for this procedure are described below, and shown in Figure 3.22d:

1. A cross-section of the backfill and back of the retaining wall is drawn to scale.
2. A *weight line* at an angle ϕ' from the horizontal, and a *construction line* at an angle $(\alpha - \delta)$ from the weight line, are drawn. The weight line defines the direction along which the various weight vectors for the trial wedges will lie. The construction line is parallel to the direction that the P' vectors assume in the rotated force polygons.
3. A trial failure surface, such as that shown in Figure 3.22d, is drawn. The weight of the trial failure wedge is determined and scaled onto the *weight line*. The line defining the trial failure surface also serves to define the direction of the F' vector in the rotated force polygon.
4. Through the end of the weight vector, a line is drawn parallel to the *construction line*, the direction of which is that of the P' vector in the rotated force polygon. The intersection of this line and the trial failure surface (which defines the direction of the F' vector) determines the magnitude of the P' vector for the particular assumed failure surface.
5. Steps 3 and 4 are repeated for several additional trial failure surfaces.
6. A curved line connecting the ends of the P' vectors is then drawn. This line, shown in Figure 3.22d, is the locus of the ends of all such P' vectors within the range of trial failure surface inclinations considered. A line parallel to the weight line is tangent to the curved

locus of the ends of the P' vectors at the maximum such vector which is the active earth pressure P'_A. This procedure is the same as that described in step 9 in Example 3.5. The upper end of the P'_A vector in the rotated force polygon also defines the failure surface predicted by the Coulomb analysis.

We must recall that the Culmann procedure is simply a useful graphical construction of the Coulomb wedge analysis. We shall illustrate its use by considering the same problem described in Example 3.5.

Example 3.6

The 15-ft-high rigid retaining wall shown in Figure 3.23a supports a cohesionless backfill with the geometry and properties previously described in Example 3.5 and shown in Figure 3.21a. In this case we wish to determine the active earth pressure, using the Culmann construction.

Solution. The solution using the Culmann graphical procedure is illustrated in Figure 3.23b. In this case four trial wedges with the accompanying force polygons are shown. These trial failure surfaces are the same as those in Example 3.5. Thus if the force polygons used in the preceding example (see Figure 3.21c) were rotated until the weight vectors coincided with those in Figure 3.23b, the force polygons would appear exactly the same. As in the preceding example, the magnitude of P'_A is scaled as 4.8 kips per lineal foot along the long axis of the wall. The failure surface is determined by this method to be inclined at $60°$ from the horizontal, and is shown in Figure 3.23b.

The Culmann procedure may be used to advantage if there are surface loads applied to the wedge. If they are vertical, they are simply added to the weight of the wedge. If the resultant of the boundary loads on the trial wedge is not vertical, it must be drawn at an angle from the weight line equal to the angle which it makes with the vertical. Examples of the application of the Culmann procedure to cases where boundary loads are applied to the backfill are given below.

Example 3.7

We now consider the case where the retaining wall and backfill described in Example 3.6 are supporting a line load of 5 kips per lineal foot two feet behind the crest of the fill as shown in Figure 3.24a. We wish to determine the magnitude of the active earth pressure force on the wall for this case.

Solution. The solution using the Culmann graphical procedure is illustrated in Figure 3.24b. The same four trial wedges examined in Example 3.6 are shown in the figure. For the first trial failure surface, the weight vector is the same as in Example 3.6 because the line load applied to the right of the trial failure surface is assumed to have no influence on the forces within the wedge.

The line load acts directly on the second trial failure surface. Thus, immediately to the left of this surface, the weight vector is not influenced by the magnitude of the

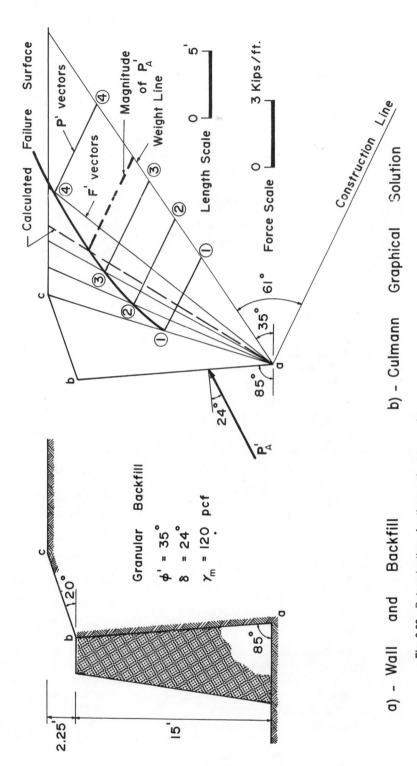

Granular   Backfill

$\phi' = 35°$
$\delta = 24°$
$\gamma_m = 120$ pcf

Length Scale

Force Scale

a) - Wall   and   Backfill   b) - Culmann   Graphical   Solution

Fig. 3.23—Determination of active earth pressure force on
retaining wall using Culmann graphical solution, Example 3.6.

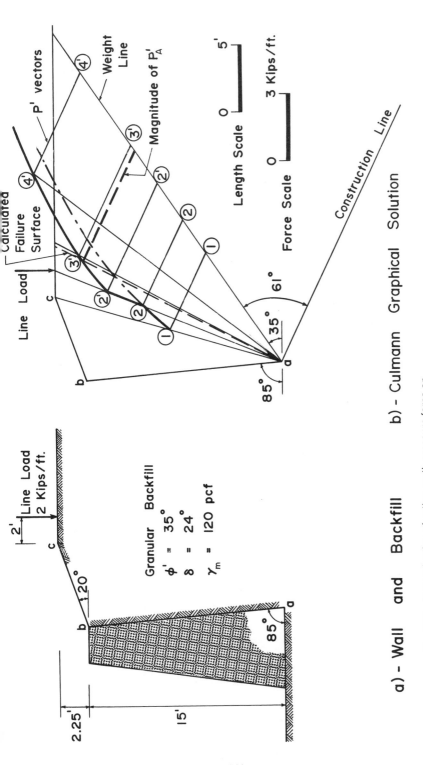

Line Load

P' vectors

Calculated Failure Surface

Weight Line

Magnitude of P'$_A$

Length Scale

0   5'

Force Scale

0   3 Kips/ft.

Construction Line

85°   35°   61°

a) - Wall and Backfill

Line Load 2 Kips/ft.

2'

c

20°

b

Granular Backfill

$\phi' = 35°$

$\delta = 24°$

$\gamma_m = 120$ pcf

85°

2.25'

15'

b) - Culmann Graphical Solution

Fig. 3.24—Determination of active earth pressure force on retaining wall using Culmann graphical solution, Example 3.7.

141

line load. Immediately to the right of this trial surface the weight vector is increased by the magnitude of the line load (2 kips per lineal foot). For all trial failure surfaces to the right of the second one, the weight vector is increased by this amount. The corresponding increase in the P' vectors may be seen by comparing the solid curved line in Figure 3.24b, which is the locus of the ends of the P' vectors in which the line load effect has been considered, and the dashed curved line, which is the locus considering only the weight of the soil. The magnitude of P'_A is scaled at 5.6 kips per lineal foot. We note that the effect of the line load was to increase the magnitude of the active earth pressure force as well as to move the calculated failure surface to the left of the position determined for the case without the line load. This may be seen by comparing Figures 3.23b and 3.24b.

Example 3.8

The retaining wall and fill described in the preceding examples are now assumed to be supporting a *strip* load of uniform intensity equal to 250 psf, as shown in Figure 3.25a. We wish to determine the active earth pressure force on the wall.

Solution. Again the Culmann graphical procedure for applying the Coulomb wedge analysis is used to determine P'_A. The solution is illustrated in Figure 3.25b. The same four trial failure wedges considered in the earlier example are shown in this figure. In the case of the distributed load, however, the effect of surface loading on the vertical vector increases gradually from an initial zero value, where the trial failure surface just intersects the left hand side of the strip load, to a maximum increment at the right-hand side of the loading. The locus of the ends of the P' vectors is shown as a heavy solid curved line. In this case the magnitude of P'_A is determined to be 5.5 kips per lineal foot. Note that, in this case, because the strip loading extends to the right, it leads to a failure plane which lies to the *right* of the failure plane determined for the case of the unloaded backfill.

Magnitude of the Coefficient of Wall Friction

The limiting value of the coefficient of wall friction, tan δ, depends upon the properties of the soil backfill and the texture and material of the wall. Laboratory tests (Meyerhof, 1961; Brumund, 1965) have demonstrated that, the rougher the wall surface, the higher is the value of tan δ. In addition, the greater the relative coarseness of the wall surface compared to the granular backfill, the greater is the interlocking of the fill and wall surface. In tests where a concrete element with a roughened surface was slid across a fine sand, the coefficient of sliding equaled the angle of internal friction of the sand, suggesting that the interlocking at the concrete-soil interface was so great that the sliding occurred within the soil rather than at the boundary (Brumund, 1965). It also appears that the coefficient of sliding and the angle of shearing resistance ϕ' depend upon the same mechanical properties of a granular soil, such as relative density and angularity of the grains (Terzaghi, 1934; Meyerhof, 1961). Static values of δ for cohesionless soils are listed in Table 3.3 for a variety of wall materials in terms of the ratio δ/ϕ'.

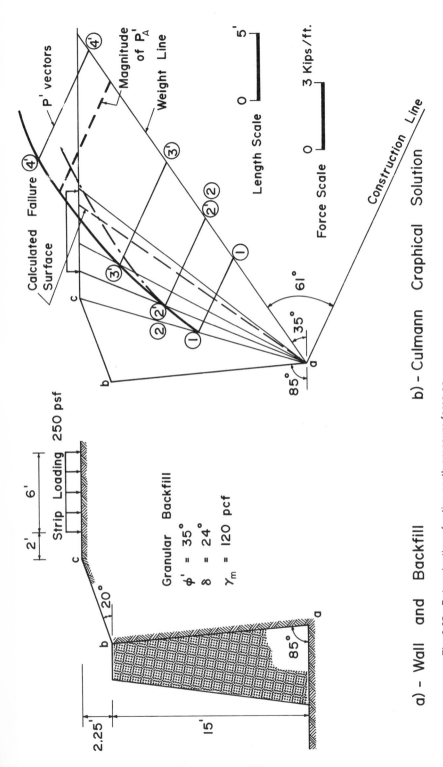

Fig. 3.25—Determination of active earth pressure force on retaining wall using Culmann Graphical Solution, Example 3.8.

a) - Wall and Backfill

Strip Loading 250 psf

2' 6'

c

20°

b

85°

a

Granular Backfill

$\phi' = 35°$

$\delta = 24°$

$\gamma_m = 120$ pcf

2.25'

15'

b) - Culmann Graphical Solution

Calculated Failure Surface

P' vectors

Magnitude of P'_A

Weight Line

Length Scale

0   5'

Force Scale

0   3 Kips/ft.

Construction Line

61°

35°

85°

a

b

c

143

TABLE 3.3
Ultimate Static Values of δ/ϕ'

Wall Material	δ/ϕ'
Smooth concrete and masonry	0.8–1.0
Rough concrete	0.9–1.0
Smooth steel	0.5–0.7
Rough steel	0.8–0.9
Wood	
parallel to grain	0.7–0.9
normal to grain	0.9–1.0

SOURCE: Modified from Meyerhof (1961).

Similar values were reported by Leonards and Brumund (1965) for steel and mortar rods sliding through sand. Terzaghi (1934), however, demonstrated that in a dense cohesionless soil dilatancy may induce δ to reach a maximum value and then decrease as the soil dilates. Hence, there does not seem to be any justification at the present time for the use of a δ value greater than the *lower* limiting values reported by Meyerhof (Table 3.3), with the additional restriction that ϕ' (for the purpose of computing δ) be taken as no greater than 30°.

We must also remember that, just as a granular soil must strain to mobilize its maximum strength, so it is essential that sliding take place between the wall and soil to mobilize the limiting value of tan δ.

Effect of Submergence

As we saw when considering the Rankine analysis, partial or total submergence of a backfill results in an increase in the total pressure on the wall if an active state of limiting equilibrium exists, or a decrease in the total pressure if a passive state of limiting equilibrium obtains. In applying the wedge analysis to the case of submerged backfill, we must again separate the effective stresses from the applied water pressure. In the case where the groundwater table is horizontal and no seepage forces exist, this may be effected by using the *effective* unit weight in place of the *total* unit weight in computations involving materials below the groundwater table.

We can show that this is the case by considering the situation illustrated in Figure 3.26. A wall and backfill with one trial failure wedge are shown in Figure 3.26a. Ther groundwater table submerges a part of the backfill. The forces on the wedge, indicated in Figure 3.26b, are:

1. *Weight of soil within wedge.* This weight consists of:
   a. The total weight of soil above the groundwater table, W_1.
   b. The effective weight of the soil below the groundwater table, W_2,

$$W'_2 = \frac{h^2}{2}(\cot \alpha + \cot i)\gamma' \qquad (3\text{--}32)$$

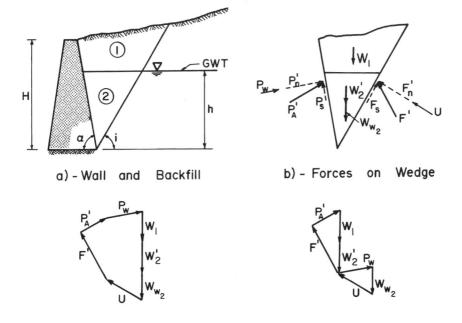

a) - Wall and Backfill

b) - Forces on Wedge

c) - Complete Force Polygon

d) - Equivalent Force Polygon

Fig. 3.26—Effect of partial submergence on active
pressure determined using wedge analysis.

in which h is the height of the groundwater table above the base
of the wall, and γ' is the submerged unit weight.

c. The weight of water below the groundwater table, W_{w_2},

$$W_{w_2} = \frac{h^2}{2}(\cot \alpha + \cot i)\gamma_w \qquad (3-33)$$

2. *Forces on trial failure surface.*

a. The normal force on the failure surface, F_N, may be replaced by
the summation of an effective normal force F'_N and the force due
to water pressure along the failure surface, U,

$$F_N = F'_N + U \qquad (3-34)$$

in which

$$U = \frac{h^2 \gamma_w}{2 \sin i} \qquad (3-35)$$

b. The shear force, F_s, and the *effective* normal force, F'_N may be
replaced by their resultant, F', acting at an angle ϕ' from the
normal to the failure plane in the direction opposing motion of
the soil wedge.

4. *Forces on wall-soil interface.*
   a. The normal force, P_N, may be replaced by the summation of the effective normal force P'_N and the resultant water pressure force P_w,

$$P_N = P'_N + P_w \qquad (3\text{–}36)$$

   in which

$$P_w = \frac{h^2 \gamma_w}{2 \sin \alpha} \qquad (3\text{–}37)$$

   b. The effective earth pressure force P'_A, acting at an angle δ from the normal to the wall in a direction opposing the movement of the soil wedge with respect to the wall, is the resultant of the shear and effective normal forces.

The directions of all forces acting on the wedge are known, as are the magnitudes of all but P'_A and F'. Consequently a force polygon (Figure 3.26c) may be constructed. We note, however, that the water forces are self-equilibrating. That is, summing forces in the vertical direction, we find:

$$W_w - U \cos i - P_w \cos \alpha =$$
$$\frac{h^2}{2}(\cot \alpha + \cot i)\gamma_w - \frac{h^2 \gamma_w}{2}\frac{\cos i}{\sin i} - \frac{h^2 \gamma_w}{2 \sin \alpha}\cos \alpha = 0 \qquad (3\text{–}38)$$

Similarly, summing forces in the horizontal direction:

$$P_w \sin \alpha - U \sin i = \frac{h^2 \gamma_w}{2} - \frac{h^2 \gamma_w}{2} = 0 \qquad (3\text{–}39)$$

Or, we may redraw the force polygon to separate the water forces, which simply form a closed polygon from the remaining effective forces, as shown in Figure 3.26d. Thus we see that the *effective* earth pressure force may be calculated using the effective unit weight of the soil. Furthermore, the water-pressure force acting on the wall is independent of the failure surface selected, and may be computed as a hydrostatic pressure when there are no seepage forces.

Effect of Direction of Relative Movements Between Wall and Soil

In applying the Coulomb earth pressure analysis, we must assume the direction of the earth pressure force, and therefore the direction of the shear force acting between the wall and the backfill. The direction of the shear stress which actually develops depends, of course, on the direction of the relative movements between the failure zone and the wall. For example, as shown in Figure 3.27a, when the backfill moves *down* with respect to the wall, the shear stress acting at the interface is directed *down* on the wall and *up* on the soil. The resultant active earth pressure force is therefore directed *down* on the wall and *up* on the soil wedge at an angle δ from the normal as

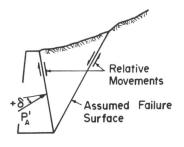
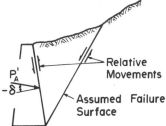

a) - Wall and Fill, Positive δ Case   b) - Wall and Fill, Negative δ Case

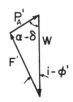
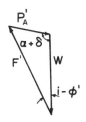

c) - Force Polygon, Positive δ Case   d) - Force Polygon, Negative δ Case

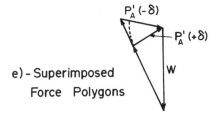

e) - Superimposed
Force Polygons

Fig. 3.27—Effect of direction of relative motion between the wall and backfill on the active earth pressure.

indicated in Figure 3.27a. This is called a *positive-δ case*, and is the most common situation for the active-pressure condition.

If, however, the wall settles faster than the backfill, the wall moves down with respect to the soil wedge. In this case, indicated in Figure 3.27b, the shear stress applied at the wall-soil interface is directed *up* on the wall and *down* on the soil wedge. We refer to this condition as the *negative-δ case.*

The solution procedure for both cases is the same. However, the magnitude and direction of P'_A are different. The Coulomb method of analysis also predicts a different failure surface for the two cases. Nonetheless, it is useful to compare the magnitudes of the active earth pressure force for each case by examining a single trial failure surface for both positive- and negative-δ cases. The force polygons for these two cases and the trial failure surface of Figure 3.27a are shown in Figures 3.27c and d respectively. The active earth pressure force for the trial wedge computed for the positive-δ case is

$$P'_{+\delta} = \frac{W \sin (i - \phi')}{\sin [(i - \phi') + (\alpha - \delta)]} \tag{3-40}$$

and for the negative-δ case is

$$P'_{-\delta} = \frac{W \sin (i - \phi')}{\sin [(i - \phi') + (\alpha + \delta)]} \tag{3-41}$$

If we restrict our attention to practical values of α' and those cases for which $(i - \phi')$ is always positive, we observe from Equations 3-40 and 3-41 that the P'_A in the negative-δ case is larger than for the positive-δ case, irrespective of which trial failure surface is considered. This is illustrated graphically by superimposing the two force polygons in Figure 3.27e. An indication of the difference in the magnitudes of P'_A in the two cases is given in Figure 3.28. In this figure is shown the ratio of the *horizontal component* of the active earth pressure force from an unloaded horizontal backfill on a vertical wall to $\gamma H^2/2$ as determined from Equations 3-40 and 3-41. The actual earth pressure force is determined by multiplying the coefficient obtained in Figure 3.28 by $\gamma H^2/2 \cos \delta$. It is apparent from this figure that for many practical combinations of δ and ϕ', the difference between the positive- and negative-δ cases may be more than 50 per cent. We might also note that the case of $\delta = 0$ leads to earth pressure coefficients which are the same as determined by the Rankine analysis. Thus we see that including the shear stress in our calculation for the positive-δ case leads to a reduction in the computed active earth pressure of approximately 10 per cent.

Although the negative-δ case is not common for active earth pressures, its impact on the design of a retaining structure makes it imperative that the

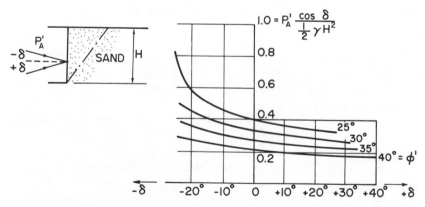

Fig. 3.28—Coefficients for determination of active earth pressure force on a vertical wall, retaining an unloaded horizontal cohesionless backfill. (After Terzaghi and Peck, 1967.)

designer consider the potential for development of this condition. This is dramatically illustrated by the crib wall failure described in the article reprinted in this chapter.

Effect of Inclination of Wall and Backfill

For the more general case of a wall inclined at an angle α to the horizontal, supporting a backfill with its surface at an angle β to the horizontal, Coulomb (1776) obtained the result

$$P'_A = \tfrac{1}{2}\gamma H^2 \left[\frac{\sin^2(\alpha + \phi')}{\sin^2\alpha \sin(\alpha - \delta)\left(1 + \sqrt{\dfrac{\sin(\phi' + \delta)\sin(\phi' - \beta)}{\sin(\alpha - \delta)\sin(\alpha + \beta)}}\right)} \right]^2 \quad (3-42)$$

in which the sign of δ is determined by the relative movement of the wall and soil (that is, see Figure 3.28).

3.6   RECONSIDERATION OF CRIB WALL FAILURE

We are now in a position to investigate in detail the crib wall failure described in the accompanying article. Its author, G. P. Tschebotarioff attributed the failure to sliding within the crib wall along the surface between the double-thick lower portion and the single-thick upper part. That the slide was able to take place within the wall was due to the fact that the shell of the wall was constructed of individual units which were not bonded to one another. Thus sliding could, and did, take place between the rows of headers and through the interior fill.

To evaluate the likelihood of sliding* in a case such as the one we are considering, let us investigate the equilibrium of a free body of a crib wall (Figure 3.29). The forces acting on the wall are:

1. The weight of the wall, W.
2. The active earth pressure force, P'_A (a positive-δ case is illustrated in the figure).
3. An effective normal force, N', acting on the base of the wall. From the equilibrium requirements normal to the base of the wall,

$$N' = W\cos\theta + P'_A \cos(\alpha - \theta - \delta) \quad (3-43)$$

4. A shear force, F_R, on the base of the wall. If equilibrium obtains, summation of the forces parallel to the base of the wall leads to the required magnitude,

$$F_R = P'_A \sin(\alpha - \theta - \delta) - W\sin\theta \quad (3-44)$$

---

* The design process for a retaining structure requires investigating the potential of other modes of failure than sliding on the base or within the wall. We shall discuss these in detail, along with other design considerations, in Chapter 13.

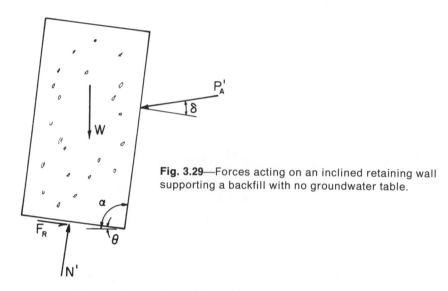

Fig. 3.29—Forces acting on an inclined retaining wall supporting a backfill with no groundwater table.

The *maximum* value of the shear force on the base of the wall is

$$F_{R_{\max}} = N' \tan \mu \qquad (3\text{--}45)$$

in which $\tan \mu$ is the coefficient of friction between the wall and soil along the sliding surface at the base.

If the maximum shearing resistance at the base of the wall, $F_{R_{\max}}$, is greater than the required shearing resistance, F_R, the wall will not fail by sliding along its base.* A correct design insures that F_R is suitably less than $F_{R_{\max}}$. The ratio of these two is defined as the *factor of safety* against sliding,

$$FS = \frac{F_{R_{\max}}}{F_R} \qquad (3\text{--}46)$$

Substituting Equations 3–42 to 3–44 into Equation 3–45 leads to the expression for the factor of safety against sliding:

$$FS = \frac{[W \cos \theta + P'_A \cos (\alpha - \theta - \delta)] \tan \mu}{P'_A \sin (\alpha - \theta - \delta) - W \sin \theta} \qquad (3\text{--}47)$$

In the original design of the crib wall, an analysis of sliding along the surface upon which failure subsequently occurred, indicated a factor of safety of 2.82 (see Figure 6 of the case article, page 97). We note, however, that the designers assumed that the backfill would settle with respect to the wall, thus estimating the active earth pressure force on the basis of a positive δ condition.

---

* Of course some *small* amount of relative displacement between the base and the soil is necessary in order to mobilize the shearing resistance between them.

If, as observation of the wall suggested, the structure had settled with respect to the backfill, then the direction of the shear stresses on the wall would have been reversed. A Coulomb analysis of the earth pressure for both positive- and negative-δ cases, using the Culmann graphical solution, is given in Figure 3.30. In the analysis the following data were used:

1. For the backfill, $\phi' = 30°$. Although the angle of shearing resistance for the fill was assumed in the design to be $33°$, the accompanying article suggests that as a result of poor compaction procedures during construction the actual *in situ* ϕ' was less than this.
2. The equivalent coefficient of friction between the back of the wall and the backfill, $\delta = 27°$. This value was assumed on the basis of the data given in Table 3.2. The magnitude of δ assumed for design of the wall was lower than this.
3. The slope of the backfill, $\beta = 12°$. This value was reported from observation of the fill after construction.
4. The unit weight of the soil, $\gamma = 118$ pcf. An estimate of the unit weight was made from the area and weight of the soil filled wall used in the original design.

The results of the wedge analysis shown in Figure 3.30 indicate that the magnitude of the active earth pressure force is 5000 pounds/lineal foot along the long dimension of the wall for the positive-δ case, and 10,700 pounds/ lineal foot for the negative-δ case.

The forces acting on the wall itself can then be computed from Equations 3–43 to 3–45, and are shown graphically in Figure 3.31 for both positive- and negative-δ cases. We note that as a result of the change in the direction of the shear force applied to the back of the wall, not only was the required shearing resistance along the base increased, but there was a simultaneous decrease in the maximum shearing resistance along the base due to a reduction in a normal force. The factors of safety for these two cases, as determined from Equation 3–46, are

$$FS_{+\delta} = \frac{(17.2)(0.532)}{1.8} = 5.08 \qquad (3-48)$$

$$FS_{-\delta} = \frac{(10.1)(0.532)}{7.0} = 0.77 \qquad (3-49)$$

From these results it is quite apparent that had the wall settled with respect to the fill an instability of the upper section of the wall with respect to sliding should be expected.

There are three reasons to believe that the wall did settle a greater amount at the rear than at the front, thus causing the back of the wall to move down with respect to the soil in the backfill, that is, a negative-δ condition:

1. The fill material *within* the wall was placed in freezing weather when adequate compaction was not possible. When this loose material

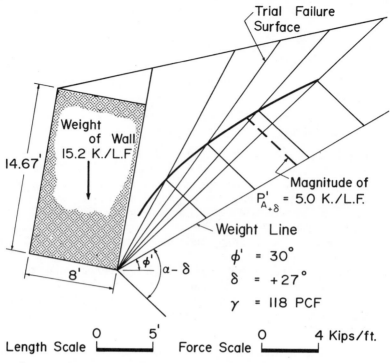

Weight
of Wall
15.2 K./L.F.

14.67'

8'

ϕ' $\alpha - \delta$

Trial Failure
Surface

Magnitude of
$P'_{A_{+\delta}}$ = 5.0 K./L.F.

Weight Line

ϕ' = 30°

δ = +27°

γ = 118 PCF

Length Scale 0 ____ 5'   Force Scale 0 ____ 4 Kips/ft.

a) - +δ Case, Soil Wedge Sliding Down Relative
to Back of Wall

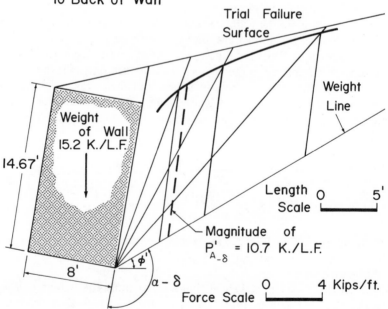

Trial Failure
Surface

Weight
Line

Weight
of Wall
15.2 K./L.F.

14.67'

8'

ϕ' $\alpha - \delta$

Length
Scale 0 ____ 5'

Magnitude of
$P'_{A_{-\delta}}$ = 10.7 K./L.F.

Force Scale 0 ____ 4 Kips/ft.

b) - -δ Case, Back of Wall Sliding Down Relative
to Soil Wedge

Fig. 3.30—Culmann graphical solution for
active earth pressure force on crib wall.

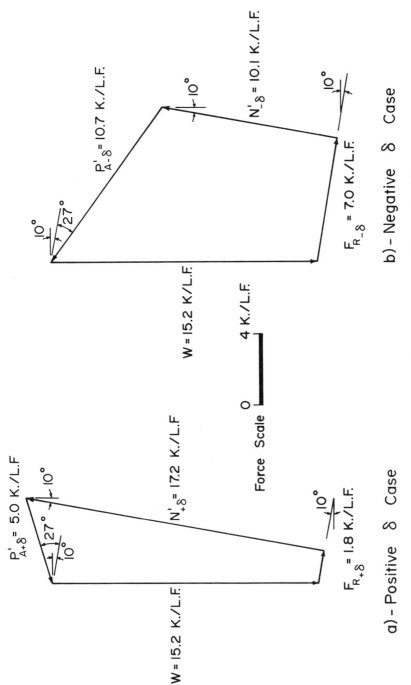

Fig. 3.31—Equilibrium force polygon for upper portion of crib wall.

thawed it tended to settle within the wall and produce a downdrag on the interior of the structural members. Because the wall was tilted toward the backfill, the rear wall of the upper section was more heavily loaded than the forward wall, by both shear and normal stresses. A numerical estimate of the shear stresses applied to the interior wall (see article, page 98) indicated that cracking of some of the header units could be expected. Excavation of material from within three cells disclosed the predicted cracks.

2. The lower section of the wall was wider than the upper section. As a result of the geometry of the wall a significant weight of backfill rested on the lower section producing a higher load intensity at the rear of the wall than at the front.

3. The foundation of the wall included a "substantial" reinforced concrete footing at the *toe*, with only three smaller stretcher units at the rear of the main portion of the wall (page 98). This factor served to contribute further to a higher intensity of pressure underneath the small stretcher units at the back of the main section of the wall, which would be expected to lead to larger settlements in that area.

Tschebotarioff's conclusion (page 99) that the shear stress between the soil and wall corresponded to the negative-δ case and caused the instability of the wall seems to have been borne out. The factors which caused this reversal in direction appear to have stemmed from a misunderstanding of how the wall and soil would interact, combined with poor construction practices.

Among other useful lessons to be learned from this example, we note that construction practice can be a major factor in determining the degree to which design assumptions are reasonable representations of field conditions. Thus it is incumbent upon the designer to insure, through specifications and field inspection when possible, that construction methods do not inadvertently produce unanticipated results.

3.7 SUMMARY OF KEY POINTS

Here, in summary, are the key points of the present chapter. The sections in which they were discussed are noted so that reference can be made to the original, more detailed discussion:

1. In order to determine the stresses between a retaining structure and the backfill which it supports, we have investigated the state of stress throughout a mass of cohesionless material. Determination of this state of stress was based upon the assumption of a material continuum subject to (Section 3.2):
   a. The requirements of equilibrium (Equation 3–3).
   b. Appropriate boundary conditions.
   c. Assumption of a suitable constitutive law describing the nature of the material.

2. One useful class of solutions is obtained by assuming that the medium is semi-infinite, and in a state of limiting equilibrium, defined by the Mohr-Coulomb failure criterion (Equation 3–8). This approach is called the Rankine earth pressure theory. These assumptions led to two solutions for the limiting *active* (minimum lateral pressure) and *passive* (maximum lateral pressure) cases of a semi-infinite cohesionless soil mass bounded by a uniformly loaded plane surface (Equations 3–9). For the special case of a horizontal boundary, subjected to normal loading only, the horizontal and vertical planes are principal planes. In this special case failure surfaces are planar and oriented $\pm(45° + \phi'/2)$ with respect to the major principal plane (Section 3.2).

3. The analysis of conditions at limiting equilibriums implies the assumption that the strains required to mobilize the full strength at each point have occurred. Experimental observations indicate that small strains within a soil mass are required in order to develop the active state of limiting equilibrium. Much larger strains are required for the passive case (Section 3.2).

4. When no lateral strains are permitted within an earth mass, the lateral pressure, earth pressure at rest, is intermediate between the active and passive earth pressures. The ratio of horizontal to vertical effective normal stress, defined as the coefficient earth pressure at rest K_0, has been found experimentally to depend upon the strength of the soil and its stress history (Equation 3–30). No reliable method is presently available for determining the *in situ* value of K_0 (Section 3.3).

5. The use of the Rankine theory to predict pressures between retaining structures and a backfill assumes that the stresses acting on the wall are identical with those which would occur in the semi-infinite medium along a plane corresponding to the position of the wall. Although a more generalized limiting equilibrium analysis in which the assumptions of interaction between the wall and the soil is not so restrictive is available, the complexity of the solution procedures has led to adoption in practice of alternative approaches (Section 3.4).

6. The use of *wedge analysis* requires the assumption that when the backfill is on the verge of failure:
    a. Sliding will occur along a specific single surface through the soil.
    b. Conditions at each point along this surface can be described by the Mohr-Coulomb failure criterion.
    c. Sufficient information concerning the distribution of forces on the failure wedge is known so that the equilibrium of forces on the wedge can be examined.
    Location of the most likely failure surface of the assumed shape must be determined by trial (Section 3.5).

7. A special case of the wedge analysis is the case in which the failure surface is assumed to be a plane. This approach, called the Coulomb earth pressure theory, requires assumption of the location and direction of the resultant force exerted by the wall on the backfill. The

primary advantage of this method is its capability to incorporate a variety of factors into the analysis, including:
   a. Nonplanar backfill surface.
   b. Complicated two-dimensional surface loading.
   c. Shear stresses between the wall in the soil such that the magnitude of the shear stress is a function of the backfill and wall materials, and the direction of the shear stress depends upon the relative movement of the wall and backfill (Section 3.5).

8. The magnitude of the earth pressure force exerted by the soil on the wall is influenced markedly by direction of the shear stress between the wall and the soil. In the active case, a relative movement between the wall and the soil which produces a shear force acting down on the soil, a negative-δ case, leads to an increase of the earth pressure force on the wall (Section 3.5).

9. Drainage conditions in the backfill are of fundamental importance to the net force acting on the wall. The lateral force on a wall supporting a submerged backfill in the active condition is likely to be more than double that resulting from a well-drained soil mass (Sections 3.2 and 3.5).

PROBLEMS

3.1. The boundary of a semi-infinite soil mass is horizontal and free of load. The material *in situ* (in the field) has a ϕ' angle of 34° at a relative density of 0.60 and is 40-percent saturated. Its minimum and maximum void ratios have been determined to be 0.32 and 0.90 respectively. The specific gravity of solids is 2.62.
   (a) Determine the distribution of the potential minimum and maximum horizontal stresses to a depth of 10 ft.
   (b) Determine the distribution of stresses to a depth of 10 ft on a plane oriented at 45° from the horizontal, for the minimum and maximum states of stress.

3.2. The soil mass described in Prob. 3.1. is now determined to be saturated, $\gamma_{sat} = 128$ pcf, with the free-water surface at the ground surface.
   (a) Describe the distribution of the potential minimum and maximum horizontal effective and total stresses to a depth of 10 ft.
   (b) Describe the distribution of stresses to a depth of 10 ft on a plane oriented at 45° from the horizontal for the minimum and maximum states of stress.

3.3. A retaining wall supporting a *dry* cohesionless backfill is inclined at 10° from the vertical. Two cases are to be considered, as shown. In the first case the wall is inclined away from the fill; in the second case the wall is inclined toward the fill. The backfill has a horizontal boundary. Its unit weight is 110 pcf and $\phi' = 35°$. (See figure on page 157.)
   (a) Using the Rankine analysis, determine for both cases the magnitude, direction, and location of the resultant force on the wall.
   (b) Although the Rankine analysis considers neither strain nor displacements explicitly, what can you say about the relative displacements of the wall and fill for these two cases?

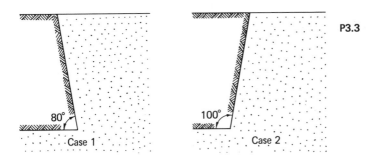

P3.3

80°   · · ·Case 1   100°   Case 2

3.4. Assuming now that the backfill described in Prob. 3.3 is submerged, with the groundwater table at the surface:

(a) Answer part a of Prob. 3.3.

(b) What general conclusion can you draw from these results?

3.5. A vertical retaining wall 10 ft high is supporting a cohesionless fill (γ = 118 pcf, ϕ' = 33°). The backfill is inclined at an angle of 15° up from the horizontal, with a uniform vertical loading of 150 psf acting over the entire ground surface. Using the Rankine method of analysis, determine the active earth pressure force acting on the wall.

3.6. Assuming the Rankine analysis is applicable, can the ratio of the shear force to the normal force on a retaining wall be greater than the magnitude of tan ϕ'? The wall may be inclined at any angle from the vertical. The upper boundary of the fill may be inclined from the horizontal. Substantiate your answer!

3.7. A vertical retaining wall 20 ft high is supporting a horizontal cohesionless backfill. The backfill is made up of two different layers of soil. The

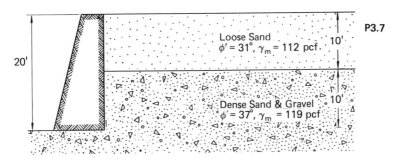

P3.7

Loose Sand
ϕ' = 31°, γ_m = 112 pcf.

10'

Dense Sand & Gravel
ϕ' = 37°, γ_m = 119 pcf

10'

20'

boundary between the two layers is horizontal and at the mid-elevation of the wall. The properties of the two soil strata are shown in the sketch.

(a) By the Rankine method of analysis, determine the magnitude and distribution of normal stress on the wall.

(b) What is the implication of assuming the Rankine theory on the relative movement of the two soil types at their interface? Explain your answer.

3.8. A compacted dry sand fill ($\phi' = 35°$) is placed adjacent to a reinforced concrete building:

(a) Estimate reasonable upper and lower limits for the earth pressure on the building at a depth of 10 ft (assume $\gamma_m = 110$ pcf). Justify your estimates.

(b) Would your estimates differ if the building were replaced by a conventional gravity retaining wall? Justify your answer. If your answer is yes, indicate the limits for this case.

3.9. A vertical retaining wall 10 ft high is supporting a clean dry sand backfill ($\gamma = 100$ pcf, $\phi' = 35°$) with a horizontal surface. Plot the magnitude of the *horizontal component* of the earth pressure force, P'_A, as a function of the angle of wall friction, δ, for $0 \leqslant \delta \leqslant 35°$, determined by the Coulomb method. Show also the magnitude of the active pressure force determined by the Rankine analysis.

3.10. The concrete retaining wall shown has been designed to support a horizontal force of 10.0 kips/ft. Determine the height x to which the water can be allowed to rise before the allowable force on the wall is exceeded.

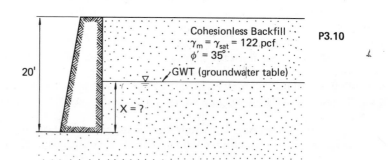

3.11. The inclined retaining wall shown is supporting a cohesionless fill. The fill is subjected to a vertical line load 5 ft behind the top of the wall. The fill was placed in an unsaturated condition with a unit weight of 118 pcf and a $\phi' = 30°$. Assuming the angle of wall friction to be 24°, determine the active earth pressure force on the wall.

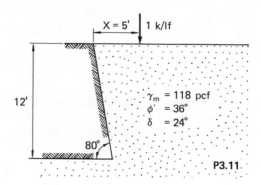

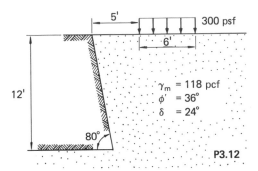

P3.12

3.12. An inclined concrete gravity retaining wall is supporting a cohesionless backfill. The fill is subjected to a strip loading shown in the accompanying sketch.

    (a)   Determine the active earth pressure force on the wall.

    (b)   To avoid excessive settlements under the strip loading the wall has been designed so it will not displace a significant distance horizontally. How would you estimate the earth pressure force on the wall?

3.13   It is desired to relocate the line load described in Prob. 3.11 above so that it will not affect the earth pressure on the wall. How far from the wall $(x = ?)$ must the load be placed for this to be the case?

4

Stresses Within
an Earth Mass

4.1 INTRODUCTION

The article on pages 161 to 166, "Surcharging a Big Warehouse Floor Saves $1 Million," from *Engineering News-Record* of March 31, 1966, presents an example of innovative design which was the solution to a problem of a quite different nature from those considered in the previous chapters. Heretofore we have investigated a variety of problems by considering stresses within masses of cohesionless soil *at the point of failure*. The article introduces an important category of problems involving the deformation of fine-grained, rather than coarse-grained soils, subjected to stress levels significantly less than that required to bring about failure. Such fine-grained materials generally exhibit a variety of properties which are absent or unimportant in connection with the behavior of granular materials. One of the most important of these is the tendency for such soils to possess strength when the effective normal stress on the failure plane is zero. This property, called *cohesion* leads to the designation of these materials as *cohesive soils*. The descriptive names given to such soils were originally based primarily on grain size (see Table 2.4). The two broad categories are *silt* and *clay*. There is, of course, a continuous spectrum of materials between coarse silt and the finest-grained clays. A detailed discussion of classification of cohesive soils is given in Chapter 9.

When a mass of cohesive soil is subjected to stresses, even at levels well below that required to cause failure, it undergoes both time-independent and time-dependent changes in shape and volume of significant magnitude. These effects are called *elastic* (or *immediate*) *distortion, creep, consolidation* or *compression*, depending upon the physical basis involved. We shall consider these phenomena in this and subsequent chapters.

Surcharging a Big Warehouse Floor Saves $1 Million

Faced with installing a foundation below the floor of an old building, at a cost that would exceed that of an entirely new building, the Port of New York Authority resorted to surcharging —a procedure common in roadbuilding but unusual in structural work.

As a result, a job that would have cost more than $1 million if done by conventional methods, is costing less than one-tenth of that.

Building 301 is a 550 x 255 ft column-and-girder structure built at Port Newark, N. J., by the U. S. Navy in 1941 as a plate shop. While its interior and exterior columns are supported on 85-ft untreated wood piles driven to support 20 tons each, its asphaltic concrete floor was merely placed on compacted sand or cinders. This was satisfactory, for while tremendous loads moved through the building, none of them was ever supported by the floor (which in spite of this, settled more than a foot). The materials entered the structure on pile-supported craneways that delivered them directly to pile-supported machines and tools, and which carried processed parts to loading docks outside.

When the PNYA acquired the property from the Navy in 1963, it was faced with doing something about the floor. Analysis of the material beneath the floor and tests on it indicate that the mere addition of material needed to level the floor would result in settlement. Since the PNYA intends to rent the building as a general warehouse mainly for the storage of shipborne cargo, the floor capacity was far from adequate.

At the same time, studies on the possibility of installing a pile-supported floor capable of sustaining a minimum live load of 500 psf indicated that the job could well require six to eight months to complete and could easily cost $500,000 just for piles alone, exclusive of the concrete for the floor. Thus, the pile costs would be about $5 psf of building—too high for the existing value of the structure.

The authority figured that some 1,500 piles, each about 84 ft long, would be needed. Because the craneway tracks are only 32 ft above the floor, the piles would have to be driven in at least two sections and probably three. The sectional requirement for the piles precluded the use of wood or precast concrete, leaving only timber-concrete composite, cast-in-place concrete or steel H-pile sections as possibilities. Each of these would require expensive and time-consuming splices. These conditions, coupled with the problem that moving a pile driver around the 25 x 56 ft bays would be slow and awkward, would put the cost of a piled floor beyond reason.

Since all the column foundations were designed to carry not only the dead loads of the building but also the extremely heavy live loads on the craneways, a surcharge could be used safely and economically.

The fill was brought into the building in 15-cu-yd rear dump trucks at the rate of 7,000 cu yd per day, by Tryon Constructors, of Hackensack, N.J,. under a $36,000 contract

The sand was removed by Orion Contracting Co., of Mountain Lakes,

Reprinted from *Engineering News-Record*, March 3, 1966.

This warehouse doesn't have a low roof; its floor carries a 55,000-cu-yd surcharge.

N.J., at an average rate of 2,700 cu yd per 9 hour day. On its best day Orion moved 3,500 cu yd. To do the job the contractor used three elevating scrapers each with a capacity of 15 cu yd struck; three dozers, one each in the large, medium and small class; and two loaders with 1-cu-yd buckets. The haul, exclusive of travel within the building was roughly 1,000 ft, and Orion completed the job in 20 working days.

The PNYA will soon let contracts for paving the floor, which will comprise 6 in. of stone screenings compacted to 95% modified AASHO maximum density, a 2-in. layer of plant-mixed asphalt macadam and a 1½-in. top course of asphaltic concrete.

The PNYA had several important factors working for it when it decided to surcharge the floor. To begin with, the structure has no utilities under the floor that could be damaged by settlement. Water and sewer lines do not enter the building but terminate in an attached structure. Electric power lines enter the building close to the roof and all feeders are overhead. Lines to electric units close to the floor, such as switches and fuse boxes, are encased in conduit and are rigidly attached to columns or walls. Where only the conduits were involved, the contractor simply backfilled against them. In places where there are wall- or column-mounted switches, outlets or fuse boxes, the conduit and wiring were disconnected and the boxes protected by wood sheeting before any fill material was brought in.

The only other major protective measure involved placement of a 4- to 5-ft-high berm around the outside of the perimeter of the building. This berm both assisted in consolidation of the soil near the wall and prevented the material inside from exerting destructive pressures against the wall.

An even more important factor in the success of the operation is the Port Authority's long experience with the method. The Port Authority has been increasingly surcharging much of its Port Newark areas since it took over

True roof height shows when sand is removed. Arrow indicates top elevation of surcharge.

Preload in place, sand almost covers the girts.

Sand out, girts are 13 ft above the floo

the port in 1948 on a 50-year lease from the City of Newark.

The entire port area, bounded on the west by the New Jersey Turnpike and Newark Airport, and on the east by Newark Bay, is geologically fairly uniform, although the thicknesses of individual strata vary by as much as 25%. A good deal of this variation is attributable to mud waves resulting from the sporadic dumping over the years of fills of varying densities.

Red shale underlies the area with its upper face about 110 ft below the high tide level. Above this there is 70 ft of red clay. Next comes about 26 ft of black organic silt, which is the last of the naturally placed materials. The materials above the silt were placed by man and vary, depending on what was available and inexpensive at the time of placement. In some areas it is garbage, in others, hydraulically placed sand. In the area of the warehouse, it happens to be cinders. Throughout the en-

tire area however, the stratum of organic silt has posed foundation problems.

For many years the loadbearing walls, columns or floors of all buildings in the port were piled. When the PNYA moved into the area in 1948, it continued this practice for about three years. Then in 1950 the Authority's soils division, under the direction of Martin Kapp started a soils study program. As a result of these studies the authority began a surcharging program that has affected every structure built in the port since.

In the ensuing years, the Port Authority has contracted to have brought in more than 31 million cu yd of sand for filling and preloading areas of Newark Airport, Port Newark and the adjacent Elizabeth-Port Authority Marine Terminal. For the most part, this sand has been dredged from New York's Lower Bay and has been placed hydraulically at various sites. Since then the

164

Line shows amount of settlement.

material has been shifted by various means from site to site in advance of construction work.

However, regardless of the actual final method of placing surcharges on virgin swamp, the authority has found it pays to start the operation by hydraulically placing a 4 to 5 ft thick blanket of sand over the area. This reduces the possibility of mud waves and facilitates the subsequent movement of equipment.

Initially surcharging was used only to consolidate parking areas and roadways. By 1951 Authority engineers had enough experience with the procedure to eliminate piles from beneath floors designed to support live loads of 500 to 800 psf—one case went to 1,000 psf.

Since 1959 there have been no piles driven under even load-bearing columns and walls in any of the one-story cargo buildings that make up the majority of the port's structures.

This has been accomplished by use of a pole-type construction that anticipates and allows for minor and uniform settlements. The design has saved millions of dollars in the construction of some 40 buildings the Port Authority has erected in the last 15 years.

In the case of Building 301, the authority filled it with 55,000 cu yd of material being removed from another site. This resulted in a 12-ft high surcharge that applied a load of 1,300

Elevating scrapers were used to remove the surcharge.

psf to the floor area. The material was placed by trucks in October, 1964, and was removed by pans 14 months later in January, 1966. The final settlement was 17 in. with only a $\frac{1}{8}$ in. rebound when the surcharge was removed.

Herbert Frank, resident engineer for the Port Authority, states that on the basis of experience with the method the new floor may settle an additional $1\frac{1}{2}$ in. during the first five years of use, but because of the use for which the structure is intended, this settlement will pose no problems. If it does, the Authority will bring the floor up to grade by adding an asphaltic top coat.

Meantime the Port Authority continues to save money in foundation costs by its surcharging program. The time required for surcharging a site has ranged from three months to 18 months, depending on the economics of an individual situation. The authority's laboratory first makes a soils analysis and decides how much settlement will occur at a particular site under the design loads. Design curves are developed that show the required surcharge period for various surcharge heights. The planners then select the most desirable design for a particular site.

Frederick F. Fontanella is Manager of the Marine Planning and Construction Division in the Port Authority's Marine Terminal Department; John S. Wilson is Engineer of Design for Marine Terminals in the bi-state agency's Engineering Department. John M. Kyle is Chief Engineer of the Authority.

In 1963 the Port of New York Authority acquired an old column-and-girder structure, Building 301, in the Port Newark, New Jersey, area, with the objective of converting it to a general warehouse. The stratigraphic profile beneath the building is shown in Figure 4.1. The data for this figure were obtained from the log of a *test boring* taken at the site. It was determined that the estimated floor loads transmitted to the underlying soft soil shown in Figure 4.1 would lead to settlement of the floor of more than one foot. The stratum which was primarily responsible for the anticipated settlement was the soft organic silt layer. This material was expected to undergo large volume changes resulting from the in-service stress transmitted from the floor of the structure through the overlying strata.

In such circumstances the indicated solution is often to bypass the unsuitable materials with a *deep foundation* that derives its support from more competent underlying strata. One example of this is described in the article "Foundation Design," reproduced in Chapter 1 (pages 4–9). In the case of Building 301, however, such a solution was deemed prohibitively expensive because of the large expanse of heavily loaded floor. This led to the decision to *surcharge* the floor. The objective of the surcharge load on the floor was to prestress the soil to levels greater than in-service conditions would create, and cause the shape and volume changes to occur before the new floor was placed.

We shall investigate the design problem posed by the accompanying article, and after the development of suitable background, shall carry out the analysis which led to design of the appropriate surcharge load, eliminating the necessity for an expensive deep foundation. In order to do this, we

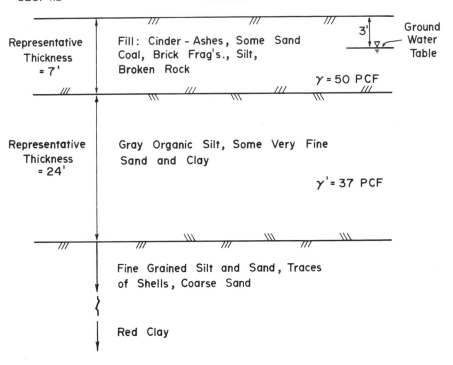

Fig. 4.1—Representative stratigraphic profile (not to scale) under New York Port Authority Building 301, Port Newark, N.J. (Data from Kapp, 1968.)

must consider two important items:

1. The way in which stresses are distributed throughout an earth mass as a function of boundary conditions and the properties of the soil.
2. The manner in which the soil deforms in response to these imposed stresses.

The remainder of this chapter will be devoted primarily to the first point. The second point will be considered in part in this chapter, and more completely in Chapters 5, 7 and 11.

4.2 STRESS

In Sections 2.5 and 3.2 we established the rules which govern the way in which the stress acting on a plane at a point changes as the orientation of the plane through that point changes, and as we move from point to point. The only addition to these results required here is to generalize the equations obtained in Section 3.2 from two dimensions to three dimensions. Withou additional detailed derivation, the *equations of equilibrium* for three dimen sions are

$$\frac{\partial \sigma_x}{\partial x} + \frac{\partial \tau_{xy}}{\partial y} + \frac{\partial \tau_{xz}}{\partial z} - X = 0$$

$$\frac{\partial \tau_{xy}}{\partial x} + \frac{\partial \sigma_y}{\partial y} + \frac{\partial \tau_{yz}}{\partial z} - Y = 0 \qquad (4-1$$

$$\frac{\partial \tau_{xz}}{\partial x} + \frac{\partial \tau_{yz}}{\partial y} + \frac{\partial \sigma_z}{\partial z} - Z = 0$$

The situation with regard to solving these equations is much less favorabl than in the case of two dimensions, for the three Equations 4–1 are expresse in terms of six unknowns. In the case of the two-dimensional stabilit problem we simply introduce a third equation (the Mohr-Coulomb failur law), which provided a third equation without introducing additiona unknowns. Now, however, we shall not assume failure conditions, but rathe investigate the *deformations* experienced by the soil mass.

4.3 DISPLACEMENT AND STRAIN

Definitions of Displacement and Strain

We consider the geometry of a small parallelepiped within a body which in its initial undeformed configuration, has sides parallel to the three co-ordinate axes, as shown in Figure 4.2a. Concomitant with the application of stress to the body, points within the body, such as point A in Figure 4.2a move. This movement, for example from A to A', is termed a *displacement* Let us consider two points, A and B on the initial undeformed element These points will displace to A' and B'. If the distance $A'B'$ is the same as AB, and if similar lines in any direction do not change length, the body has undergone a *rigid-body displacement*. Of more interest to us is the case where $A'B'$ is of different length from AB; that is, the points A and B have undergone a *relative* displacement. This is termed a *deformation*. It is such deformations which accompany and produce stress in material bodies and which are a subject of major concern in the investigation of the response of earth masses to loads imposed upon them.

To clarify the picture of deformations experienced by the small element under consideration, we shall examine the changes in the face initially parall l to the xz-plane in Figure 4.2b. Point A is shown initially with coordinates (x, y, z). During deformation it displaces to a point A'. The component of the displacement from A to A' in the x direction is $u(x, y, z)$, and in the z-direction is $w(x, y, z)$. The displacement in the y-direction, $v(x, y, z)$, is not shown in this two-dimensional view.* The deformation, that is, the change in length, of line AB is simply the difference between the length of $A'B'$ and AB. If we make the *assumption* that this change in length is a very small quantity† both in actual magnitude and relative to the length of the line AB, we may say as an approximation that

$$A'B' - AB = u(x + \Delta x, y, z) - u(x, y, z)$$

Clearly, the magnitude of the deformation experienced by a line representing one side of the small element depends upon the size of that element. This leads us to the description of deformation relative to the size of the element. Thus we define the *normal* or *direct strain*, for example the strain of line AB in the x- direction, as

$$\varepsilon_x = \lim_{AB \to 0} \frac{A'B' - AB}{AB} = \lim_{\Delta x \to 0} \left[\frac{u(x + \Delta x, y, z) - u(x, y, z)}{\Delta x} \right] \quad (4\text{--}2)$$

Note that this is the definition of the partial derivative, so that

$$\varepsilon_x = \frac{\partial u}{\partial x} \quad (4\text{--}3a)$$

It is evident that we can carry out a similar procedure to define the normal strains in the y- and z-directions respectively:

$$\varepsilon_y = \frac{\partial v}{\partial y} \quad (4\text{--}3b)$$

$$\varepsilon_z = \frac{\partial w}{\partial z} \quad (4\text{--}3c)$$

If we examine Figure 4.2b it becomes clear that the definitions of three normal strains, Equations 4–3a, b, and c, are not sufficient to define the deformed configuration of the body. We require in addition some description of the change in shape. We can obtain this by describing the change in angle between respective sides from the right angles in the initial undeformed configuration. In Figure 4.3, the rotation of line $A'B'$ is θ_{AB}, which, for

---

* The symbolism employed here implies that the displacement is a function of the coordinates of the initial undeformed configuration. This is not the only way in which such displacement can be viewed. In this development and hereafter, we shall consider displacements, deformations, and strains to be defined with regard to the initial undeformed configuration of the body. As long as strains are "small," this is entirely equivalent to defining these quantities with respect to the deformed configuration. However, as strains become large, the distinction may be important. For a more detailed discussion of this point, reference may be made to Fung (1965).

† This assumption is very important because of the mathematical simplification it introduces into the formulation and solution of *boundary value problems*. Although the assumption is not always justified, it is almost always applied in the investigation of stress distribution problems in earth masses.

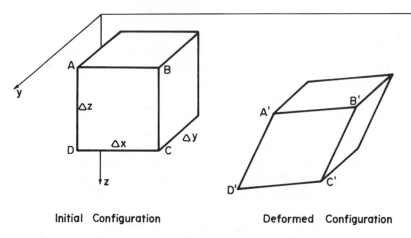

Initial Configuration   Deformed Configuration

a) - Three - Dimensional View

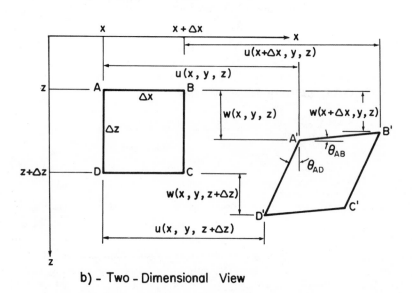

b) - Two - Dimensional View

Fig. 4.2—Displacement of small element of continuous material.

"small" deformations, is defined by

$$\tan \theta_{AB} = \frac{w(x + \Delta x, y, z) - w(x, y, z)}{\Delta x}$$

Similarly, the rotation of line AD is defined as

$$\tan \theta_{AD} = \frac{u(x, y, z + \Delta z) - u(x, y, z)}{\Delta z}$$

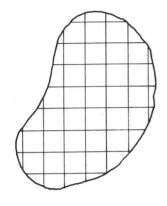

Fig. 4.3—Continuous and discontinuous deformations of a body.

a) - Undeformed   Configuration

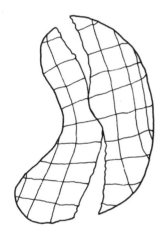

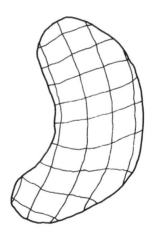

b) - Deformed   Configuration -   c) - Deformed   Configuration -
    Discontinuous   Deformations   Continuous   Deformations

If these angle changes are small, consistent with our assumption of small strains, then $\tan \theta_{AB} \simeq \theta_{AB}$ and $\tan \theta_{AD} \simeq \theta_{AD}$. The change in angle of the initially right angle BAD is then simply $\theta_{AB} + \theta_{AD}$. Taking the limit as $\Delta x \to 0$ and $\Delta z \to 0$, this angle change is defined as the *shear strain*, γ_{xz}:

$$\gamma_{xz} = \frac{\partial w}{\partial x} + \frac{\partial u}{\partial z} \qquad (4\text{-}3d)$$

A similar development for the sides of the element in Figure 4.2a parallel to the yz plane and the xy plane, respectively, leads to similar definitions for the other two independent shear strains:

$$\gamma_{yz} = \frac{\partial w}{\partial y} + \frac{\partial v}{\partial z}$$ (4–3

$$\gamma_{xy} = \frac{\partial v}{\partial x} + \frac{\partial u}{\partial y}$$ (4–3

We have now described the geometry of deformation in a continuous bod
when the body is assumed to undergo "small" strains. However, we ha
done this at the cost of introducing nine additional unknowns: three di
placements and six strains. Because we have introduced only six addition
equations (Equations 4–3) our situation with regard to solution of th
problem appears less favorable than before: we now have fifteen unknow
(six stresses, three displacements, and six strains) and only nine equatior
(three equilibrium equations and six equations defining the strains in tern
of displacements). It is evident that additional information is required. I
the next section we shall consider the form which such information migl
take. However, before doing this, it is necessary to observe an importar
constraint imposed upon the deformations of a continuous body.

Compatibility

Equations 4–3 show that the six components of strain are uniquely define
by the three displacements u, v, w. However, these three displacemen
may also be determined from the six strains. In order to insure that th
deformations resulting from these displacements are continuous, both math
matically and physically, we must recognize that the strains are not entirel
independent of each other. The reason for this is illustrated in Figure 4.
A material body is shown in Figure 4.3a with a grid of lines scribed upon i
In Figure 4.3b the body is shown in a deformed configuration in whic
arbitrary assumption of strains has produced discontinuous deformatio
that is, breaks in the body. Strains which produce such discontinuities ar
termed incompatible. Alternatively, incompatible strains could produc
multivalued deformations, that is, two or more material points might mutu
ally occupy the same point in space. This is clearly inadmissible. Thus w
see that the deformations must be finite and single-valued.

In order to insure that deformations are continuous, it is necessary tha
all components of strain and their derivatives must exist and be continuou
to at least the second order. Examining these derivatives, we will the
discover the required relationship between them. For example, differen
tiating Equation 4–3a twice with respect to y yields

$$\frac{\partial^2 \varepsilon_x}{\partial y^2} = \frac{\partial^3 u}{\partial x\, \partial y^2},$$

differentiating Equation 4–3b twice with respect to x gives

$$\frac{\partial^2 \varepsilon_y}{\partial x^2} = \frac{\partial^3 v}{\partial y\, \partial x^2},$$

and taking the mixed partial derivative of Equation 4–3f leads to

$$\frac{\partial^2 \gamma_{xy}}{\partial x\, \partial y} = \frac{\partial^3 v}{\partial y\, \partial x^2} + \frac{\partial^3 u}{\partial x\, \partial y^2}$$

Combining these, we see that

$$\frac{\partial^2 \varepsilon_x}{\partial y^2} + \frac{\partial^2 \varepsilon_y}{\partial x^2} = \frac{\partial^2 \gamma_{xy}}{\partial x\, \partial y} \qquad (4\text{–}4a)$$

A similar procedure for various of the other derivatives will lead to five more equations:

$$\frac{\partial^2 \varepsilon_y}{\partial x^2} + \frac{\partial^2 \varepsilon_z}{\partial y^2} = \frac{\partial^2 \gamma_{yz}}{\partial y\, \partial z} \qquad (4\text{–}4b)$$

$$\frac{\partial^2 \varepsilon_z}{\partial x^2} + \frac{\partial^2 \varepsilon_x}{\partial z^2} = \frac{\partial^2 \gamma_{xz}}{\partial z\, \partial x} \qquad (4\text{–}4c)$$

$$2\frac{\partial^2 \varepsilon_x}{\partial y\, \partial z} = \frac{\partial}{\partial x}\left(-\frac{\partial \gamma_{yz}}{\partial x} + \frac{\partial \gamma_{xz}}{\partial y} + \frac{\partial \gamma_{xy}}{\partial z}\right) \qquad (4\text{–}4d)$$

$$2\frac{\partial^2 \varepsilon_y}{\partial z\, \partial x} = \frac{\partial}{\partial y}\left(\frac{\partial \gamma_{yz}}{\partial x} - \frac{\partial \gamma_{xz}}{\partial y} + \frac{\partial \gamma_{xy}}{\partial z}\right) \qquad (4\text{–}4e)$$

$$2\frac{\partial^2 \varepsilon_z}{\partial x\, \partial y} = \frac{\partial}{\partial z}\left(\frac{\partial \gamma_{yz}}{\partial x} + \frac{\partial \gamma_{xz}}{\partial y} - \frac{\partial \gamma_{xy}}{\partial z}\right) \qquad (4\text{–}4f)$$

Equations 4–4 are called the *equations of compatibility*. They represent necessary and sufficient conditions that the deformations will be continuous, finite and single-valued (Love, 1944).

4.4 CONSTITUTIVE RELATIONS

The requirements of equilibrium, the definition of strain, and the compatibility relations apply equally well to any material experiencing small deformations, which can be approximated as a continuum: steel, putty, even water. It is evident that the mechanical response of materials constituted so differently must surely be different. Thus our physical intuition, combined with the need for six additional independent relations, suggests consideration of equations which incorporate the mechanical properties of the material into the analysis. These properties determine the relationship between stress, strain, time, temperature, and other factors that might affect the mechanical response. As already pointed out, because these relations reflect the constitution of the material, they are referred to as *constitutive relations*.

If we could define them, the constitutive relations for all real materials would be very complicated. In general, the response of a given material to, say, an imposed stress will depend upon the type of imposed stress, its magnitude, the time over which it acts, the previous stress and strain history experienced by the material, the location of the stress on the body, the temperature, and even possibly the electrical, magnetic, and chemical environments. Unfortunately, recognition of all of these relevant factors in the mathematical description of the constitutive relations is so complex that even though the number of equations equals the number of unknowns, they cannot be solved by presently available methods. Consequently we are led to idealize the constitutive relations to a form which leads to a mathematically tractable formulation, and which still, hopefully, provides reasonable and useful results. Thus in general, the mechanical response of a material may be expressed in the form of a set of equations such as

$$\varepsilon_x = f(\sigma_x, \sigma_y, \sigma_z, \tau_{xy}, \tau_{xz}, \tau_{yz}, \varepsilon_x, \varepsilon_y, \varepsilon_z, \gamma_{xy}, \gamma_{xz}, \gamma_{yz}, x, y, z, t, T) \quad (4\text{-}5)$$

in which f is some function, t is time, and T is temperature. Similar equations for the other strain components in which the functional relations may be different would provide the required additional six equations.

The first simplification which is usually made is to assume that the function f is *linear*. This implies that, if deformations and rigid-body rotations are suitably small, the *principle of superposition* applies, that is, the sum of the response of the material to two different sets of stresses is equal to the response of the material to the sum of those stresses. Thus the strain components cannot depend upon their own magnitude, and, for constant temperature, the strain at a given point becomes simply a function of the stress and time:

$$\varepsilon_x = f(\sigma_x, \sigma_y, \sigma_z, \tau_{xy}, \tau_{xz}, \tau_{yz}, t) \quad (4\text{-}6)$$

There are a number of special cases that have been studied extensively, and for which the solutions for a variety of problems have been obtained. Among these are:

1. *Linear-elastic material.* An elastic material is one for which the relation between stress and strain is independent of time and is the same for *unloading* as it is for loading. For an elastic material to be linear, the stress-strain relationship will be a straight line. For this case the six required expressions are called Hooke's law,* although they were formulated in terms of stress by Cauchy (1828).

$$\varepsilon_x = C_{11}\sigma_x + C_{12}\sigma_y + C_{13}\sigma_z + C_{14}\tau_{xy} + C_{15}\tau_{xz} + C_{16}\tau_{yz}$$
$$\varepsilon_y = C_{21}\sigma_x + \cdots \qquad\qquad\qquad + C_{26}\tau_{yz}$$
$$\vdots \qquad\quad \vdots \qquad\qquad\qquad\qquad\qquad \vdots \qquad (4\text{-}7)$$
$$\gamma_{yz} = C_{61}\sigma_x + \cdots \qquad\qquad\qquad + C_{66}\tau_{yz}$$

---

* Robert Hooke first described the concept of a linear-elastic body in 1676, in the form of a Latin anagram, CEIIINOSSSTTUV. This was later arranged (Hooke, 1678) to spell UT TENSIO SIC VIS, which means "the extension is in proportion to the force."

in which the C's are constants. Hence if the stress is a function of time, the strain will be the same function of time.

In the event that the material is *homogeneous*, that is, the material properties represented by the constants C do not vary from point to point, as well as *isotropic*, that is, the material properties at a point are the same in all directions, there will be only the two independent constants shown below:

$$\varepsilon_x = C_{11}\sigma_x + C_{12}(\sigma_y + \sigma_z)$$
$$\varepsilon_y = C_{11}\sigma_y + C_{12}(\sigma_z + \sigma_x)$$
$$\varepsilon_z = C_{11}\sigma_z + C_{12}(\sigma_x + \sigma_y)$$
$$\gamma_{xy} = 2(C_{11} - C_{12})\tau_{xy} \qquad \text{(4-8)}$$
$$\gamma_{xz} = 2(C_{11} - C_{12})\tau_{xz}$$
$$\gamma_{yz} = 2(C_{11} - C_{12})\tau_{yz}$$

By considering the special case of a uniaxial compression or tension test, we see that

$$C_{11} = \frac{1}{E}, \qquad C_{12} = -\frac{\mu}{E}$$

in which E is Young's modulus and μ is Poisson's ratio.

And so for an isotropic, homogeneous, linear-elastic material the constitutive relations are the generalized Hooke's law:

$$\varepsilon_x = \frac{1}{E}\left[\sigma_x - \mu(\sigma_y + \sigma_z)\right]$$

$$\varepsilon_y = \frac{1}{E}\left[\sigma_y - \mu(\sigma_z + \sigma_x)\right]$$

$$\varepsilon_z = \frac{1}{E}\left[\sigma_z - \mu(\sigma_x + \sigma_y)\right]$$

$$\gamma_{xy} = \frac{2(1 + \mu)}{E}\tau_{xy} = \frac{1}{G}\tau_{xy} \qquad \text{(4-9)}$$

$$\gamma_{xz} = \frac{2(1 + \mu)}{E}\tau_{xz} = \frac{1}{G}\tau_{xz}$$

$$\gamma_{yz} = \frac{2(1 + \mu)}{E}\tau_{yz} = \frac{1}{G}\tau_{yz}$$

where G is the *shear modulus*.

2. *Linear-viscoelastic material.* A linear-viscoelastic material exhibits a *time-dependent* component of response in addition to a time-independent elastic response. If a linear-viscoelastic material is also homogeneous and isotropic, only two independent parameters, analogous to C_{11} and C_{12} in Equations 4-8, are required to describe the mechanical behavior. Assuming that the material is unstressed and unstrained at $t = 0$, and only normal stresses are applied, these

lead to constitutive relations of the form

$$\varepsilon_x(t) = \int_0^t \frac{d\sigma_x(\theta)}{d\theta} J_{11}(t - \theta)\, d\theta$$
$$+ \int_0^t \left[\frac{d\sigma_y(\theta)}{d\theta} + \frac{d\sigma_z(\theta)}{d\theta} \right] J_{12}(t - \theta)\, d\theta \qquad (4\text{--}10)$$

in which $\varepsilon_x(t)$ is the strain in the x direction at time t, $\sigma_x(\theta)$, $\sigma_y(\theta)$, and $\sigma_z(\theta)$ are the time-dependent stresses in the x, y, z directions respectively, and θ is the dummy time variable of integration. This parameter $J_{11}(t - \theta)$ is called the *creep compliance*, and represents the normal strain in the x direction at time t due to a unit normal stress in the x direction applied at some earlier time θ. Similarly, $J_{12}(t - \theta)$ is the normal strain in the x direction at time t due to a unit load applied in either the y or z directions at time θ. Figure 4.4 illustrates a comparison of the response of typical viscoelastic and elastic materials to a constant imposed stress. When the creep compliances are constant, they become simply the constants C_{11} and C_{12} in Equations 4–8, that is, $1/E$ and $-\mu/E$ respectively.

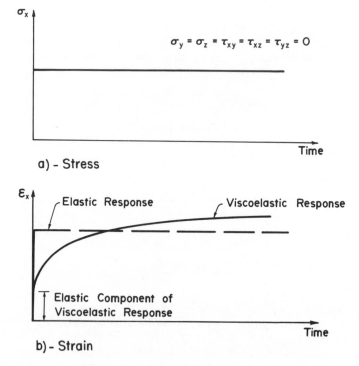

Fig. 4.4—Comparison of elastic and viscoelastic response to constant stress.

Any number of linear-viscoelastic materials may be described by specifying particular creep compliance functions. The decision about what function to use can be made either by inference from observations of small-scale or prototype models of the particular problem of concern, or from laboratory tests. The form of Equation 4–10 suggests that solutions of a boundary-value problem for a linear-viscoelastic material will be generally more complicated than for linear-elastic materials.

Real soils generally exhibit nonlinear and time-dependent response. However, until recently, the difficulty of solving nonlinear differential equations has precluded consideration of realistic constitutive relations. Now, the use of numerical methods has become practical because of the availability of high-speed computers, and successful solutions have been obtained for certain important problems utilizing the nonlinear response of soil bodies.* The main impediment to further advances in this area is the difficulty in establishing a set of constitutive relations which is applicable to a wide range of soils.

Thus, restricting our attention to linear constitutive relations, we now have fifteen linear differential and algebraic equations in fifteen unknowns. In principle, it is only necessary, in order to solve any desired boundary-value problem, to specify the appropriate boundary conditions from which the various constants of integration can be evaluated.

The category of boundary-value problem which is most often of interest to us is the case of various distributed loadings on the surface of a half-space. Investigation of a number of such cases for both elastic and various viscoelastic constitutive relations has led to the following conclusions:

1. The distribution of stress throughout a half-space resulting from loads applied to the surface is relatively insensitive to the form of the constitutive relationship. In fact, for linear constitutive laws, the vertical stress resulting from vertical surface loadings is independent of the material parameters, as are the other stresses in the case of a plane strain problem. In more general problems, the stresses may be a function of the material parameters. Again, however, the variability of natural deposits from point to point, combined with the difficulty of obtaining and testing appropriate samples of the material to yield representative results for the material parameters, introduces uncertainties, the effect of which exceeds the difference in the stresses computed using viscoelastic rather than elastic constitutive relations.

---

* The most useful technique presently available for solving boundary value problems with nonlinear constitutive relations is the *finite element method* (Zienkiewicz, 1967). Excellent predictions of displacements of a compacted earth dam (Clough and Woodward, 1967), settlement of a model footing on a cohesive soil mass, and the force-displacement relationship of a retaining wall pressed into a sand mass (Girijavallabhan and Reese, 1968) have been made. It is likely that this approach will find increasing use in design.

2. Surface displacements predicted by linear-elastic analysis, using an appropriate "equivalent" elastic modulus and Poisson's ratio, are sufficiently accurate for most soil engineering applications in the case of saturated cohesive soils, even though they fail to account for certain time-dependent effects.

Consequently it is customary to compute the stresses within an earth mass using the assumptions of linear elasticity, and then to determine the displacements of the earth mass by calculating the response of the real nonlinear material to these stresses. The way in which this is done will be discussed in Chapter 5.

4.5 SOLUTION TO LINEAR-ELASTIC PROBLEMS

Boussinesq Problem

A variety of methods exist for solving the equations of equilibrium, compatibility, and Hooke's law for various boundary conditions. However, the simplest method is to solve one problem and then apply the principle of superposition to the results obtained, to yield the solution of other problems. The most important original solution was given by Boussinesq (1885) for the distribution of stresses within a linear-elastic half-space resulting from a point load applied normal to the surface, as illustrated in Figure 4.5. The results obtained were

$$\sigma_z = \frac{3P}{2\pi} \frac{z^3}{R^5} = \frac{3P}{2\pi z^2} \frac{1}{\left[1 + \left(\dfrac{r}{z} \right)^2 \right]^{5/2}}$$

$$\sigma_r = \frac{P}{2\pi} \left[\frac{3zr^2}{R^5} - \frac{1 - 2\mu}{R(R + z)} \right]$$

$$\sigma_\theta = \frac{P}{2\pi} (1 - 2\mu) \left[\frac{1}{R(R + z)} - \frac{z}{R^3} \right] \tag{4-11}$$

$$\tau_{rz} = \frac{3P}{2\pi} \frac{z^2 r}{R^5}$$

$$\tau_{\theta z} = \tau_{r\theta} = 0$$

where μ is Poisson's ratio and the other quantities in the equation are defined in Figure 4.4. These stresses are those that would occur in a weightless, linear-elastic medium. Preexisting stress due to the weight of the material must be superimposed upon these.

A very useful result from Equations 4–11 is that the vertical normal stress, σ_z, is independent of the magnitude of the elastic constants. In fact, for a linear-viscoelastic medium, subject to the usual small-strain assumptions, the vertical stress is *also* independent of the material parameters. The result

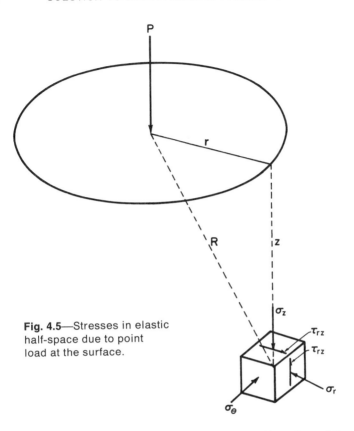

Fig. 4.5—Stresses in elastic half-space due to point load at the surface.

is *not* independent, however, of the assumptions of small strain and linearity leading to the expression.

Because of the independence of the vertical stress on the material properties, the expression may be written

$$\sigma_z = I \frac{P}{z^2} \tag{4-12}$$

in which I is an *influence value* that depends only upon the geometry, that is, the location of the point at which the stress is desired relative to the point load:

$$I = \frac{3}{2\pi} \frac{1}{\left[1 + \left(\dfrac{r}{z}\right)^2\right]^{5/2}} \tag{4-13}$$

For this reason the influence value I may be plotted as a function or r/z once, and is then available for calculation of the vertical stress due to any point load on a linear-elastic half-space. This has been done in the Appendix. (Figure A.1). The use of this diagram is illustrated by Example A.1.

Solutions by Superposition

The result of the Boussinesq point-load problem may be used to advantag in calculating the stresses at a point due to normal stresses distributed on the surface of an elastic half-space. As an example, consider the case of a uniformly distributed normal pressure applied over a circular area, illustrated in Figure 4.6. The resultant force, dP, acting on a differential area, dA, may be expressed as

$$dP = p \, dA = pr \, d\theta \, dr$$

in which p is the magnitude of the uniform pressure. Thus the first o Equations 4–11 may be used directly to determine the incremental stress a a depth z beneath the center of the circle due to the load on this differentia area:

$$d\sigma_z = \frac{3p}{2\pi z^2} \frac{r \, d\theta \, dr}{\left[1 + \left(\frac{r}{z}\right)^2\right]^{5/2}} \tag{4-14}$$

The stress due to the entire distributed load is obtained by integrating the incremental stress over the total area:

$$\sigma_z = \frac{3p}{2\pi z^2} \int_0^{2\pi} \int_0^a \frac{r \, dr \, d\theta}{\left[1 + \left(\frac{r}{z}\right)^2\right]^{5/2}} = p \left\{ 1 - \frac{1}{\left[1 + \left(\frac{a}{z}\right)^2\right]^{3/2}} \right\} \tag{4-15}$$

This expression may also be written in terms of an influence value,

$$\sigma_z = Ip \tag{4-16}$$

where

$$I = 1 - \frac{1}{\left[1 + \left(\frac{a}{z}\right)^2\right]^{3/2}} \tag{4-17}$$

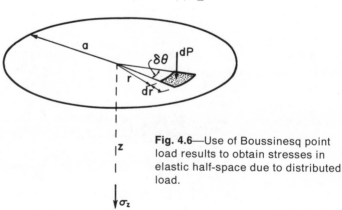

Fig. 4.6—Use of Boussinesq point load results to obtain stresses in elastic half-space due to distributed load.

This influence value has been plotted as a function of z/a in Figure A.2 (Appendix A) as the curve corresponding to a uniform loading, and in Figure A.3 as the curve for which $r/a = 0$.

Equation 4–14 may be used in a different way in order to permit evaluation of the effect of loads distributed over irregular-shaped areas. That is, if the equation is integrated over small portions of the circular area such that the influence value corresponding to each of these small areas is the same, and the loaded area is sketched to an appropriate scale, the stress can be determined by simply counting the number of such "unit" influence areas covered by the load. This was done by Newmark (1942) and is presented in Figure A.4. Use of this diagram is illustrated in Example A.1.

It is evident that the stress resulting from any number of surface load distributions can be determined by a similar process of superimposing point loads acting on a differential area. In all such cases the vertical stress can be expressed as a function of the applied normal stress on the surface multiplied by an appropriate dimensionless influence value depending only upon the geometry. Other particularly useful influence values for loads on rectangular areas are given in Figures A.5–A.7.* In general, the stresses resulting from any variety of loaded areas of arbitrary shape may be determined by programming the analysis for the computer.

Distribution of Stresses

It is important to develop physical intuition about the way in which stresses distribute through an earth mass, and the size of the zone which is affected by a loaded area on the ground surface. An appreciation of the theoretical stress distribution in an elastic medium assists the development of such intuition. It is frequently of interest in this connection to observe the *stress profile*, that is, the magnitude of stress plotted as a function of depth below selected points under a loaded area. As an illustration, Figure 4.7a shows the distribution of vertical stress beneath the centerline of a uniformly loaded circular area, and beneath a point a distance of one radius outside the loaded area. The stress is plotted so that the vertical distance below the surface is the depth, expressed as a proportion of the radius, and the abscissa of the curve is the magnitude of the stress as a proportion of the load p. One of the most important features of this diagram is the difference in the attenuation of stress with depth depending upon the location of the vertical section. Under the centerline, the stress is a maximum immediately below the loaded area, and at a depth equal to twice the width of the loaded area has attenuated to less than 10 per cent of the initial vertical stress. Beneath the point outside the loaded area the stress is zero at the ground surface and increases to a maximum at the depth of approximately 1.2 times the width of the loaded

---

* Extensive listings of equations for stresses due to a variety of surface loads applied to elastic media are given by Scott (1963) and Harr (1966).

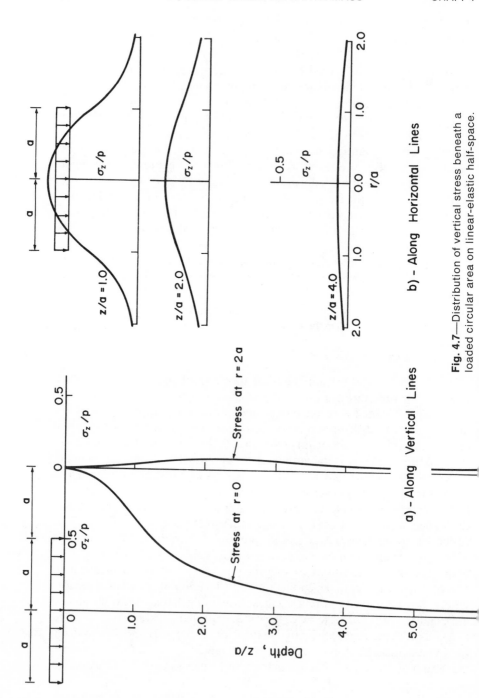

Fig. 4.7—Distribution of vertical stress beneath a loaded circular area on linear-elastic half-space.

area. However, the magnitude of this stress changes very little with depth below this point.

The distribution of vertical stress along selected horizontal lines beneath the center of the loaded area is shown in Figure 4.7b. It is clear that the stress is more highly concentrated beneath the area at shallow depths and reduces in magnitude as its effect spreads at increasing depths. Naturally, the area under the curve on each horizontal plane must equal the total applied load.

An alternative useful way to view the distribution of stresses is to consider *contours* of equal vertical stress, called *isobars*, shown in Figure 4.8 for a loaded circular area. These contours delineate the zone of influence of the

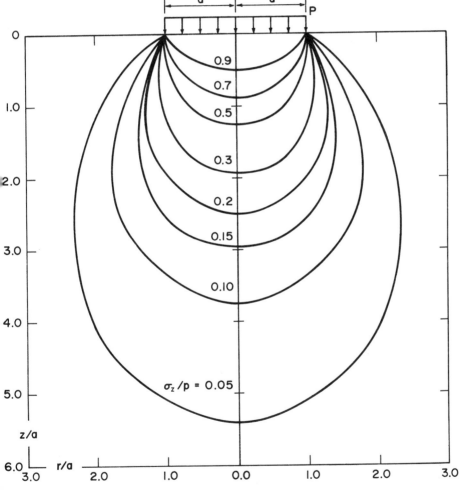

Fig. 4.8—Contours of vertical normal stress beneath uniformly loaded circular area on linear-elastic half-space.

footing in the sense that the area contained within a given contour experiences stresses larger than the stress level indicated by that contour. For example, the zone within the stress contour for which $\sigma_z = 0.05p$ contains all of the material subjected to stresses (resulting from the loaded area) of that magnitude or greater. Because of the shape of this zone it is often referred to as the *pressure bulb*. Note that for a loaded area of a given shape on the surface of a linear-elastic half-space, the size of the pressure bulb is proportional to the size of the loaded area. The importance of this to the foundation engineer cannot be overrated, and we shall see many examples in which it is essential to remember this fact.

Effects of Layered Systems

It frequently happens that the compressible stratum in which we wish to determine the stresses underlies one or more strata of different mechanical properties. Concern about the potential influence of a layered system on the distribution of stresses and displacements has led to analyses of elastic systems consisting of layers of differing elastic properties.* Typical results of such analyses are given in Figure 4.9. This figure shows a uniformly distributed load acting on a circular area on the surface of a two-layer elastic system. For illustrative purposes the thickness of the upper layer has been chosen equal to the radius of the loaded area. The vertical stress distribution under the center line of the loaded area is shown as a function of depth for various ratios of Young's modulus. The results indicate that when the upper layer is significantly stiffer than the lower layer the stress in the lower layer is reduced to an important degree. Thus it appears, from elastic analysis, that the presence of a stiff layer near the surface may mitigate the stresses distributed to an underlying compressible layer. Conversely, the stress in a layer underlain by a very stiff layer is increased markedly, as shown by the stress at the interface between the layers when the lower one is rigid. Thus on the basis of this result, one might expect that the stresses in a compressible layer underlain by a stiff granular material, or rock, would be higher than those predicted by the results of the Boussinesq problem.

A few experimental investigations have been conducted to elucidate this point. The Waterways Experiment Station (1951) measured stresses and deflections within a carefully prepared homogeneous clay-silt test section due to an applied uniform circular load on the surface. The results of their experiments indicated that the distribution of vertical stress resulting from the applied surface loading corresponded very closely to that determined by superimposing the results of the Boussinesq problem, as shown in Figure 4.7.

---

* The first attempt to consider such variations was made by Marguerre (1933), who investigated the distribution of stress resulting from point and line loads. Burmister (1943) developed expressions for stresses and displacements of 2- and 3-layer flexible elastic systems subjected to a uniform stress acting over a circular area of the surface. Numerical values of these results have been presented in tabular form by Acum and Fox (1951), and Jones (1962), and in graphical form by Burmister (1958, 1967) and Peattie (1962).

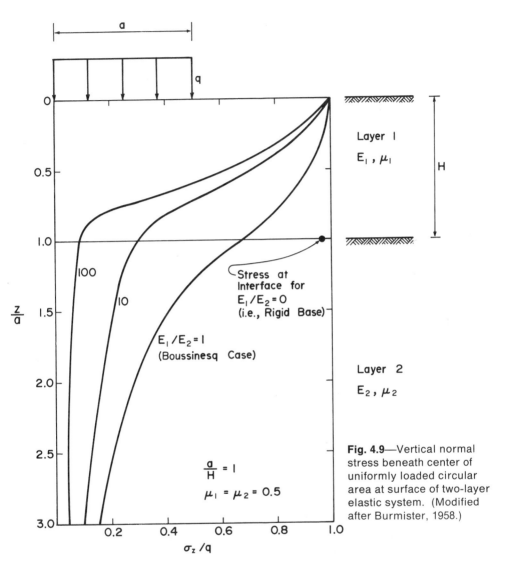

Fig. 4.9—Vertical normal stress beneath center of uniformly loaded circular area at surface of two-layer elastic system. (Modified after Burmister, 1958.)

An extensive experimental investigation of the distribution of stresses within layered systems was conducted by Sowers and Vesic (1962). They found that except in the case of very rigid lower strata, that the results were described more closely by elastic theory assuming the uniform half-space than by a multilayer theory. Burmister (1967) has challenged the validity of these latter results because of the change in conditions which results from the introduction of instruments for the measurement of stresses. It is true that the presence of a foreign body at the very point where the stresses are being measured can change the stresses significantly and modify the influence of the layered system.

Although the information about this point is not conclusive at the present time, it seems clear that the vertical stress intensity transmitted by a stiff layer overlying a much less stiff layer is less in the zone below the load in the lower layer than would be predicted by assuming a linear-elastic half-space. Except for such cases, however, stress computations for soil engineering problems are customarily made assuming a linear-elastic half-space.*

Approximate Stress Distribution

In cases where several strata appear in a profile, it may be of interest to consider the effect of a load applied at the surface of the profile on one or more of the underlying strata. In such circumstances it is often useful to approximate the distributing effect of the material overlying the surface in question by assuming that the total load from the surface is distributed over an area of the same shape as the loaded area on the surface, but with dimensions which increase by an amount equal to the depth below the surface. This is illustrated in Figure 4.10, which shows a rectangular area of dimensions $B \times L$ at the surface. At a depth z, the total load is assumed to be uniformly distributed over an area $(B + z)$ by $(L + z)$.

The relationship between the approximate distribution of stress determined by this method and the exact distribution is illustrated in Figure 4.11. In this figure, the vertical stress distribution at a depth B beneath a uniformly loaded square area of width B, along a line which passes beneath the center of the area is shown. Also shown is the assumed uniform distribution at depth B determined by the approximate method described. The discrepancy between these two methods decreases as the ratio of the depth considered to the size of the loaded area increases.

Embankments and Excavations

In many cases, the loads imposed on a soil mass by a structure may be represented by a simple system of boundary stresses without producing a significant effect on the calculated distribution of stress throughout the region of interest. In the case of loads applied due to the weight of an additional contiguous earth mass, such an approximation may lead to important differences in calculated stresses. One important example of this is *earth embankments*. The effect of an earth embankment may be approximated in a linear-elastic analysis by considering the situation shown in Figure 4.12a. This figure shows a "long" symmetric elastic embankment continuous with its foundation. The embankment material is assumed to have a unit weight

---

* When the stiffness of a load-bearing stratum exceeds that of an underlying soft soil by more than approximately 10 times, the load-distributing effect may be accounted for, approximately, by calculating stresses in the lower layer assuming the upper stiff layer to be increased in thickness. An increase of 15 per cent in the thickness of the upper layer has been used successfully.

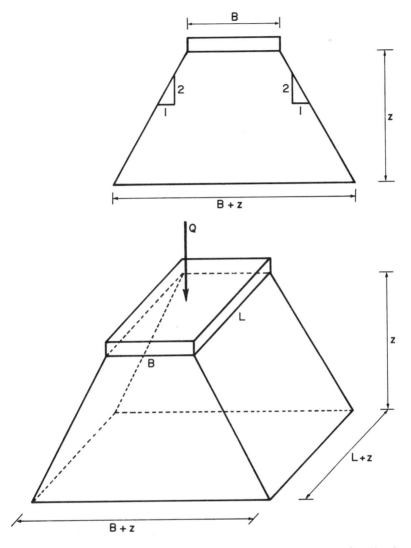

Fig. 4.10—Approximate distribution of vertical stress due to surface load.

γ. In the past, it has been customary to approximate the stresses transmitted to the foundation from this embankment as a distributed stress normal to the surface, with a magnitude equal to γ times the height of the embankment at each point. This is illustrated in Figure 4.12b, where it is referred to as the "normal loading approximation," or the Boussinesq problem. Such an approach neglects the shear stresses which develop between the embankment and its foundation. A more recent approach considers the embankment and foundation as a single body loaded only by its own weight.

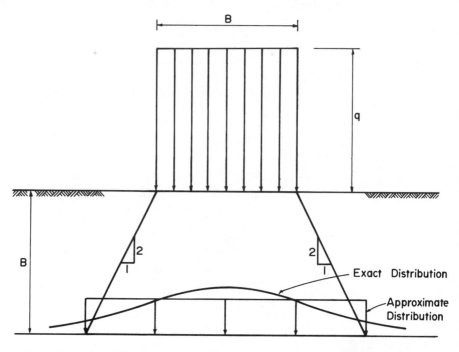

Fig. 4.11—Relationship between vertical stress below a square uniformly loaded area as determined by approximate and exact methods.

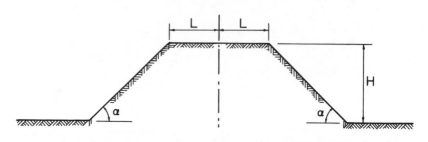

a)- Long Symmetric Elastic Embankment Continuous With Foundation

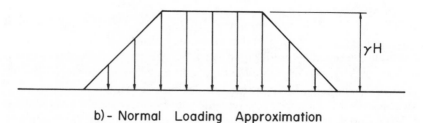

b)- Normal Loading Approximation

Fig. 4.12—Elastic embankment problem.

188

An example of the discrepancies between the results obtained by these two different methods is shown in Figure 4.13. This figure shows the distribution of vertical normal stress beneath the centerline and at the toe of a symmetric embankment with side slopes of 45° and Poisson ratio of 0.3.* The various solid curves show the vertical stress for different embankment shapes as

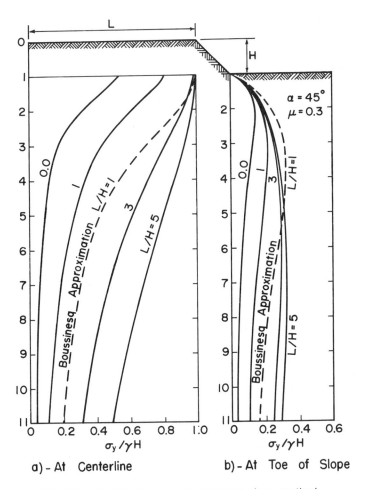

a) - At Centerline   b) - At Toe of Slope

Fig. 4.13—Distribution of vertical stress along vertical sections for varying L/H ratios. (After Perloff et al., 1967.)

---

* The distribution of stresses in a linear elastic material due to surface loadings under plane strain conditions, when weight effects can be neglected, is independent of the material parameters. When the stresses arise from gravity forces, however, the results do depend upon the magnitude of Poisson's ratio. For the cases described herein, a Poisson's ratio of 0.3 has been used. The magnitude is consistent with equivalent values calculated from field measurements of earth dam movements (Gould, 1968).

measured by the ratio L/H (these parameters are defined in Figures 4.12 and 4.13). The dashed curves shown in the figure are the results obtained for the Boussinesq approximation for $L/H = 1$. The difference between these two results is obvious.

It is believed that the data for the elastic embankment are more realistic than those for the normal loading approximation because the analysis considers the effect of the material itself on the distribution of stress.* Consequently the influence diagrams for stresses due to embankment loadings given in the Appendix, in Figures A.12–A.26, are those obtained from the elastic embankment analysis.

Another circumstance in which the stresses due to gravity forces are frequently of interest is that of unloading due to an excavation. For the case of a "long" excavation in a linear elastic material the problem can be represented as shown in Figure 4.14a. It has been conventional to represent the *stress release* due to such an excavation by an *upward* load equal in magnitude to the unit weight of the material removed times the depth of the excavation, applied at the level of the base of the excavation, which was assumed to be the surface of an elastic half-space.

As in the case of embankments, there is an important discrepancy in the stress release calculated using the excavation analysis and that determined by the Boussinesq approximation. This is illustrated in Figure 4.15 which shows the net reduction in vertical stress as a function of depth beneath the center line and edge of an excavation for which $L/H = 1$. Influence diagrams for the excavation analysis are given in Appendix A as Figures A.29–A.49.

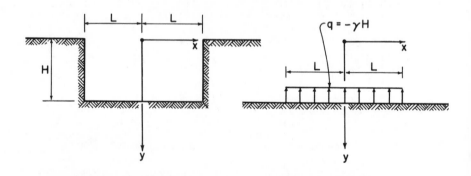

a) - Excavation   Analysis   b) - Boussinesq   Approximation

Fig. 4.14—Illustration of excavation problem.

* It is well known that shear distortions occur at the interface between an embankment and the underlying foundation. Such distortions are predicted by the elastic embankment analysis, but not by the normal loading approximation. Some embankments on very soft foundations have been strengthened by placing reinforcing mats at the fill-foundation interface.

4.6 APPLICATION TO BUILDING 301

In order to calculate the effects of the surcharge load in Building 301, we must determine the stresses produced by the surcharge in the underlying strata of interest. Because the organic silt stratum is much more compressible than any of the materials above and below it, we shall be concerned primarily with the vertical stresses produced within this 24-ft-thick layer (Figure 4.1).

For this case, the depth of interest varies from approximately 7 to 31 ft. The dimensions of the structure, and surcharged area, are 255×550 ft. Thus, for all practical purposes the additional vertical stress in the organic silt due to the surcharge is equal to the magnitude of the surcharge load (see Figure A.5, or even Figure 4.15).

The height of the surcharge fill in the building was the maximum which could be placed within the available clearance, 12 ft. Thus the increment in vertical stress beneath the center of the building, assuming for the fill $\gamma = 108$ pcf, was

$$\Delta \sigma z = (108)(12) = 1300 \text{ psf} \qquad (4\text{--}17)$$

In Chapter 5 we shall see how this stress increment is used to calculate settlements.

4.7 SUMMARY OF KEY POINTS

Summarized below are the key points which we have discussed in the previous sections. The section in which they were described is noted so that reference can be made to the original more detailed discussion:

1. The equations of equilibrium in three dimensions (Equations 4–1) constitute the rule which governs the way in which stresses acting at a point in a continuum change from point to point. The number of unknowns introduced by these equations necessitates considerations of additional aspects of the stress distribution problem (Section 4.2).
2. a. The relative displacements between two points in a continuous body along the line joining them is defined as *deformation.*
   b. *Normal* or *direct strain* describes the deformation in the limit as the length of the deformed line approaches zero (Equations 4–3a, b, c). Shear strain is also defined by a limiting process which describes the change of shape at a point (Equations 4–3d, e, f).
   c. In order to insure that a continuous body remains continuous after deformation, the displacements must satisfy constraints described by the compatibility relations (Equations 4–4). The above topics are discussed in Section 4.3.
3. The nature of the material is introduced into a boundary value problem by the constitutive relationship. It is most common to

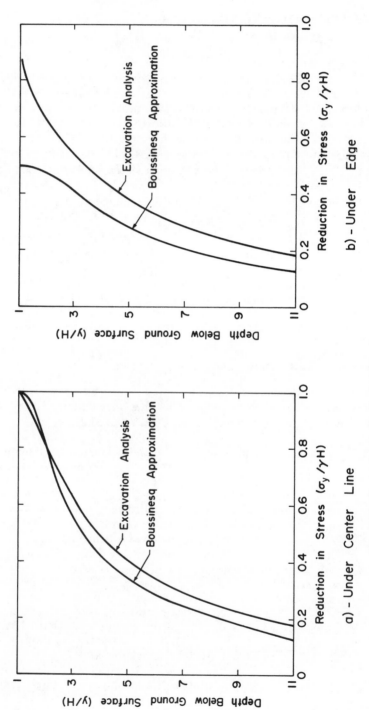

Fig. 4.15—Reduction in vertical stress due to excavation in elastic material.

calculate the distribution of stresses assuming that the material body exhibits linear-elastic behavior. Displacements may then be calculated considering the nonlinear nature of an earth mass (Section 4.4).

4. a. The distribution of stresses throughout a linear-elastic mass subjected to boundary loads has been calculated for a wide variety of boundary conditions, and many solutions are found in chart form or computer programs (Section 4.5, Appendix A).

    b. The calculated stress distribution when the mass is considered to be a layered system may be significantly different from that for a half-space. This effect can be accounted for approximately in the case in which a stiff layer overlies a more flexible stratum below (Section 4.5).

    c. The average vertical stress on a horizontal plane below a loaded surface area may be approximated by assuming that the total load is distributed over an area with dimensions exceeding the original area by an amount equal to the depth below the load (Section 4.5).

5. Stresses which result from the weight of a superimposed earth mass, or from excavation of existing material, may be significantly different from those conventionally determined on the basis of an assumed simple boundary load. Charts for calculation of stresses in the case of long excavations or embankments are available (Section 4.5, Appendix A).

PROBLEMS

4.1. Two footings, shown in plan view, are each subjected to a uniform vertical pressure q. Plot the distribution of vertical stress σ_z/q as a function of depth beneath points A and C for each of the following cases:

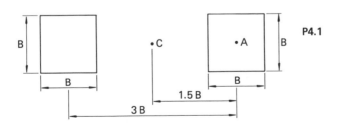

(a) Both footings are assumed to be point loads of magnitude qB^2 located at the footing centroids.

(b) The footing at point A is assumed as a uniformly loaded rectangular area and the other one is assumed as a point load of magnitude qB^2 located at the centroid of the footing.

(c) Both footings are assumed to be uniformly loaded rectangular areas.
Having completed this exercise, what conclusions can you draw from it?

4.2. Plot the distribution with depth of vertical total, neutral, and effective (overburden) stress, for the soil profile shown, to a depth of 25 ft.

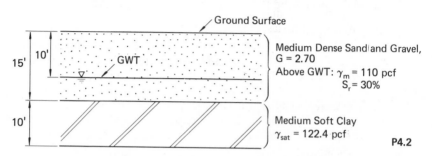

P4.2

4.3. A small water-storage tank is resting on a 20 ft diameter circular footing loaded uniformly to a vertical pressure of 1.5 tsf. The footing is located at the surface of the soil profile shown in the sketch for Prob. 4.2 above. Determine the increment in vertical stress produced by the tank at the top, middle, and bottom of the clay layer—
(a)   Beneath the center of the footing.
(b)   Beneath the edge of the footing.
(c)   Beneath a point one radius (10 ft) outside the footing.

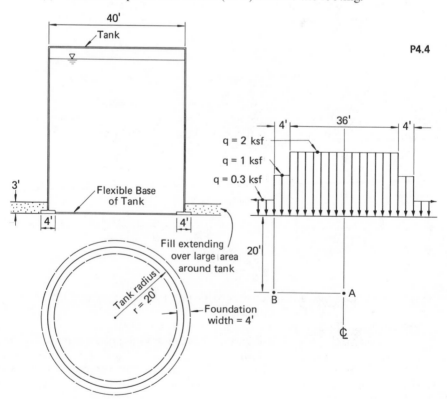

P4.4

4.4. A small emergency water tank is constructed as illustrated. The equivalent loading from the tank foundation, water within the tank, and fill surrounding the tank is shown in the right-hand sketch. Determine the vertical stress increment due to the combination of these loadings at points *A* and *B*, assuming that elastic theory can be used for this purpose.

5

Compressibility and Settlement of Cohesive Soils

5.1 INTRODUCTION

The transfer of loads from the foundation of a structure to the underlying soil produces deformation of the soil and settlement of the ground surface under, and adjacent to the structure. As a result of these foundation settlements, the value and function of the structure may be impaired even though the supporting soil has not failed. The article reprinted in Chapter 4 (pages 161–166) described the fact that settlements of more than a foot were predicted for Building 301 in Port Newark, were it to be used without modification as a warehouse. Settlements of this magnitude were considered by the designers to be intolerable, and the building site was surcharged to precompress the underlying soil. With the removal of the preload it was estimated that there would still be an additional $1\frac{1}{2}$ in. of settlement during the first five years of use of the building as a warehouse. After considering the consequences of the additional settlement on the structure and its function, the designers concluded that this magnitude of settlement during the time stated would be acceptable.

In the design of other structures, the tolerable limits on settlements may be quite different. For a building with close column spacings, $1\frac{1}{2}$ in. of settlement might lead to damage due to the *structural distortions** which would develop, whereas settlements of 17 in. under oil storage tanks might be acceptable. The pattern of settlement and its significance to a structure depend not only upon the underlying soil profile and properties, but also upon the type of structure, its response to the settlements, its interaction

---

* Structural distortion is the differential settlement between two points on the structure divided by the distance separating them, that is, it is the rotation of the line joining two points.

with the soil, and its intended function. Therefore, in considering settlement, we shall be concerned with both the total magnitude of the settlement and with the magnitude of relative displacements and distortions. In this chapter, we shall limit our attention to the behavior of the soil. The structural response is discussed in Chapter 11.

Components of Settlement

We find it useful to consider that the vertical settlement ρ of a foundation, due to the loads transmitted to the underlying soils, may be described as the sum of three components,

$$\rho = \rho_d + \rho_c + \rho_s \qquad (5-1)$$

in which ρ_d is the *immediate* or *distortion* settlement, ρ_c is the *consolidation* settlement, and ρ_s is the *secondary compression* settlement. The immediate component is that portion of the settlement which occurs concomitantly with the applied loading primarily as a consequence of *distortion** within the foundation soils. The distortion settlement is generally not elastic, although it is calculated using elastic theory when the seat of the settlement is in cohesive material.

The remaining components result from the gradual expulsion of water from the voids and the contemporaneous compression of the soil skeleton. The distinction between consolidation and secondary compression is made on the basis of the physical processes which control the time rate of settlement. Consolidation settlement arises from a hydrodynamic process called *primary consolidation* in which the time rate of settlement is controlled by the rate at which water can be expelled from the void spaces in the soil. In the case of *secondary compression*, the speed of settlement is controlled by the rate at which the soil skeleton itself yields and compresses. A typical time-settlement curve for a foundation, in which the loads are applied instantaneously without impact, is shown in Figure 5.1. The settlement components are indicated in the figure. The time at which primary consolidation ceases and secondary compression is deemed to begin is denoted by t_{100}.

Because the response of soils to applied loads is not linear, the superposition implied by Equation 5-1 is not generally valid. However, no consistent alternative method for predicting settlements is available; and experience indicates that reasonable predictions of settlement of foundations on many sub-soil types can be made with this approach.

The time-settlement relationship shown in Figure 5.1 is applicable to all soils if we recognize that the time scale and relative magnitudes of the three components may differ by orders of magnitude for different soil types. For example, water flows so readily through most clean granular soils that the

---

* An element of material undergoes a pure distortional deformation if it changes shape without changing volume.

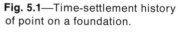

Fig. 5.1—Time-settlement history of point on a foundation.

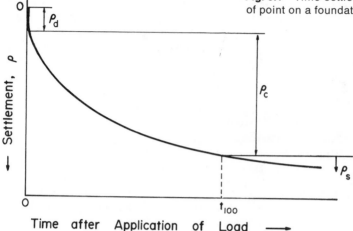

Time after Application of Load ⟶

expulsion of water from the pores is, for all practical purposes, instantaneous. Furthermore, the mechanical behavior of cohesionless soils subjected to static loadings is essentially time-independent so that they do not exhibit secondary compression effects. Therefore foundations on granular soils settle almost simultaneously with the application of load. Although these settlements occur rapidly, they are not elastic and cannot be reasonably estimated by an elastic analysis. The rationale and procedure for predicting settlements of foundations on a cohesionless mass are discussed in Chapter 11. Secondary compression is also considered in Chapter 11. Our discussion in Chapter 5 will be limited to methods for predicting the immediate and consolidation settlements of the surface of a *saturated cohesive* soil stratum. In the following sections these individual components are considered in detail.

5.2   IMMEDIATE SETTLEMENT

In Chapter 4 we examined the distribution of stresses within a soil mass resulting from the application of loads to the boundary of the medium. Concomitant with the development of these stresses, the soil will exhibit distortional and/or volumetric strain. In the case of a saturated cohesive soil, the water and soil solids are essentially incompressible so that volume change can occur only as water is squeezed out of the voids. Because time is required to permit expression of water from the voids of a saturated cohesive soil, the immediate settlement of such materials cannot involve volume change. For this reason the immediate settlement is distortional in nature.

Analytical evaluation of immediate settlements is a problem which requires satisfaction of the same set of conditions as the determination of stresses in continuous media. In fact, we could view the process as one of determining the stresses at each point in the medium, evaluating the vertical strains and integrating these vertical strains over the depth of the material. Thus the solution must satisfy:

1. The equations of equilibrium (Equations 4–1).
2. The strain-displacement relation (Equations 4–3), subject to the constraints of the requirements of compatibility (Equations 4–4).
3. Constitutive relations (Equations 4–6).
4. The boundary conditions.

To calculate immediate settlements we shall assume that the soil behaves as an isotropic linear-elastic material, and is homogeneous throughout the region of interest. We consider first the immediate settlements determined by assuming that the soil mass is an elastic half-space.

Immediate Settlement of Surface of a Half-space

The vertical settlement of any point of the surface of an elastic half-space uniformly loaded over a rectangular area is (Schleicher, 1926)

$$\rho_d(x, y) = C_s qB \left(\frac{1 - \mu^2}{E} \right) \tag{5-2}$$

in which C_s is a geometric factor which accounts for the shape of the rectangle and the position of the point for which the settlement is being calculated (Figure 5.2):

$$
\begin{aligned}
C_s = \frac{1}{2\pi} \Bigg[&\left(\frac{L}{B} - \frac{2y}{B} \right) \ln \frac{\sqrt{\left(1 - \frac{2x}{B}\right)^2 + \left(\frac{L}{B} - \frac{2y}{B}\right)^2} + \left(1 - \frac{2x}{B}\right)}{\sqrt{\left(1 + \frac{2x}{B}\right)^2 + \left(\frac{L}{B} - \frac{2y}{B}\right)^2} - \left(1 + \frac{2x}{B}\right)} \\
+ &\left(\frac{L}{B} + \frac{2y}{B} \right) \ln \frac{\sqrt{\left(1 - \frac{2x}{B}\right)^2 + \left(\frac{L}{B} + \frac{2y}{B}\right)^2} + \left(1 - \frac{2x}{B}\right)}{\sqrt{\left(1 + \frac{2x}{B}\right)^2 + \left(\frac{L}{B} + \frac{2y}{B}\right)^2} - \left(1 + \frac{2x}{B}\right)} \\
+ &\left(1 - \frac{2x}{B} \right) \ln \frac{\sqrt{\left(1 - \frac{2x}{B}\right)^2 + \left(\frac{L}{B} - \frac{2y}{B}\right)^2} + \left(\frac{L}{B} - \frac{2y}{B}\right)}{\sqrt{\left(1 - \frac{2x}{B}\right)^2 + \left(\frac{L}{B} + \frac{2y}{B}\right)^2} - \left(\frac{L}{B} + \frac{2y}{B}\right)} \\
+ &\left(1 + \frac{2x}{B} \right) \ln \frac{\sqrt{\left(1 + \frac{2x}{B}\right)^2 + \left(\frac{L}{B} - \frac{2y}{B}\right)^2} + \left(\frac{L}{B} - \frac{2y}{B}\right)}{\sqrt{\left(1 + \frac{2x}{B}\right)^2 + \left(\frac{L}{B} + \frac{2y}{B}\right)^2} - \left(\frac{L}{B} + \frac{2y}{B}\right)} \Bigg]
\end{aligned}
\tag{5-3}
$$

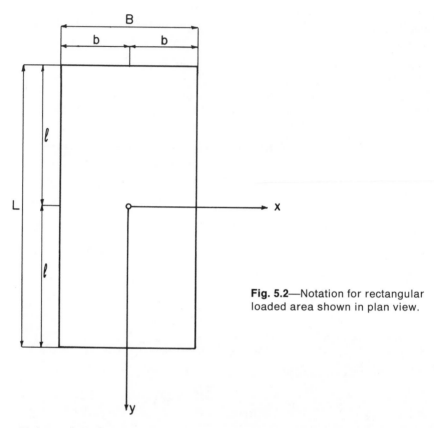

Fig. 5.2—Notation for rectangular loaded area shown in plan view.

Values of C_s for several points on uniformly loaded rectangular areas of various shapes are given in Table 5.1. We note in this table that the settlement at the corner of a uniformly loaded rectangle is only one-half of that at the center. The "dish-shaped" pattern of settlement is illustrated in Figure 5.3a.

Equation 5–2 may also be used to describe the vertical settlement of a concentrically loaded rigid rectangular area, such as a concrete footing, described by the boundary condition of *uniform displacement* rather than uniform loading. Only the form of C_s is different for the rigid case. Similarly, Equation 5–2 applies to uniformly loaded and rigid circular areas if the appropriate expression for C_s is used (Schleicher, 1926).

Data for these cases are also given in Table 5.1. We see that the settlement of a rigid area is approximately 95 per cent of that under the center of the uniformly loaded area carrying the same total load. As might be expected, the distributions of contact pressure between the loaded area and the elastic foundation are quite different for these limiting cases. The uniformly loaded area produces a uniform reaction, whereas the contact pressure on the rigid

TABLE 5.1
Values of Shape and Rigidity Factor C_s at Various Points of Elastic Half-Space

Shape	Center	Corner	Middle of Short Side	Middle of Long Side	Average
Circle	1.00	0.64	0.64	0.64	0.85
Circle (rigid)	0.79	0.79	0.79	0.79	0.79
Square	1.12	0.56	0.76	0.76	0.95
Square (rigid)	0.99	0.99	0.99	0.99	0.99
Rectangle (length/width)					
1.5	1.36	0.67	0.89	0.97	1.15
2	1.52	0.76	0.98	1.12	1.30
3	1.78	0.88	1.11	1.35	1.52
5	2.10	1.05	1.27	1.68	1.83
10	2.53	1.26	1.49	2.12	2.25
100	4.00	2.00	2.20	3.60	3.70
1,000	5.47	2.75	2.94	5.03	5.15
10,000	6.90	3.50	3.70	6.50	6.60

Source: Schleicher (1926).

foundation increases from the center to a theoretically infinite stress at the edges, as illustrated in Figure 5.3b. In practice such a foundation does induce large stresses at the edges, although they cannot exceed that stress which will produce yielding of the soil beneath the foundation. For footings with shapes other than those described, or for the determination of settlements at points other than those indicated, superposition may be used as in the analysis of stresses.

The applicability of elastic theory to the computation of settlement of structures on clay soils has always been a subject of controversy. However, for relatively uniform clay strata, E and μ can be considered more or less

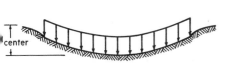

a) - Settlement of Uniformly Loaded Flexible Area

Fig. 5.3—Settlement pattern and contact pressure of loaded areas on linear elastic half-space.

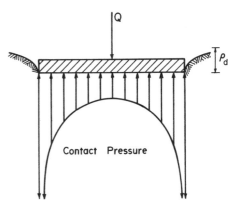

b) - Contact Pressure Distribution on Rigid Foundation

constant, and at sufficiently low stress levels, elastic theory is thought to lead to reasonable estimates of the immediate settlement.

Example 5.1

A structure is to be supported on a mat* foundation whose dimensions are 70 by 180 ft. The load on the mat is presumed to be uniformly distributed; its magnitude is 1250 psf. The mat rests on a deep saturated-clay deposit with these properties:

Unconfined compressive strength: $q_u = 2500$ psf

Young's modulus: $E = 800 \times 10^3$ psf

Estimate the immediate settlement at the center and corner of the mat.

Solution. For a saturated soil there is no elastic volume change thus Poisson's ratio is assumed to be 0.5. From Table 5.1, with

$$\frac{L}{B} = \frac{180}{70} = 2.57$$

the shape factors are determined by interpolation to be

At the center: $C_s = 1.67$

At the corner: $C_s = 0.83$

Thus from Equation 5-2 the immediate surface settlement is

$$\rho_{d_{center}} = (1.67)(1250)(70)\left(\frac{1 - 0.5^2}{800 \times 10^3}\right) = 0.137 \text{ ft} = 1.6 \text{ in.}$$

$$\rho_{d_{corner}} = (0.83)(1250)(70)\left(\frac{1 - 0.5^2}{800 \times 10^3}\right) = 0.068 \text{ ft} = 0.82 \text{ in.}$$

A mat is neither completely flexible nor completely rigid. Because of its large size the distribution of contact pressure may be as assumed, nearly uniform over the center portion of the mat. At the corners, the rigidity of the mat (owing to the thickness of the mat and amount of reinforcing) may be significant, and settlements are likely to be less than predicted.

Immediate Settlement of the Surface of Layered Systems

In practice, the foundation engineer must deal with soil profiles that are nonhomogeneous and/or multilayered. If the thickness of the uppermost layer is large relative to the dimensions of the loaded area, the soil model may be taken as a homogeneous layer of infinite depth with the immediate surface displacements calculated as previously described. If, however, the upper layer is relatively thin, ignoring the effect of layering may have an appreciable effect on the magnitude of the calculated immediate settlement. We shall consider first the settlement of the surface of a layer of elastic

---

* A *mat* is a single foundation element which extends the full length and width of the structure.

material of *finite* thickness underlain by a *rigid* base and compare it with the surface displacement of a semi-infinite mass of the same material. This special case of a layered system is of interest when soft compressible strata are underlain by rock, or very hard or dense soils. From Figure 4.9 we note that the stresses in the upper layer are increased and therefore, we might expect an increase in the surface displacement. However, there are no strains below the rigid boundary and therefore, there is a small depth over which to sum the vertical strains contributing to the surface displacement. Thus we note two effects, one acting to increase the surface displacement, the other acting to decrease it. As we shall see, the net result is a reduction in the surface displacement if a rigid boundary exists at some finite depth (Burmister, 1943, 1956). The displacements may be again evaluated by Equation 5–2 in which the shape factor C_s' now accounts for the effect of layering, and E and μ are the elastic moduli of the *compressible* layer. Values for the shape factor for this have been tabulated (Harr, 1966) and are given in Table 5.2 for the settlement under the center of the uniformly loaded area at the upper surface of a compressible layer underlain by a rigid boundary. These factors depend upon the geometry and dimensions of the loaded area as illustrated in Figure 5.4.

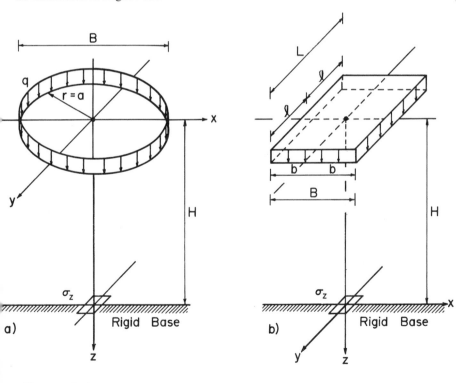

Fig. 5.4—Finite compressible material underlain by a rigid base.

As indicated in Table 5.2, the solution depends upon the boundary conditions assumed at the interface between the elastic and rigid layers. Results are given for two cases:

1. $\tau = 0$. The horizontal shear stresses acting at the boundary are assumed zero, that is, the upper elastic material can slide freely in a horizontal direction at the boundary of the rigid material. The vertical stresses on either side of the boundary are equal.

2. $u = 0$. There is no horizontal displacement of any point on the boundary of the two layers, that is, a sufficient shear resistance is developed at the interface which prevents slipping of the upper layer with respect to the lower. Neither is there a displacement in the vertical direction at the interface.

In computing the values of C'_s Poisson's ratio was assumed to be 0.3 for the case of no displacement ($u = 0$). For the case of no shear ($\tau = 0$), μ need not be specified. These two cases provide limiting values of C'_s. That is, values of C'_s for cases of partial slippage along the interface should lie between the tabulated values.

Examination of Table 5.2 indicates that the shallower the compressible layer, the smaller is the magnitude of the displacements at the surface. For $H/B = 0.5$ the reduction in surface displacement relative to that for a half-space is 50 per cent or greater. When the thickness of the compressible layer is equal to the diameter or width of the loaded area ($H/B = 1$) the reduction in displacement is nearly 30 per cent. We also can see that the effect of the interface boundary condition is greatest for small thicknesses of the compressible layer but for these small thicknesses the reduction in displacement is so great that the difference between the two cases is of small significance.

We now consider the case of a stiff layer of finite thickness underlain by a less rigid stratum of great depth. Such systems are exemplified by pavements as well as precompressed soil strata overlying less highly prestressed materials. Referring to Figure 4.9, we see that the presence of a stiffer upper layer reduces the vertical stresses in the underlying material below those which would occur in a homogeneous system. Thus, we would expect a corresponding reduction in surface displacements.

Surface displacements at the center of a uniformly loaded circular area on a stiff elastic layer underlain by an infinite depth of less stiff elastic material have been calculated (Burmister, 1965). The results are most conveniently described in terms of the surface settlement of a homogeneous system:

$$\rho_{d_i} = C''_s \, \rho_{d_0} \qquad (5\text{--}4)$$

in which ρ_{d_i} is the settlement at the center of a uniformly loaded circular area at the surface of a layer with elastic modulus E_1, Poisson's ratio μ_1, and thickness H, underlain by an infinite depth of material with elastic modulus E_2 and Poisson's ratio μ_2; ρ_{d_0} is the calculated settlement at the

TABLE 5.2
Values of Shape Factor C_s' for Settlement of Center of Uniformly Loaded Area on Elastic Layer Underlain by Rigid Base

| | Circle of Diameter B | | Rectangle | | | | | | | | | | | | Infinite Strip | |
| | | | L/B = 1 | | L/B = 1.5 | | L/B = 2 | | L/B = 3 | | L/B = 5 | | L/B = 10 | | L/B = ∞ | |
H/B	$\tau = 0$	$u = 0$	$\tau = 0$	$u = 0$	$\tau = 0$	$u = 0$	$\tau = 0$	$u = 0$	$\tau = 0$	$u = 0$	$\tau = 0$	$u = 0$	$\tau = 0$	$u = 0$	$\tau = 0$	$u = 0$
0.0	0.00	0.00	0.00	0.00	0.00	0.00	0.00	0.00	0.00	0.00	0.00	0.00	0.00	0.00	0.00	0.00
0.1	0.10	0.08	0.10	0.08	0.10	0.08	0.10	0.08	0.10	0.08	0.10	0.08	0.10	0.08	0.10	0.08
0.25	0.26	0.22	0.26	0.21	0.25	0.21	0.25	0.21	0.25	0.21	0.25	0.21	0.25	0.21	0.25	0.21
0.5	0.50	0.45	0.51	0.44	0.51	0.43	0.51	0.43	0.51	0.43	0.51	0.43	0.51	0.43	0.51	0.43
1.0	0.72	0.68	0.77	0.72	0.85	0.77	0.87	0.78	0.88	0.78	0.88	0.78	0.88	0.78	0.88	0.78
1.5	0.81	0.78	0.88	0.84	1.00	0.94	1.07	0.99	1.12	1.02	1.13	1.02	1.13	1.02	1.13	1.02
2.5	0.89	0.87	0.98	0.95	1.14	1.10	1.24	1.19	1.36	1.29	1.44	1.34	1.45	1.34	1.45	1.34
3.5	0.92	0.90	1.02	1.00	1.20	1.17	1.32	1.29	1.47	1.42	1.60	1.52	1.64	1.54	1.65	1.54
5.0	0.94	0.93	1.05	1.04	1.25	1.23	1.39	1.36	1.56	1.53	1.75	1.69	1.87	1.77	1.88	1.77
∞	1.00	1.00	1.12	1.12	1.36	1.36	1.52	1.52	1.78	1.78	2.10	2.10	2.53	2.53	∞	∞

SOURCE: Harr (1966).

center of the uniformly loaded circular area on the surface of a homogeneous half-space with parameters E_2 and μ_2, and C_s'' is the correction factor relating the two settlements.

Values for C_s'' for various ratios H/B and E_1/E_2 are given in Table 5.3. In the case considered, $\mu_1 = \mu_2 = 0.4$ and no slip at the interface between the two layers have been assumed. Note that the introduction of the stiff layer can produce a significant reduction in the surface displacement. If we consider the ratio $E_1/E_2 = 100$ to approximate a rigid layer over a flexible layer, we see that a relatively small thickness of the rigid layer $H/B = 0.25$ reduces the settlement by 57 per cent. Even with a small difference in the stiffness, $E_1/E_2 = 2$, there is a reduction of 25 per cent when the thickness of the stiffer material is half the diameter of the loaded area.

TABLE 5.3
Values of Elastic Distortion Settlement Correction Factor C_s'', at Center of Circular Uniformly Loaded Area on Elastic Layer E_1 Underlain by Less Stiff Elastic Material E_2, of Infinite Depth

H/B	Value of E_1/E_2				
	1	2	5	10	100
0	1.000	1.000	1.000	1.000	1.000
0.1	1.000	0.972	0.943	0.923	0.760
0.25	1.000	0.885	0.779	0.699	0.431
0.5	1.000	0.747	0.566	0.463	0.228
1.0	1.000	0.627	0.399	0.287	0.121
2.5	1.000	0.550	0.274	0.175	0.058
5.0	1.000	0.525	0.238	0.136	0.036
∞	1.000	0.500	0.200	0.100	0.010

NOTE: $\mu_1 = \mu_2 = 0.4$.
SOURCE: Data from Burmister, 1965.

Approximate Evaluation of Immediate Settlement of Layered Systems

Analytical and/or numerical methods for the determination of stresses and displacements in multilayered systems are available for other cases than those described above.* However, most solutions require that a numerical evaluation employing a digital computer be performed for each problem.† Except for special cases, the use of such layered system-analyses is not

---

* A bibliography of "Solutions to Boundary Value Problems of Stresses and Displacements in Earth Masses and Layered Systems" (Hampton et al., 1969) is available. The bibliography is *not* limited to analysis of media which are elastic and/or isotropic and thus provides a very useful listing. However, reference is made only to works published prior to 1966.

† A solution by Burmister (1965) for profiles containing three elastic strata are available with numerical results in graphical form. Because the results were obtained for application to pavement systems, they are limited to profiles in which the stiffness of each stratum is greater than, or equal to that of the underlying adjacent layer.

justified, for these reasons:

1. The procedures are not readily available in a form applicable for immediate computation.
2. Material parameters are generally not obtainable with the degree of accuracy required to justify a sophisticated analysis.
3. The boundary conditions under the load and at the interface between strata cannot be clearly defined.
4. The approximations required to "fit" the geometry of the real problem to that of the problem for which the solution is available are usually inconsistent with the precision of the solution procedure.

Thus in many situations an approximate analysis of the immediate settlement is appropriate. The following example illustrates one approach:*

Example 5.2 ————————————————————————————————

The mat foundation described in Example 5.1, 70×180 ft supporting a uniform normal load of 1250 psf, is founded on the soil profile shown in Figure 5.5. The profile indicates a layer of stiff clay over a more compressible clay which is in turn underlain by shale. Given the elastic modulus for each material, estimate the immediate surface settlement at the center of the mat.

Solution. Assume the shale acts as a rigid base and that above it there is a *single* stratum of thickness, $H = 50$ ft. Thus,

$$\frac{H}{B} = \frac{50}{70} = 0.175$$

$$\frac{L}{B} = \frac{180}{70} = 2.57$$

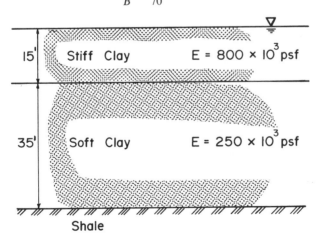

Fig. 5.5—Soil profile for Example 5.2.

* Alternative approaches to incorporating the effect of layered systems are described in connection with calculation of immediate settlement of Building 301 (Section 5.4).

The shape factors C_s' obtained from Table 5.2 by linear interpolation are:

$$C_s' = 0.66 \quad \text{for } \tau = 0,$$
$$C_s' = 0.58 \quad \text{for } u = 0$$

Without information concerning the true interface condition, we select an intermediate value,

$$C_s' = \tfrac{1}{2}(0.58 + 0.66) = 0.62$$

Substituting this value into Equation 5–2 leads to a calculated surface settlement for an assumed $\mu = 0.5$, of

$$\rho_d = (0.62)(1250)(70)\left(\frac{1 - 0.5^2}{E}\right) = \frac{41 \times 10^3}{E} \text{ ft} \tag{5-5}$$

We can bound the potential immediate surface displacement by estimating the settlement using the different moduli of the two compressible strata:

$$\frac{41 \times 10^3}{800 \times 10^3} < \rho_d < \frac{41 \times 10^3}{250 \times 10^3}$$

or

$$0.051 < \rho_d < 0.164$$

To approximate the actual immediate surface settlement, we use an equivalent Young's modulus, weighted by the relative thicknesses of the two strata:

$$E' = \frac{15(800 \times 10^3) + 35(250 \times 10^3)}{50} = 415 \times 10^3 \text{ psf}$$

Thus the estimated immediate settlement is

$$\rho_d = \frac{41 \times 10^3}{415 \times 10^3} = 0.099 \text{ ft} \tag{5-6}$$

We should expect the settlement predicted by Equation 5–6 to exceed that determined by a more exact approach because the load distributing effect of the upper stiff layer is not accounted for in the weighted average E'.

Evaluation of Elastic Constants

The magnitude of calculated elastic distortion settlements depends directly on the values of the elastic constants, Young's modulus and Poisson's ratio. Because cohesive soils are not linear-elastic materials, the objective of evaluating the "elastic constants" is to determine those values which, when substituted into Equation 5–2, will lead to a correct determination of the elastic distortion settlement.

For saturated clay soils, which deform at constant volume during the limited time required to develop the elastic distortion settlement, a Poisson's ratio $\mu = 0.5$, corresponding to an incompressible medium, is usually assumed. Although this assumption may not be correct,* the magnitude of the computed settlement is not especially sensitive to small changes in Poisson's ratio.

Determination of the appropriate value of Young's modulus E, is a much more critical problem. This parameter is commonly evaluated from the initial tangent modulus of the stress-strain curve from triaxial compression or *unconfined compression* tests† (Figure 5.6). Alternatively, a secant modulus, determined for the stress level estimated to obtain under field loading conditions, is used. There is, however, laboratory and field evidence to indicate that the values so obtained are too low, often only a small fraction of the field value (Ladd, 1964; Hanna and Adams, 1968; Soderman et al., 1968). There are two primary reasons for this discrepancy:

1. During the process of extracting a soil specimen from the ground (*sampling*) and preparation of a specimen for laboratory testing, the natural structural arrangement of the soil skeleton is disturbed.

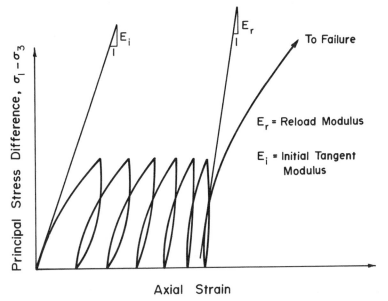

Fig. 5.6—Definitions of moduli.

* Recent measurements of horizontal movements of conduits imbedded within earth embankments (Gould, 1968) suggest that a more correct prediction of such movements is obtained by using an equivalent Poisson's ratio for the soil equal to 0.3–0.35. This apparently anomalous situation probably arises from the natural anisotropy of many cohesive deposits, particularly those which have been prestressed.

† The unconfined compression test is a special case of the triaxial compression test in which there is no confining pressure, and is useful only for specimens of cohesive material. We shall discuss this test in greater detail in Chapter 8.

Such disturbance induces increased pore water pressure with a corresponding decrease in effective stress. This leads to a reduced undrained stiffness and strength. The modulus E is one of the most sensitive parameters to disturbance effects. Thus a modulus in a *laboratory* test may not be representative of the analogous parameter in the *field*.

2. Defects in the form of *fissures* occur in a great many soils. These inhomogeneities are often unimportant to the settlement of a structure because they are small in size relative to the foundation dimensions. Such defeats may pass completely through a laboratory specimen, and produce a spuriously low measured modulus in a laboratory test in which confining pressures are low or absent.

There is very limited evidence available to serve as a guide in applying the results of laboratory tests to determine the field E. On the basis of this limited information (Ladd, 1964; Hanna and Adams, 1968; Soderman et al., 1968) at this time, the best approach appears to be the following two-step procedure:

1. Using the best available *undisturbed* specimen, conduct a triaxial compression test in which the specimen is allowed to drain freely when subjected to an all-round confining pressure equal to the vertical effective overburden stress for that specimen in the field. After equilibrium has been reached, prevent further drainage and apply additional axial loadings as described in step two below. This type of triaxial compression test is called a *consolidated-undrained* (CU) test, and is described more fully in Chapter 8.

2. Gradually increase the axial stress to that magnitude estimated for the field loading conditions, and then reduce it to zero. Repeat this loading cycle as many times as necessary. For each cycle determine the average, or secant modulus. A plot of this modulus as a function of the number of cycles will approach an asymptotic value called the *reload modulus*, E_r.

The reload modulus is shown in Figure 5.6. Its determination is illustrated in Figure 5.7 for a sample of Tilbury clay till cycled at a stress level equal to one-third the principal stress difference at failure. From this diagram, and other tests, it appears that five or six cycles is sufficient to determine the reload modulus. The fact that E_r is larger than the modulus determined upon initial loading indicates that the material is strain hardening. The physical reason for this, and the relationship between E_r and the field modulus, are not known. However, by combining the two steps above, the laboratory estimated value of E is likely to be closer to that which would be determined in the field by settlement measurements on the prototype structure.

In some cases it may be possible to conduct a field plate loading test. In this test, loads are applied to a one foot square or a one foot diameter circular plate and the accompanying deflections measured. From the results

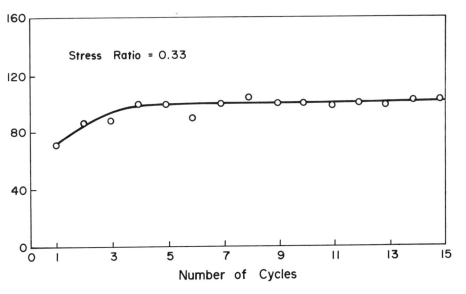

Fig. 5.7—Variation in reload modulus with number of load cycles. (From Soderman, et al., 1968.)

of such a test, all of the parameters in Equations 5–2 are known except the factor $(1 - \mu^2)/E$ which can then be determined. It has been suggested (Soderman et al., 1968) that the plate load test yields the best estimate of elastic parameters used in predicting foundation settlements. However, load tests on small plates cannot be extrapolated to the prediction of settlements of large foundations if the seat of the settlement of the footing is different from that of the foundation. A difference between small- and large-scale results may occur because of the presence of strata which intercept the pressure bulb under the foundation and not that under the small plate.

Indirect estimate of the elastic modulus may also be made by relating it to the strength measured in an undrained test. Experimental evidence (Bjerrum, 1964; Hanna and Adams, 1968; Soderman et al., 1968) suggests that

$$E = (125 \text{ to } 500)(\sigma_1 - \sigma_3)_f \qquad (5-7)$$

in which $(\sigma_1 - \sigma_3)_f$ is the principal stress difference at failure in an undrained compression test. Equation 5–7 provides, at best, a crude estimate of the elastic modulus. It may, however, be used to determine whether immediate distortion settlements will be of sufficient magnitude to justify a more accurate determination of E.

In the majority of cases, elastic distortion settlements of clay strata are small compared to settlements resulting from the volume compressibility of the material. In the case of large structures on relatively deep deposits of clay, such settlements may be significant and should not be overlooked.

One such example is the case of a large industrial structure near Montreal, Canada, supported on a mat foundation overlying a deep deposit of the Laurentian clay. It was estimated that nearly half of the total observed settlement of 6 in. was due to elastic distortion settlement of the clay (Leonards, 1966).

5.3 CONSOLIDATION SETTLEMENT—ONE-DIMENSIONAL CASE

When foundation loads are transmitted to cohesive subsoils, there is a tendency for a volumetric strain which, in the case of a saturated material is manifested in an increase in pore water pressure. With sufficient elapsed time, water flows out of the soil pores, permitting the excess pressure to dissipate. The analysis of the volumetric strains which result, and the vertical settlements accompanying them, is simplified if we assume that such strains occur only in the vertical direction. Such an assumption may not be unreasonable when the geometric and boundary conditions in the field are such that vertical strains dominate. For example, as illustrated in Figure 5.8a, when the dimensions of the loaded area are large relative to the thickness

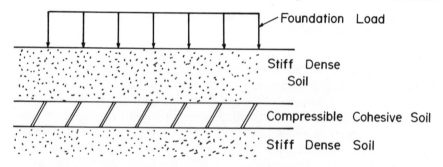

a) - Field Conditions Leading to Approximately One - Dimensional Compression

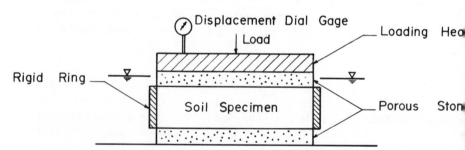

b) - Laboratory One - Dimensional Consolidation Test Simulating Field Condition

Fig. 5.8—One-dimensional compression.

of the compressible stratum and/or when the compressible material lies between two stiffer soils whose presence tends to reduce the magnitude of horizontal strains, an approximately *one-dimensional compression* of the soil will occur. We shall for the present follow conventional practice by accepting the assumption of one-dimensional compression in the discussion which follows.

A laboratory test which produces one-dimensional compression of a soil specimen to simulate the conditions illustrated in Figure 5.8a is shown schematically in Figure 5.8b. This useful test, to which we shall make reference in the discussion which follows, is called the *one-dimensional consolidation test*. As shown in the figure, a specimen of soil is constrained with a rigid ring and loaded top and bottom with porous stones which permit water to flow into and out of the soil specimen. Consolidation tests on saturated cohesive materials are usually conducted submerged to prevent drying of the soil. Loads are applied to the porous stone on top of the specimen and vertical displacement of the top surface is measured.

One type of consolidation test apparatus is illustrated in Figures 5.9 and 5.10 along with auxiliary apparatus for trimming the soil specimen to the appropriate size.

Fig. 5.9—Geonor consolidation apparatus; consolidation frame on the right, sample cutting apparatus on left.

Fig. 5.10—Close-up view of the Geonor consolidation apparatus. Left, loading head. Above, sample container; from left to right—base plate cylindrical container for water within which sample is immersed, lower porous stone, guide ring, sample cylinder fastener ring, upper porous stone and piston.

It is assumed that the test conditions produce a homogeneous state of stress and strain within the soil specimen when it has reached equilibrium under the applied loading. Because deformations are one-dimensional in nature, the vertical strains and volumetric strains in the specimen are equal, and can be related to the measured settlement of the top of the specimen by

$$\rho_c = \varepsilon_z H_0 = \varepsilon_v H_0 \tag{5-8}$$

in which the vertical strain, ε_z, is

$$\varepsilon_z = \frac{dH}{H_0} \tag{5-9}$$

and the volumetric strain, ε_v, is

$$\varepsilon_v = \frac{dV}{V_0} \tag{5-10}$$

and H_0 and V_0 are the thickness and volume of the specimen prior to load application. Because the volume change occurs only in the voids, the volumetric strain can be related to the void-ratio change by

$$\frac{dV}{V_0} = -\frac{de}{1 + e_0} \qquad (5\text{-}11)$$

in which de is the change in void ratio due to the applied load increment and e_0 is the initial void ratio. Thus we see that the settlement of the soil specimen, its vertical or volumetric strain, and its void-ratio change are all proportional in the case of one-dimensional compression. We find it convenient, therefore, to use the void-ratio change to describe the effect of load application in a one-dimensional situation.

The relationship between void ratio and applied vertical effective stress for one-dimensional compression of a cohesive soil is shown in Figure 5.11. The slope of the relation shown in the figure is called the *coefficient of compressibility*, a_v, defined as

$$a_v = \frac{-de}{d\sigma'_c} \qquad (5\text{-}12)$$

in which de is the void-ratio change consequent on application of a vertical effective stress increment $d\sigma'_c$. The coefficient of compressibility is an inverse measure of the stiffness of the soil. Thus from Figure 5.11 we observe an important characteristic of most cohesive soils: namely that the material becomes stiffer as the effective consolidation pressure increases.

If we consider the case in which the effective vertical stress is increased beyond the maximum value previously experienced by the soil and is then

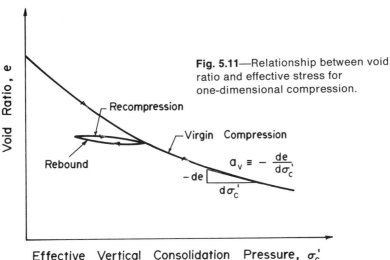

Fig. 5.11—Relationship between void ratio and effective stress for one-dimensional compression.

reduced, the specimen *swells*, or *rebounds*; water flows into the sample. It is evident in Figure 5.11 from the portion of the curve labeled "rebound" that the soil does not respond in an elastic manner, as evidenced by the permanent deformation that occurs. We also note that there is a significant reduction in a_v in the rebound and subsequent *recompression* cycle as compared to the initial compression curve. After exceeding the previous maximum effective stress, the curve rapidly approaches the relation which would have been followed had the rebound-recompression cycle not been introduced. The void-ratio–effective-stress relationship for stress levels exceeding the maximum past pressure experienced by the soil is called the *virgin compression curve*, indicated in Figure 5.11.

It is more useful to present the relationship of Figure 5.11 in a form in which the effective stress is plotted to a logarithmic scale as shown in Figure 5.12. It is apparent from this figure that the virgin compression curve is a

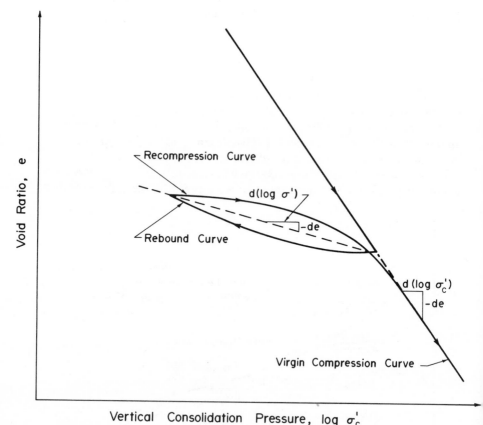

Fig. 5.12—Laboratory void-ratio–Effective-stress curve for one-dimensional compression.

straight line with a slope C_c, called the *compression index*:

$$C_c \equiv \frac{-de}{d(\log \sigma_c')} \tag{5-13}$$

in which the logarithm is taken to the base 10. This portion of the curve corresponds to the condition in which the existing vertical effective stress is the maximum such stress experienced by the soil. In accordance with the nomenclature in Chapter 3, we refer to such a material as "normally consolidated." Thus, for a normally consolidated soil in the field, the overburden pressure is equal to the maximum vertical effective stress experienced by this material. That is, if we define this maximum past pressure as the *preconsolidation pressure*, σ_p', then for a normally consolidated soil,

$$\sigma_p' = \sigma_0' \tag{5-14}$$

in which σ_0' is the *in situ* overburden pressure.

When the vertical effective stress has been rebounded, the preconsolidation pressure exceeds the existing consolidation stress. In such a condition, the soil is considered to be *over-consolidated*. In the field, an over-consolidated soil is one for which

$$\sigma_p' > \sigma_0' \tag{5-15}$$

Subsequent reloading leads to a hysteresis loop, as shown in Figure 5.12. The average slope of this portion of the e–$\log \sigma_c'$ curve, described as the *expansion* or *rebound* or *recompression index*, is

$$C_e \equiv -\frac{de}{d(\log \sigma_c')} \tag{5-16}$$

Effect of Sample Disturbance

The curve shown in Figure 5.12 is that which might be obtained in the laboratory by consolidating a slurry of cohesive soil to some pressure at which it is rebounded and then reloaded. It also depicts the one-dimensional compression response of a relatively recent uncemented field deposit. Unfortunately, the void-ratio–pressure relationship determined in the laboratory for a specimen extracted from a field deposit is not the same as would be experienced by that element of soil in the field in response to applied loads. The difference in the laboratory and field response is due to disturbance of the soil structure during sampling and testing, as well as other features of the testing procedure. Because we wish to know the *in situ* void-ratio–effective-stress relationship to estimate consolidation settlements in the field, we must give attention to methods for interpreting laboratory test data in such a way as to permit reconstruction of the field parameters. This problem is analogous to the one we faced in determining elastic parameters for estimation of elastic distortion settlements.

In the field, an element of soil is subjected to a vertical effective stress $\sigma_v' = \sigma_0'$ and a horizontal effective stress $\sigma_h' = K_0 \sigma_v'$, in which K_0 is the at-rest earth pressure coefficient, as defined in Equation 3–29. In general, K_0 is not unity: less than one for normally and lightly over-consolidated soils and greater than one for heavily over-consolidated soil. When a soil sample is removed from the ground, the external confining pressure is removed. The tendency of the saturated specimen to swell is resisted by the development of negative pore water pressure due to *capillary tension.**
If air does not come out of solution, the sample volume will not change, and the effective confining pressure is equal to the magnitude of the pore water pressure,

$$\sigma_v' = \sigma_h' = -u \tag{5-17}$$

Thus the ratio σ_h'/σ_v' is changed by an amount which depends upon the value of K_0. The resulting strains cause disturbance of the soil structure, that is, the soil is partially remolded. More detailed discussion of this effect is given by Skempton and Sowa (1963), Ladd and Lambe (1963), and Ladd (1964).

The influence of sampling, and other disturbance effects on the compressibility of the soil is illustrated in Figure 5.13. The loading history for a specimen in a normally consolidated deposit is shown in Figure 5.13a. The field virgin compression curve is shown as a solid line down to the point which represents the *in situ* conditions, for which $\sigma_0' = \sigma_p'$. An additional increment of load on this deposit will produce a void ratio change corresponding to the dashed continuation of the field virgin compression curve. Due to disturbance however, the effective consolidation pressure for a specimen brought into the laboratory is reduced as shown in the figure, even though the void ratio remains constant. When the specimen is reloaded in the laboratory, a void-ratio decrease occurs due to the disturbance effect, and the solid laboratory curve shown in the figure results.

In the case of an over-consolidated clay (Figure 5.13b), the *in situ* stress history is represented by the solid field virgin compression curve to the point at which the maximum past pressure, σ_p', was reached, after which the load was reduced to the existing overburden pressure σ_0'. The solid rebound curve shows the *in situ* void-ratio–log pressure relation during the stress release. Subsequent reloading in the field would produce the dashed field recompression curve which, when the preconsolidation pressure was exceeded, would rejoin the field virgin compression curve. Again the effect of disturbance is to reduce the effective consolidation pressure at constant void ratio leading to the laboratory curve shown as a solid line in the figure.

To reconstruct the field compressibility relations from those in the laboratory, it is necessary to determine three parameters: the preconsolidation pressure, σ_p'; the compression index, C_c; and the recompression index, C_e.

---

* Capillarity is discussed further in Chapter 6.

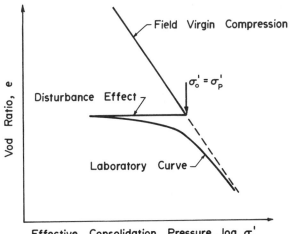

a) - Normally Consolidated Clay $(\sigma_p' = \sigma_o')$

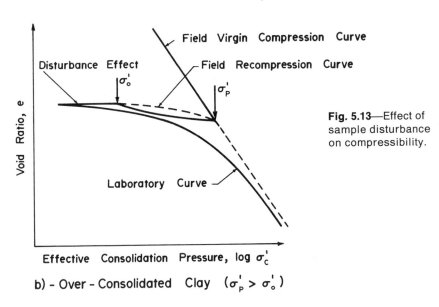

Fig. 5.13—Effect of sample disturbance on compressibility.

b) - Over - Consolidated Clay $(\sigma_p' > \sigma_o')$

Determination of the Preconsolidation Pressure

The preconsolidation pressure is the most important parameter to reconstruct from the void-ratio–log pressure relation. This is the case because of the important difference in compressibility, and accompanying settlements under load, between a normally consolidated and over-consolidated clay. In order to determine whether, and to what degree, a soil is over-consolidated, we must find the preconsolidation pressure and compare it to the existing overburden stress. Based on the results of many laboratory experiments, Casagrande (1936) suggested an empirical procedure which has become accepted in practice. This graphical procedure, illustrated in Figure 5.14, is described below:

1. Plot the void-ratio–$\log_{10}$ effective consolidation pressure relation obtained from a one-dimensional consolidation test.
2. At the point of maximum curvature on the e–$\log \sigma'_c$ curve (point A, Figure 5.14), construct a horizontal line.
3. At this same point draw a tangent to the e–$\log \sigma'_c$ curve.
4. Construct the bisector of the angle α between the horizontal and tangent to the curve at point A.
5. Extend upward the straight line portion of the virgin compression curve, as shown by the dashed line in the figure.

The intersection of the extended virgin compression curve and the bisector of the angle α defines the magnitude of the preconsolidation pressure σ'_p.

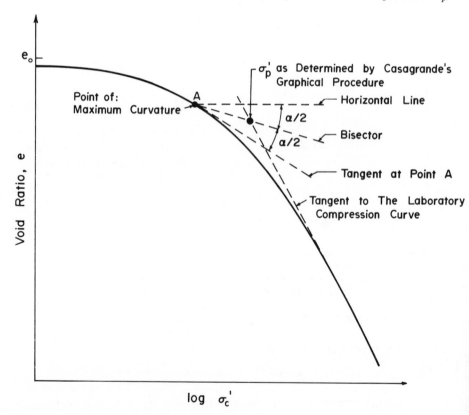

Fig. 5.14—Casagrande graphical procedure for the determination of σ'_p.

The preconsolidation pressure defined by the Casagrande empirical procedure is considered to represent the most probable value. It is often useful, however, to estimate the possible error involved in this procedure by establishing the range in the determined preconsolidation pressure. This range may be established by determining the maximum possible and minimum probable values of σ_p', as illustrated in Figure 5.15. The maximum possible preconsolidation pressure is the magnitude of the pressure required for the laboratory void-ratio–log pressure curve to become a straight line delineating the virgin compression curve. This pressure is greater than the actual preconsolidation pressure by an amount that depends upon the degree of disturbance as suggested by Figures 5.13 and 5.15. If a clay were normally consolidated, and no disturbance effects existed, then the maximum possible preconsolidation pressure would coincide with the actual preconsolidation pressure (Figure 5.13a). For an over-consolidated soil (Figure 5.13b), the maximum possible σ_p' determined by this method is greater than the actual value, even for the best possible undisturbed sample.

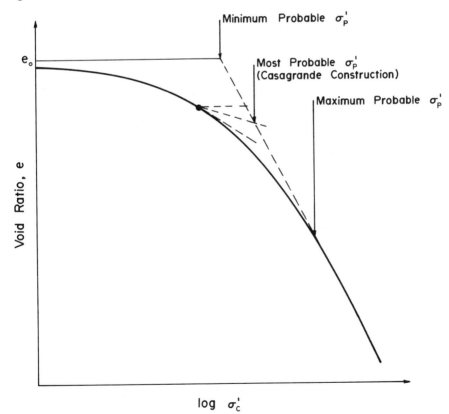

Fig. 5.15—Range in preconsolidation pressure determined from laboratory consolidation test.

As disturbance increases, the difference between the maximum and most probable values increases.

The minimum probable preconsolidation pressure is estimated by extrapolating from the *laboratory* virgin compression curve until it intersects the initial void ratio, e_0. Referring again to Figures 5.13 and 5.15, we see that the sample disturbance leads to a lower value of C_c, which produces a minimum probable σ'_p below the actual value. In subsequent discussion, we shall refer to the value determined by the Casagrande method as the preconsolidation pressure, keeping in mind the fact that this is only the most probable value within a range of possible values.

Prediction of Field Settlement—Normally Consolidated Clays

If the magnitude of σ'_p determined from the laboratory test data is equal to the *in situ* overburden pressure for the particular sample tested, the clay is *normally consolidated*. The field virgin compression curve, which is a straight line on a semilogarithmic plot is then reconstructed by a method developed by Schmertmann (1955). Referring to Figure 5.16, the steps required to reconstruct the virgin curve are these:

1. Point *B* is plotted. This point represents the *in situ* effective stress and void ratio at the sample location. For the normally consolidated case, this stress is the preconsolidation pressure as well as the *in situ* overburden pressure.
2. Point *C* is determined as that point at which the laboratory virgin compression curve intersects a void ratio equal to $0.42e_0$. On the basis of many laboratory tests, Schmertmann (1955) found that the laboratory compression curve for varying degrees of disturbance intersected the field virgin compression curve at approximately this point.
3. Line *BC* is then constructed. This line is the estimated field virgin compression curve.

In calculating settlements of normally consolidated soils, the slope of the *field* virgin compression curve, C_c, is used rather than the slope of the *laboratory* virgin compression curve.

In order to determine settlement of a deposit in the field, Equation 5-13 is rewritten in terms of a finite increment in effective vertical stress $\Delta\sigma'_c$ and the corresponding void-ratio change Δe:

$$-\Delta e = C_c \log\left(\frac{\sigma'_0 + \Delta\sigma'}{\sigma'_0}\right) \qquad (5\text{--}18)$$

in which σ'_0 is the *in situ* overburden pressure.* If it is assumed that, over a

---

* Alternatively the Δe corresponding to a given pressure increment can be obtained directly, that is, graphically, from the reconstructed field virgin compression curve.

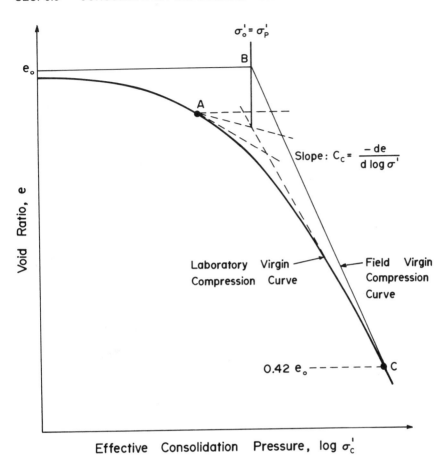

$\sigma_o' = \sigma_P'$

B

e_o

A

Slope: $C_c = \dfrac{-de}{d\log\sigma'}$

Void Ratio, e

Laboratory   Virgin
Compression   Curve

Field   Virgin
Compression
Curve

$0.42\ e_o$ ------- C

Effective   Consolidation   Pressure, $\log \sigma_c'$

Fig. 5.16—Construction of the virgin field compression curve—normally consolidated cohesive soil. (After Schmertmann, 1955.)

small depth dz, the stresses and strains may be assumed homogeneous, then Equation 5–18 can be substituted into Equation 5–11, and then into Equation 5–8, to yield

$$d\rho_c = \frac{C_c}{1 + e_0} \log\left(\frac{\sigma_0' + \Delta\sigma'}{\sigma_0'}\right) dz \qquad (5\text{–}19)$$

in which $d\rho_c$ is the small increment in settlement due to compression of the small thickness dz. The total settlement of the stratum, ρ_c, is the integral of Equation 5–19 taken over the thickness of the compressible stratum:

$$\rho_c = \int_{z=0}^{H} \frac{C_c}{1 + e_0} \log\left(\frac{\sigma_0' + \Delta\sigma}{\sigma_0'}\right) dz \qquad (5\text{–}20)$$

The integration in Equation 5–20 is frequently awkward because the applied pressure increment is usually determined from an influence diagram at discrete points rather than as a continuous function of depth. Therefore it is generally more convenient to consider the compressible stratum to consist of a number of layers within each of which the initial and applied pressure increment are assumed constant. Equation 5–20 can then be written as a sum of finite settlements,

$$\rho_c = \sum_{i=1}^{n} \Delta \rho_{c_i} = \sum_{i=1}^{n} \frac{C_c}{1 + e_{0_i}} \log \left(\frac{\sigma'_0 + \Delta \sigma}{\sigma'_0} \right)_i H_i \qquad (5\text{–}21)$$

in which $\Delta \rho_{c_i}$ is the settlement in the ith layer, the summation is taken over all n layers, e_{0_i} is the initial void ratio of the ith layer, and H_i is the thickness of the ith layer. It is generally unnecessary to use increments of uniform thickness when subdividing a compressible stratum into layers. With increasing depth, the error introduced by assuming an average σ'_0 and $\Delta \sigma'$ throughout the depth of a layer decreases because the rate of change of the applied pressure increment with respect to depth decreases as the depth increases. Thus it is usual to increase the thickness of the subdivision of a stratum as the depth increases.

If the clay is normally consolidated and the stratum is approximately homogeneous, the magnitude of C_c will not vary greatly throughout the stratum, and a single value may reasonably be used as representative for the entire stratum. The initial void ratio e_0 may also be taken as constant with depth without much loss in accuracy unless contrary data are available. If water content data have been obtained at several depths, the variation in initial void ratio with depth can be determined readily and may be used. An example follows.

Example 5.3 ————————————————————————

A warehouse 40 × 100 ft in plan, is to be constructed on the soil profile shown in Figure 5.17. The structure will be supported by a slab which is assumed to distribute the building loads uniformly to the soil. Determine the maximum differential settlement at the ground surface due to the consolidation of the clay layer.

Solution. For a uniformly distributed loading, the maximum differential settlement will occur between the middle and corners of the warehouse. The initial effective overburden pressure and the increase in the vertical effective stress within the clay stratum under the middle and corner of the structure are shown in Figure 5.18. In this case, the overburden pressure results only from the effective unit weight of the soil.

An e–$\log \sigma'$ curve obtained in the laboratory is shown in Figure 5.19. The soil was determined to be normally consolidated and the virgin curve drawn. The compression index, C_c, was found to be 0.45.

Dividing the clay into three layers, the average value of σ'_0 and $\Delta \sigma'$ for each layer is determined beneath the center from Figure 5.18a, and beneath the corner from Figure 5.18b. Using the value of C_c obtained from the laboratory sample, the settlement of each layer was determined. The numerical work is tabulated below.

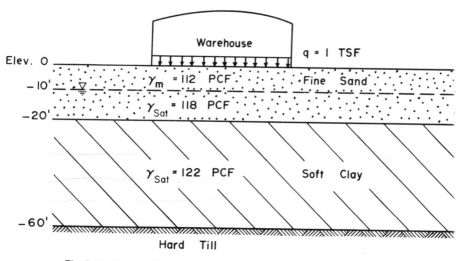

Fig. 5.17—Soil profile and loading for Example 5.3.

STRESS CALCULATIONS
1. Overburden pressure, σ'_0

z (ft)	$z\gamma' = \sigma'_0$ (psf)
0	0
-10	$10(112) = 1120$
-20	$1120 + 10(118 - 62.4) = 1676$
-60	$1676 + 40(122 - 62.4) = 4060$

2. Increase in effective stress due to warehouse, $\Delta\sigma'$, determined using influence diagram Figure A.5
   (a) Center

z	$L/z = m$	$B/z = n$	I	$4I$	$\Delta\sigma'$(psf)
-20	2.5	1.0	0.202	0.808	1606
-25	2.0	0.800	.181	.724	1428
-30	1.67	0.667	.161	.644	1288
-35	1.43	0.571	.143	.572	1144
-40	1.25	0.500	.127	.508	1016
-50	1.00	0.400	.101	.404	808
-60	0.833	0.333	0.078	0.312	624

   (b) Corner

z	$L/z = m$	$B/z = n$	I	$\Delta\sigma'$(psf)
-20	5.0	2.0	0.239	476
-25	4.0	1.6	.232	464
-30	3.33	1.33	.222	444
-35	2.86	1.14	.212	424
-40	2.50	1.00	.202	404
-50	2.00	0.80	.181	362
-60	1.67	0.667	0.162	324

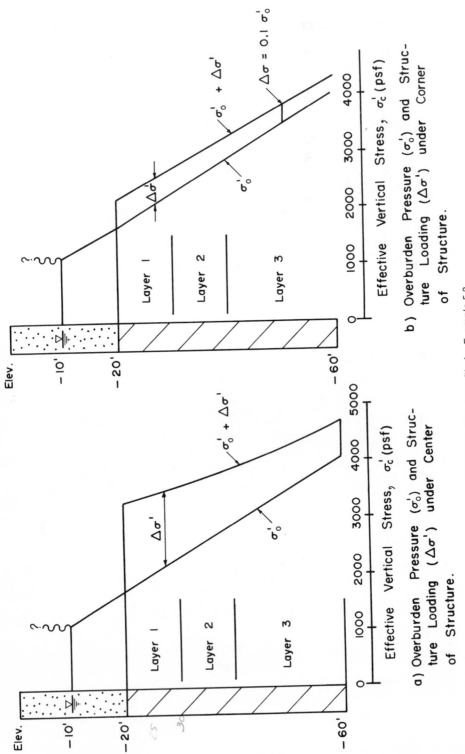

a) Overburden Pressure (σ_o') and Structure Loading ($\Delta\sigma'$) under Center of Structure.

b) Overburden Pressure (σ_o') and Structure Loading ($\Delta\sigma'$) under Corner of Structure.

Fig. 5.18—Stress profile for Example 5.3.

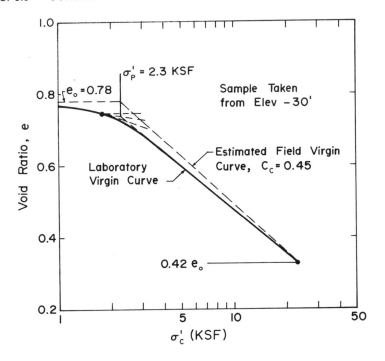

Fig. 5.19—Compressibility data for soft clay, Example 5.3.

SETTLEMENT CALCULATIONS

$$e_0 = 0.78; \frac{C_c}{1 + e_0} = 0.253$$

(a) Center of Structure

Layer	z' Middle (ft)	H_0 (ft)	σ'_0 (psf)	$\Delta\sigma$ (psf)	$\log \dfrac{\sigma'_0 + \Delta\sigma'}{\sigma'_0}$	$\Delta\rho_c$ (ft)
1	25	10	1970	1430	0.237	0.600
2	35	10	2550	1140	0.149	0.377
3	50	20	3480	810	0.0917	0.463

Total settlement at center, $\rho_{c_{center}} = 1.44$ ft

(b) Corner of Structure

Layer	z' Middle (ft)	H_0 (ft)	σ'_0 (psf)	$\Delta\sigma$ (psf)	$\log \dfrac{\sigma'_0 + \Delta\sigma'}{\sigma'_0}$	ΔH
1	25	10	1970	464	0.0906	0.229
2	35	10	2550	424	0.0660	0.167
3	45.5	11	3180	370	0.0477	0.133

Total settlement at corner, $\rho_{c_{corner}} = 0.53$ ft

Differential settlement $= \rho_{c_{center}} - \rho_{c_{corner}} = 1.44 - 0.53 = \underline{0.91 \text{ ft}}$

Note that at the corner the depth over which the clay was assumed to consolidate was limited to <u>31 ft.</u> It is normally assumed that if $\Delta\sigma' \leqslant 0.10\sigma'_0$ there is no need to calculate additional settlement. We shall see that this assumption is generally likely to be conservative. An exception occurs when the compressibility of a deep layer is unusually high, that is, $C_c \gg 0.5$.

Over-consolidated Clays

If the preconsolidation pressure, σ'_p, is found to be larger than the present overburden pressure σ'_0, the sample is over-consolidated. In this case, the steps required to reconstruct the laboratory data are (Figure 5.20):

1. Point D is plotted. This point represents the *in situ* effective stress and the void ratio at the sample location. The effective stress is

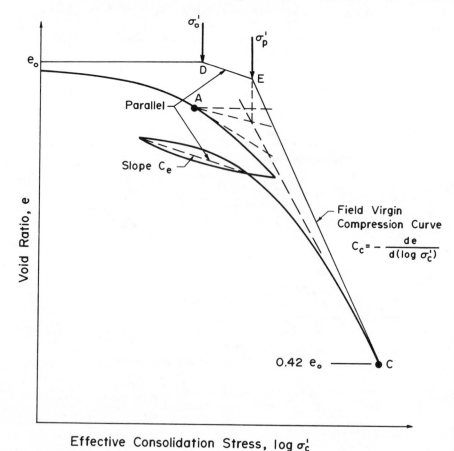

Fig. 5.20—Reconstruction of the field virgin compression curve for over-consolidated clay. (After Schmertmann, 1955.)

usually the computed overburden pressure. If the clay is not permitted to swell during storage, the field void ratio is the computed saturated initial void ratio, e_0, of the consolidation test sample.

2. The preconsolidation pressure σ'_p is determined, using the Casagrande procedure. Other methods are available for estimating σ'_p, (Schmertmann, 1955; Burmister, 1942, 1951). These may be of particular importance if the laboratory sample has been greatly disturbed and the curvature near σ'_p is small.

3. Draw a line through point D, parallel to the mean slope of the rebound curve. Extend this line until it intersects the value of σ'_p. The straight line DE will be assumed to be the field *recompression* curve with slope C_e. It is important to determine C_e from a rebound-recompression cycle, rather than the *initial* recompression curve of the laboratory specimen. This is so because sample disturbance leads to an apparent C_e for the initial recompression curve which may be as much as fifty times the *in situ* C_e. The value determined from a rebound-recompression cycle is likely to be closer to the field value. Because the average slopes of the rebound and recompression curves increase as the soil is further consolidated, it is advisable to determine C_e by conducting the rebound-recompression cycle at a stress slightly below σ'_p.

4. The point C is determined as the point at which the laboratory virgin compression curve intersects a void ratio equal to $0.42e_0$.

5. The line EC is then constructed. This line is the estimated field virgin compression curve.

The minimum probable preconsolidation pressure for an over-consolidated clay is determined by extending the laboratory virgin compression curve back until it intersects the estimated field rebound curve. This intersection is the minimum probable σ'_p.

Having reconstructed the *in situ* recompression and virgin compression curves, we are now able to calculate the settlement of over-consolidated cohesive strata. If the increase in effective stress $\Delta\sigma'$ is greater than $(\sigma'_p - \sigma'_0)$, the consolidating soil will undergo both recompression and virgin compression. The total void ratio change can be calculated as the sum of two components. The first, $\Delta e'$, is the recompression due to the change in stress from the present overburden pressure σ'_0, to the preconsolidation pressure σ'_p, Figure 5.21a:

$$\Delta e' = C_e \log \frac{\sigma'_0 + (\sigma'_p - \sigma'_0)}{\sigma'_0} = C_e \log \frac{\sigma'_p}{\sigma'_0} \qquad (5\text{--}22a)$$

The second component is the virgin compression, $\Delta e''$, due to the stress increment from σ'_p to $(\sigma'_0 + \Delta\sigma')$ (Figure 5.21a):

$$\Delta e'' = C_c \log \frac{\sigma'_0 + \Delta\sigma'}{\sigma'_p} \qquad (5\text{--}22b)$$

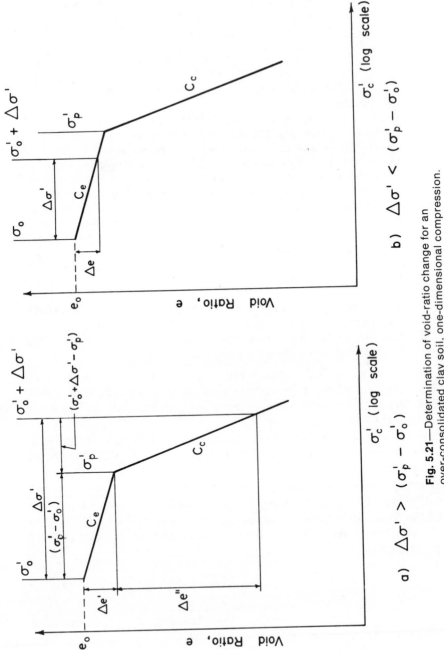

Fig. 5.21—Determination of void-ratio change for an over-consolidated clay soil, one-dimensional compression.

a) $\Delta\sigma' > (\sigma_p' - \sigma_o')$

b) $\Delta\sigma' < (\sigma_p' - \sigma_o')$

The total void-ratio change is thus

$$\Delta e = \Delta e' + \Delta e'' = C_e \log \frac{\sigma'_p}{\sigma'_0} + C_c \log \frac{\sigma'_0 + \Delta\sigma'}{\sigma'_p} \qquad (5\text{-}23)$$

If the preconsolidation pressure is not exceeded, that is, $\Delta\sigma' < (\sigma'_p - \sigma'_0)$, the soil will undergo recompression only as shown in Figure 5.21b. The change in void ratio is

$$\Delta e = C_e \log \frac{\sigma'_0 + \Delta\sigma'}{\sigma'_0} \qquad (5\text{-}24)$$

The total consolidation settlement can be calculated by integrating Equation 5–23 over that portion of the stratum for which $\Delta\sigma' > (\sigma'_p - \sigma'_0)$ and Equation 5–24 over the depth for which $\Delta\sigma' < (\sigma'_p - \sigma'_0)$. It is generally more useful, however, to divide the stratum into several layers, determine the average consolidation settlement for each layer, and sum them to determine the settlement of the stratum. Thus, for $\Delta\sigma'_i > (\sigma'_p - \sigma'_0)_i$, the settlement of the n layers so affected is

$$\rho'_c = \sum_{i=1}^{n} \frac{H_i}{1 + e_0} \left[C_e \log \frac{\sigma'_p}{\sigma'_0} + C_c \log \frac{\sigma'_0 + \Delta\sigma'}{\sigma'_p} \right]_i \qquad (5\text{-}25)$$

in which the subscript i implies those values of the parameters corresponding to the ith layer. For $\Delta\sigma'_i < (\sigma'_p - \sigma'_0)_i$, the settlement of the m layers involved is

$$\rho''_c = \sum_{i=1}^{m} \frac{H_0}{1 + e_0} C_e \log \frac{\sigma'_0 + \Delta\sigma'}{\sigma'_0} \qquad (5\text{-}26)$$

The total consolidation settlement is then

$$\rho_c = \rho'_c + \rho''_c \qquad (5\text{-}27)$$

Example 5.4

We shall reconsider the problem of Example 5.3 for the case in which the clay stratum is over-consolidated. Laboratory one-dimensional consolidation tests reveal that the magnitude of σ'_p increases approximately uniformly with depth and is about 1550 psf greater than σ'_0. A review of the local geologic history indicates that wind blown sand dunes had passed over the area at one time, accounting for the uniform increase in σ'_p. Values of C_c, C_e, and e_0 determined from laboratory tests on three samples obtained from different depths are:

Sample	Depth (ft)	C_c	C_e	e_0
1	25	0.62	0.060	0.81
2	35	0.60	0.072	0.76
3	45	0.66	0.074	0.68

Determine the maximum differential settlements under the warehouse described in Example 5.3 due to consolidation of the clay stratum.

Solution. We may again assume that the maximum differential settlement will occur between the middle and corners of the warehouse. The clay was divided into three layers and the settlement calculated for each layer. The numerical work is given in Tables 5.4 and 5.5. The overburden pressure, preconsolidation pressure and increase in vertical stress within the clay layer due to the structure are shown along lines beneath

TABLE 5.4
Settlement Calculation for Example 5.4 Below Center of Structure

Layer No. i	Depth at Middle (ft)	H_i (ft)	e_{0_i}	C_e	$\dfrac{C_e H_i}{1 + e_{0_i}}$	σ_0' (ksf)	$\sigma_0' + \Delta\sigma'$ (ksf)	$\log\dfrac{\sigma_0' + \Delta\sigma'}{\sigma_0'}$	$\Delta\rho_{c_i}$ (ft)
1	25	10	0.81	0.060	0.33	1.97	3.40	0.238	0.07
2	35	10	0.76	0.072	0.41	2.55	3.69	0.161	0.06
3	50	20	0.68	0.078	0.93	3.48	4.29	0.091	0.08

TABLE 5.5
Settlement Calculation for Example 5.4 Below Corner of Structure

Layer No. i	Depth at Middle (ft)	H_i (ft)	e_{0_i}	C_e	$\dfrac{C_e H_i}{1 + e_{0_i}}$	σ_0' (ksf)	$\sigma_0' + \Delta\sigma'$ (ksf)	$\log\dfrac{\sigma_0' + \Delta\sigma'}{\sigma_0'}$	$\Delta\rho_{c_i}$ (ft)
1	25	10	0.81	0.078	0.33	1.97	2.43	0.0906	0.03
2	35	10	0.76	0.072	0.41	2.55	2.97	0.0660	0.02
3	45.5	11	0.68	0.060	0.39	3.18	3.55	0.0477	0.01

the center and corner of the structure in Figure 5.22. It is evident from this figure that the clay stratum will experience only recompression, that is, at all depths $\Delta\sigma' < (\sigma_p' - \sigma_0')$. From Table 5.4, the settlement at the center of the building is

$$\rho_{c\,\text{center}} = \sum\rho_{c_i} = 0.230\,\text{ft}$$

and, from Table 5.5, at the corner,

$$\rho_{c\,\text{corner}} = \sum\rho_{c_i} = 0.085\,\text{ft}$$

The differential settlement $= \rho_{c\,\text{center}} - \rho_{c\,\text{corner}} = 0.230 - 0.085 = \underline{0.145}\,\text{ft}$.

Note that the over-consolidation of the clay was sufficient to reduce the settlement by more than 80 per cent when compared to that for Example 5.3.

We now consider an example which illustrates the importance of the magnitude of the preconsolidation pressure in the prediction of settlements.

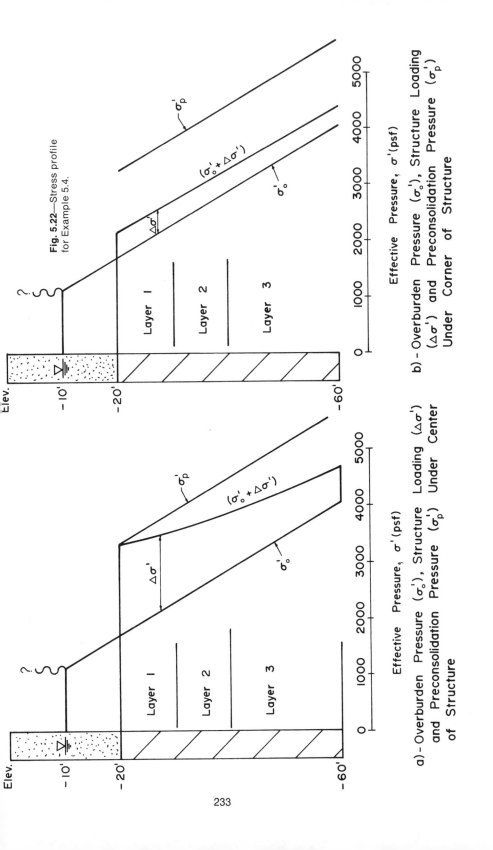

Fig. 5.22—Stress profile for Example 5.4.

a) - Overburden Pressure (σ_o'), Structure Loading $(\Delta\sigma')$ and Preconsolidation Pressure (σ_p') Under Center of Structure

b) - Overburden Pressure (σ_o'), Structure Loading $(\Delta\sigma')$ and Preconsolidation Pressure (σ_p') Under Corner of Structure

Example 5.5 ───

A structure is to be constructed at the surface on the soil profile shown in Figure 5.23. *Spread footings* 10 ft square are to support column loads of 200 kips each. Determine the settlement of the footings due to consolidation of the clay stratum, assuming they are far enough apart so that one does not influence the others. (This is not generally the case, but simplifies the example.)

Solution. Data obtained from the field exploration and laboratory testing program indicate the clay stratum is over-consolidated. Pertinent results from one-dimensional consolidation tests on samples obtained at different depths are:

Depth (ft)	e_0	σ_p' (ksf)	C_c	C_e
8	1.61	1.00	0.62	0.071
12	1.58	1.00	.63	.068
18	1.57	1.20	.59	.073
24	1.52	1.45	.56	.076
32	1.47	1.80	0.67	0.072

Preconsolidation pressures are shown as a function of depth in Figure 5.24. The large point is the most probable value with the ends of the horizontal line through each point indicating the minimum and maximum probable values. This representation is helpful in constructing a curve through the values.

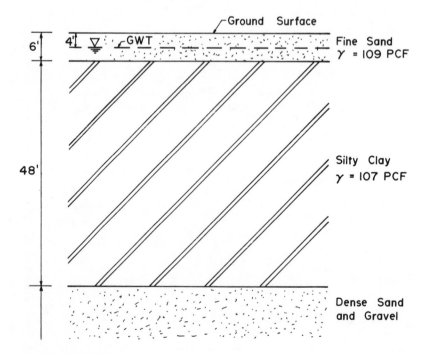

Fig. 5.23—Subsurface profile for Example 5.5.

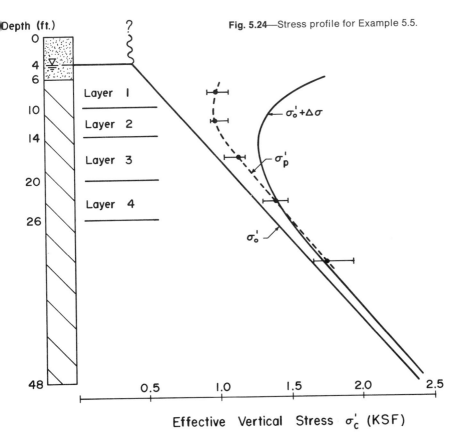

Fig. 5.24—Stress profile for Example 5.5.

Effective Vertical Stress σ_c' (KSF)

The effective overburden pressure and the increase in effective stress due to the footing loads are tabulated below and shown as a function of depth in Figure 5.24:

Depth (ft)	σ_0' (ksf)	$\Delta\sigma'$ (ksf)	$\Delta\sigma' + \sigma_0'$ (ksf)
6	0.529	1.223	1.75
8	0.618	0.900	1.52
10	0.707	0.668	1.38
14	0.885	0.404	1.29
18	1.06	0.264	1.32
30	1.60	0.083	1.68
48	2.40	0.048	2.45

The clay stratum has been divided into four layers between depths of 6 and 26 ft. Below 26 ft the increase in effective stress is less than 10 per cent of the effective overburden pressure. Because the material below this depth will undergo recompression only, we can assume the consolidation settlement below this depth to be negligible.

Making use of Equation 5-25, the consolidation settlement may now be determined. Recompression settlements are:

Layer Number, i	H_i (ft)	Depth at Middle (ft)	e_{0_i}	C_e	$\dfrac{C_e H_i}{1 + e_{0_i}}$	σ'_0 (ksf)	σ'_p (ksf)	$\log \dfrac{\sigma'_p}{\sigma'_0}$	$\Delta\rho_{c_i}$
1	4	8	1.61	0.071	0.109	0.618	1.00	0.218	0.024
2	4	12	1.58	0.068	0.105	0.797	1.00	0.098	0.010
3	6	17	1.57	0.073	0.171	1.06	1.17	0.041	0.007
4	6	23	1.52	0.076	0.181	1.33	1.41	0.025	0.005

$$\sum \text{recompression settlements} = 0.046 \text{ ft}$$

The magnitudes of σ'_p have been estimated from the curve in Figure 5.24 connecting the values obtained in the laboratory.

Layer Number, i	H_i (ft)	Depth at Middle (ft)	e_{0_i}	C_c	$\dfrac{C_c H_i}{1 + e_{0_i}}$	σ'_p (ksf)	$\sigma'_0 + \Delta\sigma'$ (ksf)	$\log\left(\dfrac{\sigma'_0 + \Delta\sigma'}{\sigma'_p}\right)$	$\Delta\rho_{c_i}$ (ft)
1	4	8	1.61	0.62	0.947	1.00	1.52	0.182	0.172
2	4	12	1.58	0.63	0.977	1.00	1.32	0.120	0.117
3	6	17	1.57	0.59	1.342	1.17	1.30	0.046	0.062
4	6	23	1.52	0.56	1.332	1.41	1.42	0.004	0.005

$$\sum \text{virgin compression settlement} = 0.356 \text{ ft}$$

Total consolidation settlement $= 0.046 + 0.356 = 0.40 \text{ ft} = \underline{4.8 \text{ in.}}$

This magnitude of settlement would lead to intolerable distortion of many structures and would require consideration of alternative foundation types. Others might withstand it if the settlement occurred during many years.

On the basis of the previous two examples it is apparent that in an overconsolidated clay stratum the magnitude and distribution of the preconsolidation pressure, σ'_p, is the most influential factor in determining consolidation settlements. In fact, calculated settlements will be within tolerable limits for most structures only if the increment of stress due to foundation loads, $\Delta\sigma$, is less than $\sigma'_p - \sigma'_0$. If this is the case, the soil will undergo only recompression. The importance of this will become more apparent when we consider allowable settlements of structures in Chapter 11. If there is no virgin compression, the consolidation settlement is then proportional to the recompression index C_e. It is to be emphasized again that this value should be taken only from a rebound section of the consolidation test, never from the initial slope, and that this rebound should be initiated at a pressure near σ'_p, to avoid overestimating the consolidation settlements.

Preconsolidation Profiles

The over-consolidation of a cohesive soil due to changes in stress may be caused by a variety of factors:

1. Rise in the groundwater table and a concomitant reduction in the effective overburden.
2. Drying of a surface exposed to the air (desiccation) with a resultant decrease in void ratio by capillary forces.
3. Erosion of pre-exisiting surface materials.
4. Prior loads which have been removed. Examples are glaciers, wind-blown sand dunes, old structures.
5. Seepage forces.
6. Tectonic forces due to movements of the earth's crust.

Useful inferences concerning the degree and causes of over-consolidation of a deposit can be drawn from consideration of a plot of preconsolidation pressure as a function of depth. Various such *preconsolidation profiles* are illustrated in Figure 5.25. The case of a normally consolidated stratum for which $\sigma'_p = \sigma'_0$ is illustrated in Figure 5.25b.

Over-consolidation resulting from desiccation is indicated in Figure 5.25c. The large negative capillary stresses induce an equally large effective stress in the region of drying. Because the groundwater level is lowered during this process, the underlying material is lightly over-consolidated.

Because erosion occurs over a relatively wide area, the release in stress due to removal of the overburden is approximately uniform at all depths, resulting in a uniform $(\sigma'_p - \sigma'_0)$, indicated in Figure 5.25d. A similar profile may also result from a general rise in the groundwater table accompanied by a uniform reduction in effective stresses.

We must keep in mind that in judging the potential of each of the aforementioned factors to over-consolidate a given stratum of soil we must also consider drainage conditions for the compressible material and whether sufficient time was available for the transient process of consolidation to have been completed. We shall discuss the time rate of settlement in Chapter 7.

It is clear that a knowledge of the geologic and engineering history of an area may be invaluable in deciding whether a soil *in situ* has been over-consolidated. Such a decision is often critical to the determination of foundation type and design.

Apparent Preconsolidation

Factors other than the magnitude of the *in situ* vertical stress may cause a presumably normally consolidated clay soil to exhibit an *apparent preconsolidation pressure* in excess of the actual applied stress. All such factors operate to increase the interparticle resistance to relative motion of individual particles. For example, a change in pore fluid to one in which the electrochemical attractive forces between particles are increased, leads to an increase

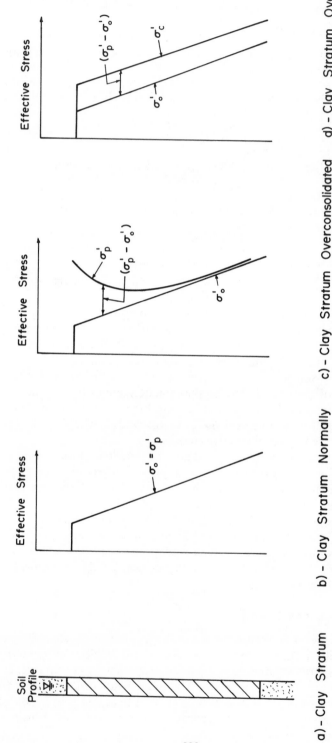

a) - Clay Stratum

b) - Clay Stratum Normally Consolidated, $\sigma_o' = \sigma_p'$

c) - Clay Stratum Overconsolidated as a Result of Drying

d) - Clay Stratum Over-Consolidated as a Result of Surface Erosion

Fig. 5.25—Preconsolidation profiles in a clay stratum.

in the observed preconsolidation pressure. An illustration of this phenomenon is shown in Figure 5.26, in which compressibility curves are shown for two undisturbed specimens of Mexico City clay sampled from the same elevation. In one specimen, the natural aqueous pore fluid was replaced with carbon tetrachloride, a nonpolar material which produces a net increase in the electrochemical attractive forces between particles. The increased resistance of the skeleton to deformation, as manifested by the apparent additional preconsolidation pressure, is evident. *In situ* changes in pore fluid composition can be produced by weathering or leaching. We shall examine electrochemical forces and their influence on clay behavior in Chapter 9.

Apparent preconsolidation pressures in excess of the actual maximum past pressure are also produced by cementation at particle contacts due primarily to precipitation of cementitious materials from pore fluid. It appears however, that the most common reason for apparent preconsolidation pressures exceeding the existing overburden stress in *normally consolidated* soils is the gradual reduction in void ratio, accompanied by an increase in interparticle attractive forces, which occurs at essentially constant effective stress over long periods of time. Although this time-dependent volume change at constant effective stress, called *secondary compression,* is not well understood, its effects have been documented. Referring to the magnitude of the apparent preconsolidation pressure as the *quasi-preconsolidation pressure* σ'_{qp}, the results of many determinations of the quasi-preconsolidation

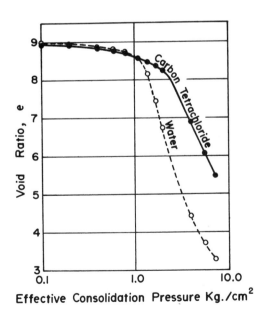

Fig. 5.26—Effect of pore fluid on compressibility of Mexico City clay. (After Leonards and Girault, 1961.)

pressure for natural undisturbed, laboratory sedimented, and remolded soils are shown in Figure 5.27. Note that from this figure the ratio $\sigma'_{qp}/\sigma'_p \simeq 1.45$.

The physical basis for the development of additional interparticle bonding forces which produce the quasi-preconsolidation pressure is not clear at present. A general discussion of such forces is given in Chapter 9.

In estimating consolidation settlements, we need not differentiate between the real preconsolidation pressure and the quasi-preconsolidation pressure if σ'_{qp} can be detected in the laboratory. Although it may be masked to some extent by disturbance effects, even a small magnitude of quasi-preconsolidation can be of considerable significance in reducing the consolidation settlement due to compression of normally consolidated clays. The reduction in settlements is of particular importance for deep deposits of compressible soil where the ratio of the applied pressure increment to the overburden pressure is likely to be small throughout most of the deposit. As a result of

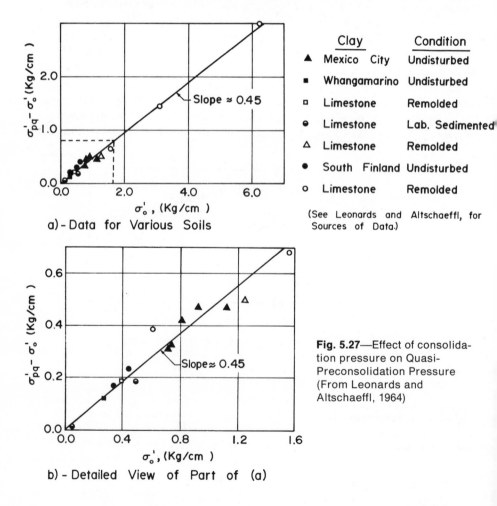

Clay	Condition
▲ Mexico City	Undisturbed
■ Whangamarino	Undisturbed
□ Limestone	Remolded
◉ Limestone	Lab. Sedimented
△ Limestone	Remolded
● South Finland	Undisturbed
○ Limestone	Remolded

(See Leonards and Altschaeffl, for Sources of Data.)

a) - Data for Various Soils

b) - Detailed View of Part of (a)

Fig. 5.27—Effect of consolidation pressure on Quasi-Preconsolidation Pressure (From Leonards and Altschaeffl, 1964)

the quasi-preconsolidation pressure, compression settlements within major portions of such a deposit will be reduced markedly.

Effect of Pressure-Increment Ratio in the Laboratory Test

The *pressure-increment ratio* is defined as the ratio of the applied pressure increment, $\Delta\sigma'$, to the present consolidation pressure, σ'_c. The magnitude of the pressure-increment ratio used in consolidation testing is influential in a number of important respects. Current attempts at standardizing laboratory practice for the one-dimensional consolidation test (ASTM, 1965) recommend a pressure-increment ratio equal to one, that is, the applied pressure is doubled for each load increment. The use of such a large pressure-increment ratio may make it difficult to delineate σ'_{qp} in the laboratory. Unfortunately, as discussed in Chapter 11, the magnitude of the pressure-increment ratio also influences the shape of the time-settlement curve. Thus the use of a small pressure-increment ratio which may be desirable for determining the quasi-preconsolidation pressure, may also create difficulty in establishing the time at which the consolidation settlement is essentially complete and secondary compression settlements begin.

Consequently, the appropriate loading sequence in a consolidation test depends upon the specific cohesive soil being tested; one cannot decide *a priori* what magnitude of pressure-increment ratio to use. It is generally a good idea to test one specimen with a relatively high pressure-increment ratio, say $\Delta\sigma'/\sigma'_c = 1$, and then test a second specimen using reduced pressure-increment ratios in the vicinity of the quasi-preconsolidation pressure.

Under-consolidated Soils

The situation occasionally arises in which a stratum of cohesive soil is found to exhibit a preconsolidation pressure less than the calculated existing overburden pressure. This circumstance occurs when a deposit has not reached equilibrium under the applied overburden stresses, as in areas of recent landfill, or where the groundwater table has recently been lowered.

When an additional load, due for instance to a foundation, is applied to the soil, settlements will occur not only in response to the additional load but to the existing load under which equilibrium has not yet been attained. This is illustrated in Figure 5.28, which shows the laboratory void-ratio–log pressure curve for an under-consolidated clay. Note that the preconsolidation pressure determined on the laboratory specimen is less than σ'_0. As indicated in the diagram, the consolidation settlement is the sum of two components: the settlement which would occur if no additional load were placed on the soil; and the settlement in response to the additional stress increment,

$$\rho_c = \sum_{i=1}^{n} C_{c_i} \frac{H_i}{1 + e_0} \log\left(\frac{\sigma'_c + (\sigma'_0 - \sigma'_c) + \Delta\sigma'}{\sigma'_c}\right)_i \tag{5-28}$$

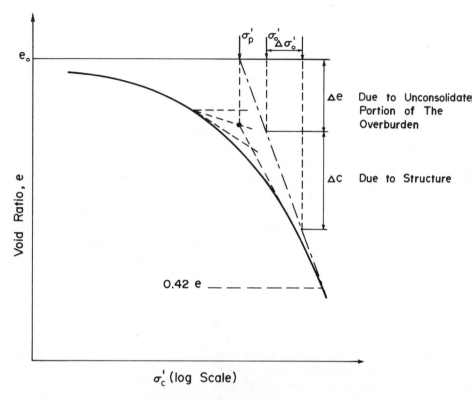

Fig. 5.28—Compressibility of under-consolidated cohesive soil.

or

$$\rho_c = \sum_{i=1}^{n} C_{c_i} \frac{H_i}{1 + e_0} \log \left(\frac{\sigma'_4 + \Delta\sigma'}{\sigma'_c} \right)_i \tag{5-29}$$

in which σ'_c is the current *effective* vertical stress in the soil.

Although under-consolidated soils occur infrequently, it is important that they be recognized as such when they are encountered. Failure to do so may lead to observed settlements far in excess of those calculated on the basis of a simple normally consolidated deposit.

Sampling and Testing Artifacts

Two artifacts may have an important influence on the relationship between the compressibility curve determined in the laboratory and that *in situ*. The most important of these is *sample disturbance*, illustrated in Figure 5.29. The difference between the observed compressibility curve and the *in situ* relation increases as the amount of disturbance increases. We have already considered the effect of sampling due both to changes in the

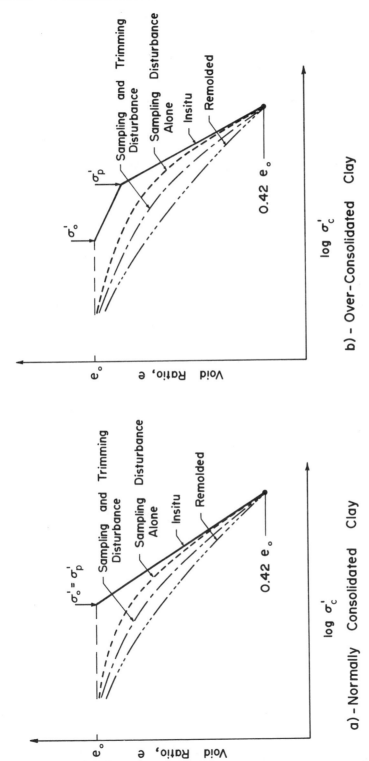

Fig. 5.29—Effect of sampling and specimen preparation on the laboratory e–log σ'_c curve.

a) - Normally Consolidated Clay

b) - Over-Consolidated Clay

effective stress state as well as a certain amount of unavoidable remolding occurring due to the sampling process. Additional disturbance is introduced, however, when the sample is removed from the sampling tube, trimmed to fit into the consolidometer ring and handled during this procedure.

Cutting and trimming remolds the surface of the sample. The degree of disturbance diminishes proceeding from the outside of the sample toward the interior. Thus the minimum trimming disturbance occurs when the ratio of the surface area to volume is a minimum for the specimen. For a cylindrical specimen, this occurs as the height to diameter ratio, H/D, approaches one.

The influence of increasing degrees of disturbance induced by the various processes to which the specimen is exposed is illustrated in Figure 5.29a for normally consolidated clays, and 5.29b for over-consolidated clays. As we saw before (Figure 5.13), the effect of such disturbance is to shift the compressibility curve and mask the preconsolidation pressure. In the case of the completely remolded sample, the "memory" of the stress history has been completely erased. Although disturbance effects cannot completely be avoided, they can be minimized by suitable care in sampling and testing.

Another important artifact is the friction force which develops between the wall of the consolidation ring and the specimen. The total load applied normal to the specimen is reduced by the magnitude of this friction force. The average value of σ'_c is then the applied σ' less the frictional force divided by the area of the specimen. Thus the effect of side friction is to cause the void-ratio–log pressure relation to appear to lie to the right of the curve which represents the actual relationship. This is illustrated in Figure 5.30

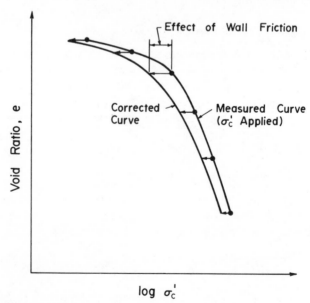

Fig. 5.30—Effect of wall friction on the observed e–log σ'_c curve.

where the difference between the two curves is the effect of sidewall friction. Clearly, the relative magnitude of the frictional effect is reduced as H/D is reduced. The effect may be further lessened by:

1. Allowing the top and bottom porous stone to move into the ring simultaneously (called *floating* the ring) rather than by maintaining the ring and bottom stone fixed and allowing only the top stone to displace.
2. Using antifriction materials to line the ring, (e.g., teflon), or coating the interior of the ring with lubricants such as molybdenum disulfide, or silicone grease.

Recent developments in low friction lining and lubricants have reduced, though not eliminated, the effects of sidewall friction.

Because the H/D required to minimize trimming and friction effects are different, current practice is to establish a compromise between these. At the present time, $H/D = 0.3$ is common.

5.4 APPLICATION TO BUILDING 301

We are now prepared to estimate the settlements of New York Port Authority warehouse Building No. 301, that resulted from the surcharge loadings. We recall that the structure and surcharged area are 255×550 ft subjected to a surcharge fill 12 ft in height which applied an increment in vertical stress of 1300 psf throughout the depth of the 24-ft-thick compressible organic silt stratum (Figure 4.1).

Immediate Surface Settlement

We shall assume that the immediate surface settlement results primarily from distortions within the organic silt layer which is considerably more flexible than the overlying cinder fill and underlying layers of sand and clay. If the silt extended from the ground surface to the top of the underlying sands, a depth of 31 ft, the parameters required to determine the shape factor C'_s from Table 5.2 are:

$$\frac{H}{B} = \frac{31}{255} = 0.121, \qquad \frac{L}{B} = \frac{550}{255} = 2.16$$

from which, for an intermediate interface condition, $C'_s = 0.11$. Estimating Young's modulus to be 80×10^3 psf with Poisson's ratio equal to 0.5, the elastic distortion under the center is, from Equation 5-2,

$$\rho_{d_{31}} = C'_s qB \left(\frac{1 - \mu^2}{E} \right) = \frac{(0.11)(1300)(255)(0.75)}{80 \times 10^3} = 0.34 \text{ ft} \quad (5\text{--}30)$$

To account for the real depth of organic silt, we shall calculate the settlement which would occur in a seven-foot-thick layer of the silt overlying a rigid base, and subtract that from the settlement determined above for the equivalent 31-ft-thick layer. For the 7-ft-thick layer and an intermediate interface condition,

$$\frac{H}{B} = 0.0264, \qquad C_s' = 0.024$$

from which

$$\rho_{d_7} = \frac{(0.024)(1300)(255)(0.75)}{80 \times 10^3} = 0.08 \text{ ft} \qquad (5\text{-}31)$$

The distortion settlement due to the organic silt stratum is then

$$\rho_{d_{\text{silt}}} = \rho_{d_{31}} - \rho_{d_{10}} = 0.26 \text{ ft} = 3.1 \text{ in.} \qquad (5\text{-}32)$$

Alternatively, because the vertical stress increment is nearly constant with depth, we could have arrived at approximately the same result by assuming that the settlement of the silt stratum was proportional to its thickness, that is,

$$\rho_{d_{24}} = \rho_{d_{31}}\left(\frac{24}{31}\right) = 0.26 \text{ ft} \qquad (5\text{-}33)$$

Of course, neither approach accounts for the potential reduction in the stresses and deformations with depth due to the relative rigidity of the upper layer of cinder fill (recall Figure 4.9).

Consolidation Settlement

The results of laboratory consolidation tests conducted on the organic silt provided by the Port of New York Authority (Kapp, 1968) are shown in Figures 5.31 through 5.34. The four specimens tested were obtained at different elevations from a test boring taken in the near vicinity of Building 301. Initial void ratios, compression and recompression indices, and the magnitude of the preconsolidation pressure in excess of the overburden pressure, as interpreted by the authors from these laboratory data, are given in Table 5.6. These results suggest that the material within this stratum is approximately homogeneous, and over-consolidated by approximately 370 psf.

To calculate consolidation settlement, the silt stratum is divided into two layers of 8- and 16-ft thickness. It is assumed that the values of e_0, C_c, C_e, and $(\sigma_p' - \sigma_0')$ are uniform within the silt. These values, interpreted from the test results in Table 5.6, are:

$$e_0 = 1.64$$
$$C_c = 0.69$$
$$C_e = 0.05$$
$$(\sigma_p' - \sigma_0') = 370 \text{ psf}$$

Table 5.6
Results of Laboratory Consolidation Tests on
Samples Taken from Below Building 301

Sample	Depth (ft)	e_0	C_c	C_e	$(\sigma_p' - \sigma_0')$ (psf)
$7U$	12	1.65	0.68	0.10	400
$9U$	17	1.64	.70	.04	370
$11U$	20	1.52	.61	.05	360
$13U$	24	1.875	0.90	0.10	370

PERTINENT STRESS DATA.

Layer Number	Depth at Mid-height (ft)	σ_0' (ksf)	$(\sigma_0' + \Delta\sigma')$ (ksf)	σ_p' (ksf)
1	11	0.65	1.95	1.02
2	23	1.09	2.39	1.46

CALCULATION OF RECOMPRESSION SETTLEMENT

Layer Number	H_i (ft)	$C_e H_i/(1 + e_0)$ (ft)	$\log(\sigma_p'/\sigma_0')$ (ft)	$\Delta\rho_i$ (ft)
1	8	0.15	0.196	0.029
2	16	0.30	0.127	0.038

Recompression settlement = <u>0.067 ft</u>

CALCULATION OF VIRGIN COMPRESSION SETTLEMENT

Layer Number	H_i (ft)	$C_c H_i/(1 + e_0)$ (ft)	$\log \dfrac{\sigma_0' + \Delta\sigma'}{\sigma_p'}$	$\Delta\rho_i$ (ft)
1	8	2.09	0.281	0.587
2	16	4.18	0.214	0.896

Virgin Compression settlement = <u>1.48 ft</u>

The ultimate consolidation settlement is thus,

$$\rho_c = 0.07 + 1.48 = 1.55 \text{ ft}$$

and the combined distortion and consolidation settlement is

$$\rho = \rho_d + \rho_c = 0.26 + 1.55 = 1.81 \text{ ft} = 22 \text{ in.}$$

The predicted settlement is greater than the 17 in. of settlement observed at the end of the surcharge period (pages 161–166). In order to appreciate the reason for this discrepancy, we must recognize that the consolidation settlement calculated from the compressibility curve is an equilibrium value reached only at the completion of a transient consolidation process. This transient outflow of water from the voids generally occurs so slowly that it is

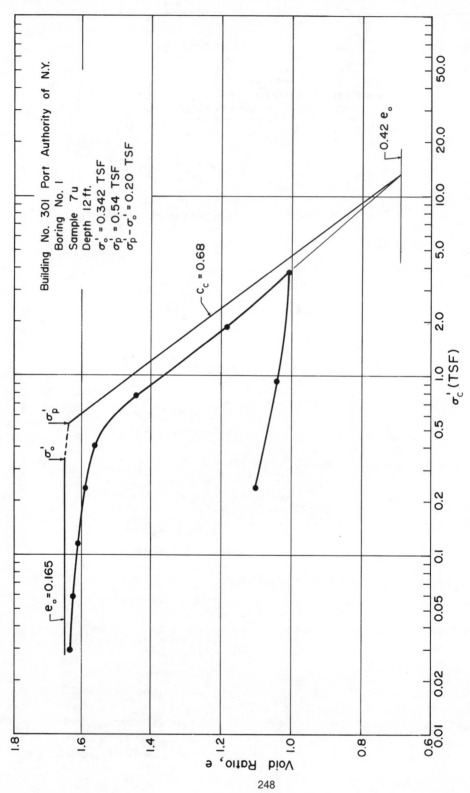

Fig. 5.31—Results of consolidation tests on samples taken rrom the vicinity of Building No. 301. (From Kapp, 1962.)

Building No. 301 Port Authority of N.Y.
Boring No. I
Sample 7 u
Depth 12 ft.
$\sigma_o' = 0.342$ TSF
$\sigma_p' = 0.54$ TSF
$\sigma_p' - \sigma_o' = 0.20$ TSF

$C_c = 0.68$

$0.42\ e_o$

σ_p'

σ_o'

$e_o = 0.165$

σ_c' (TSF)

Void Ratio, e

248

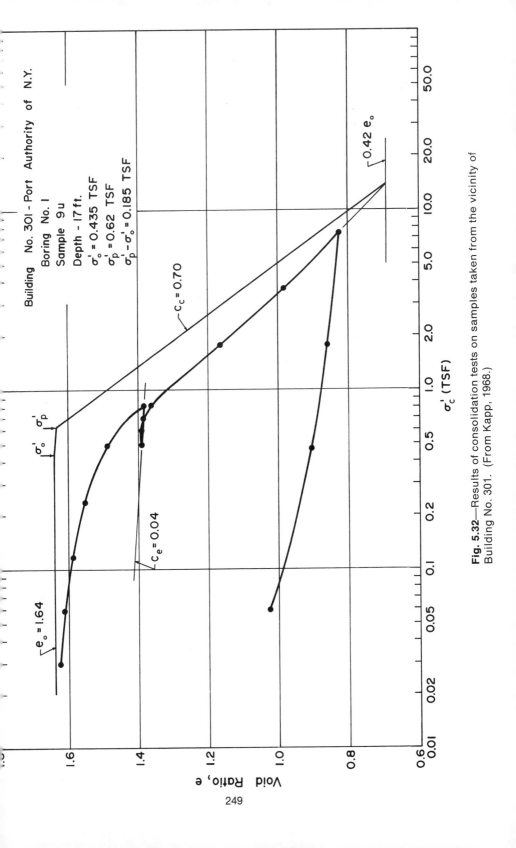

Fig. 5.32—Results of consolidation tests on samples taken from the vicinity of Building No. 301. (From Kapp, 1968.)

Building No. 301 - Port Authority of N.Y.

Boring No. I
Sample 9u
Depth - 17 ft.
$\sigma_o' = 0.435$ TSF
$\sigma_p' = 0.62$ TSF
$\sigma_p' - \sigma_o' = 0.185$ TSF

$C_c = 0.70$

$C_e = 0.04$

σ_o' σ_p'

$e_o = 1.64$

$0.42\ e_o$

Void Ratio, e

σ_c' (TSF)

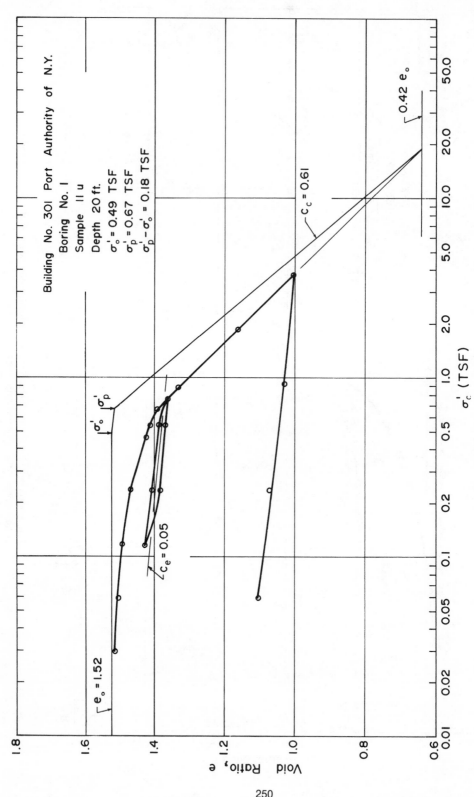

Fig. 5.33—Results of consolidation tests on samples taken from the vicinity of

Building No. 301 Port Authority of N.Y.
Boring No. I
Sample II u
Depth 20 ft.
$\sigma_o' = 0.49$ TSF
$\sigma_p' = 0.67$ TSF
$\sigma_p' - \sigma_o' = 0.18$ TSF

$C_c = 0.61$

$0.42\ e_o$

$C_e = 0.05$

$e_o = 1.52$

σ_o' σ_p'

Void Ratio, e

σ_c' (TSF)

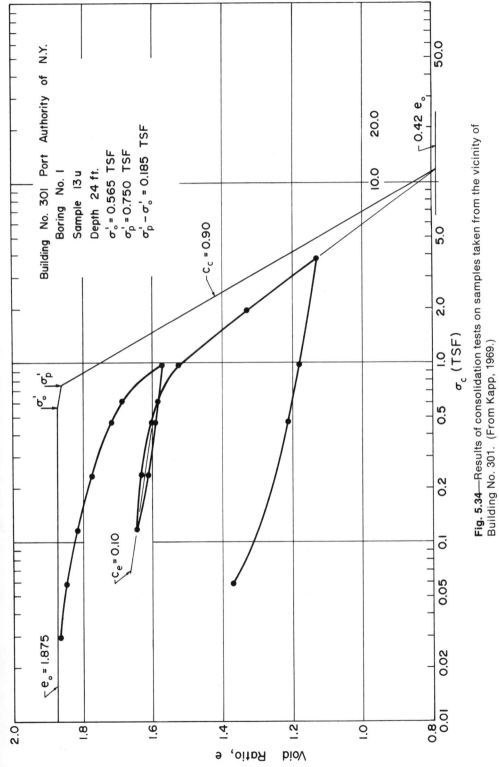

Fig. 5.34—Results of consolidation tests on samples taken from the vicinity of Building No. 301. (From Kapp, 1969.)

251

incorrect to assume that the full consolidation settlement has occurred during some time of interest. Thus in order to determine whether our calculation of consolidation settlement would be expected to overestimate that magnitude of settlement which has occurred during the 14-month surcharge period, it is necessary to investigate the principles which govern the transient flow of fluids through porous media such as soils. We shall consider the fundamental principles governing all such fluid flow in Chapter 6, and concentrate on the one-dimensional transient-flow problem in Chapter 7 after which we shall have developed a sufficient background to consider the time-dependent settlement of the surcharge load at Building 301.

5.5 SUMMARY OF KEY POINTS

Summarized below are the key points which we have discussed in the previous sections. Reference is made to the place in the text where the original more detailed description may be found:

1. The settlement of a foundation may be conveniently described as the sum of three components (Equation 5–1):

   ρ_d—the immediate distortion settlement,
   ρ_c—the consolidation settlement, and
   ρ_s—the settlement due to secondary compression.

2. Immediate distortion settlement due to saturated cohesive deposits is usually calculated on the basis of linear elastic theory:

$$\rho_d(x, y) = C_s q B \left(\frac{1 - \mu^2}{E} \right) \tag{5-2}$$

   For a saturated material, Poisson's ratio μ is generally assumed 0.5. The factor C_s depends upon the shape of the loaded area and the position of the point whose settlement is being calculated as well as:
   a. The stiffness of the loaded area when the foundation is considered as an elastic half-space;
   b. The depth of the elastic medium below the loaded area, and restraint conditions between the medium and the underlying base, when the foundation is considered to consist of an elastic material of limited depth underlain by a rigid base;
   c. The depth of the elastic material immediately below the loaded area relative to the dimensions of the loaded area, when the foundation is assumed to consist of a stiff elastic area overlying a deep less stiff material (Section 5.2).
3. Determination of the appropriate "elastic constants" from laboratory tests is difficult, and conventional methods are apt to underestimate the Young's modulus (Section 5.2).
   a. Limited available evidence suggests that the best estimate of E from laboratory tests can be made using the reload modulus E_r, obtained from a CU test on a specimen consolidated to the

vertical overburden stress and cycled at a representative stress level for the field loading conditions.

b. The elastic parameters can be back-figured from field plate loading tests in those cases where the seat of settlement of the test plate is the same as that of the prototype foundation.

c. A crude estimate of Young's modulus for preliminary analyses may be made by using empirical relations between E and the shearing resistance (Equation 5–7).

4. Analyses of consolidation settlements of cohesive soils ρ_c, resulting from superimposed load generally assumes one-dimensional compression. The relationship between void ratio change and effective stress in one-dimensional compression is

$$de = -C\, d(\log \sigma_c')$$

in which the material parameter C is called the *compression index* C_c if the applied effective stress exceeds the preconsolidation pressure (Equation 5–13), and is called the *recompression* or *expansion index* C_e if the applied effective stress is less than the preconsolidation pressure (Equation 5–16). C_c is of the order of 10 times C_e.

5. Because of the great difference between C_e and C_c, and the consequent calculated settlements, determination of the preconsolidation pressure σ_p' is most important. The Casagrande empirical approach is commonly used (Figures 5.14, 5.15).

a. Determination of whether a portion of a deposit is normally or over-consolidated is made by comparing the preconsolidation pressure determined in the laboratory on a specimen from the zone of interest, to the existing *in situ* overburden pressure.

b. Apparent preconsolidation of normally consolidated soils, when present, can lead to greatly reduced settlements as compared to calculations based upon the assumption of normal consolidation. The use of small pressure-increment ratios in the vicinity of preconsolidation pressure is useful in identifying the presence of apparent preconsolidation (Section 5.3).

c. An under-consolidated deposit, for which the preconsolidation pressure is less than the existing overburden stress, will settle in response to existing stresses without any increment in applied load. Such additional settlements must be considered.

d. A knowledge of the geological history of a deposit is helpful in interpreting results of laboratory measurement of the preconsolidation pressure. A vertical profile of preconsolidation pressure is useful for this purpose (Figure 5.25).

6. The compressibility curve obtained from the laboratory test will in general differ from that which is obtained in the field due to disturbance effects and other artifacts of the sampling and testing process. Realistic settlement predictions require reconstructing the field curve from the laboratory results. The best current method for doing this is that developed by Schmertmann (Figures 5.16, 5.20).

7. Settlements at the surface of one or more compressible strata are determined by dividing the strata into smaller layers for which

average properties and vertical stresses are appropriate, and summing the settlements of the individual layers (Section 5.3).

PROBLEMS

5.1. A raft foundation 80×200 ft is to be built on the soil profile shown. An excavation is to be made to provide a single basement of 10-ft depth. The uppermost cohesive stratum is highly over-consolidated and quite stiff. The two successive underlying strata are also cohesive but softer.

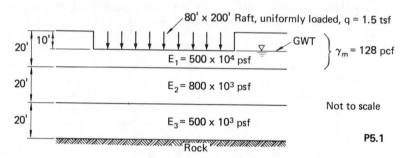

P5.1

Estimate the immediate settlement under the center of the raft. Do not forget to account for the excavation.

5.2. The 40-ft clay layer shown is loaded uniformly through its depth by a $\Delta\sigma' = 1000$ psf. Water-content data suggest the void ratio is constant throughout.

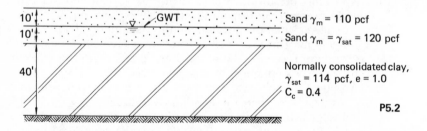

P5.2

(a) Determine the settlement of the clay stratum regarding it as one layer in the calculations.
(b) Determine the settlement assuming the clay stratum is subdivided into four equal layers.
(c) Determine the exact settlement of the entire clay stratum.

5.3. It is proposed to build a small circular oil-storage tank on the surface of an area, with the profile shown. The tank is 40 ft in diameter and will exert a pressure of 1500 psf on the ground surface. Results of a consolidation test on a sample taken from a depth of 40 ft below the ground surface are given in the accompanying table. From a triaxial compression test, Young's modulus E was estimated to be 2.5×10^5 psf.

(a) Estimate the immediate settlement of the clay layer.

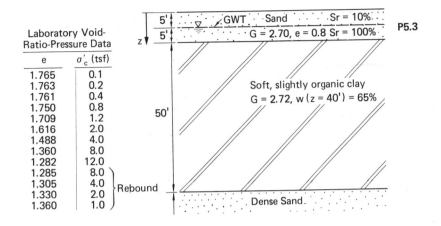

Laboratory Void-Ratio-Pressure Data	
e	σ'_c (tsf)
1.765	0.1
1.763	0.2
1.761	0.4
1.750	0.8
1.709	1.2
1.616	2.0
1.488	4.0
1.360	8.0
1.282	12.0
1.285	8.0
1.305	4.0
1.330	2.0
1.360	1.0

(b) Is the clay normally consolidated? What is C_c for the clay?

(c) What will be the ultimate consolidation settlement at the center of the tank, assuming that the center of the clay stratum is representative of the entire layer?

(d) What will be the answer to part c if the clay stratum is considered to be divided into *five* 10-ft-thick layers for analysis and only the one given set of e–log σ' data is given? What additional information might you have requested?

5.4. A building 30 × 100 ft in plan is to be constructed at the surface of the profile shown. It will apply uniform vertical stress at the ground surface

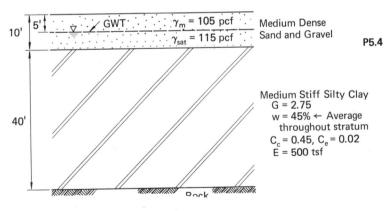

equal to 0.6 tsf. The clay is determined to be over-consolidated by an amount equal to 0.3 tsf at all depths, i.e., $\sigma'_p - \sigma'_o = 0.3$ tsf. Determine the settlement at the center of the building due to settlement of the clay stratum.

5.5. A small circular oil-storage tank is to be constructed at the surface of the soil profile shown. The oil will apply a uniform vertical stress to the ground surface of 1800 psf; the weight of the tank itself can be neglected. Shown on the sketch are soil parameters for the two clay strata obtained from

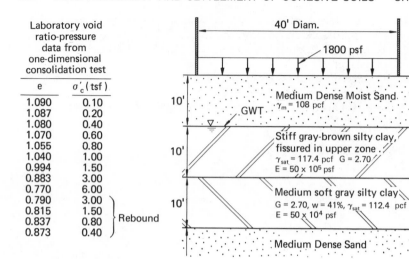

Laboratory void ratio-pressure data from one-dimensional consolidation test

e	σ'_c (tsf)
1.090	0.10
1.087	0.20
1.080	0.40
1.070	0.60
1.055	0.80
1.040	1.00
0.994	1.50
0.883	3.00
0.770	6.00
0.790	3.00
0.815	1.50
0.837	0.80
0.873	0.40

0.790, 0.815, 0.837, 0.873 rows marked: } Rebound

tests on samples taken at the mid-depth of each layer. Also given are results of a one-dimensional consolidation test made on a specimen taken from an "undisturbed" sample from the middle of the softer clay layer.

(a) Suggest a geologic history for this deposit which would be compatible with the information available to you. Justify your answer briefly.

(b) Calculate the settlement of the *center* and the *edge* of the tank due to deformation of the softer clay stratum.

(c) Identify *three* factors which will tend to make the actual settlements due to the softer clay less than that calculated. Indicate qualitatively the probable relative importance of these factors.

5.6. A long highway embankment is to be constructed on the profile shown in schematic. The preconsolidation pressure of the soft-to-medium

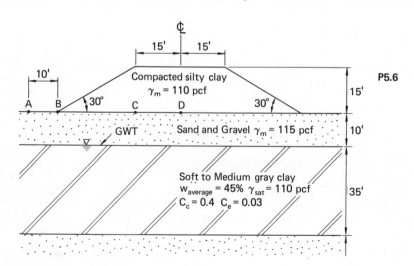

clay layer was determined from consolidation tests to be 1650 psf at the top, and to vary, approximately linearly, to a maximum of 3300 psf at the bottom.

(a) Determine settlement of points A, B, C, D due to consolidation of the clay layer.

(b) Discuss the effects which these settlements may have on the embankment and its performance.

Note: We might also wish to inquire concerning the stability of the embankment.

6

Fundamentals of Fluid Flow Through Porous Media

6.1 INTRODUCTION

The need to predict the time rate of settlement resulting from the transient consolidation process at Building 301 has led us to consider the flow of fluids through porous media. The consolidation problem is simply a special case of the general problem of seepage of fluids through porous solids.

Flow of water through soil creates problems of vital importance to the soil engineer. Such flows include seepage through and around structures designed to retain water, flow into wells, excavations and behind earth retaining structures. Concern about these problems results in part from the importance of maximizing the yield of wells or other sources of water, and reducing the losses from reservoirs and other storage facilities. Of more pressing concern to the soil engineer, however, is the effect of the flow on the effective stresses and deformations of the soil through which the fluid flow occurs. These are of interest in two connections:

1. The changes in effective stress which occur during transient flow are accompanied by volume changes which are in turn reflected in surface settlements. This process is called *consolidation.*
2. The flowing fluid applies body forces to the soil which influence the stability of the soil mass through which the flow is occurring, as well as the stability of nonsoil structures within the flow domain.

Because the principles which underlie the analysis of fluid flow through porous media are the same for all of these cases, we shall consider them in this chapter. In Chapter 7 we shall apply them to the consolidation problem, deferring consideration of other important applications to Chapter 14.

A *porous medium* is one in which the void spaces, or pores, are mostly continuous so that continuous flow of fluid through the pore spaces is possible. We shall limit our attention to a region which is completely saturated. Although the soil mass may be compressible, we shall assume that the solids and pore fluid are themselves incompressible and that only the volume of the pores can change.

In classical fluid mechanics, the analysis of flow of an incompressible fluid derives from three fundamental principles:

1. Conservation of mass
2. Conservation of energy
3. Conservation of linear momentum (Newton's second law of motion extended to continuous media)

Fluid flow through a porous medium is also controlled by these three basic relationships.

To model analytically the flow domain in a soil, we might wish to visualize the soil medium as a series of pipes, each pipe with a varying cross-section simulating the changing cross-sectional area of the soil voids. Application of the three principles stated above to this problem would lead to differential equations describing the flow: the Navier-Stokes equations for a viscous fluid; and Euler's equation for a nonviscous fluid. Unfortunately, these differential equations have been solved for only a limited number of special cases. Most of the problems of interest to the soil engineer have proven too complicated to consider by application of the Navier-Stokes equations because of the complexity of the void spaces in which the fluid flows. Although we shall gain some insight concerning the nature of flow through a soil mass by examining a special case of the Navier-Stokes equations, we shall generally require a different form of the equations of motion for a fluid within a soil mass. The analytical model which we shall use is also applied to other flows resulting from energy gradients such as the conduction of heat. We begin our discussion by considering conservation of mass.

6.2 CONSERVATION OF MASS

Because we assume the soil to be saturated, and the fluid and solid constituents to be incompressible, conservation of mass requires that any net flow of fluid into or out of an element of soil must be accompanied by an equal magnitude of volume change. We can visualize this process by considering the small parallelepiped shown in Figure 6.1. The center of the element is located at the point (x, y, z), with sides of length $\Delta x, \Delta y, \Delta z$ in the respective coordinate directions shown in the figure. Note that in this figure we have assumed the positive z direction to be upward in contrast to our previous discussion. This change in convention will be useful for description of fluid flow problems and will be used only in that context.

The arrows shown in the figure represent components of flow rate (in terms of volume of flow per unit time) entering and leaving the element. The notation identifies the direction and position of the flow. For example,

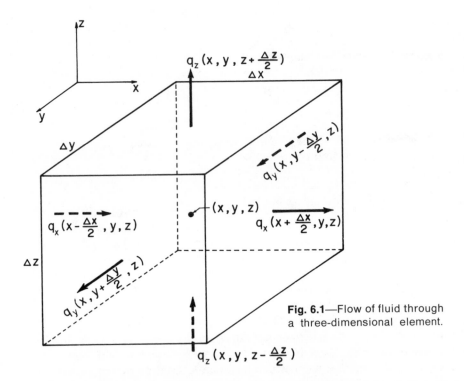

Fig. 6.1—Flow of fluid through a three-dimensional element.

$q_x(x - \Delta x/2, y, z)$ is the rate of flow in the x direction passing through and normal to the face located at the position $(x - \Delta x/2, y, z)$. We shall assume that the element is small enough so that we may represent the flow passing through a face by a vector at the center of that face.

The component of the net rate of flow out of the element due to flow in the x direction is

$$\Delta q_x = q_x \left(x + \frac{\Delta x}{2}, y, z \right) - q_x \left(x - \frac{\Delta x}{2}, y, z \right) \qquad (6\text{--}1)$$

in which Δq_x is the incremental net flow rate out of the element due to the flow in the x direction. Multiplying and dividing the right hand side of Equation 6–1 by Δx, and letting the Δx approach, but not reach, zero, we can express the net flow as a differential,

$$dq_x = \frac{\partial q_x}{\partial x} dx \qquad (6\text{--}2a)$$

A similar procedure leads to the net flow out of the element due to fluid flowing in the y and z directions respectively:

$$dq_y = \frac{\partial q_y}{\partial y} dy \qquad (6\text{--}2b)$$

$$dq_z = \frac{\partial q_z}{\partial z} dz \qquad (6\text{--}2c)$$

The sum of these components is then the net loss in volume,

$$dq_x + dq_y + dq_z = -\frac{\partial V}{\partial t} \qquad (6\text{--}3)$$

or

$$\frac{\partial q_x}{\partial x} dx + \frac{\partial q_y}{\partial y} dy + \frac{\partial q_z}{\partial z} dz = -\frac{\partial V}{\partial t} \qquad (6\text{--}4)$$

in which $-\partial V/\partial t$ is the net rate of volume loss.

We find it useful to define a quantity called the *discharge velocity* defined, for example in Figure 6.1, as

$$v_x = \frac{q_x}{\Delta y \, \Delta z}$$

$$v_y = \frac{q_y}{\Delta z \, \Delta x} \qquad (6\text{--}5)$$

$$v_z = \frac{q_z}{\Delta x \, \Delta y}$$

in which v_x, v_y, v_z are the *discharge* velocities in the x, y, z directions respectively. That is, the component of the discharge velocity is defined as the component of the flow rate in that direction divided by the area normal to the flow rate as we allow that area to approach a limiting small value. This quantity is obviously artificial because the area normal to the flow rate consists of both void spaces in which the flow actually occurs and solids. Thus the discharge velocity is not an actual velocity of flow but rather a parameter equivalent to the flow rate.*

---

* A parameter which characterizes the average velocity within the pores is the *seepage velocity*, v_s, with components

$$v_{s_x} = \frac{q_x}{A_{v_x}}, \qquad v_{s_y} = \frac{q_y}{A_{v_y}}, \qquad v_{s_z} = \frac{q_z}{A_{v_z}} \qquad (6\text{--}A)$$

in which A_{v_x}, A_{v_y}, A_{v_z} are the areas of the *voids* normal to the respective flow directions. Recalling the definition of the porosity n,

$$n = \frac{V_v}{V} \qquad (6\text{--}B)$$

the void areas of the faces normal to the component directions in the element shown in Figure 6.1 are determined as

$$A_{v_x} = n \frac{\Delta x \, \Delta y \, \Delta z}{\Delta x} = n \, \Delta y \, \Delta z$$

$$A_{v_y} = n \, \Delta z \, \Delta x \qquad (6\text{--}C)$$

$$A_{v_z} = n \, \Delta x \, \Delta y$$

Substituting Equations 6–C into Equations 6–A we find that the seepage velocity is related to the discharge velocity by

$$v_{s_x} = \frac{v_x}{n}, \qquad v_{s_y} = \frac{v_y}{n}, \qquad v_{s_z} = \frac{v_z}{n} \qquad (6\text{--}D)$$

If we then divide Equations 6–4 by the volume of the element V_0,

$$V_0 = \Delta x\, \Delta y\, \Delta z \qquad (6\text{–}6)$$

we have

$$\frac{\partial v_x}{\partial x} + \frac{\partial v_y}{\partial y} + \frac{\partial v_z}{\partial z} = -\frac{1}{V_0}\frac{\partial V}{\partial t} \qquad (6\text{–}7)$$

We note that the right side of Equation 6–7 is the negative of the volumetric strain rate. From Equation 5–11 we recall that the volumetric strain, when the soil grains and pore fluid are incompressible, is

$$\frac{dV}{V_0} = \frac{de}{1 + e_0} \qquad (6\text{–}8)$$

So, Equation 6–7 may be rewritten as

$$\frac{\partial v_x}{\partial x} + \frac{\partial v_y}{\partial y} + \frac{\partial v_z}{\partial z} = -\frac{1}{1 + e_0}\frac{\partial e}{\partial t} \qquad (6\text{–}9)$$

This expression is one form of the *continuity equation* and is the statement of conservation of mass. When the flow is steady-state so that the velocity at each point is independent of time, the rate of volume strain must also be zero and Equation 6–9 reduces to

$$\frac{\partial v_x}{\partial x} + \frac{\partial v_y}{\partial y} + \frac{\partial v_z}{\partial z} = 0 \qquad (6\text{–}10)$$

6.3 FREE ENERGY

Free energy is the capacity to do work. Fluid flow occurs only when a difference in free energy exists between two points; the flow is in a direction from higher to lower free energy. In fact we shall see that the flow rate in a porous medium is related to the space gradient of free energy just as strain in a continuous solid is related to stress.

For incompressible flow through a porous medium, the most common energy gradient is mechanical, although energy gradients may be thermal, electrical, or chemical as well. In the case of the West Branch Dam, described in Chapter 1 (pages 10–16), an electrical-energy gradient was used to increase substantially the rate of consolidation. In this section we shall restrict our attention to flow in response to mechanical-energy gradients. In Section 6.4 we shall consider capillarity, which arises in response to a molecular free-energy gradient; and in Section 9.6 we shall investigate the effect of electrical-energy gradients.

Conservation of Energy—Bernoulli's Equation

For steady flow of an incompressible fluid the requirements for the conservation of energy are embodied in the Bernoulli equation:

$$h(x, z, t) - \frac{u(x, z, t)}{\gamma_w} - z - \frac{v^2(x, z, t)}{2g} = \text{const} \qquad (6\text{-}11)$$

in which $h(x, z, t)$ is the free energy of the fluid per unit weight, that is, ft-lb/lb or g-cm/g. The free energy per unit weight is called the *total head*.

$u(x, z, t)/\gamma_w$ is called the *pressure head*, where u is the pressure in the water.

z is the vertical position of the point relative to an arbitrary datum, and is called the *elevation head*.

$v^2(x, z, t)/2g$ is the kinetic energy per unit weight of the fluid, where v is the flow velocity and g is the acceleration of gravity.

Thus we see that the free energy of the fluid differs by a constant from the sum of the potential energy resulting from pressure and positional potential and the kinetic energy. The constant is arbitrary depending upon the selection of a datum from which to measure the energy.

In the case of fluid flow through porous media, the velocity component is usually small enough so that the kinetic energy term is negligible compared to the potential energy term. Thus Equation 6-11 may be rewritten as

$$h = \frac{u}{\gamma_w} + z + \text{const} \qquad (6\text{-}12)$$

The difference in free energy (total head) between any two points is

$$\Delta h_{12} = h_2 - h_1 = \frac{u_2 - u_1}{\gamma_w} + z_2 - z_1 \qquad (6\text{-}13)$$

We note that the head difference is independent of the constants in Equation 6-12 and the position of the elevation datum. Because it is the differences in head which lead to flow, the selection of the elevation datum is arbitrary, and the constant may be set to zero.

The space rate of head dissipation, called the *hydraulic gradient*, is defined by

$$\mathbf{i} = -\lim_{\Delta s \to 0} \frac{\Delta h}{\Delta s}\bigg|_{t = \text{const}} = -\frac{\partial h}{\partial \mathbf{s}} \qquad (6\text{-}14)$$

in which $\Delta \mathbf{s}$ is the vectorial path along which the energy Δh is dissipated. When the seepage is steady state, that is, independent of time, the gradient may be written as a total derivative:

$$\mathbf{i} = -\frac{dh}{d\mathbf{s}} \qquad (6\text{-}15)$$

Thus we see that the gradient is a *vector* quantity with components in the x

and z directions respectively,

$$i_x = -\frac{\partial h}{\partial x}$$
$$i_z = -\frac{\partial h}{\partial z} \qquad (6-16)$$

It is often convenient to approximate the hydraulic gradient by the average gradient across a finite distance:

$$i_x \simeq -\frac{\Delta h}{\Delta x}$$
$$i_z \simeq -\frac{\Delta h}{\Delta z} \qquad (6-17)$$

We shall illustrate some features of the relationship between head loss, flow and its relationship to effective stresses in the following example:

Example 6.1

A long trench is to be cut into a water-bearing soil stratum, as shown in Figure 6.2. Before the trench is constructed, the *groundwater table** is at the elevation indicated

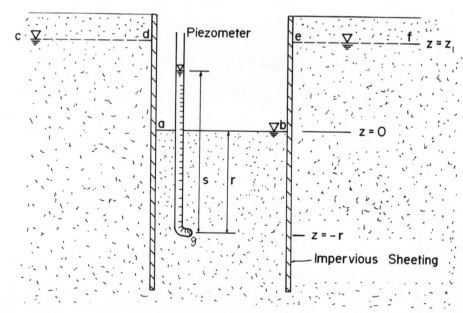

Fig. 6.2—Flow into a trench, Example 6.1.

* We recall that the *groundwater table*, or *free-water surface* is that surface in a body containing water at which the pressure in the water is equal to atmospheric pressure.

in the figure by lines $\overline{cd}$ and $\overline{ef}$. The trench is to be constructed by driving interlocking sheet piling, which are impervious to water, as shown in the figure. The trench is then excavated between the sheet piling which may be supported by a variety of bracing systems not shown. In order to be able to work in the excavation, water is pumped out of the trench so that the elevation of the free water surface within the trench is at the base of the cut, line $\overline{ab}$ in Figure 6.2. In order to monitor groundwater conditions beneath the excavation, a *piezometer** has been inserted in the ground with its tip at a depth r beneath the base of the excavation (Figure 6.2). The column of water in the piezometer is of height s.

(a) Show that the arrangement described above will lead to a flow of water into the trench.
(b) Determine the vertical distribution of vertical effective stress from the base of the cut to the depth of the piezometer tip.

Solution. (a) As shown in Figure 6.2, we have selected the arbitrary elevation datum, $z = 0$, to be at the base of the trench, line ab. Taking the constant in Equation 6–12 as zero, the total head at ab is

$$h_{ab} = \frac{u}{\gamma_w} + z = 0 + 0 = 0$$

Along the boundaries cd and ef, at which the elevation is z_1, the head is

$$h_{cd} = h_{ef} = 0 + z_1 = z_1$$

Because the free energy along the boundaries cd and ef is greater than that along the base of the trench, ab, water will flow from the former to the later and therefore into the trench.

The magnitude of the head difference between cd and ab is a measure of the "driving force" tending to cause the flow. This head difference is

$$\Delta h_{cd, ab} = h_{ab} - h_{cd} = -z_1$$

Note that the head difference is equal to the elevation difference between the free water surfaces.

(b) The profile of vertical effective stress can be obtained by determining the profiles of the total vertical stress and pore water pressure and subtracting the second from the first. The total vertical stress distribution, shown schematically in Figure 6.3a, is

$$\sigma_z = -z\gamma_{sat} = -z(\gamma' + \gamma_w) \tag{6–18}$$

in which　　　　　　γ_{sat} is saturated unit weight of the soil
　　　　　　　　　γ' is submerged unit weight of the soil
　　　　　　　　　γ_w is unit weight of water

As shown in Figure 6.2, the piezometer indicates a pore water pressure at $z = -r$ of

$$u(-r) = s\gamma_w \tag{6–19}$$

---

* A piezometer is a device for observing the water pressure at a point beneath the ground surface. Although piezometers are manufactured with a variety of operating principles, the simplest is an open standpipe, illustrated schematically in Figure 6.2, with a porous tip which permits water to flow into the piezometer but prevents the entrance of soil. The height of water in such a piezometer indicates the water pressure at the tip.

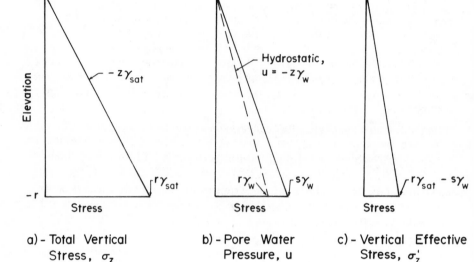

a) - Total Vertical
Stress, σ_z

b) - Pore Water
Pressure, u

c) - Vertical Effective
Stress, σ'_z

Fig. 6.3—Stress profiles for Example 6.1.

We shall assume that the pore water pressure between elevations $z = 0$ and $z = -r$ is linearly distributed,* (Figure 6.3b), so that

$$u = -z\left(\frac{s}{r}\right)\gamma_w \qquad -r \leqslant z \leqslant 0 \tag{6-20}$$

The distribution of vertical effective stress is now obtained by subtracting Equation 6-20 from Equation 6-18,

$$\sigma'_z = -z\left[\gamma_{sat} - \left(\frac{s}{r}\right)\gamma_w\right] \tag{6-21a}$$

or

$$\sigma'_z = -z\left[\gamma' - \left(\frac{s-r}{r}\right)\gamma_w\right] \qquad -r \leqslant z \leqslant 0 \tag{6-21b}$$

The expression for the vertical effective stress distribution with depth is illustrated in Figure 6.3c.

We note from Equation 6-21b that the quantity $(s - r/r)\gamma_w$ is the magnitude of the pore water pressure excess over hydrostatic pressure. Thus as the pore water pressure at the piezometer tip, represented by s, increases,

---

* As we shall see in Chapter 14, this assumption is not entirely correct. However for the boundary conditions shown, it is a reasonable approximation and will serve well for our present discussion.

the effective stress beneath the excavation decreases. By setting the left side of Equation 6–21b to zero, we find that when

$$s = r\left(1 + \frac{\gamma'}{\gamma_w}\right)$$

the vertical effective stress is reduced to zero. In the case of cohesionless soils, such as sands and gravels, the strength under such circumstances would also be reduced to zero. A sand in this condition is incapable of supporting loads, and acts simply as a very viscous liquid. Upward flows producing this condition may arise, due to construction of the sort described in Example 6.1 above or through natural artesian conditions. Soils found in this condition are commonly called *quicksand* and are said to be in a *quick* condition. Hence the presence of quicksand beneath an excavation or constructed work may have disastrous consequences. We recall that in the case of the Fort Peck slide, the rapid and extensive movements of the failure mass occurred because the increased pore water pressure produced a reduction in effective stress, not to zero, but probably to a magnitude less than that required to produce failure. We shall consider the quick condition again in Chapter 14.

It is useful to express the requirements for producing the quick condition in terms of the vertical hydraulic gradient required to produce zero effective stress. The elevation head at the piezometer tip is $-r$; the pressure head is s. Thus the total head at point g is

$$h_g = s - r$$

Recalling that the head at the base of the excavation is zero, the head difference between point g and the base of the cut is

$$\Delta h_{g,\,ab} = -(s - r)$$

Because the distribution of pore water pressure between the piezometer tip and the base of the excavation has been assumed linear, the head loss is also linear with depth, and the hydraulic gradient is a constant equal to the average hydraulic gradient over the distance:

$$i_z = \frac{s - r}{r}, \qquad -r \leqslant z \leqslant 0 \qquad (6\text{--}22)$$

Substituting this expression into Equation 6–21b, we see that the effective stress can be expressed in terms of the hydraulic gradient as

$$\sigma'_z = -z(\gamma' - i_z \gamma_w) \qquad (6\text{--}23)$$

Thus, in order to produce a quick condition in cohesionless soils ($\sigma'_z = 0$), the upward vertical hydraulic gradient must acquire a critical magnitude of

$$i_{\text{crit}} = \frac{\gamma'}{\gamma_w} \qquad (6\text{--}24)$$

By noting that the submerged unit weight of cohesionless soil is approximately equal to the unit weight of water, we see that the critical hydraulic gradient for cohesionless materials is approximately unity.

Seepage Force

The work required to produce flow in a porous medium results from a complex interaction between the fluid and the particle. It is intuitively evident, however, that a significant portion of this work results from frictional drag as the fluid passes over and around the individual particle. Thus one might say that the fluid exerts a *seepage force* on the porous medium through which it flows. To relate this seepage force to the hydraulic gradient we shall consider the vertical equilibrium of a mass of soil beneath a trench, bounded by the base of the trench and the horizontal surface at the elevation of the piezometer tip. The vertical forces acting on the wedge from Example 6–1, shown in Figure 6.4, include:

1. Summation of the water pressure acting on the base of the trench $= 0$,
2. Weight of the soil within the wedge $= Br\gamma_{\text{sat}}$,
3. Summation of the water pressure at the base of the soil mass $= Bs\gamma_w$, and
4. Summation of the vertical effective stress on the base of the element $= B\sigma_z'$.

Note that we have assumed no vertical shear stress on the sides of the soil mass; we shall consider the rationale for this assumption below. If the wedge is in equilibrium,

$$B\sigma_z' + Bs\gamma_w = Br\gamma_{\text{sat}} \qquad (6\text{--}25)$$

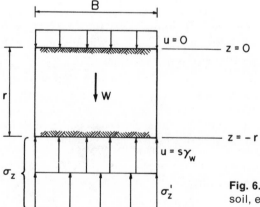

Fig. 6.4—Vertical forces on mass of soil, example 6.1.

or

$$Bo'_z + Bs\gamma_w = Br(\gamma' + \gamma_w) \qquad (6-26)$$

Rearranging terms, we find

$$Bo'_z = Br\left[\gamma' - \left(\frac{s-r}{r}\right)\gamma_w\right] \qquad (6-27)$$

Recalling that the vertical component of the hydraulic gradient is

$$i_z = \frac{s-r}{r}$$

we can rewrite Equation 6–27 as

$$Bo'_z = Br(\gamma' - i_z\gamma_w) \qquad (6-28)$$

Examining the terms inside the parenthesis in this expression, we note that γ' is the magnitude of a vector quantity, the submerged weight per unit volume of the soil mass. The quantity $i_z\gamma_w$ is the vertical component of another *body force* acting on the soil skeleton. This force per unit volume is called the *seepage force*. In general, the seepage force is a vector quantity $i\gamma_w$ which acts in the direction of the hydraulic gradient $\mathbf{i}$. Thus when it acts vertically upward, as in the case of the example described, it *reduces* the vertical effective stress, (Equation 6–28). Were the flow downward, it would have increased the vertical effective stress by an amount $i_z\gamma_w$.

We note from Equation 6–28 the previously obtained result (Equation 6–24) that, when the hydraulic gradient becomes sufficiently large,

$$i_z = i_{crit} = \frac{\gamma'}{\gamma_w} \qquad (6-29)$$

the upward seepage force equals the submerged unit weight, and the vertical effective stress becomes zero.

We now return to the question of why the shear forces on the side of the element in Figure 6.4 were neglected. There are two reasons:

1. If the domain in which flow is occurring is wide relative to its length, the magnitude of the shear forces will be negligible in comparison with the other forces considered.
2. In considering the stability of the wedge as the hydraulic gradient approaches i_{crit}, it is apparent that the shear forces on the side must approach zero even in the case of a narrow flow domain.

Thus we see that it is frequently justified, and always conservative from the point of view of evaluating the stability of the soil mass, to have made the assumption discussed.

6.4   CAPILLARITY

Surface Tension

The groundwater table within a soil mass is the surface at which the pore water pressure equals atmospheric pressure. In nearly all soils, however, there exists water at elevations above the free surface. The soil within this zone of *capillary rise* may be either fully or partially saturated. The water is drawn into the soil voids by the attraction between water molecules and those of the soil particles at the air-water interface. The finer the grain size, the greater is the rise of water above the free surface.

This phenomenon may be demonstrated by immersing a glass tube with a small inside diameter into water. Upon contact, the water rises in the tube to a height h_c, the *height of capillary rise* (Figure 6.5). A curved surface, the *meniscus*, is formed at the air-water interface, its edge making an angle of α with the tube. Clearly, the water has not been drawn by a mechanical-energy gradient. There must therefore be another source of energy to produce the capillary rise. This is a molecular free energy which develops at the interface of a liquid with a gas and/or solid. The nature of this free energy is not clearly understood but its physical effects may be predicted by the use of a simple membrane analogy. We shall assume there exists a *surface tension*, or water skin under tension, at the air-water interface. As shown in Figure 6.6:

T_s is the tension in the membrane, a force per unit length of the membrane perimeter.

α is the angle between T_s and the tube wall.

d is the diameter of the capillary tube.

r_m is the radius of the meniscus.

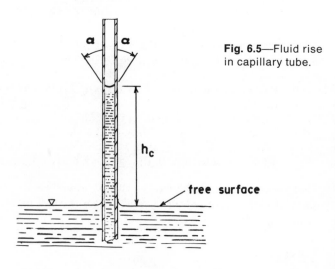

Fig. 6.5—Fluid rise in capillary tube.

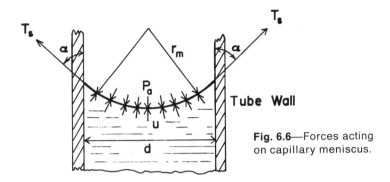

Fig. 6.6—Forces acting on capillary meniscus.

Although this is a fictitious analogy, experimental results indicate that the concept of a surface membrane not only allows us to visualize the phenomenon of capillary rise, but also leads to accurate quantitative predictions.

Capillary Pressure

If the curved surface of a thin flexible membrane is under tension, equilibrium dictates that there be a different pressure on either side of the membrane. Considering the membrane as a free body and summing forces vertically (Figure 6.6),

$$T_s \, d\pi \cos \alpha = p_a \frac{\pi d^2}{4} - u \frac{\pi d^2}{4} \qquad (6\text{-}30)$$

where u is the pressure in the water behind the membrane. Assuming the air pressure above the membrane is atmospheric, ($p_a = 0$), the pore water pressure is

$$u = \frac{-4T_s \cos \alpha}{d} \qquad (6\text{-}31)$$

The magnitude of T_s is a characteristic of the particular fluid, whereas the angle α depends upon both the fluid and the tube material. For clean glass and water $\alpha = 0°$ and $T_s = 0.075$ gm/cm. For mercury and glass α is less than $90°$ and there is a capillary depression. Because T_s and α are fixed for a given combination of materials, it is apparent from Equation 6-31 that, if the full height of capillary rise can occur, the magnitude of negative pressure is limited only by the diameter of the capillary tube, the tensile strength of water, and the compressive strength of the tube walls.

Example 6.2

The lower end of a glass capillary tube with an inside diameter of 0.02 cm is immersed in a container of water. What is the maximum negative water pressure which can develop within the tube.

Solution. From Equation 6-31, with $\alpha = 0$ and $T_s = 0.075$ gm/cm,

$$u = \frac{-4(0.075)}{0.02} = -15.0 \text{ gm/cm}^2 = -0.213 \text{ psi}$$

An inside diameter of 0.02 cm corresponds to a pore size for a medium-to-coarse sand. Had we considered a tube with an inside diameter of 0.002 mm, corresponding to the pore size of a fine silt, we would have found $u = -3060$ psf! From this we find that the pore water pressure may be negative and of large magnitude. Common evidence of this is seen in the high effective stresses generated during the shrinkage process in fine-grained soils. This subject is discussed below.

The water pressure which we have calculated is the largest negative magnitude that will occur within the capillary tube. It is easy to show that the pressure at any position within the tube at any elevation z above the free-water surface is proportional to the elevation:

$$u = -z\gamma_w \tag{6-32}$$

This may be shown either by making use of the fact that the total head within a body of water at equilibrium is constant and substituting into Equation 6-21, or by considering the vertical equilibrium of the cylinder of water above the height desired. We use this second approach to determine the maximum height of capillary rise below.

Height of Capillary Rise

It is useful to know the height to which capillary water will rise in a soil mass. A preliminary approach to the problem can be obtained by considering capillary rise in a simple tube immersed in water as shown in Figure 6.7. The horizontal scale has been exaggerated for illustrative purposes. At the elevation of the free-water surface within the tube, the water pressure is atmospheric. Considering vertical equilibrium of the cylinder of water in the tube above this section, we find

$$T_s \pi d \cos \alpha - p_a \frac{\pi d^2}{4} = \gamma_w \frac{\pi d^2 h_c}{4} - p_a \frac{\pi d^2}{4} \tag{6-33}$$

in which

$$\gamma_w \frac{\pi d^2 h_c}{4}$$

is the weight of the cylinder of water. Because we are measuring pressures with respect to atmospheric pressure, $p_a = 0$. We then solve for the height of capillary rise, h_c,

$$h_c = \frac{4T_s \cos \alpha}{d\gamma_w} \tag{6-34}$$

Thus we see for a given combination of materials, the height of capillary rise depends only upon the diameter of the tube.

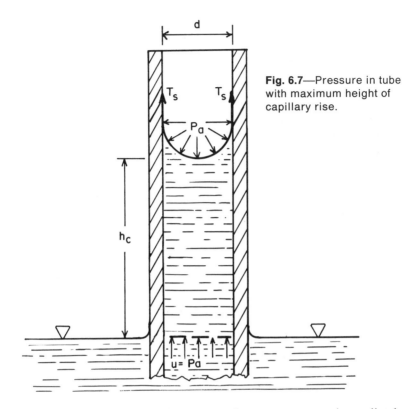

Fig. 6.7—Pressure in tube with maximum height of capillary rise.

From Equation 6–32, or Equation 6–12, the water pressure immediately below the meniscus is

$$u = -h_c \gamma_w \tag{6-35}$$

In the event that the capillary tube extends to a height above the free-water surface h_{cr}, which is less than the maximum height of capillary rise h_c, the pore water pressure immediately below the meniscus will be

$$u = -h_{cr} \gamma_w \tag{6-36}$$

Substituting this expression into Equation 6–31, we find that

$$T_s \cos \alpha = \gamma_w h_{cr} \, d \tag{6-37}$$

Because T_s is a property only of the water, it is apparent that $\cos \alpha$, that is, the orientation of the surface-tension force, must change, as illustrated in Figure 6.8a. We recall, however, that angle α is a function only of the properties of the tube and fluid, not of the tube geometry. This apparent anomaly can be understood by considering a close-up view of the top of the tube wall where the meniscus meets the tube, as illustrated in Figure 6.8b. From this figure we see that, for the glass tube and water, angle α is still zero, but that the meniscus meets the tube wall at the corner in such a way

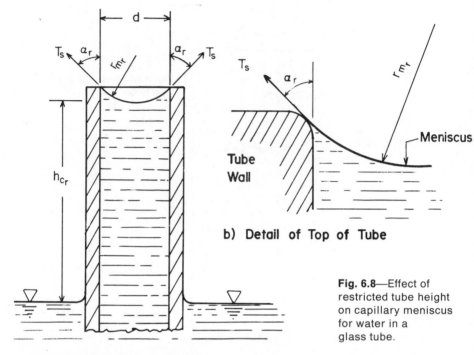

b) Detail of Top of Tube

Fig. 6.8—Effect of restricted tube height on capillary meniscus for water in a glass tube.

a) Sketch of Tube and Water

that the surface-tension force is oriented to satisfy the requirements of equilibrium. This apparent α, resulting from the restricted height of the tube, is designated α_r in the figure. As a consequence, the radius of the meniscus is larger than $d/2$ and is found from

$$r_{mr} = \frac{d}{2 \cos \alpha_r} \qquad (6-38)$$

Because the pore diameter varies from point to point within a soil mass it is worthwhile to extend our simple tube analogy to the case in which the tube diameter varies along the length of the tube. The effect of varying tube diameters is discussed in the following example.

Example 6.3

A clean glass capillary tube with variable inside diameters is sketched in Figure 6.9a. Determine the height to which water will rise in the tube under the following circumstances:

(a)   The bottom of the tube is inserted into a standing body of water until the top of the tube is 48 cm above the free-water surface.

(b)   The tube is completely immersed in a standing body of water and then extracted until the top of the tube is 30 cm above the free-water surface.

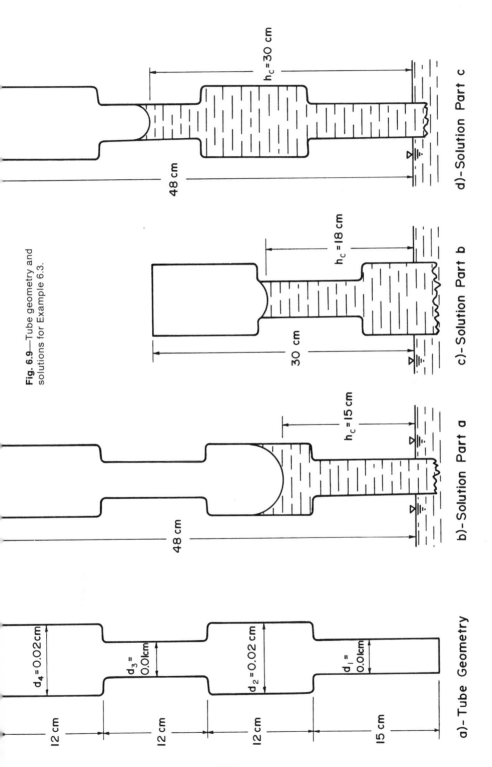

Fig. 6.9—Tube geometry and solutions for Example 6.3.

a)- Tube Geometry

$d_4 = 0.02\,cm$

$d_3 = 0.01\,cm$

$d_2 = 0.02\,cm$

$d_1 = 0.01\,cm$

12 cm

12 cm

12 cm

15

b)- Solution Part a

48 cm

$h_c = 15\,cm$

c)- Solution Part b

30 cm

$h_c = 18\,cm$

d)- Solution Part c

48 cm

$h_c = 30\,cm$

(c)   The tube is immersed in a standing body of water and then extracted until the top of the tube is 48 cm above the free-water surface.

Solution. We shall first determine the maximum height of a column of water which can be supported by each of the tube diameters. From Equation 6–34, for water in a glass tube,

$$h_c = 30 \text{ cm} \quad \text{for } d = 0.01 \text{ cm}$$
$$h_c = 15 \text{ cm} \quad \text{for } d = 0.02 \text{ cm}$$

(a) When the tube is inserted into the water, water rises in the lower section of the tube. Because h_c for this section is greater than the 12-cm length of this section above the free-water surface, the water rises into a second section of the tube, for which $d_2 = 0.02$ cm. The water rises in this section of the tube until it is at a height $h_c = 15$ cm above the free-water surface. This solution is illustrated in Figure 6.9b.

(b) When the tube is immersed and then extracted from the water, the meniscus remains at the top of the tube until the top of the tube is 15 cm above the free-water surface. The tube is pulled out further, the meniscus remains at $h_c = 15$ cm until the tube has been extracted a distance of 27 cm. At that point, the top of section 3, for which $d_3 = 0.01$ cm, is 15 cm above the free-water surface. As the tube is pulled out further, the meniscus remains at the top of section 3 and the radius of the meniscus decreases in size to accommodate the increased height of water supported. When the top of the tube is 30 cm above the free-water surface, the top of section 3 is 18 cm above the free-water surface, which is less than h_c for the 0.01-cm-diameter tube. Thus the meniscus remains at the height $h_{cr} = 18$ cm. This is illustrated in Figure 6.9c.

(c) If the tube is extracted further from the free-water surface, the meniscus at the top of section 3 develops further until the top of the tube has raised to a point 42 cm above the free-water surface. At this point, the top of section 3 is then 30 cm above the free-water surface and the meniscus in this section is fully developed so that $2r_m = d_3 = 0.01$ cm. As the tube is raised still further, the meniscus remains at its maximum height of capillary rise, $h_c = 30$ cm. This result is shown in Figure 6.9d.

From this example, it is evident that the height of capillary rise in a medium with variable pore size will depend not only upon the pore sizes but also in part upon the source of the water and the fluctuation of the free-water surface.

Because of the irregular nature of the pores in a real soil, the height of capillary rise in a soil mass cannot be predicted as easily as in a capillary tube. Terzaghi has suggested (Meinzer, 1942) that an equivalent pore diameter is proportional to the void ratio of the soil and the effective grain size D_{10}, so that the height of capillary rise is

$$h_c = \frac{C}{e D_{10}} \tag{6–39}$$

in which the constant C is a function of the shape of the grains and the contact angle α. Unfortunately, it is impossible at the present time to determine the value of C without measuring h_c. Furthermore, the height of capillary rise is different in a dry soil as compared to a slightly moistened soil (Meinzer, 1942), presumably owing to a difference in the contact angle. The height of capillary rise has been measured however in a variety of soils with a capillary front *rising* from the free-water surface, and some measured values of capillary rise are given in Table 6.1.

TABLE 6.1
Typical Values of Height of
Capillary Rise

Soil Type	Height of Capillary Rise h_c (cm)
Coarse sand	2– 5
Sand	12– 35
Fine sand	35– 70
Silt	70–150
Clay	200–400 and greater

SOURCE: Silin-Bekehurin (1958).

As a result of the variability of the void sizes from one point to the next in both the horizontal and vertical directions in a soil mass, the soil is generally found at different degrees of saturation within the capillary zone. Above the free-water surface there is a zone of soil saturated by capillary water. The height of this zone will be at least h_c for the *maximum* void sizes (Equation 6–39). If, however, the free-water surface has been dropping, the height of capillary rise in this saturated zone will be influenced primarily by the smallest void size. Because the voids are continuous horizontally as well as vertically, there will be a zone of nearly saturated soil above the saturated zone. The capillary water will bypass the larger voids, leaving them unsaturated, so that the water in this zone will be continuous but not all inclusive. The height above the free-water surface of this zone of nearly saturated soil will also depend on the smallest pore sizes. Above the zone of nearly saturated soil the capillary water will be discontinuous. Water behind a falling capillary front or that brought in by infiltration from the ground surface will remain at the contact points of the individual particles as shown in Figure 6.10. Evaporation from the ground surface also influences the extent of this discontinuous capillary water.

6.5   EFFECT OF CAPILLARITY ON SOIL BEHAVIOR

Effective Stresses

The negative pore water pressure due to capillarity, that is, the presence of an air-water interface, produces a concomitant increase in effective stresses. The increased strength imparted by these stresses to cohesionless soils can

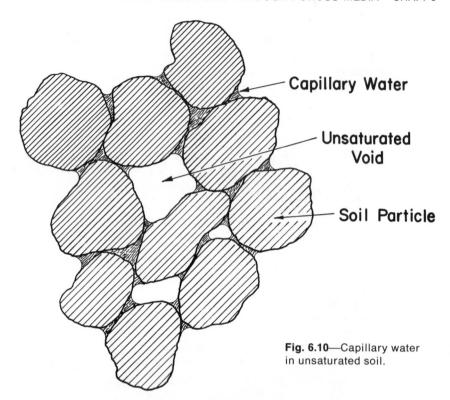

Capillary Water

Unsaturated Void

Soil Particle

Fig. 6.10—Capillary water in unsaturated soil.

be observed in many applications from children's sand castles at the beach to nearly vertical temporary excavations in moist cohesionless sand. The loss of the capillary effect, either by drying or adding free water, results in a loss of strength, and the collapse of the sand castles or excavated slopes. In the case of saturated soils the magnitude of negative pore water pressure, and thus the effective stress is proportional to the height above the free-water surface. When the soil is unsaturated and the voids are discontinuous as shown in Figure 6.10, the pore water pressure is independent of position, and cannot be predicted at the present time. The pore pressure depends on the radius of the meniscus (Figure 6.11), which depends on, among other factors, the amount of water present.

The effective stresses induced by capillary moisture in cohesionless materials causes these soils to resist volume change. This phenomenon, known as *bulking* makes field *compaction* of such materials difficult unless they are nearly dry or nearly saturated.

Sampling Disturbance of Cohesive Soils

In obtaining "undisturbed" samples of cohesive soils, we depend upon capillary stresses to prevent swelling of the soil as it is extracted from the

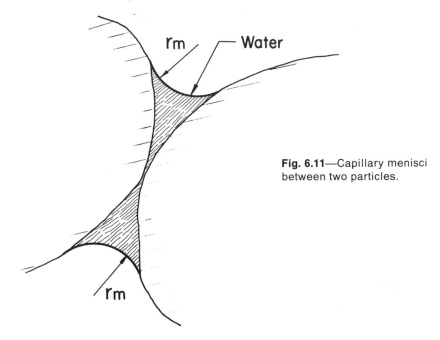

Fig. 6.11—Capillary menisci between two particles.

ground. When the overburden stress is removed from the sample, there is a tendency to expand in response to the reduced stresses. If no source of water is available to a saturated specimen, capillary pressures develop in opposition to the expansive tendency. The volumetric strain required to produce this negative pore water pressure is negligible. Thus the effective stresses due to the overburden are replaced in part by those due to capillary tensions. Unfortunately, some disturbance of the soil structure must occur because the ratio of horizontal to vertical effective stress is changed during the sampling process. *In situ*, the ratio of horizontal to vertical effective stress is (recall Equation 3–29):

$$\frac{\sigma'_x}{\sigma'_z} = K_0 \qquad (6\text{--}40)$$

in which K_0 is the coefficient of earth pressure at rest. As in the case of cohesionless soils, the horizontal effective earth pressure for *normally consolidated* clays may be approximated (Jaky, 1948) by:

$$K_0 = 1 - \sin \phi' \qquad (6\text{--}41)$$

Except for very heavily over-consolidated clays, K_0 is less than unity.

After sampling, when the total stresses are

$$\sigma_x = \sigma_z \simeq 0 \qquad (6\text{--}42)$$

the effective stresses are

$$\sigma'_x = \sigma'_z = -u \qquad (6\text{--}43)$$

and

$$\frac{\sigma'_x}{\sigma'_z} = 1 \tag{6-44}$$

Thus both vertical and effective stresses are changed, leading to distortional strains of the skeleton. This effect, termed *perfect sampling* disturbance, occurs for even the most carefully obtained specimens (Ladd and Lambe, 1963; Skempton and Sowa, 1963; Noorany and Seed, 1965). The influence of these distortions appears in laboratory test results as discussed in Chapter 5.

Shrinkage of Cohesive Soils

Shrinkage of cohesive soils is the process of volume reduction in response to the capillary stresses induced by evaporation of water from the soil mass. Thus, the existence of an air-water interface is essential to the notion of shrinkage. Before evaporation of water begins, the soil is in equilibrium with the forces around it. If there are external forces acting on the soil, the inter-particle repulsive forces are just sufficient to counteract these external forces. As evaporation begins, the menisci develop in the water in every pore where there is an air-water interface. As evaporation continues, the radius of the meniscus, r_m, decreases. Since the pore diameter varies, r_m will be initially the radius corresponding to the largest pore diameter, as in Figure 6.12a. This is so because equilibrium requires that r_m be the same everywhere at the exterior of the soil mass. For fine-grained soils, all possible variations in pore diameter will be encountered very close to the surface. Thus, as evaporation continues, the menisci will quickly retreat a small distance to the point where the pore diameter is a minimum, as shown in Figure 6.12b. When all menisci are fully developed, the maximum possible capillary pressure has been exerted on the soil structure. At this stage, the soil is still saturated, and is at the *shrinkage limit*. Any further evaporation must result in a gradual retreat of the menisci into the interior of the soil and a loss of full saturation. At this point, the color of the soil usually becomes much lighter.

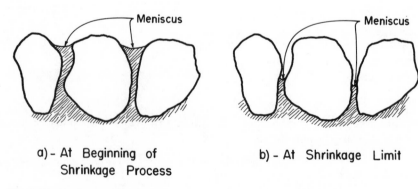

a) - At Beginning of Shrinkage Process

b) - At Shrinkage Limit

Fig. 6.12—Reduction in meniscus size during the shrinkage process.

The importance of the shrinkage process stems primarily from the fact that significant volume changes may occur in response to the increased effective stresses. In the case of building foundations on shrinking soils, this may result in undesirable movements of the foundation.* In embankments and earth structures, differential movement and cracking may result. The reason for this volume change can be seen by considering the interplay of forces involved in the soil during the shrinking process. These are illustrated in Figure 6.13. Recall that for a particular soil at a particular initial state, there is a more or less unique relationship between the void ratio and the all-round effective pressure. A typical void-ratio–pressure curve for a cohesive soil is illustrated in Figure 6.13a, as the solid curve. Furthermore, we recognize that at every void ratio, there will be some pore diameter which will be the minimum diameter in the soil. If we assume that it is possible to determine what this pore diameter is corresponding to each void ratio, then we can plot the relationship between void ratio and the smallest pore diameter, as in Figure 6.13b. Knowing the smallest pore diameter corresponding to a particular void ratio, we may compute the maximum capillary tension corresponding to that pore diameter and therefore to that void ratio. Such a curve is shown as a dashed line in Figure 6.13a.

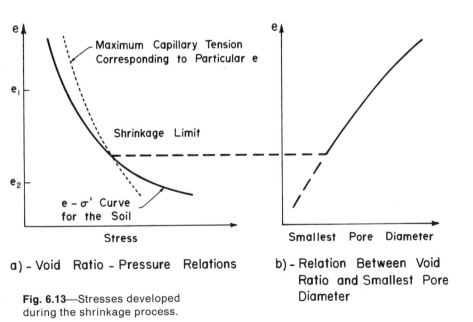

a) - Void Ratio - Pressure Relations

b) - Relation Between Void Ratio and Smallest Pore Diameter

Fig. 6.13—Stresses developed during the shrinkage process.

---

* Examples of structural damage accompanying foundation movements due to shrinkage are given by Bozozuk (1962), and Hammer and Thompson (1966) for cases in which the shrinkage was caused by removal of water by adjacent large trees. Zaporozhchenko and Volodin (1964) describe a case in which the shrinkage occurred during an unusually dry summer.

At a void ratio such as e_1 in Figure 6.13a, the maximum capillary tension that can be developed is larger than the effective stress required to compress the soil to a void ratio e_1 as the meniscus radius decreases. On the other hand, at a void ratio such as e_2, the maximum capillary tension which can be developed is less than the effective stress required to bring the soil to that void ratio. Thus, the soil could never be compressed to void ratio e_2 by the shrinkage process alone. At the point where the maximum capillary tension is just equal to the effective stress required to produce a particular void ratio, no further reduction in void ratio will occur as evaporation continues. This is the shrinkage limit.

The shrinkage limit can be determined experimentally by continuously measuring the volume and weight of a saturated soil as it shrinks. The results of such observations are given in Figure 6.14. If we measure the volume in cubic centimeters and the weight in grams, then shrinkage of a saturated soil should produce losses in volume and weight which are numerically equal, leading to a 45° line on the figure. At the shrinkage limit, a continued loss in weight causes no reduction in volume and the slope of the line should break sharply and become horizontal. The weight will continue to decrease until the soil is completely dry at which point its weight is that of the solid particles. Thus, the shrinkage limit can be determined from the point at which this line breaks. In practice, the shrinkage limit does not appear as a point, but lies on a curved portion, as shown by the dashed line in Figure 6.14. The reason for the curved portion is not fully understood, although it may result in part from deformation of the particles themselves, as well as some particle rearrangement.

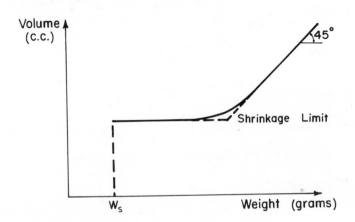

Fig. 6.14—Idealization of
the shrinkage process.

6.6 DARCY'S LAW—CONSERVATION OF LINEAR MOMENTUM

Darcy's Law

In Section 6.2 we developed the continuity equation to describe the conservation of mass for an incompressible fluid flowing through a porous medium. In Section 6.3 we discussed Bernoulli's equation expressing conservation of energy. We shall now consider conservation of linear momentum expressed in the form of an empirical constitutive law for flow.

As an outgrowth of experiments associated with the design of rapid sand filters for the city of Dijon, France, Henry Darcy developed an empirical relation describing flow of water through a sand medium (Darcy, 1856). Based on these one-dimensional experiments, illustrated schematically in Figure 6.15, Darcy concluded that the flow rate through a sand mass was directly proportional to the uniform cross-sectional area A and the difference in head Δh, and inversely proportional to the length of the flow path L:

$$q \propto \frac{A\,\Delta h}{L} \tag{6-45}$$

or

$$q = -k\,\frac{A\,\Delta h}{L} \tag{6-46}$$

in which the constant of proportionality k is called the *coefficient of permeability*. This coefficient is a measure of the ease with which the specific fluid considered, water, can flow through the soil. The negative sign in Equation 6-46 indicates that the flow is moving in a direction from higher head to lower head. This expression is more conveniently represented in terms of

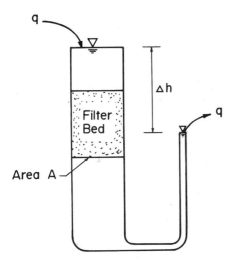

Fig. 6.15—Schematic diagram of Darcy's sand filtration experiment.

the discharge velocity,

$$v = -k\frac{\Delta h}{L} \tag{6-47}$$

Darcy's law has been generalized for the case where the cross-sectional area is not constant and the flow path may not be straight, so that

$$\mathbf{v} = -k\frac{\partial h}{\partial \mathbf{s}} \tag{6-48}$$

or

$$\mathbf{v} = k\mathbf{i} \tag{6-49}$$

in which $\mathbf{i} = -\partial h/\partial \mathbf{s}$ is the hydraulic gradient. Equations 6–46 through 6–49 are all forms of Darcy's law.

Although Darcy's law provides a useful constitutive relation for flow in a form which is amenable to analytical treatment, we must understand that it describes a fictitious velocity which does not correspond in general to the true velocity of the fluid at any given point within the medium.

Range of Validity of Darcy's Law

There appear to be both upper and lower limits to the magnitude of hydraulic gradient for which Darcy's law has been verified for a variety of soils and fluids. In investigating the upper limit, it has been found that Darcy's law accurately represents the flow through a porous medium if the flow is approximately laminar. For the normal *maximum* gradients or velocities encountered in soils, Muskat (1946) suggests that laminar flow can be expected in soils up to a range in size of medium to coarse sands. However, Anandakrishnan and Veradarajulu (1963) found nonlaminar flow in fine and medium sands under moderate hydraulic gradients, and suggested a nonlinear form of Darcy's law:

$$v = (k'i)^{1/n} \tag{6-50}$$

in which k' is a modified coefficient of permeability and n an exponent between 1 and 2. Because their data indicate that these values are not constant over a wide range of hydraulic gradients, and the methods of analysis in use at the present time in civil engineering practice do not permit the use of a nonlinear relationship between the velocity and hydraulic gradient, it is suggested that the linear form of Darcy's law be used in analyses, and that the coefficient of permeability be measured under hydraulic gradients of the same magnitude as anticipated in the field.

Laboratory permeability tests have also been conducted in which low gradients were applied to fine-grained saturated soils. As a result, some experimenters have reported finding a threshold or initial gradient which had to be exceeded before flow would take place (Oakes, 1960; Miller and Low,

1963). Others (Hansbo, 1960; Swartzendruber, 1962a,b, 1963) have suggested that an initial gradient does not exist but under a very low gradient, deviations from Darcy's law may be observed. More recent work (Olsen, 1965; Mitchell and Younger, 1966) suggests that the deviations from the predicted flow are not due to non-Darcian flow, but rather to a combination of experimental artifacts and changes in the soil fabric as flow is taking place. In either case these deviations from predicted flow using Darcy's law are likely to be most important in the prediction of the time rate of consolidation of cohesive soils in the field, where low hydraulic gradients occur. It is not clear, at this time, what magnitude of error results from using Darcy's law and a constant coefficient of permeability for such small gradients.

Permeability—Kozeny-Carman Equation

The coefficient of permeability in Darcy's law depends upon the properties of both the soil and the fluid flowing through it. Although the coefficient of permeability can be measured directly by a variety of methods, it is useful to identify the independently measurable soil parameters which influence its magnitude. Unfortunately, there is no theoretical model of a soil mass which leads to a correct prediction of the permeability based upon such independently measurable parameters. However, considerable insight into the physical nature of permeability can be gained by considering the soil mass to consist of a series of tubes of variable cross section.

We begin our investigation of this problem by examining the flow of an incompressible viscous fluid between a pair of closely spaced parallel plates. This is illustrated in Figure 6.16a, which shows the plates inclined at an angle α to the horizontal and a small element of fluid between them. The forces which act on the small element are indicated in Figure 6.16b. We shall assume that flow is viscous and laminar, that is, all fluid "particles" move along smooth paths which do not intersect, and that viscous forces are governed by Newton's law of viscosity,

$$\tau = -\eta \frac{\partial v'_s}{\partial r} \tag{6-51}$$

in which τ is the shear stress parallel to the flow direction, η is the viscosity, and v'_s is the actual velocity in the flow direction at a point in the fluid. If we assume that the time rate of change of velocity is small enough so that inertial forces may be neglected, the forces acting on the small element of fluid shown in Figure 6.16b must satisfy the requirements of equilibrium. Thus, for a unit thickness into the page,

$$\left[u\left(r, s + \frac{\Delta s}{2}\right) - u\left(r, s - \frac{\Delta s}{2}\right) \right] \Delta r + \Delta W \sin \alpha$$
$$+ \left[\tau\left(r + \frac{\Delta r}{2}, s\right) - \tau\left(r - \frac{\Delta r}{2}, s\right) \right] \Delta s = 0 \tag{6-52}$$

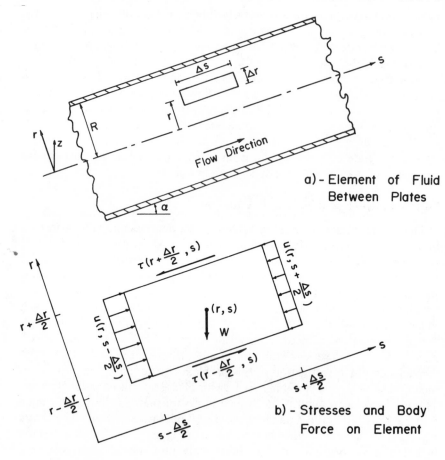

Fig. 6.16—Fluid flow between inclined parallel plates.

Recognizing that

$$\sin \alpha = \frac{\Delta z}{\Delta s}$$

and

$$\Delta W = \gamma_w \, \Delta r \, \Delta s$$

and dividing through by the volume of the element, Equation 6–52 becomes

$$\frac{u\left(r, s + \dfrac{\Delta s}{2}\right) - u\left(r, s - \dfrac{\Delta s}{2}\right)}{\Delta s} + \gamma_w \frac{\Delta z}{\Delta s}$$

$$+ \frac{\tau\left(r + \dfrac{\Delta r}{2}, s\right) - \tau\left(r - \dfrac{\Delta r}{2}, s\right)}{\Delta r} = 0 \qquad (6\text{–}53)$$

Taking the limit of Equation 6–53 as $\Delta r \to 0$ and $\Delta s \to 0$ leads to

$$\frac{\partial u}{\partial s} + \gamma_w \frac{dz}{ds} + \frac{\partial \tau}{\partial r} = 0 \qquad (6\text{--}54)$$

Substituting

$$h = \frac{u}{\gamma_w} + z$$

into this expression gives

$$\frac{\partial \tau}{\partial r} = -\gamma_w \frac{\partial h}{\partial s} \qquad (6\text{--}55)$$

We note that the right side of Equation 6–55 has been shown previously to be a seepage force. It is evident now that this can be interpreted as a drag, or shear force exerted by the flowing fluid on the solid material. Substituting Equation 6–51 into Equation 6–55,

$$\frac{\partial^2 v_s'}{\partial r^2} = \frac{\gamma_w}{\eta} \frac{\partial h}{\partial s} \qquad (6\text{--}56)$$

Because the parallel plates are assumed to extend indefinitely in the s-direction v_s' is independent of s, and the left side of Equation 6–56 becomes an ordinary derivative. The assumption of flow parallel to the plate also requires that h be independent of r. Thus the right side of Equation 6–56 must be independent of r and the equation may be integrated directly. Assuming symmetry of flow about the midplane (defined by the x axis) and no slip at the surface of the plates, integration of Equation 6–56 leads to

$$v_s' = \frac{\gamma_w}{2\eta}(r^2 - R^2)\frac{\partial h}{\partial s} \qquad (6\text{--}57)$$

The average flow velocity between the plates is

$$\bar{v}_s' = \frac{1}{2R}\int_{-R}^{R} v_s' \, dr = -\frac{\gamma_w}{3\eta} R^2 \frac{\partial h}{\partial s} \qquad (6\text{--}58)$$

It is often convenient to write this expression in terms of the hydraulic radius, R_H, defined as

$$R_H = \frac{\text{cross-sectional area of flow channel}}{\text{wetted perimeter}}$$

For the case of flow between parallel plates, in which the fluid is simply a strip, $R_H = R$, and

$$\bar{v}_s' = -\frac{\gamma_w}{3\eta} R_H^2 \frac{\partial h}{\partial s} \qquad (6\text{--}59)$$

This expression may be generalized for individual flow channels of any cross-section shape as

$$\bar{v}_s' = -\frac{\gamma_w}{\bar{C}_s \eta} R_H^2 \frac{\partial h}{\partial s} \qquad (6\text{–}60)$$

$\bar{C}_s$ is a shape factor, corresponding to a particular-shaped channel. Shape factors for a variety of channel types are given in Table 6.2. Equation 6–60 is often called the generalized *Hagen-Poiseuille equation.*

TABLE 6.2
Values of Shape Factor $\bar{C}_s$ for Generalized Hagen-Poiseuille Equation

Flow-Channel Shape	$\bar{C}_s$
Strip	3.00
Square	1.78
Circle	2.00
Equilateral triangle	1.68

The Hagen-Poiseuille equation applies to viscous flow in a single channel. In order to relate this expression to flow in a soil mass consisting of many sinuous channels of variable shape and size we take note of the following:

1. The term $-\partial h/\partial s$ in Equation 6–60 is the free-energy gradient along the actual flow path. The hydraulic gradient to which we generally refer in a soil mass is the gradient along the average flow direction. That is, even though flow will occur in the x direction, for instance, a given flow channel may wind about and at any given point in the soil mass may not be oriented parallel to the average flow direction. Thus, considering the x component as an example,

$$i_x = -\frac{\partial h}{\partial x} = -\frac{\partial h}{\partial s}\frac{ds}{dx}$$

2. In a similar fashion, the average flow velocity within each flow channel, $\bar{v}_s'$, will be different from the seepage velocity component which acts in the direction of "average" flow. That is,

$$v_{s_x} = \bar{v}_s' \frac{dx}{ds}$$

3. We define the tortuosity T_0 as

$$T_0 = \frac{ds}{dx}$$

for flow in the x direction, and in an analogous manner for flow in the other directions. This parameter is presumably a measure of the degree of sinuous flow in which ds is the length of the actual flow path and dx is the flow length in the direction of "average" flow.

4. The average hydraulic radius for the entire soil mass may be written as

$$R_H = \frac{e}{S_s}$$

in which S_s is the *specific surface* defined by

$$S_s = \frac{\text{surface area of solids}}{\text{volume of solids}}$$

The specific surface of a soil mass depends upon the grain sizes and shapes. It has a particularly important effect on the properties of fine-grained soils, and is discussed in more detail in Chapter 9. At the present time we shall note simply that for perfect spheres of a single diameter D the specific surface is

$$S_{\text{spheres}} = \frac{6}{D} \qquad (6\text{-}61)$$

5. The discharge velocity can be related to the seepage velocity, say in the x direction, from Equation 6–D (page 261) by

$$v_{s_x} = v_x \left(\frac{1 + e}{e} \right)$$

Incorporating these factors into Equation 6–60, the discharge velocity, for example for flow in the x direction, becomes

$$v_x = \frac{1}{\bar{C}_s T_0^2 S_s^2} \frac{e^3}{(1 + e)} \frac{\gamma_w}{\eta} i_x \qquad (6\text{-}62)$$

This is known as the Kozeny-Carman equation (Carman, 1956). Comparing this expression with Equation 6–49 for Darcy's law, we see that the coefficient of permeability is represented by

$$k = \frac{1}{\bar{C}_s T_0^2 S_s^2} \frac{e^3}{(1 + e)} \frac{\gamma_w}{\eta} \qquad (6\text{-}63)$$

Thus the Kozeny-Carman equation provides an opportunity to evaluate the soil parameters which are suggested by our model to be important in determining the coefficient of permeability.

Equation 6–63 implies that the permeability of a soil depends upon physical properties of the fluid (γ_w and η), and geometric properties of the soil solids and the pore spaces between them ($\bar{C}_s$, T_0, e, and S_s). If we compare the relative significance of the terms, we find:

1. For a given fluid such as water and a moderate range in temperatures, γ_w and η remain approximately constant.

2. Table 6.2 states that the shape factor varies from a maximum value of 3 for flow between two parallel plates to approximately 60 per cent of that value when the flow channels are equilateral triangles. Although more complex cross section shapes will produce still smaller shape factors, the variation is of the order of a factor of two. It has been found that for a range of grain-size distributions, the variation in the tortuosity is also modest. Because the tortuosity tends to increase as the range of grain sizes in a specimen increases, the product $\bar{C}_s T_0^2$ tends to remain approximately constant. It has been found that for sand and silt-size particles $\bar{C}_s T_0^2 = 5$ is a good approximation (Carman, 1956; Wyllie and Spangler, 1952).

3. The influence of void ratio can be seen by considering a soil mass with a given distribution of particles so that the specific surface remains constant. The term

$$\frac{e^3}{1 + e}$$

supports our intuitive view that a decrease in void ratio, producing a decrease in spacing between particles, is reflected in a corresponding decrease in permeability. The magnitude of this effect for a cohesionless soil can be estimated using data from Figure 2.13. For the grain-size distribution corresponding to point A, the maximum void ratio is 0.8 and the minimum is 0.35. The ratios of the corresponding coefficients of permeability is approximately 10:1.

4. The specific surface is influenced most significantly by the particle size. Thus the permeability of two samples of uniform size spheres at the same void ratio, one with grain diameter of 1.0 mm (sand size), the other with grain diameter 0.01 mm (silt size) will be in the ratio of 10,000:1.

So, we see that the most significant factor in determining the permeability of a soil is the particle size.

This was recognized by Hazen (1911), who proposed an empirical expression for permeability of loose *clean* sands and gravels:

$$k = C_1 D_{10}^2$$

in which D_{10} is the effective grain size (see Section 2.4). When D is expressed in centimeters and k in centimeters per second, C_1 is approximately 100.

Typical values for the coefficient of permeability are given in Table 6.3. It may seem surprising that k varies over a range of nearly 100 billion times. This is a much greater range than occurs for most physical properties with which we are familiar. Values of k for various soils are shown in Table 6.4.

The observations discussed above depend upon the validity of Equation 6–63 in describing k for a real soil. For sandy and silty soils, this equation has been shown to be approximately correct (Carman, 1956). In the case of clay soils however, Equation 6–63 does not adequately predict the permeability. There are a number of reasons for this (Leonards, 1962) but the

TABLE 6.3
Permeability and Drainage Characteristics of Soils

Coefficient of Permeability k in cm per sec (log scale)

$10^2 \quad 10^1 \quad 1.0 \quad 10^{-1} \quad 10^{-2} \quad 10^{-3} \quad 10^{-4} \quad 10^{-5} \quad 10^{-6} \quad 10^{-7} \quad 10^{-8} \quad 10^{-9}$

Drainage

Good (from $\sim10^1$ to $\sim10^{-3}$) — Poor (from $\sim10^{-3}$ to $\sim10^{-6}$) — Practically Impervious (below $\sim10^{-6}$)

Soil types

- Clean gravel
- Clean sands, clean sand, and gravel mixtures
- Very fine sands, organic and inorganic silts, mixtures of sand silt and clay, glacial till, stratified clay deposits, etc.
- "Impervious" soils, e.g., homogeneous clays below zone of weathering
- "Impervious" soils modified by effects of vegetation and weathering

Direct determination of k

- Direct testing of soil in its original position-pumping tests. Reliable if properly conducted. Considerable experience required.
- Constant-head permeameter. Little experience required.

Indirect determination of k

- Falling-head permeameter. Reliable. Little experience required.
- Falling-head permeameter. Unreliable. Much experience required.
- Falling head permeameter. Fairly reliable. Considerable experience necessary.
- Computation from grain-size distribution, i.e., Hazen's formula. Applicable only to clean cohesionless sands and gravels.
- Computation based on results of consolidation tests. Reliable. Considerable experience required.

Source: After Casagrande and Fadum (1940).

TABLE 6.4
Coefficients of Permeability of Common
Natural Soil Formations

Formation	k (cm/sec)
River deposits	
Rhone at Genissiat	up to 0.4
Small streams, eastern Alps	0.02–0.16
Missouri	0.02–0.20
Mississippi	0.02–0.12
Glacial deposits	
Outwash plains	0.05–2.00
Esker, Westfield, Mass.	0.01–0.13
Delta, Chicopee, Mass.	10^{-4}–0.015
Till	$< 10^{-4}$
Wind deposits	
Dune sand	0.1 –0.3
Silt (loess)	0.001 $\pm$
Silt (loess loam)	$10^{-4}\pm$
Lacustrine and marine offshore deposits	
Very fine uniform sand, $C_u = 5$ to 2	10^{-4}–0.0064
Silt (Bull's liver), Sixth Ave.,	
N.Y.C., $C_u = 5$ to 2	10^{-4}–0.0050
Silt (Bull's liver), Brooklyn, $C_u = 5$	10^{-5}–10^{-4}
Clay	$< 10^{-7}$

Source: Modified from Terzaghi and Peck (1967).

most important by far stems from the arrangement or *fabric* of clays. We recall that in the derivation of Equation 6–63 it was assumed that the hydraulic radius for the pores could be computed as an average value by using the specific surface for the entire mass of soil. In the case of clay soils, however, this assumption is frequently unjustified because the particles are packed into many clusters, and the pore space between particles *within* the clusters is very small while the pore space *between* clusters is much larger. It is intuitively evident that the presence of a relatively few large pores will have an overwhelming influence on the ease of flow. In fact we recognize that the flow rate is approximately proportional to the fourth power of the pore radius. Olsen (1960) has shown that the discrepancies between Equation 6–63 and values of k measured in the laboratory could be explained by incorporating the effects of the variation in pore sizes. Thus the factors incorporated in Equation 6–63 are significant, but, in the case of clays, not sufficient to predict k.

Measurement of the Coefficient of Permeability

A variety of methods have been developed to measure or infer the magnitude of the coefficient of permeability. We shall mention the most commonly used methods briefly here. Permeability tests fall into two broad categories:

laboratory tests performed on a representative specimen of soil; and field tests performed on the *in situ* material, usually involving much larger masses of soil. We consider these separately:

1. *Laboratory tests.* Tests performed in the laboratory on representative specimens of soil may also be divided into two categories: direct tests which require no assumptions other than the validity of Darcy's law to permit determination of k; and indirect tests in which some additional assumptions are imposed to produce a theory from which k may be inferred.

    a. Direct tests. The constant-head permeability test. The constant-head test is conceptually the same as the experiment from which Darcy inferred the relation that bears his name. It is illustrated schematically in Figure 6.17a. If, in this test, the flow is maintained long enough so that the flow rate is constant, the hydraulic gradient across the sample length must be constant:

$$i_z = -\frac{\partial h}{\partial z} = -\frac{\Delta h}{L} \qquad (6\text{-}64)$$

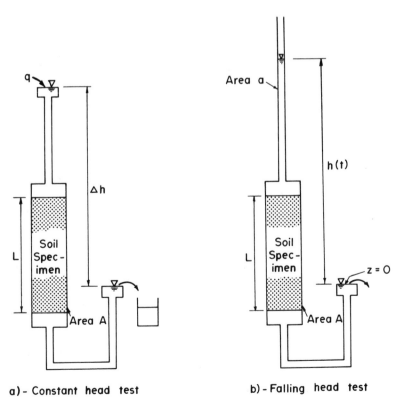

a) - Constant head test   b) - Falling head test

Fig. 6.17—Schematic representations of laboratory permeability tests.

Then k is found from

$$k = -\frac{qL}{A\,\Delta h} \tag{6-65}$$

Unless excess air pressure is applied to the top of the inlet reservoir to increase the head loss markedly, the constant-head permeability test is restricted to relatively permeable soils, as indicated in Table 6.3. For soils for which $k < 10^{-3}$ cm/sec the time involved in the usual constant-head permeability test is impractically long.

b. Indirect tests. The indirect laboratory tests required additional assumptions about some feature of the flow problem in order to develop a theory for determining k.

(1) Falling-head permeability test. The falling-head test is illustrated schematically in Figure 6.17b. If we define the elevation datum to be at the surface of the outlet reservoir, then the total head loss across the sample at any time t is

$$\Delta h(t) = 0 - h(t) \tag{6-66}$$

in which $h(t)$ is the height of the water in the standpipe above the outlet at time t (Figure 6.17b). If we *assume* that the change in position of the water in the standpipe is sufficiently slow, then we may approximate the average hydraulic gradient in the sample as a constant at any time,

$$i_z(t) = \frac{h(t)}{L} \tag{6-67}$$

From which, the flow rate through the specimen is

$$q(t) = k\frac{h}{L}A \tag{6-68}$$

where A is the cross-sectional area of the specimen. The flow rate can also be written as

$$q(t) = -\frac{dh}{dt}a \tag{6-69}$$

in which a is the cross-sectional area of the standpipe above the specimen. Equating these two expressions and rearranging terms we have

$$-\frac{dh}{h} = \frac{k}{L}\frac{A}{a}dt \tag{6-70}$$

If we note the position of the water in the standpipe as h_0 and h_1 at times t_0 and t_1, respectively, we can integrate Equation 6-70 between these limits to obtain an expression for k:

$$k = \frac{a}{A} \frac{L}{(t_1 - t_0)} \ln \left(\frac{h_0}{h_1} \right) \qquad (6\text{--}71)$$

Detailed discussion of the conduct of such tests, and potential experimental errors are given in various laboratory manuals (see, for example, Lambe, 1951, or Bowles, 1970).

(2)   Consolidation tests. The coefficient of permeability can be inferred from the results of consolidation tests when interpreted using an appropriate theory. The method for doing this is discussed in detail in Chapter 7, and we shall defer consideration until then.

2.   *Field tests.* There are a variety of field tests which permit *in situ* determination of the average coefficient of permeability of a large mass of soil. All such tests require some assumptions concerning the nature of the flow patterns throughout a large mass of soil leading to the solution of a boundary value problem which approximates the flow regime imposed or observed during the test. A number of the most important methods are described succinctly by Golder and Gass (1963) and Cedergren (1967).

Field tests have the advantage that the material being tested is the *in situ* stratum, and that the difficulty of obtaining a useful representative specimen for permeability tests in the laboratory is obviated. Field testing has the disadvantages, however, of being generally more expensive and frequently more time-consuming than laboratory testing.

An extensive list of references about permeability testing is given by Johnson and Richter (1967).

Example 6.4

A falling-head permeability test is conducted on a saturated sample of clean fine to medium sand. The soil specimen, 10 cm² in cross-sectional area is 15 cm long. The water level in the standpipe above the sample falls from a height of 40 cm to 20 cm above the free-water surface outlet during an elapsed time of one minute. The cross-sectional area of the standpipe is 1 cm².

(a)   Determine the coefficient of permeability of the sand.

(b)   If k for the sand were 10 times as large as that determined in part a above, would a falling-head test be useful in determining its magnitude?

Solution. (a) Substituting the measured parameters into Equation 6–71 we can solve directly for k:

$$k = \frac{(1)(15)}{(10)(60)} \log_e \left(\frac{40}{20} \right) = 0.017 \text{ cm/sec}$$

This magnitude is appropriate for the soil type described in the problem statement.

(b) It is apparent from Equation 6–71 that the elapsed time is inversely proportional to the coefficient of permeability. So, if k were 100 times greater than determined above, the elapsed time would be 1/100th of that measured for part a, that is, 0.6 sec. Clearly, this would impose experimental difficulties which would make the test much less useful for more coarse-grained material.

Inhomogeneous and Anisotropic Permeability

Because Darcy's law applies at a point, we have not so far considered the effect of variations in the coefficient of permeability from point to point. Although this problem is treated more appropriately in Chapter 14, two special cases are of particular interest here. These are linear flow normal to or parallel to, the layers in a parallel multilayered system illustrated in Figure 6.18. Knowing the head loss across the system, it is often useful to calculate the flow rate for these special cases. This can be accomplished conveniently by determining an equivalent permeability $\overline{k}$ such that the flow rate is

$$q = -\overline{k}\frac{\Delta h_T}{L_T}A \qquad (6\text{--}72)$$

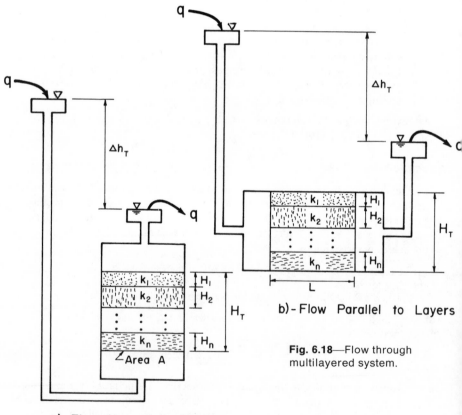

b)- Flow Parallel to Layers

Fig. 6.18—Flow through multilayered system.

a)- Flow Normal to Layers

in which Δh_T is the total head loss across the system, L_T is the length of flow path, A is the cross-sectional area of the system normal to the flow direction. We can determine the equivalent permeability for the case of flow normal to the planes of bedding (Figure 6.18a) by noting that the total head difference across the system, Δh_T, is the sum of the head loss through the individual layers:

$$\Delta h_T = \Delta h_1 + \Delta h_2 + \cdots + \Delta h_n \qquad (6\text{-}73)$$

In addition, the flow rate must be the same in all layers for steady-state flow:

$$\frac{q}{A} = -k_1 \frac{\Delta h_1}{H_1} = -k_2 \frac{\Delta h_2}{H_2} = \cdots = -k_n \frac{\Delta h_n}{H_n} \qquad (6\text{-}74)$$

Rearranging the terms in Equations 6–74 and substituting them into Equation 6–73, we have

$$\Delta h_T = -\frac{q}{A}\left(\frac{H_1}{k_1} + \frac{H_2}{k_2} + \cdots + \frac{H_n}{k_n}\right) \qquad (6\text{-}75)$$

From this expression we see that the equivalent permeability as defined in Equation 6–72, for flow *normal* to the layers, is

$$\bar{k} = \frac{H_T}{\dfrac{H_1}{k_1} + \dfrac{H_2}{k_2} + \cdots \dfrac{H_n}{k_n}} \qquad (6\text{-}76)$$

If the flow is parallel to the bedding planes, as shown in Figure 6.18b, the total flow rate must equal the sum of the flow rates in each of the layers,

$$q = q_1 + q_2 + \cdots + q_n \qquad (6\text{-}77)$$

and the head loss across all layers is equal to the total head loss across the system:

$$\Delta h_T = -\frac{q_1}{H_1} \frac{L}{k_1} = -\frac{q_2}{H_2} \frac{L}{k_2} = \cdots = -\frac{q_n}{H_n} \frac{L}{k_n} \qquad (6\text{-}78)$$

in which the cross-sectional area for the ith layer, assuming a unit distance into the page, is simply the height of the layer H_i. Rearranging terms in Equation 6–78 and substituting into Equation 6–77 gives the flow rate for a unit distance into the page:

$$q = -\frac{\Delta h_T}{L}(k_1 H_1 + k_2 H_2 + \cdots k_n H_n) \qquad (6\text{-}79)$$

From Equations 6–79 and 6–72 we find that the equivalent permeability for flow *parallel* to the strata is

$$\bar{k} = \frac{k_1 H_1 + k_2 H_2 + \cdots + k_n H_n}{H_T} \qquad (6\text{-}80)$$

Recognizing the analogy between the coefficient of permeability in Darcy's law and the conductance (the reciprocal of resistance) in Ohm's law, we can observe that fluid flow normal and parallel to the layers in a multilayered system is analogous to the flow of electric current through resistances in series and parallel, respectively. The influence of the relative magnitudes of the coefficients of permeability is illustrated in the following example.

Example 6.5 ———————————————————————————————————

For the two-layer system shown in Figure 6.19, determine the equivalent permeabilities for flow normal to and parallel to the bedding.

Solution. For flow *normal* to the bedding we find, from Equation 6–76

$$\bar{k} = \frac{2}{\dfrac{1}{10^{-4}} + \dfrac{1}{10^{-2}}} = \frac{2 \times 10^{-6}}{0.0101} = \underline{1.98 \times 10^{-4}\ \text{ft/min}}$$

We note that the equivalent permeability in this case is only twice the permeability of the silt even though the permeability of the sand is 100 times greater than that of the silt.

In the case of flow *parallel* to the bedding, Equation 6–80 gives

$$\bar{k} = \frac{(10^{-4})(1) + (10^{-2})(1)}{2} = \underline{0.51 \times 10^{-2}\ \text{ft/min}}$$

which shows that the equivalent permeability is only half of that of the sand in spite of the factor of 100 in the ratio of the permeabilities.

———

This illustrates the fact that when there are two strata in a multilayered system, one of which is much less permeable than the others, and the other of which is much more permeable, then the impermeable stratum will tend to control the flow normal to the bedding whereas the permeable stratum will tend to control the flow parallel to the bedding.

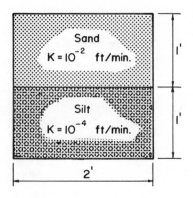

Fig. 6.19—Soil system for Example 6.5.

In our discussion of Darcy's law, we have thus far considered permeability to be a scalar property of a soil mass and the fluids flowing through it. It is very common, however, for the permeability in a mass of soil to exhibit anisotropy, that is, to be different in different directions at the same point. Anisotropic permeability in a soil mass may arise from a variety of sources. Among the most common of these is the tendency for a horizontal orientation of particles in a cohesive soil mass due to the developing overburden stresses during deposition. Another common feature is stratification due to seasonal and other changes in the deposition process. These and a variety of other features often lead to the usual circumstance that the horizontal permeability is greater than the vertical permeability. Many examples are available (Muskat, 1946; Hvorslev, 1951) for both cohesionless and cohesive materials. It is not uncommon for the ratio of horizontal to vertical coefficient of permeability to be of the order of 10 to 100, although lower values have been noted (Mansur and Dietrich, 1965). Theoretical considerations suggest that the coefficient of permeability for a general anisotropic porous mass can be written as a symmetric matrix (Scheidegger, 1957):

$$[k] = \begin{bmatrix} k_{xx} & k_{xy} & k_{xz} \\ k_{xy} & k_{yy} & k_{yz} \\ k_{xz} & k_{yz} & k_{zz} \end{bmatrix} \tag{6-81}$$

Thus in its most general form, Darcy's law in terms of components becomes

$$v_x = -k_{xx}\frac{\partial h}{\partial x} - k_{xy}\frac{\partial h}{\partial y} - k_{xz}\frac{\partial h}{\partial z}$$

$$v_y = -k_{xy}\frac{\partial h}{\partial x} - k_{yy}\frac{\partial h}{\partial y} - k_{yz}\frac{\partial h}{\partial z} \tag{6-82}$$

$$v_z = -k_{xz}\frac{\partial h}{\partial x} - k_{yz}\frac{\partial h}{\partial y} - k_{zz}\frac{\partial h}{\partial z}$$

In the case where the coordinate axes are also the principal directions of permeability so that the off-diagonal elements of the matrix are zero, the generalized form of Darcy's law appears as commonly written:

$$v_x = -k_x\frac{\partial h}{\partial x}$$

$$v_y = -k_y\frac{\partial h}{\partial y} \tag{6-83}$$

$$v_z = -k_z\frac{\partial h}{\partial z}$$

We shall assume in our further developments that the coordinate axes do correspond to principal axes and that Equations 6–83 are the appropriate form of Darcy's law.

6.7   THE GENERAL DIFFUSION EQUATION

By comparing Darcy's law with Equation 6–62, which was derived from the requirements of the conservation of energy and momentum, we see that Darcy's law is, in fact, a statement of the conservation of energy and momentum relations. We are then ready to derive the general diffusion equation by substituting requirements of energy and momentum conservation in the form of Darcy's law, Equations 6–83, into the continuity relation, Equation 6–9:

$$\frac{\partial}{\partial x}\left(k_x \frac{\partial h}{\partial x}\right) + \frac{\partial}{\partial y}\left(k_y \frac{\partial h}{\partial y}\right) + \frac{\partial}{\partial z}\left(k_z \frac{\partial h}{\partial z}\right) = \frac{1}{1 + e_0}\frac{\partial e}{\partial t} \qquad (6\text{–}84)$$

If the coefficient of permeability is independent of position within the soil mass, that is, the soil is homogeneous, Equation 6–84 may be rewritten

$$k_x \frac{\partial^2 h}{\partial x^2} + k_y \frac{\partial^2 h}{\partial y^2} + k_z \frac{\partial^2 h}{\partial z^2} = \frac{1}{1 + e_0}\frac{\partial e}{\partial t} \qquad (6\text{–}85)$$

This expression is the general diffusion equation governing flow of an incompressible fluid through a homogeneous saturated porous medium.

If our interest is restricted to the problem of steady-state seepage for which

$$\frac{\partial e}{\partial t} = 0$$

then Equation 6–85 becomes

$$k_x \frac{\partial^2 h}{\partial x^2} + k_y \frac{\partial^2 h}{\partial y^2} + k_z \frac{\partial^2 h}{\partial z^2} = 0 \qquad (6\text{–}86)$$

If, furthermore, the soil is isotropic, so that

$$k_x = k_y = k_z = k$$

the expression for steady-state seepage becomes

$$\frac{\partial^2 h}{\partial x^2} + \frac{\partial^2 h}{\partial y^2} + \frac{\partial^2 h}{\partial z^2} = 0 \qquad (6\text{–}87)$$

which is the well-known Laplace's equation.

In the following chapter we shall consider the problem of one-dimensional consolidation, Equation 6–85, reduced to one dimension, and defer discussion of the steady-state seepage problem to Chapter 14.

6.8 SUMMARY OF KEY POINTS

We summarize below the key points discussed in the preceding sections. Reference is made to the place in the text where the original more detailed description can be found:

1. Flow of an incompressible fluid through a porous medium is governed by the requirements of:
   a. Conservation of mass (Section 6.2). The requirement is described by the continuity equation (Equation 6–9).
   b. Conservation of energy (Section 6.3). This is expressed by the Bernoulli equation for a slowly flowing fluid (Equation 6–12).
   c. Conservation of linear momentum (Section 6.6). This is combined with the conservation of energy into an empirical constitutive relation for flow, Darcy's law (Equation 6–49).
   d. Combining Darcy's law with the continuity equation leads to the general diffusion equation. If the soil is homogeneous, this is Equation 6–85. If flow is steady-state, and the soil is isotropic, the resulting expression is Laplace's equation (Equation 6–87).
2. Darcy's law introduces a parameter, the coefficient of permeability k, which is a function of both soil parameters (grain-size distribution, grain shape, void ratio) and fluid properties (density, viscosity). For natural soils, k may vary from approximately 10 cm/sec for clean gravel, to 10^{-8} cm/sec for homogeneous clays (Section 6.6). The magnitude of k can be determined by a variety of laboratory and field tests.
3. The hydraulic gradient i, induces a body force, the seepage force $i\gamma_w$ (Section 6.3). This effective stress acts in a direction parallel to the hydraulic gradient at each point. When i acts vertically upward with magnitude sufficient to overcome the submerged weight of a cohesionless soil (Equation 6–24), quicksand develops.
4. Capillarity is a consequence of the attraction between soil grains and water in the presence of an air-water interface. Capillary effects can be predicted using the concept of surface tension. The height of capillary rise and/or capillary tension which can be developed in a soil depends upon the size of the void spaces, as well as whether the source of water is rising or falling (Section 6.4). Capillary effects include bulking of cohesionless soils, perfect sampling disturbance of cohesive soils, shrinkage and prestress of cohesive soils due to desiccation (Section 6.5).
5. One-dimensional steady-state flow through layered systems can be conveniently described in terms of an equivalent coefficient of permeability whose magnitude is determined by Equation 6–76 if the flow is normal to the layers, and by Equation 6–80 if the flow is parallel to the layers.

PROBLEMS

6.1. From the data given on the sketch, determine the head h in the region between the two materials, and find the discharge q in cc/sec.

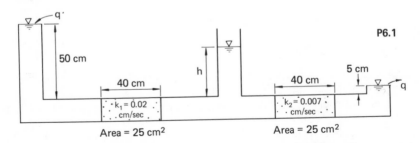

P6.1

6.2. The accompanying sketch illustrates a case in which water is seeping from a reservoir into an open trench, 1000 ft away, through a confined stratum of fine sand overlain and underlain by impervious material. The sand stratum is 20 ft deep at the reservoir and decreases to 10 ft thick about midway to the trench.

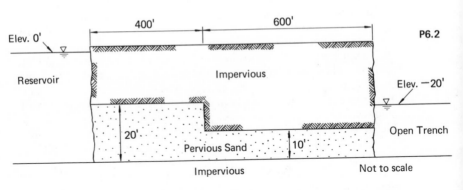

P6.2

In order to maintain the water level in the trench 20 ft below the reservoir surface elevation, it is found necessary to remove 1 ft³ of water per hour per foot of trench. What is the coefficient of permeability of the sand, in ft/min? Assume the reservoir and trench sides to be vertical, and assume the change of section of the sand stratum to be abrupt.

6.3. Two reservoirs are connected by a sand trench, as shown. In placing the fill material through water, the soil was segregated. The fines settled more slowly, resulting in a varying permeability. Measurements indicate that the permeability along the trench can be approximated as

$$k(s) = k_0 \left(1 - 0.9 \frac{s}{L} \right)$$

(a) Determine the quantity of flow between the reservoirs, per foot into the page, as a function of k_0, z_1, z_2, and B.

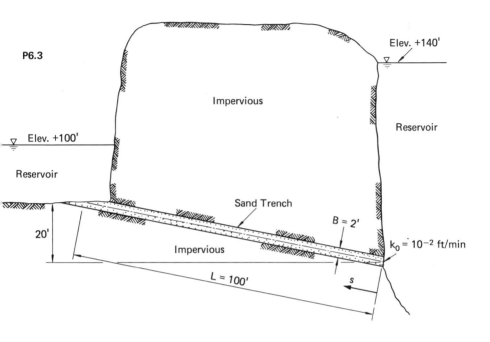

P6.3

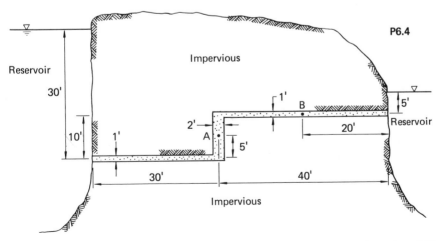

(b) Determine the factor of safety against a quick condition for the numerical values given.

Note: The trench extends a "long" distance into page.

6.4. Water is flowing through a trench of varying cross-sectional thickness, shown in profile. The trench, which is long into the page, contains a

homogeneous sand. Determine the *pressure u* in the water at points *A* and *B* *for the water levels shown.*

6.5. Water is being pumped from the sand aquifer shown. As a consequence, the water *pressure* at point A is $u_A = 62.4$ psf, and is the same at all points which are just at the top of the sand. Determine:

 (a) The quantity of flow through the silt stratum in cubic feet/minute/ square foot of plan area.

 (b) The water *pressure* at point B.

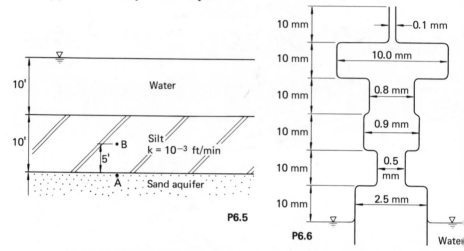

P6.5

P6.6

6.6. The capillary tube shown has inside dimensions indicated on the sketch. If it is clean and moist:

 (a) How high will distilled water rise in the tube, for the situation shown in the sketch?

 (b) Where will the water level be in the tube if the tube is inserted over its full height into the water, and then pulled out to the position shown in the sketch?

6.7.

 (a) Why would the shrinkage limit of a clay be different if the water in the voids were replaced by some other liquid with a *smaller* surface tension.

 (b) Would the sample in the second case shrink more or less? Why?

6.8. Derive an expression for the shrinkage limit in terms of the following parameters:

V_1 = volume of a saturated pat of soil

W_1 = weight of a saturated pat of soil

V_2 = volume of same pat when completely dry

W_2 = weight of same pat when completely dry

7

Consolidation and Time Rate of Settlement

7.1 INTRODUCTION

The consolidation process involves expulsion of water from the soil being compressed. The movement of water through and out of a soil mass requires time; the time rate of consolidation is controlled principally, though not entirely, by the permeability of the soil. In the case of a clay soil in the field, the time required to reach equilibrium may be of the order of days, months, or years. Thus, in reconditioning Building 301 at Newark Airport ("Surcharging a Big Warehouse Floor Saves $1 Million," pages 161–166), the Port of New York Authority engineers had to decide how long to leave the surcharge loading in place in order to reduce postconstruction settlements to an acceptable magnitude. As we shall see in Chapter 11, the waiting time for consolidation due to a surcharge can be modified by changing the magnitude of the surcharge load. In the case of Building 301, the interior building frame limited the height of the surcharge fill. In other situations the controlling factors are often the strength of the underlying soil (Chapter 8) or the cost of the fill.

In this chapter we shall concentrate our attention on the development of a mathematical description of the consolidation process, solution of the resulting equation, and the use of the solution results for predicting the time rate of consolidation settlement of a cohesive soil. The mathematical description by Karl Terzaghi of the consolidation process, which included elucidation of the principle of effective stress (Terzaghi, 1923, 1924*), marked the beginning of soil mechanics as an organized discipline within civil engineering.

---

* These pioneer works appeared in German. Terzaghi introduced the civil engineering profession in the United States to the principles of soil mechanics in a classic series of articles in *Engineering News-Record* (Terzaghi, 1925).

7.2 ONE-DIMENSIONAL CONSOLIDATION

Equation 6–84 is the general three-dimensional form of the consolidation equation. We shall restrict our use of the term *consolidation* to the hydro-dynamic process, of which Equation 6–84 is a simplified representation, and use *consolidation settlement* for the settlement resulting only from this process. Although a numerical procedure will be presented that can be extended to the solution of the three-dimensional form of Equation 6–84, it is only the solution of the one-dimensional case and some special two-dimensional cases which have found significant application to civil engineering practice. In a later section, we shall consider some of the factors which limit the use of the more general formulation.

Before beginning, let us reiterate the assumptions we have made in developing Equation 6–84:

1. The soil is fully saturated.
2. The pore fluid and soil solids are both incompressible.
3. The change in the volume of the voids is sufficiently small that small-strain theory is applicable.
4. Darcy's law is valid.

We now add the following assumption:

5. Flow takes place in only one direction.

This means that two of the terms on the left side of Equation 6–84 are zero. For flow in the vertical direction, we have then

$$\frac{\partial}{\partial z}\left(k_z \frac{\partial h}{\partial z}\right) = \frac{1}{1 + e_0}\frac{\partial e}{\partial t} \qquad (7\text{–}1)$$

In order to solve this equation, we wish to express it in terms of a single dependent variable. It will be convenient to make this variable the pore water pressure in *excess of hydrostatic pressure*, $u_e(z, t)$. The definition of this parameter is presented in Figure 7.1. A typical compressible stratum with the groundwater located above it is shown in Figure 7.1a. The hydrostatic component of pore water pressure is indicated in Figure 7.1b. Figure 7.1c shows the total magnitude of pore water pressure $u(z, t)$ after some external loading has caused an excess over hydrostatic to be developed. Also shown is the magnitude of that excess pressure. Because the hydraulic gradient under *hydrostatic* conditions is zero, that is,

$$\frac{\partial h}{\partial z} = \frac{\partial z}{\partial z} - \frac{1}{\gamma_w}\frac{\partial u}{\partial z} = 1 - 1 = 0 \qquad (7\text{–}2)$$

there is no flow for this case. When flow occurs, the hydraulic gradient depends upon the pore water pressure in excess of hydrostatic:

$$\frac{\partial h(z, t)}{\partial z} = \frac{1}{\gamma_w}\frac{\partial u_e(z, t)}{\partial z} \qquad (7\text{–}3)$$

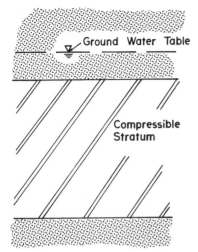

a) - Stratigraphic Profile

Fig. 7.1—Hydraulic gradients in compressible stratum.

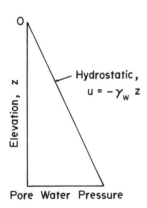

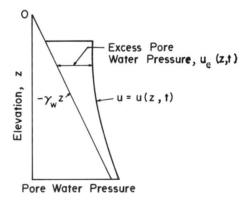

b) - Equilibrium, $\dfrac{\partial h}{\partial z} = 0$ c) - Transient Flow, $\dfrac{\partial h}{\partial z} = f(z,t)$

In most cases, the excess pore water pressure develops in response to an externally imposed load.* The relationship between the magnitude of the excess pore water pressure and an increment in vertical stress depends upon a variety of factors, and is explored in detail in Chapter 8. In the one-dimensional case, however, it appears that an increment in vertical stress $\Delta\sigma_z(z)$ produces an increment in excess pore water pressure $\Delta u_e(z)$ of equal

---

* Consolidation may also occur when the pore water pressure is reduced below hydrostatic and there is no change in total stress. This occurs, for example, when water is pumped from an aquifer immediately underlying a compressible stratum.

magnitude. Denoting the initial vertical effective stress by $\sigma_0'(z, 0)$, the principle of effective stress requires that for a stratum without initial u_e, for which

$$\Delta u_e(z) = u_e(z),$$
$$\sigma_0'(z, 0) - \gamma_w z + \Delta\sigma_z(z, t) = \sigma_z'(z, t) - \gamma_w z + u_e(z, t) \qquad (7-4)$$

If the increment in vertical stress is applied at time $t = 0$ and maintained constant,

$$\Delta\sigma_z(z, t) = \Delta\sigma_z(z, 0) \qquad (7-5)$$

and the left side of Equation 7–4 remains constant, so that

$$\frac{\partial \sigma_z'}{\partial t} = -\frac{\partial u_e}{\partial t} \qquad (7-6)$$

That is, the dissipation of excess pore water pressure is accompanied by an equal and opposite change in effective stress.

At this point we introduce three additional assumptions:

6. Compression of the soil is one-dimensional, so the void ratio change may be related to the change in vertical effective stress by

$$de = -a_v \, d\sigma_z' \qquad (7-7)$$

in which a_v is the coefficient of compressibility.

7. The coefficient of compressibility is a constant, independent of the stress level.

8. The coefficient of permeability is independent of position and changes in void ratio with depth and time, that is,

$$k_z = k = \text{const}$$

Assumptions 7 and 8 are made to facilitate obtaining a closed-form solution. We shall see later that these assumptions need not be made if a numerical procedure is used. The assumptions that a_v and k remain constant for all positions and time are clearly not correct. We know that, as the void ratio decreases, both a_v and k decrease. Because they appear in the final expression as a ratio, however, their quotient tends to remain constant even though they may vary individually.*

Substituting Equation 7–7 into the right side of Equation 7–1, assumption 7 leads to

$$\frac{\partial e}{\partial t} = a_v \frac{\partial u_e}{\partial t} \qquad (7-8)$$

---

* Analytical solutions to this problem formulated for varying permeability (Schiffman, 1958) and coefficient of compressibility (Davis and Raymond, 1965) have been presented. For pressure-increment ratios common in field applications, however, the effect of variations in these parameters is minor. Where such effects are important, a numerical solution is still useful.

Assumption 8 permits eliminating k from the derivative when we substitute Equation 7–3 into Equation 7–1, which leads to

$$\frac{k}{\gamma_w} \frac{\partial^2 u_e}{\partial z^2} = \frac{a_v}{1 + e_0} \frac{\partial u_e}{\partial t} \tag{7-9}$$

or

$$c_v \frac{\partial^2 u_e}{\partial z^2} = \frac{\partial u_e}{\partial t} \tag{7-10}$$

in which

$$c_v = \frac{k(1 + e_0)}{a_v \gamma_w} \tag{7-11}$$

is a constant factor called the *coefficient of consolidation*.

Equation 7–10 is the one-dimensional consolidation equation derived by Terzaghi. The assumptions made to this point have led to an expression which is a *linear* partial differential equation in one dependent variable, solutions to which are readily available. Such solutions, when applied to consolidation problems are referred to as the Terzaghi one-dimensional consolidation theory.* The form of the solution depends upon the boundary and initial conditions imposed. We shall develop a numerical procedure for solving the equation. A description of the classical closed-form solution procedure is given by Wu (1966).

7.3 NUMERICAL SOLUTION

Recurrence Formula

The solution of the one-dimensional consolidation equation may be viewed as a surface in a three-dimensional space, as shown in Figure 7.2. The horizontal plane is defined by the values of time t and elevation z. For each value of time and elevation there is a single value of excess pore pressure $u_e(z, t)$ which, when plotted along the vertical axis and connected with the adjacent values, defines the solution surface. Note that along the time axis the surface stretches from $0 \leqslant t < \infty$, and along the axis of depth from $0 \leqslant z \leqslant D$ where z is now measured from the top of the material being consolidated, positive downward†, and D is the thickness of the consolidating layer.

Consider now the intersection of a vertical plane parallel to the z axis, as shown in Figure 7.2, with the $u_e(z, t)$ surface. The resulting curve $u_e(z, t_j)$

---

* We note that this equation is identical to the heat conduction equation where u is analogous to the temperature and c_v is analogous to the constant $k/\rho c$, where c is the specific heat, k is the coefficient of thermal conductivity, and ρ is the density.

† Now that we need consider *excess* pore water pressure only, and the elevation head is cancelled out of the expressions, it is convenient to return to our earlier convention of defining the downward direction as positive.

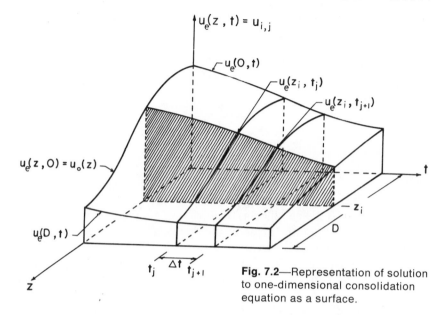

Fig. 7.2—Representation of solution to one-dimensional consolidation equation as a surface.

is the distribution of excess pore pressure with depth at a given value of time, t_j. Such a curve is called an *isochrone** of excess pore pressure. A tangent to the isochrone at any point is the partial derivative of $u_e(z, t)$ with respect to z. By definition

$$\frac{\partial u_e}{\partial z} = \lim_{\Delta z \to 0} \frac{u_e(z + \Delta z, t_j) - u_e(z, t_j)}{\Delta z} \tag{7-12}$$

where, as in Figure 7.3a, the partial derivative is the secant to the isochrone, taken at the depth z and $z + \Delta z$ in the limit as $\Delta z \to 0$. Therefore let us approximate the partial derivative by a secant

$$\frac{\partial u_e}{\partial z} \simeq \frac{u_e(z + \Delta z, t_j) - u_e(z, t_j)}{\Delta z} = \frac{u_e(z_{i+1}, t_j) - u_e(z_i, t_j)}{\Delta z} \tag{7-13}$$

where z_i and z_{i+1} are the two depths separated by a distance Δz. The approximation given by Equation 7–13 is known as a *finite difference*.†
Noting that

$$\frac{\partial^2 u_e}{\partial z^2} = \frac{\partial}{\partial z}\left(\frac{\partial u_e}{\partial z}\right) \tag{7-14}$$

---

* From the Greek: *isos*—equal, *chronos*—time.

† There are several finite difference schemes for approximating differentials. The one applied here makes use of a *forward difference* operator $\Delta f(x) = f(x + \Delta x) - f(x)$, where Δx is the step size or difference in the value of x at which the function is to be evaluated. We might also define a *backward difference* operator as $\Delta f(x) = f(x) - f(x - \Delta x)$, or a *central difference* operator $\Delta f(x) = f(x + \Delta x/2) - f(x - \Delta x/2)$. For a discussion of finite difference procedures and their application to boundary-value problems, see Beckett and Hurt (1967).

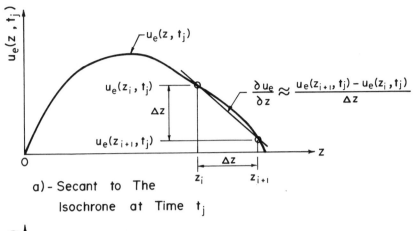

a) - Secant to The
Isochrone at Time t_j

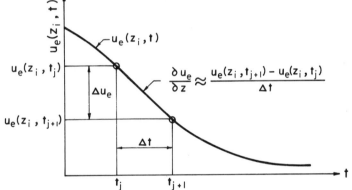

b)- Secant to The Excess Pore Pressure Function at Depth z_i

Fig. 7.3—Definition of partial derivatives as finite differences.

We shall approximate the second partial derivative as the finite difference of the forward and backward differences:

$$\frac{\partial^2 u_e}{\partial z^2} \simeq \frac{\dfrac{u_e(z_{i+1}, t_j) - u_e(z_i, t_j)}{\Delta z} - \dfrac{u_e(z_i, t_j) - u_e(z_{i-1}, t_j)}{\Delta z}}{\Delta z} \quad (7\text{--}15)$$

or

$$\frac{\partial^2 u_e}{\partial z^2} \simeq \frac{u_e(z_{i+1}, t_j) - 2u_e(z_i, t_j) + u_e(z_{i-1}, t_j)}{\Delta z^2} \quad (7\text{--}16)$$

If we now consider the intersection of a vertical plane parallel to the t axis with the solution surface $u_e(z, t)$, Figure 7.3b, we see that the resulting curve describes the dissipation of pore water pressure with time at a particular depth z_i. A tangent to this curve is the partial derivative $\partial u / \partial t$ at z_i. We shall approximate this by a secant

$$\frac{\partial u}{\partial t} \simeq \frac{u_e(z_i, t_{j+1}) - u_e(z_i, t_j)}{\Delta t} \tag{7-17}$$

in which t_{j+1} is $t_j + \Delta t$.

For convenience in notation, we will dispense with writing u_e as a function of z and t, and express Equations 7-16 and 7-17 more concisely as

$$\frac{\partial^2 u_e}{\partial z^2} \simeq \frac{u_{e_{i+1, j}} - 2u_{e_{i, j}} + u_{e_{i-1, j}}}{\Delta z^2} \tag{7-18}$$

$$\frac{\partial u_e}{\partial t} \simeq \frac{u_{e_{i, j+1}} - u_{e_{i, j}}}{\Delta t} \tag{7-19}$$

Substituting Equations 7-18 and 7-19 into Equation 7-10, and recognizing that the equalities shown are exact only in the limit,

$$c_v \frac{(u_{e_{i+1, j}} - 2u_{e_{i, j}} + u_{e_{i-1, j}})}{\Delta z^2} = \frac{(u_{e_{i, j+1}} - u_{e_{i, j}})}{\Delta t} \tag{7-20}$$

Rearranging terms, we now have a *recurrence formula*,

$$u_{e_{i, j+1}} = \alpha u_{e_{i+1, j}} + (1 - 2\alpha)u_{e_{i, j}} + \alpha u_{e_{i-1, j}} \tag{7-21}$$

in which

$$\alpha = \frac{c_v \Delta t}{(\Delta z)^2} \tag{7-22}$$

Thus to evaluate the excess pore pressure at any depth, z_i, and time, t_{j+1}, by the difference equation, Equation 7-21, we need to know:

1. The increments of time Δt and depth Δz such that the finite difference scheme will produce a mathematically stable solution.
2. Three values of the excess pore pressure, $(u_{e_{i+1, j}}, u_{e_{i, j}}, u_{e_{i-1, j}})$, above, below, and at the elevation z_i, at the previous time increment, $t_i = t_{i+1} - \Delta t$.
3. The value of c_v.

Boundary and Initial Conditions

To begin the finite difference scheme we need starting values or *initial conditions* ($t = 0$). In the case of *one-dimensional* compression of a stratum initially in equilibrium, the initial *excess* pore water pressure distribution is equal to the distribution of the applied vertical stress increment

$$u_e(z, 0) = \Delta\sigma(z) \tag{7-23}$$

Because the calculated value of excess pore pressure at any depth z_i, for $t > 0$, depends upon the magnitude of u_e at points above and below z_i, we need to know boundary conditions $u_e(0, t)$ and $u_e(D, t)$ that is, the excess pore water pressure at all times after $t = 0$, at the top and bottom of the consolidating stratum. If, at a boundary there is *free drainage*,* then $u_e = 0$ for $t > 0$. If both top and bottom boundaries are free-draining,

$$u_e(0, t) = u_e(D, t) = 0 \qquad (7\text{-}24)$$

In situ, free-draining boundaries generally exist where a sand or silt stratum borders or intersects a clay layer, because the permeability of the sand or silt is usually sufficiently greater than that of the clay to permit unimpeded flow of water from the clay to the more pervious material.

If at a boundary *no drainage* is possible, there can be no flow across this boundary. Thus, we see from Darcy's law that, at an impervious boundary,

$$\frac{\partial u_e}{\partial z} = 0 \qquad (7\text{-}25)$$

The numerical procedure used to introduce this boundary condition is discussed below.

The use of the initial and boundary conditions in the finite-difference solution is illustrated in Examples 7–1 to 7–3.†

Incrementing Time and Depth

To insure that the finite-difference solution is an acceptable approximation to the "true" result, two criteria must be satisfied:

1. The difference equation must be mathematically stable. For our purpose we shall interpret this to mean that roundoff or other errors introduced at some point in time in the solution must not increase in magnitude as t increases.
2. The approximate solution must converge to the exact solution as $\Delta z \to 0$ and $\Delta t \to 0$.

A detailed discussion of the means by which the satisfaction of these criteria is verified is beyond the scope of this text. It has been shown, however (Forsythe and Wasow, 1960), that both criteria are met if

$$\alpha = \frac{c_v \, \Delta t}{(\Delta z)^2} \leqslant \frac{1}{2} \qquad (7\text{-}26)$$

Our intuition supports this result, at least with respect to the stability criterion, because of the nature of Equation 7–21. The expression states that the excess pore water pressure at a point in space and time is a weighted

---

* Free drainage implies that, for practical purposes, no head loss occurs within the stratum serving as a drain. Gray (1945) found that one stratum would generally serve as a free drain for another if the coefficient of permeability of the first were of the order of 100 times that of the second.

† Boundary and initial conditions are also required if one is to obtain a closed-form solution (Wu, 1966).

average of the value at that and adjacent points for an earlier time. The parameter α determines the weighting factors. When $\alpha > \frac{1}{2}$, one of the weighting factors is negative!

For $\alpha = \frac{1}{2}$, Equation 7–21 becomes

$$u_{e_{i,\,j+1}} = \tfrac{1}{2}(u_{e_{i-1,\,j}} + u_{e_{i+1,\,j}}) \qquad (7\text{–}27)$$

That is, the excess pore pressure is simply the average of that at the adjacent points for the preceding time. For this reason, $\alpha = \frac{1}{2}$ is an especially convenient value to use for manual computations.

A still more accurate result is possible if $\alpha = \frac{1}{6}$. It can be shown that for this special case, the error due to the discretization into differences is minimized (Forsythe and Wasow, 1960). For $\alpha = \frac{1}{6}$, Equation 7–21 is

$$u_{e_{i,\,j+1}} = \tfrac{1}{6}(u_{e_{i+1,\,j}} - u_{e_{i-1,\,j}}) + \tfrac{2}{3}u_{e_{i,\,j}} \qquad (7\text{–}28)$$

Thus for calculations using a digital computer, $\alpha = \frac{1}{6}$ is more desirable.

We shall now illustrate solutions of some specific problems.

Example 7.1

The clay stratum described in Example 5.3 (Figure 5.17) is consolidating under a warehouse loading of 1 tsf distributed over an area of 40×100 ft^2 at the ground surface. The coefficient of consolidation c_v was determined in a consolidation test to be 1.0×10^{-4} ft^2/min. Determine the distribution of excess pore water pressure as a function of depth and time, assuming that the clay stratum is drained at both boundaries.

Solution. The finite-difference procedure previously described was used to determine the distribution of excess pore pressures. The numerical results are shown in Table 7.1. This table is useful not only for illustrating the results but also serves as a form for manual computation. For this problem the 40-ft stratum was divided into 5 uniform layers* so that $\Delta z = 8$ ft. To facilitate manual computation,

$$\alpha = \frac{c_v\,\Delta t}{(\Delta z)^2} = \frac{1}{2}$$

was used, from which

$$\Delta t = \frac{0.5(8)^2}{1.0 \times 10^{-4}} = 3.2 \times 10^5 \text{ min} = 222 \text{ days}$$

The value of depth and time and their respective index parameters i and j are shown along the top and left side of Table 7.1, respectively.

The vertical normal stress distribution $\Delta\sigma_z$, due to the loading at the ground surface, was determined using elastic theory (Figure A.5). The initial excess pore pressure distribution

$$u_e(z, 0) = \Delta\sigma(z)$$

---

* Five layers is a relatively crude approximation for actual use, and was selected for this example to simplify illustration of the principles. For manual computation in this problem ten layers would be better.

TABLE 7.1
Values of Excess Pore Pressure for Stratum Drained at Both Boundaries, obtained in Example 7.1 by Finite Differences

i	0	1	2	3	4	5
z_i (ft)	0	8	16	24	32	40

j	Time t (days)	Excess Pore Pressure u_e (z, t) (psf)				
		(Initial Conditions)				
	0^-	1620	1360	1120	930	770	600
0	0	810	1360	1120	930	770	300
1	222	0	965	1145	945	615	0
2	444	0	573	955	880	473	0
3	667	0	478	726	714	440	0
4	889	0	363	596	583	357	0
5	1110	0	298	473	476	292	0
6	1330	0	237	387	382	238	0
7	1555	0	194	310	313	191	0
8	1780	0	155	253	250	156	0

NOTE: 1. The elevation (z_i) refers to the depth below the upper boundary of the stratum.
2. The subscript indices i and j both have initial values of zero for $z = 0$ and $t = 0$, respectively.

is shown in Table 7.1 in the row corresponding to $t = 0^-$. The reason for denoting the initial time this way will become apparent.

For free drainage at both boundaries, the boundary conditions are

$$u_e(0, t) = u_e(40', t) = 0$$

These values are tabulated in the columns corresponding to $z = 0$ and $z = 40$ at all values of $t > 0$. Note, however, that at $t = 0$, for $z = 0$ and $z = 40$ ft, there is an inconsistency between the initial condition and the boundary conditions. This has been resolved by using the average of the initial and subsequent boundary values for each of $u(0, 0)$ and $u(40', 0)$ to start the recurrence relation.* This is shown in the row corresponding to $j = 0, t_j = 0$.

Using the recurrence formula, Equation 7–27, the values of the excess pore pressure were calculated row by row, that is, the values of $u_e(z, t)$ were calculated at all depths for a given value of t before proceeding to $t + \Delta t$.

A typical computation of the excess pore pressure at $z = 16$ ft $(i = 2)$ and $t = 889$ days $(j = 4)$ follows:

$$u_{e_{2,4}} = \tfrac{1}{2}(u_{e_{3,3}} + u_{e_{1,3}})$$
$$u_{e_{2,4}} = \tfrac{1}{2}(478 + 714) = 596 \text{ psf}$$

* Because of the finite increments of t and z, there would be no gradient calculated at the boundaries at time zero if the initial conditions were used unmodified. The error introduced by this approximation is unimportant after a few time increments.

The computational procedure was continued until a time of 1780 days (4.9 years) was reached. Even after this period of time the *excess* pore water pressure at the middle of the stratum was approximately 25 per cent of the initial excess pore pressure.

The preceding example illustrates the computational procedure for the finite difference solution for Equation 7–10. We now wish to investigate some special cases which are of particular interest. The first of these is the case of a uniform initial excess pore water pressure.

Example 7.2

As a result of the lowering of the groundwater table, a clay stratum 80 ft thick is consolidating under an increment of load $\Delta\sigma_z = 1000$ psf, which is uniform with depth. The clay layer is drained at both top and bottom boundaries. We wish to determine the excess pore water pressure as a function of position and time. The coefficient of consolidation of the clay is 1.0×10^{-4} ft^2/min.

Solution. The numerical solution for excess pore water pressures is given in Table 7.2, and plotted as isochrones of pore water pressure in Figure 7.4. The stratum was divided into 10 layers, and α was set equal to $\frac{1}{2}$ to facilitate manual computation. The isochrones are shown as points connected by straight line segments to indicate their approximate nature.

TABLE 7.2

Values of Excess Pore Pressure and Uniform Initial Excess Pore Pressure for Stratum Drained at Both Boundaries, Obtained in Example 7.2 by Finite Differences

i	0	1	2	3	4	5	6	7	8	9	10
z_i (ft)	0	8	16	24	32	40	48	56	64	72	80

j	t_j (days)	Excess Pore Pressure $u_e(z, t)$ (psf)									
		(Initial Conditions)									
	0^-	1000	1000	1000	1000	1000	1000	1000	1000	1000	1000	1000
0	0	500	1000	1000	1000	1000	1000	1000	1000	1000	1000	500
1	222	0	750	1000	1000	1000	1000	1000	1000	1000	750	0
2	444	0	500	875	1000	1000	1000	1000	1000	875	500	0
3	666	0	438	750	938	1000	1000	1000	938	750	438	0
4	889	0	375	688	875	969	1000	969	875	688	375	0
5	1111	0	344	625	828	938	969	938	828	625	344	0
6	1333	0	313	586	781	898	938	898	781	586	313	0
7	1555	0	293	547	742	859	898	859	742	547	293	0
8	1778	0	273	518	703	820	859	820	703	518	273	0
9	2000	0	259	488	669	781	820	781	669	488	259	0
10	2222	0	244	464	635	745	781	745	635	464	244	0

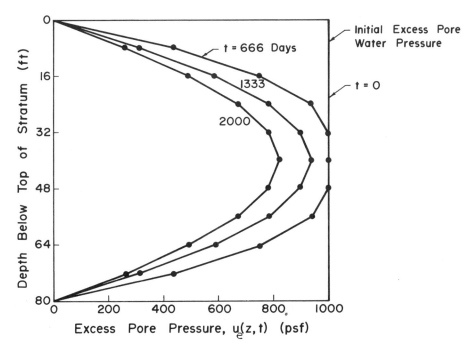

Fig. 7.4—Approximate excess pore water pressure isochrones as a function of the depth and time for a stratum drained at both boundaries, Example 7.2.

We note that in this case (a uniform initial excess pore water pressure and a doubly drained stratum) that the distribution of excess pore water pressure is symmetrical about the midplane of the stratum ($i = 5$, $z_i = 40$ ft). Because of this symmetry

$$u_{e_{6,j}} = u_{e_{4,j}} \tag{7-29}$$

And the recurrence relation, Equation 7–21, for the midplane may be rewritten as

$$u_{e_{5,j+1}} = 2\alpha u_{e_{4,j}} + (1 - 2\alpha)u_{e_{5,j}} \tag{7-30}$$

When $\alpha = \frac{1}{2}$, this reduces to the simple case

$$u_{e_{5,j+1}} = u_{e_{4,j}} \tag{7-31}$$

We can generalize the foregoing results by recognizing that the symmetrical condition requires that tangents to the isochrones at the plane of symmetry are vertical, that is,

$$\frac{\partial u_e}{\partial z} = 0 \tag{7-32}$$

which implies no flow across the plane of symmetry. So we see that, for a symmetrical problem such as the case of a uniform initial excess pore water pressure in a doubly drained homogeneous stratum divided into n layers (n even),

$$u_{e_{\frac{n}{2}},j+1} = 2\alpha u_{e_{\frac{n}{2}-1,j}} + (1 - 2\alpha)u_{e_{\frac{n}{2}},j} \tag{7-33}$$

When $\alpha = \frac{1}{2}$, Equation 7–33 takes the form

$$u_{e_{\frac{n}{2}},j+1} = u_{e_{\frac{n}{2}-1,j}} \tag{7-34}$$

Drainage at One Boundary Only

When the contiguous stratum at one boundary is impervious, there is no flow across the boundary, that is,

$$\frac{\partial u_e}{\partial z} = 0 \tag{7-35}$$

Noting Equation 7–32, we see that an impervious boundary is equivalent to the midplane of a homogeneous stratum with a symmetrical distribution of initial excess pore water pressure. Consequently, the distribution of excess pore pressure in a singly drained stratum is identical to that in one-half of the analogous symmetrically loaded doubly drained stratum.

We observe then that an impervious boundary is accounted for in the finite difference solution by using Equations 7–33 and 7–34 rewritten as

$$u_{e_{m,j+1}} = 2\alpha u_{e_{m-1,j}} + (1 - 2\alpha)u_{e_{m,j}} \tag{7-36}$$

for $\alpha \leqslant \frac{1}{2}$, and

$$u_{e_{m,j+1}} = u_{e_{m-1,j}} \tag{7-37}$$

for $\alpha = \frac{1}{2}$, in which m is the number of layers in the stratum.

Example 7.3

The 40-ft clay stratum described in Example 7.1 is found to be undrained at its lower boundary. Determine the distribution of excess pore pressures as a function of depth and time for the same initial conditions.

Solution. The stratum was divided into 5 uniform layers as before, and α was taken as $\frac{1}{2}$. The initial conditions are the same as in Example 7.1:

$$u_e(z, 0) = \Delta\sigma(z)$$

but the boundary conditions are now

$$u_e(0, t) = 0$$

and

$$\frac{\partial u_e}{\partial z}(40', t) = 0$$

The recurrence relation, Equation 7–27 was used to determine the excess pore pressures at the bottom of layers 1–4, and Equation 7–37 was used to find the excess pore pressure at the lower boundary (bottom of layer 5). The results are given in Table 7.3. Note again that the calculation proceeds row by row.

TABLE 7.3
Values of Excess Pore Pressure for Stratum Drained at
Upper Boundary Only, Obtained in Example 7.3 by
Finite Differences

i	0	1	2	3	4	5
z_i (ft)	0	8	16	24	32	40

Time						
j $\quad t_j$ (days)						

j	t_j (days)	\multicolumn{6}{c}{Excess Pore Pressure $u_e(z, t)$ (psf)}					
		\multicolumn{6}{c}{(Initial Conditions)}					
	0^-	1620	1360	1120	930	770	600
0	0	810	1360	1120	930	770	600
1	222	0	965	1145	945	765	770
2	444	0	573	955	955	858	765
3	667	0	478	764	906	860	858
4	889	0	382	692	812	882	860
5	1110	0	346	597	787	836	882
6	1330	0	298	566	716	834	836
7	1555	0	283	507	700	776	834
8	1780	0	254	492	642	767	776
⋮	⋮						
12	2670	0	201	394	522	633	644
⋮	⋮						
18	4000	0	148	290	386	469	477
⋮	⋮						
26	5780	0	99	194	259	314	320
⋮	⋮						
40	8880	0	49	96	128	156	158

Comparing the results of this example with those of Example 7.1 (Table 7.1) we see that there is a pronounced decrease in the rate of pore pressure dissipation as a consequence of the impervious lower boundary. The significance of this point, and the influence of the length of the longest drainage path are considered below.

7.4 DEGREE OF CONSOLIDATION

Equation 7–21 yields the distribution of excess pore water pressure with space and time. It is frequently of interest to determine the degree to which the excess pore water pressure at a point has been dissipated. This is described by the *degree of consolidation* defined as

$$U_z(z, t) = \frac{e(z, 0) - e(z, t)}{e(z, 0) - e(z, \infty)}$$ (7–38)

If we assume that the coefficient of compressibility a_v, is constant for the range of effective stresses occurring at a point, the changes in void ratio can be expressed as

$$e(z, 0) - e(z, t) = a_v[u_e(z, 0) - u_e(z, t)]$$ (7–39a)

$$e(z, 0) - e(z, \infty) = a_v[u_e(z, 0) - u_e(z, \infty)] = a_v u_e(z, 0)$$ (7–39b)

and the degree of consolidation at a point can be expressed in terms of excess pore water pressure,

$$U_z(z, t) = 1 - \frac{u_e(z, t)}{u_e(z, 0)}$$ (7–40)

The physical significance of the degree of consolidation, as expressed by Equation 7–40, is illustrated in Figure 7.5. This figure shows an isochrone in a doubly drained stratum at time t_j following application of a uniform initial excess pore pressure. The excess pore water pressure at a depth z_i is indicated, along with the magnitude of the pressure which has been dissipated at that depth. From this figure we see that the degree of consolidation

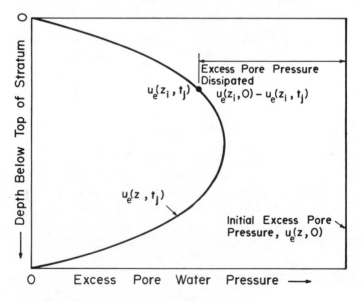

Fig. 7.5—Dissipation of excess pore water pressure in doubly drained stratum.

is the proportion of the initial excess pore pressure which has been dissipated at the time t_j.

Of more frequent interest to the designer is the *average degree of consolidation* $U(t)$, defined as

$$U(t) = \frac{\rho_c(t)}{\rho_c} \tag{7-41}$$

in which $\rho_c(t)$ is the consolidation settlement of the entire stratum which has occurred at time t and ρ_c is the ultimate value of the consolidation settlement. Thus we define the average degree of consolidation as the proportion of consolidation settlement which has occurred. From Chapter 5, we see that the one-dimensional consolidation settlement at any time can be related to the void ratio change throughout the stratum by

$$\rho_c(t) = \frac{1}{1 + e_0} \int_0^D [e(z, 0) - e(z, t)] \, dz \tag{7-42}$$

in which D is the thickness of the stratum and e_0 is the average initial void ratio. The ultimate one-dimensional consolidation settlement is

$$\rho_c = \frac{1}{1 + e_0} \int_0^D [e(z, 0) - e(z, \infty)] \, dz \tag{7-43}$$

Hence we can express the average degree of consolidation in terms of the void ratio change by

$$U(t) = \frac{\int_0^D [e(z, 0) - e(z, t)] \, dz}{\int_0^D [e(z, 0) - e(z, \infty)] \, dz} \tag{7-44}$$

Substituting Equations 7–39 into this expression, and assuming further that a_v is constant throughout the stratum, then it may be removed from within the integral and the average degree of consolidation may be expressed in terms of the excess pore water pressure:

$$U(t) = 1 - \frac{\int_0^D u_e(z, t) \, dz}{\int_0^D u_e(z, 0) \, dz} \tag{7-45}$$

Graphical interpretation of the average degree of consolidation is illustrated in Figure 7.6. The figure shows the distribution of initial excess pore water pressure with depth, and an isochrone of excess pore water pressure at time t_j. The area to the left of the initial excess pressure, denoted by area (t_0),

$$\text{Area}(t_0) = \int_0^D u_e(z, 0) \, dz \tag{7-46}$$

is a measure of the total pore water pressure which must be dissipated. The

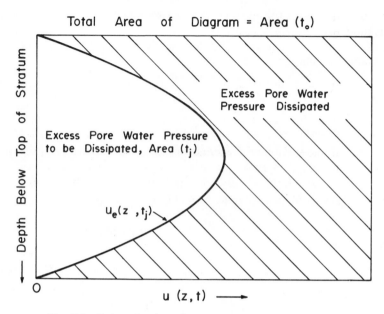

Fig. 7.6—Determination of average degree of consolidation.

area to the left of the isochrone, area (t_j),

$$\text{Area}(t_j) = \int_0^D u_e(z, t_j)\, dz \qquad (7\text{–}47)$$

is a measure of the total excess pore water pressure which remains to be dissipated after time t_j. The average degree of consolidation is then

$$U(t) = 1 - \frac{\text{area}(t_j)}{\text{area}(t_0)} \qquad (7\text{–}48)$$

which is a measure of the excess pore water pressure which has been dissipated at time t_j.

In the case of the results obtained by finite differences, numerical integration is necessary. Using the trapezoidal rule* (Beckett and Hurt, 1967):

$$\int_0^D u_e\, dz \simeq \Delta z(\tfrac{1}{2}u_{e_0} + u_{e_1} + u_{e_2} + \cdots + u_{e_{n+1}} + \tfrac{1}{2}u_{e_n}) \qquad (7\text{–}49)$$

We have

$$\text{Area}(t_0) = \Delta z[\tfrac{1}{2}u_e(0, 0) + u_e(1, 0) + u_e(2, 0) + \cdots \\ + u_e(n - 1, 0) + \tfrac{1}{2}u_e(n, 0)] \qquad (7\text{–}50)$$

$$\text{Area}(t_j) = \Delta z[\tfrac{1}{2}u_e(0, j) + u_e(1, j) + u_e(2, j) + \cdots \\ + u_e(n - 1, j) + \tfrac{1}{2}u_e(n, j)] \qquad (7\text{–}51)$$

---

* Simpson's rule is more accurate, but requires an even number of layers (Beckett and Hurt, 1967).

Area (t_0) is a constant for a given set of initial conditions and need be calculated only once for a specific problem.

Example 7.4

A 20-ft-thick clay stratum is consolidating, owing to a uniform initial excess pore water pressure,

$$u_e(z, 0) = 1000 \text{ psf}$$

The stratum, for which $c_v = 2 \times 10^{-2}$ in.^{2}/min is drained only at its top surface.

(a) Determine the average degree of consolidation using the finite difference solution.
(b) Compare the results to those obtained by an exact analysis.

Solution. (a) The numerical solution for excess pore water pressure is given in Table 7.4 for five layers and $\alpha = \frac{1}{2}$. As in the preceding example, Equation 7–27 was used to calculate the excess pore water pressures at the bottom of layers 1–4, and Equation 7–37 was applied for the bottom of layer 5 at the impervious boundary.

TABLE 7.4
Values of Excess Pore Pressure and Average Degree of Consolidation for Stratum Drained at Upper Boundary Only, Obtained in Example 7.4.

i		0	1	2	3	4	5	Average Degree of Consolidation $U(t)$ (%)	
z_i (ft)		0	4	8	12	16	20		
Time								By Finite	
j	t_j (days)	\multicolumn{6}{c}{Excess Pore Pressure $u_e(z, t)$ (psf)}		Differences ($\alpha = 1/2$)	By Exact Analysis
		\multicolumn{6}{c}{(Initial Conditions)}			
	0^-	1000	1000	1000	1000	1000	1000		
0	0	500	1000	1000	1000	1000	1000	0	0
1	40	0	750	1000	1000	1000	1000	15.0	16.0
2	80	0	500	875	1000	1000	1000	22.5	22.6
3	120	0	435	750	938	1000	1000	27.5	27.6
4	160	0	375	688	875	969	1000	31.9	31.9
5	200	0	344	625	828	938	969	35.6	35.7
10	400	0	244	464	635	745	781	50.4	50.4
15	600	0	189	359	494	580	610	61.5	61.3
20	800	0	147	279	384	452	475	70.0	69.8
30	1200	0	89	169	232	274	287	81.9	81.6
40	1600	0	54	102	141	166	174	89.0	88.8
50	2000	0	33	62	85	100	105	93.4	93.1

Data are shown for each time increment up to $j = 5$ and then at varying intervals up to $j = 50$. We note that as time increases, changes in excess pore pressure occur relatively more slowly.

The next-to-last column shows the average degree of consolidation calculated using Equation 7–48 in which Area (t_0) and Area (t_j) were determined by numerical integration (Equations 7–50 and 7–51, respectively). As an example, for $t_j = 200$ days ($j = 5$),

$$\text{Area}(t_j) = 4(0 + 344 + 625 + 828 + 938 + \tfrac{1}{2} \times 969) = 12{,}880$$

and

$$\text{Area}(t_0) = 4(\tfrac{1}{2} \times 1000 + 1000 + 1000 + 1000 + 1000 + \tfrac{1}{2} \times 1000) = 20{,}000$$

from which

$$U(200 \text{ days}) = 1 - \frac{12{,}880}{20{,}000} = 0.356 = \underline{35.6\%}$$

as shown in the next-to-last column.

(b) The average degree of consolidation calculated by an exact solution for this problem, which is discussed further below, is shown in the last column of Table 7.4. It is apparent that even for a relatively coarse division of the stratum, $\alpha = \tfrac{1}{2}$, and application of the trapezoidal rule for numerical integration, the numerical computation of $U(t)$ leads to results which compare favorably with the exact analysis.

Example 7.5

The ultimate consolidation settlement of the clay stratum considered in Example 7.4 has been predicted as 3 in. How long will be required for 1 in. of settlement to occur.

Solution. The average degree of consolidation, corresponding to 1 in. of settlement in this case, is by definition (Equation 7–41)

$$U(t) = \frac{\rho_c(t)}{\rho_c} = \frac{1}{3} = 33.3\%$$

Interpolating in either of the right-side columns in Table 7.4, we find that

$$t = 175 \text{ days}$$

is required.

The preceding examples have shown that we can determine the distribution of excess pore water pressure and the degree of consolidation at a given time (or conversely, the time required to reach a given degree of consolidation) if the geometry of the problem and initial conditions and the coefficient of consolidation are known. We now inquire whether it is necessary to solve a new problem each time one of these parameters is changed. The answer is that *many* different problems can be solved using a single solution by reformulating the problem statement in suitable dimensionless terms.

Dimensionless Parameters

We wish to express Equation 7–10 in a useful dimensionless form. To do this, we define the following dimensionless parameters:

1. *Depth factor, Z,*

$$Z \equiv \frac{z}{H} \tag{7-52}$$

in which z is the depth below the top of the compressible stratum and H is some characteristic dimension of the stratum. We shall find it convenient to define H as the thickness of a singly drained stratum and $\frac{1}{2}$ the thickness of a doubly drained stratum. The reasons for this are discussed in detail below. Thus for the depth in Equation 7–10, we can substitute

$$z = ZH \tag{7-53}$$

2. *Time factor, T,*

$$T \equiv \frac{c_v t}{H^2} \tag{7-54}$$

so that in Equation 7–10,

$$t = \frac{H^2}{c_v} T \tag{7-55}$$

3. Normalized *excess pore water pressure, $\bar{u}_e$,*

$$\bar{u}_e \equiv \frac{u_e}{u_{e_0}} \tag{7-56}$$

in which u_{e_0} is a characteristic value of the pore water pressure, usually selected as the maximum initial value. The excess pore water pressure in Equation 7–10 may be written

$$u_e = u_{e_0} \bar{u}_e \tag{7-57}$$

Substituting Equations 7–53, 7–55, and 7–57 into Equation 7–10, we have the dimensionless expression:

$$\frac{\partial^2 \bar{u}_e}{\partial Z^2} = \frac{\partial \bar{u}_e}{\partial T} \tag{7-58}$$

The form of the solution to this expression is the same as that for Equation 7–10 except that geometric and soil parameters, as well as a specific magnitude of pore water pressure, do not appear in the results.

The recurrence relation for the finite difference solution of Equation 7–58 is still Equation 7–21, in which

$$\alpha = \frac{\Delta T}{(\Delta Z)^2} = \frac{c_v \Delta t}{(\Delta z)^2} \tag{7-59}$$

is the same as before. Thus the results obtained from the finite difference solution are independent of c_v and the thickness of the stratum. We may observe that this is the case by reviewing the results in Tables 7.1 through 7.4. Once α and the number of layers have been selected, the magnitude of the excess pore water pressure is independent of the actual magnitude of the layer thickness for the time. Furthermore we see by inspection that the degree of consolidation (Equation 7–40) and the average degree of consolidation (Equation 7–45) are the same irrespective of whether we use the actual or normalized excess pore water pressure.

Recognizing that the initial excess pore water pressure can be expressed as

$$u_e(Z, 0) = u_{e_0} f(Z) \qquad (7-60)$$

it is clear that the initial *normalized* excess pore pressure depends solely upon the dimensionless depth:

$$\bar{u}_e(Z, 0) = f(Z) \qquad (7-61)$$

As a consequence of the foregoing, we see that the distribution of excess pore water pressure and average degree of consolidation determined for one stratum may be used to describe the consolidation process for any other stratum with the *same boundary conditions* and the *same form of initial excess pore pressure distribution* ($f(Z)$ in Equation 7–61).

We can make use in this connection of the results generated for Example 7.4 by expressing the variables in the appropriate dimensionless form. Thus, referring to Table 7.4, we make the following substitutions:

1. Let H be the depth of the singly drained stratum so that the dimensionless depth factor is

$$Z = \frac{z}{20}$$

from which the *incremental* depth factor ΔZ is

$$\Delta Z = 0.2$$

2. For the clay stratum described in Example 7.4 ($c_v = 2 \times 10^{-2}$ in.^{2}/min), the time factor is

$$T = \left[\frac{2 \times 10^{-2} \text{ in.}^2/\text{min}}{(20 \text{ ft})^2} \right] \left[\frac{1}{144 \text{ in.}^2/\text{ft}} \right] (1440 \text{ min/day})t = 5 \times 10^{-4} t$$

For the time increments $\Delta t = 40$ days, the time factor *increment* size is

$$\Delta T = 0.02$$

3. Letting u_{e_0} be the initial uniform value of excess pore water pressure,

$$u_{e_0} = 1000 \text{ psf}$$

the normalized initial pore water pressure is

$$\bar{u}_e(Z, 0) = 1.0$$

The results from Table 7.4 have been rewritten in Table 7.5 in terms of the dimensionless parameters described above. Consequently this table now applies to any homogeneous single-drained stratum for which the initial excess pore water pressure is uniform with depth.

TABLE 7.5

Values of Normalized Excess Pore Pressure and Average Degree of Consolidation for Stratum Drained at Upper Boundary only

i	0	1	2	3	4	5	
Z_i	0.0	0.2	0.4	0.6	0.8	1.0	
Time Factor $\dfrac{}{}$ j $\quad T_j$	Normalized Excess Pore Pressure, $u_e(Z, T)$						$U(T)$ (%)
(Initial Conditions)							
0⁻	1.0	1.0	1.0	1.0	1.0	1.0	

j	T_j							$U(T)$ (%)
0	0	0.5	1.0	1.0	1.0	1.0	1.0	0
1	0.02	0	0.750	1.0	1.0	1.0	1.0	15.0
2	0.04	0	0.500	0.875	1.0	1.0	1.0	22.5
3	0.06	0	0.438	0.750	0.938	1.0	1.0	27.5
4	0.08	0	0.375	0.688	0.875	0.969	1.0	31.9
5	0.10	0	0.344	0.625	0.828	0.938	0.969	35.6
10	0.20	0	0.244	0.464	0.635	0.745	0.781	50.4
15	0.30	0	0.189	0.359	0.494	0.580	0.610	61.5
20	0.40	0	0.147	0.279	0.384	0.452	0.475	70.0
30	0.60	0	0.089	0.169	0.232	0.274	0.287	81.9
40	0.80	0	0.054	0.102	0.141	0.166	0.174	89.0
50	1.00	0	0.033	0.062	0.085	0.100	0.105	93.4

SOURCE: Data from Example 7.4.

Example 7.6

A 30-ft-thick, single-drained, clay stratum, for which $c_v = 0.15\ \text{ft}^2/\text{day}$, is consolidating due to a uniform initial excess pore water pressure of 2000 psf. Determine:

(a) The time required for 65 per cent of the consolidation settlement to occur.
(b) The excess pore water pressure at a depth of 12 ft below the top of the stratum 1 year after the initial application of load.

Solution. Because the boundary conditions and the form of the initial excess pore pressure distribution are the same as those pertaining to Table 7.5, we can apply these results to our example:

(a) Interpolating in Table 7.5, we find that for $U(T) = 65$ per cent, $T = 0.34$. The actual time for 65 per cent consolidation is determined from Equation 7-55, in which $H = 30$ ft and $c_v = 0.15$ ft^2/day.

$$t = \frac{(30)^2}{0.15} 0.34 = 2040 \text{ days} = \underline{5.6} \text{ years}$$

We note that *if the thickness of the stratum were doubled, the time required to reach the same average degree of consolidation would be four times longer.*

(b) For $t = 1$ year (365 days), the time factor is

$$T = \frac{(0.15)(365)}{(30)^2} = 0.0609$$

The depth factor corresponding to $z = 12$ ft is

$$Z = \frac{12}{30} = 0.4$$

Interpolating in Table 7.5, we find the normalized excess pore water pressure

$$\bar{u}_e = 0.747$$

from which the actual pore water pressure is determined as

$$u_e = (2000)(0.747) = \underline{1490} \text{ psf}$$

7.5   RESULTS FROM ANALYTICAL SOLUTION

For many practical cases it is not necessary to generate a finite difference solution, because analytical solutions are available. One such solution, for a uniform initial excess pore water pressure in a doubly drained stratum is shown in Figure 7.7. Illustrated are isochrones of excess pore water pressure. The ordinate in the figure is the dimensionless depth factor below the top of the stratum, in which H is one-half the thickness of the stratum. The abscissa shown below the figure is the normalized pore water pressure factor. The abscissa above the figure is the degree of consolidation. Each of the isochrones corresponds to a specific time factor (Equation 7-54).

Results obtained analytically for this, and a variety of other cases of interest, were assembled by Terzaghi and Frölich (1936). The relationships between average degree of consolidation and time factor for a doubly drained stratum subjected to several useful initial excess pore water pressure distributions given by them are shown in Tables 7.6 and 7.7. Table 7.6 gives the results as degree of consolidation corresponding to specific time factors. To reduce interpolation, Table 7.7 gives the time factors corresponding to specific average degrees of consolidation for the same cases. The use of Figure 7.7 and Tables 7.6 and 7.7 is illustrated in the following examples.

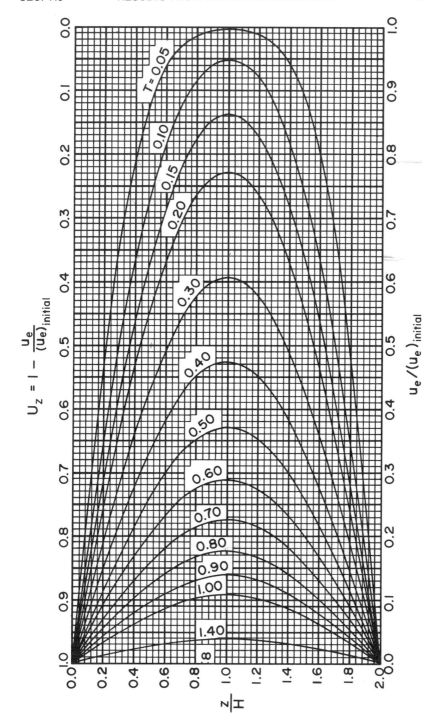

Fig. 7.7—Isochrones of excess pore water pressure for uniform initial excess pressure.

TABLE 7.6
One-Dimensional Consolidation Theory:
Solutions for Four Cases of initial Excess Pore
Water Pressure Distribution in Double-Drained
Stratum.

Average Degree of Consolidation for Various Values of T					Distributions of Initial Excess Pore Water Pressure

T	Case 1	Case 2	Case 3	Case 4	
0.004	7.14	6.49	0.98	0.80	Constant
0.008	10.09	8.62	1.95	1.60	
0.012	12.36	10.49	2.92	2.40	
0.020	15.96	13.67	4.81	4.00	
0.028	18.88	16.38	6.67	5.60	
0.036	.21.40	18.76	8.50	7.20	Linear Variation CASE 1
0.048	24.72	21.96	11.17	9.60	
0.060	27.64	24.81	13.76	11.99	
0.072	30.28	27.43	16.28	14.36	
0.083	32.51	29.67	18.52	16.51	
0.100	35.68	32.88	21.87	19.77	Half Sine Curve CASE 2
0.125	39.89	36.54	26.54	24.42	
0.150	43.70	41.12	30.93	28.86	
0.175	47.18	44.73	35.07	33.06	
0.200	50.41	48.09	38.95	37.04	
0.250	56.22	54.17	46.03	44.32	Sine Curve CASE 3
0.300	61.32	59.50	52.30	50.78	
0.350	65.82	64.21	57.83	56.49	
0.400	69.79	68.36	62.73	61.54	
0.500	76.40	76.28	70.88	69.95	
0.600	81.56	80.69	77.25	76.52	Triangular CASE 4
0.700	85.59	84.91	82.22	81.65	
0.800	88.74	88.21	86.11	85.66	
0.900	91.20	90.79	89.15	88.80	
1.000	93.13	92.80	91.52	91.25	
1.500	98.00	97.90	97.53	97.45	
2.000	99.42	99.39	99.28	99.26	

Average Degree of Consolidation, U (%)

Example 7.7

A 20-ft-thick clay layer is consolidating due to a uniform loading of 1500 psf. The clay layer is drained at both its top and bottom boundaries; $c_v = 0.015$ in.^{2}/min.

Determine:

(a) The time required to reach 50 per cent consolidation at depths of 5 ft and 10 ft from the top of the clay layer.

(b) The degree of consolidation at a depth of 6 ft from the top surface of the clay layer for $t = 365$ days. What is the excess pore pressure at this depth and time?

TABLE 7.7

One-Dimensional Consolidation Theory: Time Factor for Various Average Degrees of Consolidation Double-Drained Stratum

	Time Factor T			
U (%)	Case 1	Case 2	Case 3	Case 4
0	0	0	0	0
5	0.0020	0.0030	0.0208	0.0250
10	.0078	.0111	.0427	.0500
15	.0177	.0238	.0659	.0753
20	.0314	.0405	.0904	.101
25	.0491	.0608	.117	.128
30	.0707	.0847	.145	.157
35	.0962	.112	.175	.187
40	.126	.143	.207	.220
45	.159	.177	.242	.255
50	.197	.215	.281	.294
55	.239	.257	.324	.336
60	.286	.305	.371	.384
65	.342	.359	.425	.438
70	.403	.422	.488	.501
75	.477	.495	.562	.575
80	.567	.586	.652	.665
85	.684	.702	.769	.782
90	0.848	0.867	0.933	0.946
95	1.129	1.148	1.214	1.227
100	∞	∞	∞	∞

NOTE: See Table 7.6 for description of initial excess pore pressure distribution.

Solution. This stratum is doubly drained and loaded uniformly throughout its depth. Therefore use Figure 7.7 with $H = 10$ ft. From Equation 7–55,

$$t = \frac{(10)^2}{0.015} \times \frac{144}{1440} T = 667T \text{ days}$$

(a) At $z = 5$ ft, $z/H = 0.5$ and for $U_z = 50$ per cent, $T = 0.23$. Thus

$$t_{50} = (667)(0.23) = \underline{150} \text{ days}$$

At $z = 10$ ft, $z/H = 1.0$ and for $U_z = 50$ per cent, $T_{50} = 0.37$. Thus

$$t_{50} = (667)(0.37) = \underline{250} \text{ days}$$

(b) At $t = 365$ days, $T = 0.53$, and for $z = 6$ ft, $z/H = 0.6$. Thus $U_z = 73\%$. The excess pore pressure from Equation 7–39 is

$$u(z, t) = 1500(1 - 0.73) = \underline{410} \text{ psf.}$$

Example 7.8 ────────────────────────────────

A double-drained 40-ft-thick clay layer, for which $c_v = 0.002$ in.^{2}/min, has been consolidating for 3 years due to an initial excess pore water pressure varying linearly with depth. It is estimated that there will be 3 in. of total settlement due to consolidation of the clay layer. Determine the vertical consolidation settlement which has occurred up to the present time.

Solution. The time factor (Equation 7–54) for $t = 3$ years is

$$T = \frac{(0.002)(3)}{(20)^2} \times \frac{60 \times 24 \times 365}{144}$$
$$T = 0.0548$$

From Table 7.7 Case 1,

$$U = 26.4\%$$

The settlement at the end of 3 years is therefore

$$\rho_c(3 \text{ yr}) = 0.264(3) = \underline{0.8} \text{ in.}$$

Example 7.9 ────────────────────────────────

A 20-ft-thick clay layer drained at both boundaries is consolidating due to an excess pore water pressure which was initially uniform with depth. If $c_v = 0.002$ in.^{2}/min, how long will it take to achieve 50 per cent of the ultimate settlement:

Solution. From Table 7.8, for Case 1, corresponding to $U(T) = 50$ per cent, $T = 0.197$. Using Equation 7–55,

$$t = \frac{0.197(10)^2}{.002} \times \frac{144}{60 \times 24} = \underline{985} \text{ days} = \underline{2.7} \text{ years}$$

Example 7.10 ────────────────────────────────

If the clay stratum described in Example 7.9 were 40 ft thick instead of 20 ft, how long would it take to achieve 50 per cent of the ultimate settlement?

Solution. Using Equation 7–55,

$$\frac{t_1}{t_2} = \frac{\dfrac{T_1(H_1)^2}{c_{v_1}}}{\dfrac{T_2(H_2)^2}{c_{v_2}}}$$

as $T_1 = T_2$ and $c_{v_1} = c_{v_2}$

$$\frac{t_2}{t_1} = \left(\frac{H_2}{H_1}\right)^2$$

$$t_2 = \left(\frac{20}{10}\right)^2 t_1 = 4(985) = \underline{3940} \text{ days} = \underline{10.8} \text{ yr}$$

Boundary Drainage Conditions

We have previously noted that, when the initial excess pore pressure distribution is symmetrical about the midheight of a stratum drained at both boundaries, the distribution of excess pore water pressure will be symmetrical about this plane at all times. Thus the distribution of excess pore pressure in one-half of the stratum is the same as that in a stratum one-half the depth, drained at only one boundary. Because we can represent both depth and time in dimensionless form, the results for a doubly drained stratum with symmetrical initial distribution of u_e are valid for both cases, irrespective of the actual thickness of the stratum, or the magnitude of c_v. We can then use some of the results in Tables 7.6 and 7.7, obtained for a double-drained stratum, to describe conditions for a single-drained stratum. The initial excess pore water pressure distributions for which the double-drained case is applicable to the single-drained case are shown in Figure 7.8. Illustrated in the figure is the fact that the parameter H used in determining the time factor, is the *full* thickness of the single-drained stratum, and *one-half* the

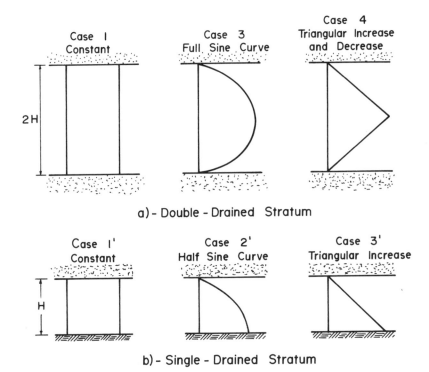

Fig. 7.8—Initial excess pore water pressure distribution for double-drained and single-drained strata for which Table 7.6 is applicable.

thickness of the double-drained stratum. Thus we may visualize H as the *length of the longest drainage path in the stratum.*

The use of Figure 7.8 and Table 7.6 to determine $U(T)$ for a consolidating stratum drained only at one boundary is illustrated in the following example.

Example 7.11

A clay stratum 5 ft thick is drained only at its upper boundary. The initial excess pore pressure increases with depth and may be approximated by a half-sine wave distribution in the area of interest. A distribution of this sort occurs outside of a loaded area. It is estimated that there will be a total of 2 in. of consolidation settlement at the point in question. If the coefficient of consolidation is 0.06 ft²/day, determine the settlement due to consolidation at the end of one year.

Solution. The time factor corresponding to one year is

$$T = \frac{(6.0 \times 10^{-2})(365)}{(5)^2} = 0.876$$

From Figure 7.8, we use Case 3 in Table 7.6. Interpolating, we find

$$U = 88.4\%$$

Thus at the end of one year

$$\rho(t) = 0.884(2) = \underline{1.8} \text{ in.}$$

Superposition of Results

Because the differential equation of consolidation, Equation 7–10, is linear, solutions to the equation may be superimposed. Thus, if an initial excess pore pressure distribution can be represented as the sum of two or more of the distributions given in Table 7.6, then the average degree of consolidation for such a case can be computed from the average degrees of consolidation corresponding to the appropriate initial excess pore pressures. This is done by summing the degrees of consolidation weighted in proportion to the relative areas of the initial distributions. This process is illustrated in Figure 7.9, in which it is desired to determine the average degree of consolidation due to the initial excess pore pressure shown in Figure 7.9a, $U_A(T)$. This initial excess pore pressure is the difference between an initial linear excess pore pressure shown in Figure 7.9b and a half-sine wave distribution shown in Figure 7.9c. Hence,

$$\text{Area}(A) = \text{Area}(B) - \text{Area}(C) \qquad (7\text{--}62)$$

From superposition, for any time factor T, the degree of consolidation due to the initial excess pore pressure distribution in Figure 7.9a will be

$$U_A(T) = \frac{U_B(T)\text{Area}(B) - U_C(T)\text{Area}(C)}{\text{Area}(A)} \qquad (7\text{--}63)$$

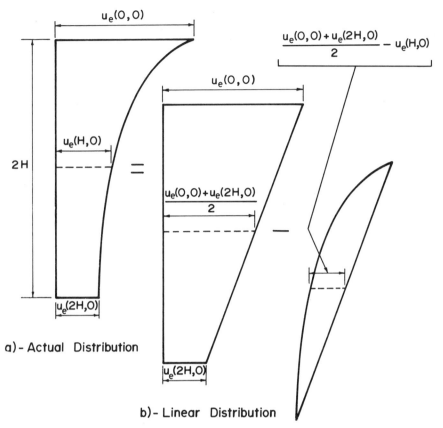

a)- Actual Distribution

b)- Linear Distribution

Fig. 7.9—Weighting of the excess pore pressure distribution for calculation of the average degree of consolidation. (After Leonards, 1962.)

c)- Sinusoidal Distribution

Example 7.12

A stratum 10 ft thick, drained only at its top boundary, is consolidating under a linearly *decreasing* excess pore pressure distribution. At the top of the layer $u_e(0, 0) = 1000$ psf, at the bottom $u_e(10', 0) = 500$ psf. For a time factor $T = 0.400$, determine $U(T)$.

Solution. The linearly decreasing $u_e(z, 0)$ is represented as shown in Figure 7.10 by cases 1' and 4' from Figure 7.8. The corresponding double-drained stratum is also shown. For the single-drained stratum,

$$\text{Area } B = 1000(10) = 10,000 \text{ lb/ft}$$

$$\text{Area } C = \frac{500}{2}(10) = 2500 \text{ lb/ft}$$

$$\text{Area } A = \frac{1000 + 500}{2}(10) = 7500 \text{ lb/ft}$$

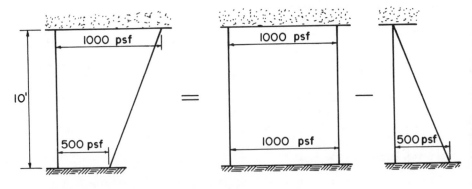

a) - Single Drained Stratum

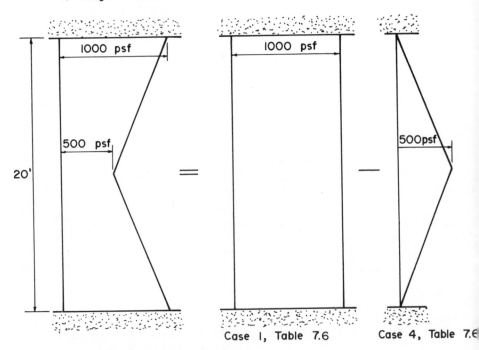

b) - Equivalent Double - Drained Stratum

Fig. 7.10—Initial excess pore water pressure distributions for equivalent double-drained and single-drained strata for Example 7.12.

For $T = 0.400$: From Table 7.7,

(Case 1) $U_B = 69.8\%$

(Case 4) $U_C = 61.5\%$

Thus from Equation 7–63

$$U_A = \frac{(10,000)(69.8) - (2,500)(61.5)}{7,500} = \underline{73\%}$$

Internal Drainage Layers

In the case where a compressible stratum contains one or more seams of pervious material which act as drainage strata, the average degree of consolidation may also be found by taking advantage of the linearity of the mathematical procedure

$$U = \frac{1}{\rho_c} (U_1 \rho_{c_1} + U_2 \rho_{c_2} + \cdots U_M \rho_{c_M}) \tag{7–64}$$

where U and ρ_c are the average degree of consolidation and ultimate consolidation settlement, respectively, for the entire compressible stratum,

$$U = \frac{\rho_c(t)}{\rho_c}$$

U_L and ρ_{c_L} are the average degree of consolidation and ultimate consolidation settlement for the Lth portion of the stratum, $L = 1, \ldots M$.

Example 7.13

A 20-ft-thick stratum drained at the top boundary is consolidating due to a uniform initial excess pore pressure distribution. A thin horizontal seam of sand acts as a drain at a depth of 10 ft from the upper boundary. The bottom of the stratum is undrained. It has been estimated that the ultimate settlement of the upper 10 ft is 3 in., and that of the lower 10 ft is 2 in. For the entire stratum, $c_v = 2.5 \times 10^{-3}$ in.^{2}/min.

(a) Determine the time at which the average degree of consolidation for the entire stratum is 50 per cent.

(b) If the sand seams were not present, how long would it take for 50 per cent of the ultimate settlement to occur?

Solution. (a) We consider the stratum to consist of two layers: the upper layer 10 ft thick and double-drained; the lower layer 10 ft thick and single-drained. Thus, $H_1 = 5$ ft; $H_2 = 10$ ft.

$$\rho_{c_1} = 3 \text{ in.} \qquad \rho_{c_2} = 2 \text{ in.} \qquad \rho_c = 5 \text{ in.}$$

From Equation 7–64,

$$U = U_1 \left(\frac{3}{5}\right) + U_2 \left(\frac{2}{5}\right) = 50\%$$

and

$$T_1 = \frac{c_v t_1}{H_1^2}, \qquad T_2 = \frac{c_v t_2}{H_2^2},$$

but $t_1 = t_2$ and $c_{v_1} = c_{v_2}$ so that

$$T_1 = \frac{c_v t}{(5)^2}, \qquad T_2 = \frac{c_v t}{(10)^2} \text{ and } T_2 = 0.25 T_1$$

or

$$T_2 = 0.25 T_1$$

We proceed by trial and error, using case 1 in Table 7.6 to obtain the results shown below:

T_1	U_1 (%)	$\frac{3}{8}U_1$ (%)	T_2	U_2 (%)	$\frac{2}{8}U_2$ (%)	U_T (%)
0.600	81.6	49.0	0.150	43.7	17.5	66.5
0.400	69.8	41.9	0.100	35.7	14.3	56.2
0.350	65.8	39.5	0.0875	33.4	13.4	52.9
0.300	61.3	36.8	0.0750	30.9	12.4	49.2

Interpolating from these results, $T_1 = 0.315$ and

$$t = \frac{0.315(5)^2}{2.5 \times 10^{-3}} \frac{(144)}{(60)(24)(365)} = \underline{0.86} \text{ years}$$

(b) With no internal drainage layer, $H = 20$ ft (only one drainage boundary). Then from Table 7.7 for Case 1 and $U = 50\%$, $T = 0.197$ and

$$t = \frac{0.197(20)^2}{2.5 \times 10^{-3}} \frac{(144)}{(60)(24)(365)} = \underline{8.6} \text{ years}$$

The effect of the sand seam acting as a drainage layer within the clay stratum is to reduce the time for a 50 per cent average degree of consolidation to one tenth the magnitude of that without the sand seam. It is apparent, therefore, that a critical aspect of the exploration of cohesive strata for which the time rate of settlement is to be estimated, is the definition of drainage layers.

Relative Usefulness of Numerical and Analytical Solutions

Until the general availability of the high-speed digital computer it has been conventional practice to simplify the description of design problems so that available analytical solutions can be used for estimating the time-rate of settlement and/or change effective stress.

Because the finite-difference solution permits ready incorporation of multilayered systems (see, for example, Harr, 1966), changes in permeability and compressibility with changing effective stress, and even variable boundary conditions, it is much more versatile than the analytical solution. Although use of numerical solution procedures has been restricted in the past primarily to research on the influence of variation in material parameters on the consolidation process (Barden and Berry, 1965), the numerical solution is finding increasing application in practice. One early example of the application of the finite-difference solution procedure to the case in which the analytical solutions available were deemed unsuitable is the design investigation of heave and settlement of a tunnel in Sweden involving cutting and filling of a 100-m-thick layer of soft normally consolidated clay underlain by artesian water-bearing sand. Settlement estimates including variations in the ground water table, changes of permeability and compressibility over a 100-year period were made (Hansen and Nielsen, 1965).

It is believed that the versatility of the numerical solutions procedure and the less restrictive assumptions required will make it an increasingly popular tool for the designer.

7.6 LABORATORY DETERMINATION OF THE COEFFICIENT OF CONSOLIDATION.

To calculate the time rate of settlement in the field one must know the magnitude of the coefficient of consolidation c_v. Thus, one of the reasons for performing a laboratory one-dimensional consolidation test is to evaluate c_v.

If, for a laboratory specimen of known thickness and known boundary conditions (usually double-drained in the consolidometer) the average degree of consolidation is known at a specific time, the corresponding time factor T may be determined from Table 7.7, Case 1. Then c_v is calculated from Equation 7-54,

$$c_v = \frac{TH^2}{t}$$

in which H is one-half the initial specimen thickness.

To determine $U(T)$ at a known time, we might plot the settlement versus time for a given increment of vertical stress in the consolidometer, measure the total settlement at the completion of the consolidation phase, and then determine from the time settlement data, the time at which some arbitrary value of the average degree of consolidation, say $U = 50$ per cent, had taken place. Unfortunately, although the consolidation phase ends when the pore pressure is zero, the sample in the consolidometer continues to deform due to secondary compression. Therefore either the pore pressure must be measured during consolidation or some interpretation of the time-settlement data must be made to determine when consolidation is essentially complete.

Due to small amounts of air in the pore water arising from cavitation upon reduction of the *in situ* pore water pressure there may also be an instantaneous settlement which is not a part of the consolidation process. The initial height or zero displacement of the sample at the beginning of consolidation must therefore also be interpreted.

Casagrande Log-Time Fitting Method

One procedure used in practice to determine c_v is that developed by Casagrande and Fadum (1940). It is often called the "Casagrande log-time fitting method." The procedure is as follows:

1. Plot the settlement versus the log of time, as in Figure 7.11,* for a given increment of loading.
2. The dial reading corresponding to complete consolidation ($U(T) = 100\%$), R_{100}, is the point of intersection of the two tangents shown in the figure. The lower tangent is the projection of the secondary compression portion of the settlement curve, the upper tangent is taken at the point of contraflexure.
3. Assume the initial portion of the curve can be represented by a parabola on a natural scale as can the theoretical $U(T)$ vs. T curve for $U(T) < 60$ per cent (Gray, 1938). Then on the semilog time-settlement plot, choose an early time t_1 and the corresponding dial reading R_1. For $t_2 = 4t_1$, read the corresponding value of R_2 from the plot. The initial dial reading R_0 is then

$$R_0 = R_1 - (R_2 - R_1) \qquad (7-65)$$

This may be carried out most easily graphically, as shown in Figure 7.11. Several trials should be made to give an average value.
4. Define the dial reading corresponding to 50 per cent consolidation, R_{50}, by

$$R_{50} = \frac{R_{100} - R_0}{2} \qquad (7-66)$$

The time to 50 per cent average degree of consolidation, t_{50}, is that time on the time-settlement curve for which $R = R_{50}$.
5. For $U(T) = 50\%$, $T = 0.197$ (Table 7.7). With the initial sample height H_0 known (for the particular load increment) the coefficient of consolidation may be computed as

$$c_v = \frac{T_{50} \left(\dfrac{H_0}{2}\right)^2}{t_{50}} \qquad (7-67)$$

---

* In practice it is generally the settlement dial reading rather than the actual settlement that is plotted. The average degree of consolidation is then proportional to settlement or change in dial reading.

Example 7.14 ——

The following dial reading data were obtained for one load increment of a laboratory consolidation test on a limestone residual clay:

Time After Load Applied	Dial Reading	Time After Load Applied	Dial Reading
6 sec	1579	15 min	1916
15	1598	30	1977
30	1618	60	2000
1 min	1650	120	2010
2	1693	240	2017
4	1752	480	2023
8	1837	740	2026
		1045	2030

Determine the coefficient of consolidation by the Casagrande log-time fitting method.

Solution. The dial reading-time data are shown in Figure 7.12 on a semilogarithmic plot. Using the procedure described above, the values of R_0 and R_{100} were determined to be 1543 and 1998 respectively. Thus R_{50} was 1770 and t_{50} was determined to be 4.2 min. The initial sample height under this increment of load was 0.196 in. Thus,

$$c_v = \frac{0.197 \left(\dfrac{0.196}{2}\right)^2}{4.2} = 0.45 \times 10^{-3} \text{ in}^2/\text{min}$$

———

This procedure "fits" the laboratory time-settlement curve to the theoretical relationship at three points: zero, 100, and 50 per cent average degree of consolidation. The two curves may not coincide at other points, in fact they generally do not, but the discrepancy between them is often quite small.

There are, however, circumstances in which the Casagrande procedure is not applicable. If secondary compression is significant for the given increment of loading, the time at which the pore pressure is essentially zero, that is, when consolidation phase is completed, may not be distinguished by the break in curvature on the dial reading-log time plot. The plot of R vs. log t may be similar to one of those shown in Figure 7.13. The type of curve observed depends upon the pressure increment ratio $\Delta\sigma/\sigma_0'$ (Leonards and Altschaeffl, 1964). If R_{100} cannot be identified on the semi-log plot of the time-settlement curve, either the pore water pressure must be measured, or another means of interpreting c_v must be found.

Taylor's Square-Root-of-Time Method

A second procedure for determining the coefficient of consolidation c_v is Taylor's square-root-of-time method (Gilboy, 1936; Taylor, 1940). The

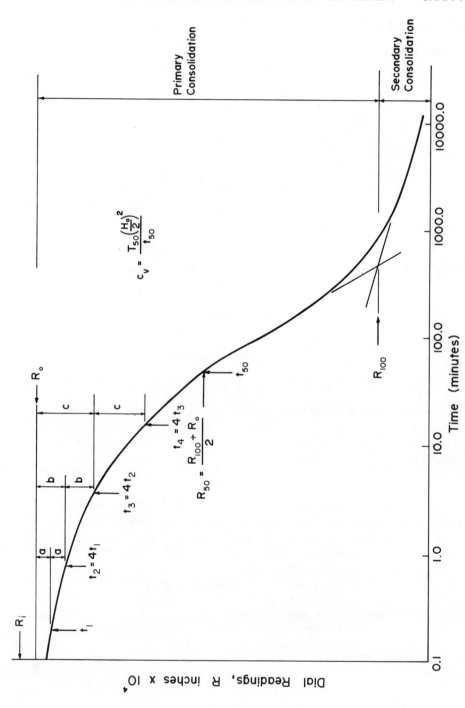

Fig. 7.11 Determination of c_v by the casagrande logtime method.

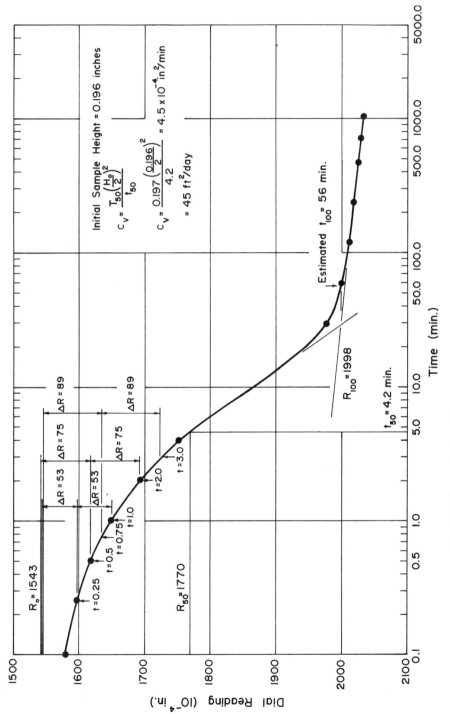

Fig. 7.12—Determination of C by the casagrande log time method, Example 7.14.

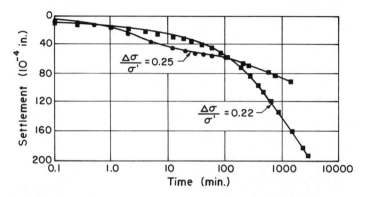

Fig. 7.13—Effect of pressure increment ratio on laboratory time-settlement curves. (Modified from Leonards and Altschaeffl, 1964.)

procedure is as follows:

1. Plot the measured settlement or displacement dial readings versus the square root of the measured time after load application, as in Figure 7.14.
2. Project the early straight line portion of the curve back to zero time; the point of intersection defines R_0. The assumed linearity of the data in the early part of the consolidation process is a consequence of the parabolic shape of the theoretical consolidation-time curve.
3. Plot a second straight line through R_0, with a slope 1/1.15 times that of the initial strength line portion, that is, for each ordinate the abscissa is 1.15 times greater than that of the initial line. The second line will intersect the laboratory data at a value of R, t, which is defined to be the (R_{90}, t_{90}) point. This method is based on the fact that the procedure applied to the theoretical time-settlement curve leads to determination of R_{90}, t_{90}.
4. From Table 7.7 for $U(T) = 90\%$, $T_{90} = 0.848$. With the initial sample height H_0 known (for the given load increment applied) the coefficient of consolidation is

$$c_v = \frac{T_{90}\left(\dfrac{H_0}{2}\right)^2}{t_{90}} \qquad (7\text{--}68)$$

The more pronounced influence of secondary compression effects at later values of time suggest that Taylor's method may not be as reliable as Casagrande's method. However, Taylor's method may be applicable when a semilog plot of the type shown in Figure 7.13 is obtained and Casagrande's method cannot be used at all. We must recognize that in such a case the

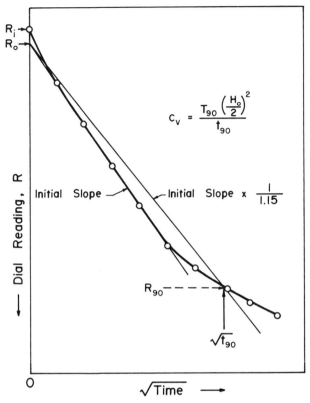

Fig. 7.14—Determination of c_v by the Taylor square-root-of-time method.

secondary compression is likely to be of considerable consequence, and the value of c_v obtained by Taylor's method will be in error.

Example 7.15 ————————————————————————————

Use the dial reading-time data from Example 7.14 to determine the coefficient of consolidation by the Taylor square root of time method.

Solution. The dial readings are shown plotted in Figure 7.15 versus the square root of time. Using the procedure described above, the initial straight line segment was projected back to time zero, to fix R_0 as 1544. A second line with slope 1/1.15 that of the initial line intersects the laboratory data at $R_{90} = 1942$, $t_{90} = 18.9$ min. The initial sample height under this increment of load was 0.196 in. Thus

$$c_v = \frac{0.848 \left(\dfrac{0.196}{2}\right)^2}{18.9} = 0.43 \times 10^{-3} \text{ in}^2/\text{min}$$

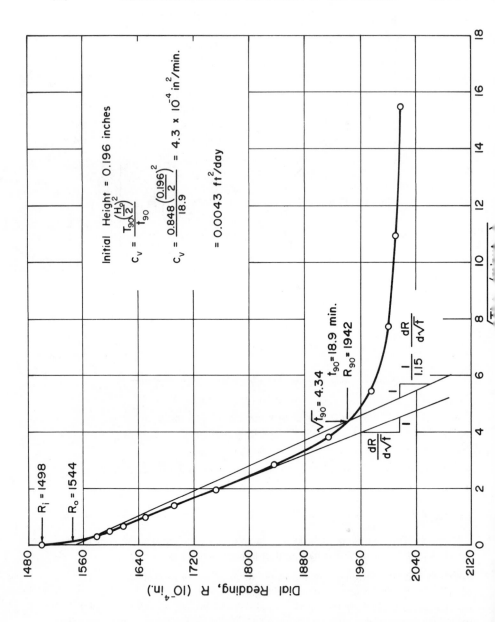

Fig. 7.15—Determination of C_v by the Taylor square-root-of-time method, Example 7.15.

7.7   APPLICATION TO BUILDING 301

At the conclusion of Chapter 5 we investigated the magnitude of settlement that would result from the surcharging of Building 301 at Newark. The immediate settlement was calculated to be 0.26 ft (3.1 in.) and the ultimate consolidation settlement was calculated as 1.55 ft (18.6 in.). Before comparing our estimate of the settlement with measured values we concluded that it would be necessary to determine whether the silt layer was still consolidating under the influence of the surcharge at the end of the 14 months and, if so, what the average degree of consolidation would be at this time.

Laboratory consolidation-test data made available by the Port Authority engineers to the authors, from four of the soil samples tested, are shown in Figure 7.16 (Kapp, 1968). Because of the large secondary compression effects on the data, they are plotted as dial reading versus the square root of time. Taylor's method was used to determine the coefficient of consolidation c_v. The coefficient of consolidation was estimated by the authors to be 0.12 ft²/day. The consolidating stratum was drained at both boundaries and assumed to have an average height of 24 ft. Using Equation 7–54, the time factor at the end of 425 days was

$$T = \frac{0.12 \text{ ft}^2/\text{day}(425 \text{ days})}{(\frac{24}{2}\text{ft})^2} = 0.35$$

From Table 7.6, case 1, the average degree of consolidation $U(T)$ was determined to be 66 per cent. Thus the silt stratum was still consolidating at the time the surcharge was removed. Knowing the ultimate consolidation settlement and the average degree of consolidation the consolidation settlement at the end of fourteen months was

$$\rho_c(t) = 18.6(0.66) = 12.3 \text{ in.}$$

Adding to $\rho_c(t)$ the immediate settlement of 3.1 inches, we determine the expected total settlement at the end of 14 months to be 15.4 in. In the article describing the surcharging of Building 301, the measured settlement was reported to be 17 in.

We must recognize that the accuracy of our estimate was in part fortuitous. The reported value of measured settlement was an average or representative value, as was our predicted value, which was based on an average depth of compressible material. Twenty-two settlement plates were located in Building 301 shortly after surcharging was begun, and settlements ranging from 6 to 18 in. were observed at the end of the 14 months.

Although a reasonable estimate of one-dimensional field settlements may be based on the methods of analysis we have discussed, for an accurate assessment of the extent to which the excess pore pressures have dissipated, it is advisable to make use of field measurements (piezometers, settlement plates) to validate the results of the consolidation analysis. We cannot

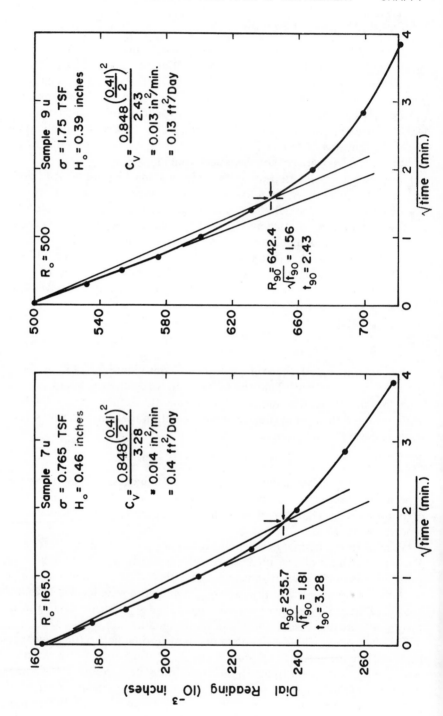

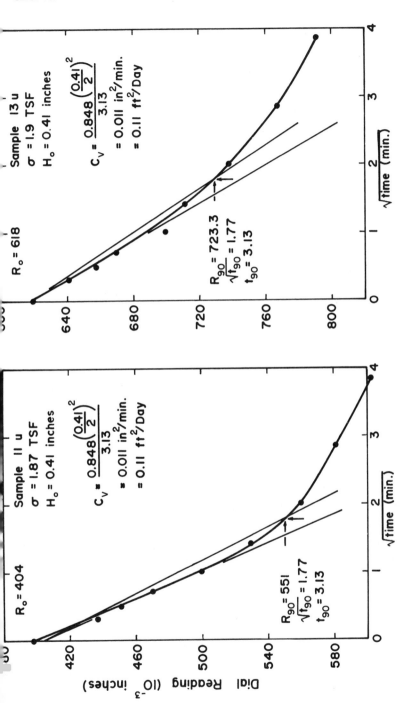

Fig. 7.16—Interpretation of the coefficient of consolidation for the settlement analysis of Building 301. (Data from Kapp, 1968.)

depend in such circumstances upon the calculated results alone because of the many possible sources of error.

In the article about Building 301, it was observed that a pile foundation to support 500 psf live loads was considered as an alternate design. With the value of 500 psf as the magnitude of $\Delta\sigma'$, the ultimate consolidation settlement should have been 6.7 in. Comparing this with the measured settlements of 17 in., we might ask: Was it then necessary to leave the surcharge on for a period of 14 months? To answer this question consider the degree of consolidation U at the center of the layer (drained at both boundaries) where it would be a minimum, at the end of the surcharge period. From Figure 7.7, with $T = 0.35$, U_z at the midheight is 46 per cent. Thus the magnitude of the pore pressure dissipated at this depth was

$$\Delta u_e = U_z \, \Delta\sigma = (0.46)(1300)$$
$$= 600 \text{ psf}$$

We may conclude that, by leaving the load on for 14 months, the entire silt stratum was consolidated to a $\Delta\sigma'$ of greater than 500 psf. Thus in service, the Warehouse would have experienced consolidation settlements due only to recompression of the silt and no virgin compression. Had the load been removed when the recorded settlement was equal to the sum of the immediate settlement and a consolidation settlement of only 6.7 in., a major portion of the stratum would not have experienced an increment in effective consolidation pressure of 500 psf. Then the settlement of the Warehouse in service would have been greater than desired, and the surcharge would not have served its purpose.

7.8 SUMMARY OF KEY POINTS

Summarized below are the key points which we have discussed in the previous sections. Reference is made to the place in the text where the original more detailed description can be found:

1. Starting with the general diffusion equation specialized for one-dimensional flow and incorporating several linearizing assumptions, the Terzaghi one-dimensional consolidation equation is derived (Equation 7–10, Section 7.2). The equation is expressed in terms of a single dependent variable, the excess pore water pressure u_e, and includes a parameter of soil behavior under one-dimensional conditions, the coefficient of consolidation c_v (Equation 7–11).
2. The solution to the equation $u_e(z, t)$ is viewed as generating a surface in a space with coordinates u_e, z, t. The family of curves obtained from the intersection of this surface with planes of $t =$ constant, called isochrones, show the distribution of u_e with depth at the specific times (Section 7.3).
    a. The solution can be obtained for simple boundary and initial conditions analytically. More general cases can be solved nu-

merically using a recurrence relation (Equation 7–21), and appropriate boundary conditions.

b. The progress of consolidation at a point can be expressed in terms of the degree of consolidation (Equations 7–38, 7–40). The progress of consolidation throughout the entire consolidating stratum is described by the average degree of consolidation (Equations 7–41, 7–45).

c. The use of dimensionless parameters, depth factor (Z), time factor (T) and normalized excess pore pressure (U_e) permits generalizing results from both numerical solutions (Section 7.4) and analytical solutions (Section 7.5) to cases with similar boundary conditions, and similar forms of initial conditions.

d. The linear nature of the Terzaghi consolidation equation permits superposition of results obtained for initial conditions which are superimposed (Section 7.5).

3. Determination of the coefficient of consolidation from laboratory one-dimensional consolidation tests may be made by "fitting" the observed time-settlement relation to the theoretical curve. Two methods are currently employed (Section 7.6):

a. The Casagrande log-time fitting method, which enforces coincidence between experimental and theoretical curves at $U = 50$ per cent.

b. The Taylor square-root-of-time method which assumes coincidence between experimental and theoretical curves at $U = 90$ per cent.

Taylor's method is more useful primarily when the 100 per cent consolidation point cannot be estimated from a semi-logarithmic plot of the laboratory time-settlement data.

PROBLEMS

7.1. The clay stratum shown in the figure is subjected to an instantaneous excess pore water pressure equal to 1.0 tsf throughout its depth. Determine the average degree of consolidation and the excess pore water pressure at point A one year after the load application if—

(a) the stratum is drained top and bottom.

(b) the stratum is drained top and bottom and a thin drainage seam occurs at mid-depth.

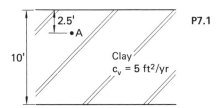

P7.1

7.2. For the situation described in Prob. 5.2, in which $c_v = 10$ ft²/year for the clay, determine the magnitude of consolidation settlement *and* the distribution of excess pore water pressure 5 years after the load is applied, in the following cases:

(a) Assuming that only the top of the clay is drained.

(b) Assuming that in addition to drainage at the top of the clay, a thin drainage layer exists at the mid-depth of the clay.

7.3. The long embankment shown is to be constructed very rapidly at the surface of the profile shown in the sketch. Determine the magnitude of the settlement at the centerline of the fill after *one year* due only to consolidation of the clay layer.

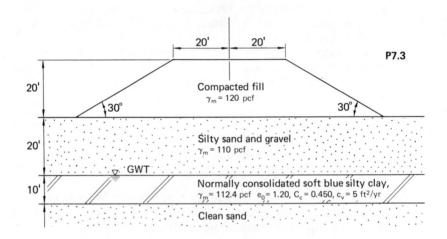

P7.3

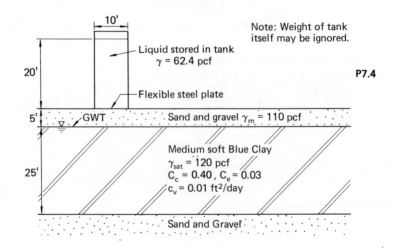

Note: Weight of tank itself may be ignored.

P7.4

Note: Values of soil parameters given for clay are typical of the whole layer, and were obtained from a sample from the center of the layer.

7.4. A small liquid-storage tank is located at the surface of the soil profile shown in the sketch below. The preconsolidation pressure in the clay stratum is found to be approximately 300 psf in excess of the overburden pressure at all depths.

Determine the magnitude of the consolidation settlement of the clay after 5 years, for the following situations:

    (a)   Assuming both sand and gravel strata are drainage layers.

    (b)   Assuming only the top sand and gravel layer is drained.

8

Shearing Resistance of Cohesive Soils

8.1 INTRODUCTION

In the various articles concerning major civil engineering projects so far considered, one soil engineering problem in particular seems to appear again and again: the problem of correctly assessing the shearing resistance of cohesive soils, and the changes which occur in that resistance with time. The failure of the West Branch Dam, described in Chapter 1 ("Electro-Osmosis Stabilizes Earth Dam's Tricky Foundation Clay"), is one example of this. As the result of an incorrect assessment of the strength of the clay soils beneath the dam and the rate at which this strength would increase due to consolidation under the weight of the dam, an expensive, difficult, and time-consuming remedial measure was required.

The Fort Peck slide, discussed in Chapter 2 ("Large Slide in Fort Peck Dam Caused by Foundation Failure"), was triggered by a failure of the weathered shale and Bentonite clay seams under the heel section of the dam. The low shearing resistance of this material was not recognized at the time the dam was built.

Progressive failure of cohesive soils has led to major failures of earth slopes and retaining structures which had been apparently stable for as much as fifty years (Skempton, 1964). We shall consider this problem in Chapter 12.

Even in those cases where compressibility of the subsoils is the primary problem ("Surcharging a Big Warehouse Floor Saves $1 Million", Chapter 4), it is a prerequisite of the design to insure that the shear strength of the soil has not been exceeded.

Many more such examples could be cited, but suffice it to say that the problem of correctly assessing the shear strength of cohesive soils at a given point in time and the changes which will occur in that strength with time is one of the most difficult and important questions which the soil engineer faces. At first we shall restrict our consideration to acquiring an under-standing of the prediction of maximum strength in a saturated cohesive soil, and the influence of water content, effective stresses, and stress history on this strength. In subsequent chapters we shall see the way in which the principles discussed below are applied, as well as cases in which additional factors such as progressive failure play an important role.

8.2 HVORSLEV HYPOTHESIS

$$\tau'_{1f} = \tau'_{3f} \tan^2\left(45° + \frac{\phi'}{2}\right)$$

$$\sin \phi' = \frac{\tau'_{1f} - \tau'_{3f}}{\tau'_{1f} + \tau'_{3f}}$$

Conditions at Failure

The Hvorslev (1936, 1960) hypothesis is an extremely useful modification of the Mohr-Coulomb failure theory for application to saturated uncemented cohesive soils.* It states that, irrespective of stress history,

$$\tau_{ff} = c_e + \sigma'_{ff} \tan \phi'_e \qquad (8-1)$$

where τ_{ff} is the shear stress on the failure plane at failure, that is, the peak shear stress

c_e is the "effective cohesion," that is, the shear strength at $\sigma'_{ff} = 0$

σ'_{ff} is the effective normal stress on the failure plane at failure

ϕ'_e is the "effective angle of shearing resistance"

Experimental results indicate that c_e is a function of the void ratio, so that a plot of Equation 8-1 using τ_{ff}, σ'_{ff} and void ratio as variables, appears as shown in Figure 8.1. Note the contrast with sand. Recall that, for sand, ϕ' is a function of the void ratio, while for clay ϕ'_e is a constant for a given clay and c_e is a function of the void ratio. The relationship shown in Figure 8.1 has been found to be valid for both normally and over-consolidated clays (Hvorslev, 1960).

The relationship between strength and effective stress at a particular void ratio is a straight line, as indicated by Equation 8-1 and shown in Figure 8.1. The way in which failure at constant void ratio corresponds to both normally and over-consolidated conditions is illustrated schematically in Figure 8.2. Consider that several specimens of a saturated normally con-solidated clay are consolidated under different pressures and then sheared to failure. The relationship between void ratio and effective stress on the failure plane at failure would be a curve such as that defined by point ①, ② and ③ in Figure 8.2a. The void ratio for point ① is designated e_1. The strength corresponding to these conditions is shown in Figure 8.2b as point ①'.

---

* More comprehensive constitutive laws which include consideration of stress-strain behavior at states preceding failure have been developed (Roscoe and Burland, 1968). However, these are beyond the scope of our present discussion.

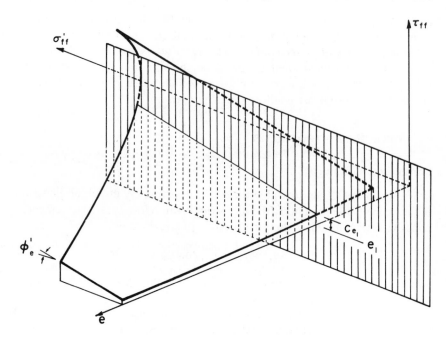

Fig. 8.1—Plot of Hvorslev failure hypothesis.
(Modified from Roscoe, et al., 1958.)

Alternatively we might consolidate a specimen to a higher pressure, say, that corresponding to point ② in Figure 8.2a, and then rebound it so that at failure it would be at void ratio e_1. These failure conditions are represented by point ④. The strength is represented by point ④' in Figure 8.2b. Consolidating the soil to still higher pressure, such as that corresponding to point ③ in Figure 8.2a, and rebounding until failure conditions are those of point ⑤ leads to a strength shown by point ⑤'.

The three points ①', ④', and ⑤' all lie on the straight line *Hvorslev failure envelope* for the void ratio e_1. The end point (point ①') applies to the normally consolidated condition, with points for the over-consolidated case lying to the left.

Equivalent Consolidation Pressure

It is convenient to define a new quantity, the *equivalent consolidation pressure*, σ'_e, which is the consolidation pressure on the virgin compression curve corresponding to the void ratio of the soil. Thus, the equivalent consolidation pressure for a *normally* consolidated soil is simply the actual consolidation pressure. But for an *over*-consolidated soil, σ'_e is the consolidation pressure which will produce the same void ratio on the virgin curve. This is illustrated in Figure 8.3, which shows an idealized void-ratio–

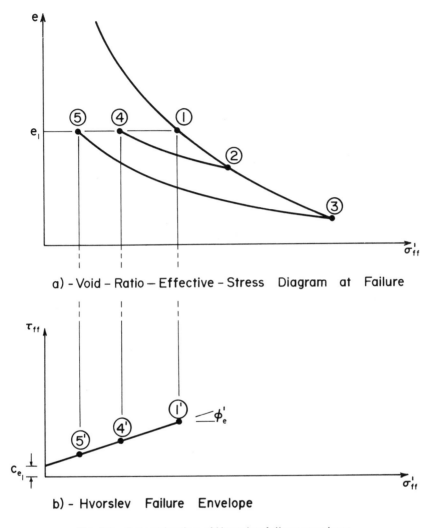

a) - Void – Ratio – Effective – Stress   Diagram   at   Failure

b) - Hvorslev   Failure   Envelope

Fig. 8.2—Determination of Hvorslev failure envelope.

pressure diagram. Point ① is on the virgin compression curve at void ratio e_1 and pressure σ'_{c_1}. Points ② and ③ are at the same void ratio, but at lower pressures. They represent over-consolidated conditions. For all three points, σ'_{c_1} is the equivalent consolidation pressure. It follows that for every void ratio there is a unique value of σ'_e.

For clays *remolded near the liquid limit, and then consolidated*, it has been shown experimentally (Hvorslev, 1960) that,

$$c_e = \kappa\sigma'_e \qquad (8\text{–}2)$$

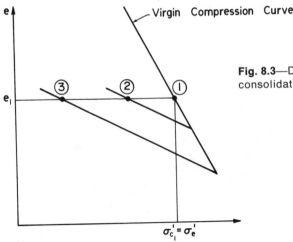

Fig. 8.3—Definition of equivalent consolidation pressure.

Effective Consolidation Pressure (log scale)

where κ is a constant. Hence the shearing resistance can be expressed as

$$\tau_{ff} = \kappa\sigma'_e + \sigma'_{ff} \tan \phi'_e \qquad (8-3)$$

For many clays, Equation 8–3 completely describes the conditions at failure. For those natural soils to which Equation 8–2 does not apply, the relationship between c_e and e can be determined experimentally.

Determination of Hvorslev Parameters

The parameters κ and ϕ'_e characterize a cohesive soil, and are called the *Hvorslev parameters*. For the purpose of evaluating them, it is convenient to express Equation 8–3 in the form

$$\frac{\tau_{ff}}{\sigma'_e} = \kappa + \frac{\sigma'_{ff}}{\sigma'_e} \tan \phi'_e \qquad (8-4)$$

This is the equation of a straight line, and is expressed in useful terms for considering the results of direct shear tests. Because the triaxial compression test is more commonly used for evaluation of the shearing resistance of cohesive soils, it is desirable to express the strength in terms of the parameters usually measured in the triaxial compression test. Manipulation of the Mohr circle at failure leads to the equivalent expression for triaxial compression test results (Skempton and Bishop, 1954),

$$\frac{(\sigma_1 - \sigma_3)_f}{2\sigma'_e} = \frac{\kappa \cos \phi'_e}{1 - \sin \phi'_e} + \frac{\sigma'_{3f}}{\sigma'_e}\left(\frac{\sin \phi'_e}{1 - \sin \phi'_e}\right) \qquad (8-5)$$

where $(\sigma_1 - \sigma_3)_f$ is the maximum principal stress difference at failure and σ'_{3f} is the minor effective principal stress at failure.

Experimental determination of the Hvorslev parameters is arduous and, fortunately, seldom necessary. An understanding of the principles involved, however, permits effective interpretation of many types of test results encountered in practice.

8.3 PORE WATER PRESSURES

In order to appreciate the usefulness of the Hvorslev hypothesis, it is necessary to understand an additional aspect of conditions at failure: the pore water pressure developed in the soil. To do this we shall visualize a small element of soil in equilibrium under a set of stresses. Let us suppose that the element is then subjected to increments in stress, $\Delta\sigma_1$, $\Delta\sigma_2$, $\Delta\sigma_3$. As a consequence, an increment in pore water pressure, Δu, is developed within the soil in the element. The increments in effective principal stresses are then

$$\Delta\sigma'_1 = \Delta\sigma_1 - \Delta u$$
$$\Delta\sigma'_2 = \Delta\sigma_2 - \Delta u \qquad\qquad (8\text{--}6)$$
$$\Delta\sigma'_3 = \Delta\sigma_3 - \Delta u$$

At this point it is useful to visualize a saturated soil as a compressible skeleton containing within its voids an elastic fluid. If the compressible skeleton is assumed to behave in a linear-elastic fashion, Hooke's law (Equation 4–9) applies. For the case of principal stresses only, the volumetric strain of the compressible skeleton $\Delta V/V$, in response to the increments in effective stress is then

$$\frac{\Delta V}{V} = \varepsilon_1 + \varepsilon_2 + \varepsilon_3 = \frac{1 - 2\mu}{E}(\Delta\sigma'_1 + \Delta\sigma'_2 + \Delta\sigma'_3) \qquad (8\text{--}7a)$$

in which ε_1, ε_2, ε_3 are the major, minor, and intermediate principal strains respectively, E is Young's modulus and μ is Poisson's ratio *for the skeleton*. Substituting for the effective stresses from Equations 8–6, the volumetric strain is

$$\frac{\Delta V}{V} = C_s[\tfrac{1}{3}(\Delta\sigma_1 + \Delta\sigma_2 + \Delta\sigma_3 - 3\,\Delta u)] \qquad (8\text{--}7b)$$

in which $C_s = 3(1 - 2\mu)/E$ is the volumetric compressibility of the soil skeleton. This parameter is the inverse of the bulk modulus. Equations 8–7 indicate that for a linear-elastic material, and in fact for all nondilatant materials, the volumetric strain depends only upon the mean stress. If however, dilatancy is important, as it is for most soils, we must introduce an empirical factor to account for it. This is conveniently done by expressing the volumetric strain as

$$\frac{\Delta V}{V} = C_s[\tfrac{1}{3}(\Delta\sigma_1 + \Delta\sigma_2 + \Delta\sigma_3 - 3\,\Delta u)] + D|\Delta\tau\;| \qquad (8\text{--}8)$$

in which D is an empirical parameter, which is a measure of the tendency of the soil to change volume in response to shearing stresses, $|\Delta\tau_{oct}|$ is the absolute value of the increment in a parameter τ_{oct}, called the octahedral shear stress. In terms of principal stresses, τ_{oct} is*

$$\tau_{oct} = \tfrac{1}{3}[(\sigma_1 - \sigma_2)^2 + (\sigma_2 - \sigma_3)^2 + (\sigma_3 - \sigma_1)^2]^{\frac{1}{2}} \qquad (8-9)$$

The octahedral shear stress is a useful measure of the shear stresses which act on the soil because it can be varied independently of the mean stress. That is, the magnitudes of the principal stress increments can be manipulated so that the mean stress remains constant while the octahedral shear stress assumes any arbitrary value. This permits separating the response of the soil to a change in volumetric effective stress as measured by C_s and a change in shearing effects as measured by D.

If the pore water is assumed to behave in a linear-elastic manner, the volumetric strain in the water is

$$\frac{\Delta V_w}{V_w} = C_w \, \Delta u \qquad (8-10)$$

in which V_w is the volume of the water in the pores and ΔV_w is the change in pore water volume. However $V_w = n \cdot V$, where n is the porosity. Furthermore, if no drainage is permitted, $\Delta V_w = \Delta V$, so that Equation 8–10 may be rewritten

$$\frac{\Delta V}{V} = nC_w \, \Delta u \qquad (8-11)$$

Equating the right sides of Equations 8–8 and 8–11, and solving for the increment in pore water pressure, we have

$$\Delta u = \frac{1}{1 + n\left(\dfrac{C_w}{C_s}\right)} \left[\frac{1}{3}(\Delta\sigma_1 + \Delta\sigma_2 + \Delta\sigma_3) + \frac{D}{3C_s}\left|3\,\Delta\tau_{oct}\right|\right] \qquad (8-12a)$$

The ratio C_w/C_s is generally less than 10^{-3} (Perloff, 1962), so that for all practical purposes the coefficient in front of the brackets in Equation 8–12a is unity. If, for convenience, we let $\alpha \equiv D/3C_s$, Equation 8–12a may be written as

$$\Delta u = \tfrac{1}{3}(\Delta\sigma_1 + \Delta\sigma_2 + \Delta\sigma_3) + \alpha\left|3\,\Delta\tau_{oct}\right| \qquad (8-12b)$$

This expression relates the pore water pressure increment to changes in the mean stress and octahedral shear stress for a saturated soil in terms of the empirical parameter α.† This parameter is in general a function of strain, so that the expression is nonlinear.

---

* Note that because τ_{oct} is not linearly related to stress, $\Delta\tau_{oct}$ *cannot*, in general, be calculated directly from the principal stress *increments*. Rather the initial and final stresses must be substituted into Equation 8–9, and the difference between the two resulting τ_{oct} values obtained.

† The relation between changes in applied stress and pore water pressure for partially saturated soils is more complex (Bishop et al., 1960; Bishop, 1961; Barden et al., 1969) and is beyond the scope of this text.

In the special case of the triaxial compression test, in which we assume

$$\Delta\sigma_a = \Delta\sigma_1$$
$$\Delta\sigma_r = \Delta\sigma_\theta = \Delta\sigma_2 = \Delta\sigma_3 \tag{8-13a}$$

then

$$\Delta\tau_{\text{oct}} = \Delta\left[\frac{\sqrt{2}}{3}(\sigma_1 - \sigma_3)\right] = \frac{\sqrt{2}}{3}(\Delta\sigma_1 - \Delta\sigma_3) \tag{8-13b}$$

Equation 8-12b becomes

$$\Delta u = \Delta\sigma_3 + (\sqrt{2}\alpha + \tfrac{1}{3})(\Delta\sigma_1 - \Delta\sigma_3) \tag{8-14a}$$

We shall find it convenient for this special but frequently applied case to define a new empirical coefficient, $A \equiv \sqrt{2}\alpha + \tfrac{1}{3}$, so that Equation 8-14a is written as

$$\Delta u = \Delta\sigma_3 + A(\Delta\sigma_1 - \Delta\sigma_3) \tag{8-14b}$$

in which A is a dimensionless number representing the ratio of magnitude of pore water pressure induced by the axial component of the applied stress, to the magnitude of the axial component itself. Because of the nonlinear behavior of the soil, the A factor for a given specimen will depend on the magnitude of the strain induced, and therefore the stress applied. Since the present concern is for conditions at failure, our current discussion of the A factor will be restricted to its magnitude *at failure*, A_f.[*]

Experimental results (Henkel, 1956; Parry, 1960; Perloff and Osterberg, 1964) indicate that, for cohesive soils, the A factor at failure is a function of the degree of over-consolidation as expressed by the *over-consolidation ratio*, OCR, defined in Chapter 3 as

$$\text{OCR} = \frac{\sigma'_p}{\sigma'_c} \tag{8-15}$$

where σ'_p is the preconsolidation pressure and σ'_c is the present consolidation pressure. Thus when the OCR = 1, the soil is normally consolidated. A large value of OCR indicates a high degree of over-consolidation.

Values of A_f as a function of OCR for two soils are shown in Figure 8.4. Note that, for large values of σ'_p/σ'_c, the A factor is negative, that is, the pore water pressures induced by the axial stress are negative. Recall that for sand, the pore water pressure at failure is a function of the void ratio and effective confining pressure. Values of A_f for a number of normally consolidated clays are given in Table 8.1.

Our discussion so far has revealed that the shearing resistance of given saturated cohesive soil consists of two components. One of these depends upon the void ratio at failure, the other is a function of the effective stress on the failure plane at failure. Corresponding to a particular void ratio, the effective stress at failure is in turn a function of the type of stress field and

---

[*] Because the factor A_f was introduced first by Skempton (1954), it is commonly called the *Skempton A-factor*.

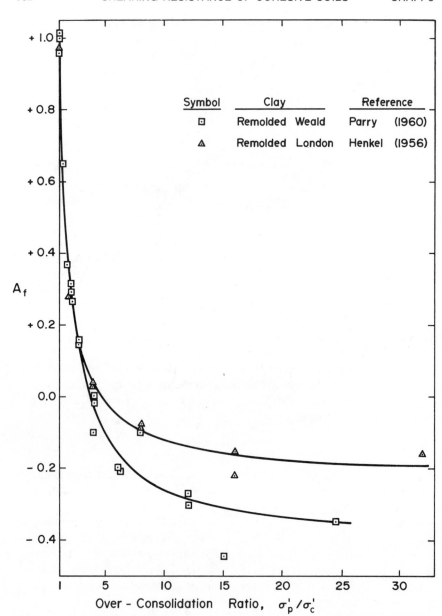

Fig. 8.4—Effect of over-consolidation ratio
on *A* at failure for two remolded clays.

TABLE 8.1
Values of A_f for Undrained
Triaxial Compression
Tests on Normally
Consolidated Clays

Clay Type	A
Drammen clay, Norway	0.85
Oslo clay, Norway	0.95
Seven Sisters, Canada	0.72
Littelton, New Zealand	0.96
Lilla Edet, Sweden	1.03

SOURCE: From Bjerrum and
Simons (1960).

the over-consolidation ratio. We shall discuss now, in some detail, the interpretation of various types of triaxial compression tests on cohesive soils.

8.4 NORMALLY CONSOLIDATED CLAYS

Interpretation of Test Results—Drained Tests

Consider the following experiment performed on a cylindrical specimen of saturated soft, *normally consolidated* clay:

1. Apply an all-round pressure, σ_3, greater than the preconsolidation pressure, and permit drainage until the soil is fully consolidated. Then $\sigma_3 = \sigma'_3 = (\sigma'_1)_{\text{initial}} = \sigma'_{c_0}$ (Figure 8.5). The void ratio is e_0.
2. Apply additional axial stress, $\Delta\sigma_1 = \sigma_1 - \sigma_3$, very slowly to failure so that at all times, $u = 0$ and $\Delta\sigma_1 = \Delta\sigma'_1$.
3. Measure the volume change so that the void ratio at failure, e_f, is known. This failure void ratio, say of magnitude e_1, corresponds to some consolidation pressure, σ'_{c_1}, as shown in Figure 8.5.

Then, at failure, $\sigma'_{3f} = \sigma_3$ and $\sigma'_{1f} = \sigma'_{3f} + \Delta\sigma'_{1f}$.

The Mohr circle at failure, in terms of effective stresses, can then be constructed. By definition, it must be tangent to the Hvorslev failure envelope for a failure void ratio equal to e_1, as shown in Figure 8.6.*

---

* If we wish to consider the shearing resistance of the soil alone, we must take into account the work done in changing the volume during the shearing process as failure is approached. Several approaches to this problem have been pursued (Skempton and Bishop, 1954; Poorooshasb and Roscoe, 1961; Rowe et al., 1964; Roscoe et al., 1965) and the precise form of the correction applied to test results which include volume change varies. However, the results of various approaches differ to a minor extent. Thus, as in Chapter 2, we shall assume for purposes of discussion that an appropriate correction to the measured strength values has been made, or is not necessary. In the case of normally consolidated clays, the rate of volume change at failure is very small and can be neglected.

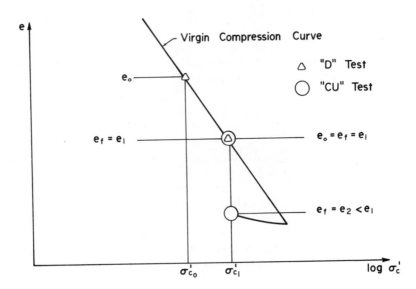

Fig. 8.5—Initial and final void ratios for drained and consolidated-undrained tests.

The experiment can be repeated on identical specimens, consolidating them under higher and higher values of σ_3, and then imposing $\Delta\sigma'_1$ to failure. *The Mohr circles for each failure condition will be tangent to the Hvorslev failure envelope corresponding to the specific void ratio at failure.* Observa-

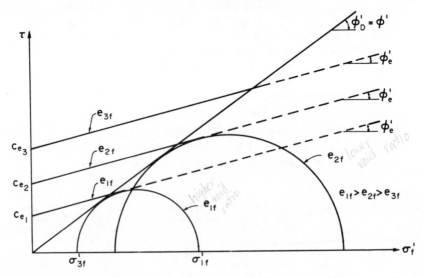

Fig. 8.6—Drained test envelope superimposed on Hvorslev envelopes.

tions of such experiments have shown that a straight line can be drawn through the origin and tangent to these Mohr circles as illustrated in Figure 8.6. This line represents a limiting line, or *envelope* of the Mohr circles at failure, for the *normally consolidated* soil, in terms of effective stresses. Note however, that each point on this failure envelope corresponds to a different void ratio *and* effective stress at failure.

The experiments described above are standard *drained* (D) tests (also called *consolidated-drained* [CD] tests), or *slow* (S) tests. The results of drained tests are very useful because the measured total stresses are also effective stresses. However, interpretation of the failure envelope, when Hvorslev envelopes are not available for comparison (as in Figure 8.5), is complicated by the fact that the *envelope expresses changes in void ratio as well as in effective stress*. In practice, the drained test is infrequently used because of the *very* low testing rates required to keep the pore water pressure close to zero.*

Interpretation of Test Results—Consolidated-Undrained Tests

Let us now consider the results of the following experiment on a cylindrical specimen of a soft saturated *normally* consolidated clay:

1. Apply an all-round pressure, $\sigma_3 = \sigma_{c_1}$, just sufficient to consolidate the specimen to the void ratio e_1 (Figure 8.5), and allow water to drain from the specimen.
2. After consolidation is complete, *prevent any further drainage* and apply an additional axial stress $\Delta\sigma_1$ to failure.
3. Measure pore water pressure so that u_f is known.

The Mohr circle at failure, in terms of effective stresses, can then be constructed. By definition, it must be tangent to the Hvorslev failure envelope for a failure void ratio equal to e_1 at the point for a normally consolidated soil (Figure 8.6). But this is the same circle for the drained test described above for which the failure void ratio was e_1.† Therefore the failure envelope for this test on normally consolidated clays, in terms of effective stresses, is coincident with the drained-test envelope.

The experiment described above is a standard *consolidated-undrained* (CU) test, also called a *consolidated-quick* (R) test. This test has the practical advantage that once consolidation is complete, and the drainage lines are closed, the specimen can be tested to failure more rapidly than in the case of the drained test. In order to obtain the effective stress envelope, however, pore water pressures must be measured. This is difficult to do, and it requires care and expertise to obtain reliable results. Therefore, pore water pressures

---

* Typical testing time for a drained triaxial compression test on a 1.4-in.-dia., 2.8-in.-long specimen of medium soft clay may be as long as 18 hours (Bishop and Henkel, 1962).

† This is a somewhat idealized picture which is satisfied by some, but by no means all, clays. clays. Nonetheless it is a useful concept for development of an understanding of the subject.

are frequently not measured, and the results of CU tests are often presented in terms of *total* stresses. The relationship between the effective and total stress envelopes for a normally consolidated clay with $A_f = 0.8$ is shown in Figure 8.7. Since, $u_f = 0.8$, $\Delta\sigma_{1f} = 0.8(\sigma_1 - \sigma_3)_f$, the Mohr circle in terms of total stresses will be displaced to the right of the circle in terms of effective stresses by an amount equal to 0.8 times the diameter of the circle. It can be seen that ϕ_{CU} is a much smaller angle than ϕ'. The equation for the two envelopes shown in Figure 8.7 are:

Effective stresses: $\tau_f = \sigma'_f \tan \phi'$ (8–16)

Total stresses: $\tau_f = \sigma_f \tan \phi_{CU}$ (8–17)

These envelopes identify the Mohr circles at failure, but do not represent conditions on the actual *plane of failure*. Hence only a single subscript f is used to denote conditions *at* failure.

The results of consolidated-undrained tests in terms of *total* stresses, for *normally* consolidated clays can be very useful *if interpreted properly*. The Hvorslev hypothesis provides the appropriate insight, permitting a more meaningful interpretation of these test results than is apparent from construction of a total stress failure envelope. For a normally consolidated clay, the equivalent consolidation pressure is equal to the actual consolidation pressure, which is also the minor principal stress in a standard CU triaxial compression test, that is,

$$\sigma'_e = \sigma'_c = \sigma_3$$

Therefore, Equation 8–5 may be rewritten as

$$\frac{(\sigma_1 - \sigma_3)_f}{2\sigma'_c} = K_1 + \frac{\sigma'_{3f}}{\sigma'_c} K_2 \qquad (8–18)$$

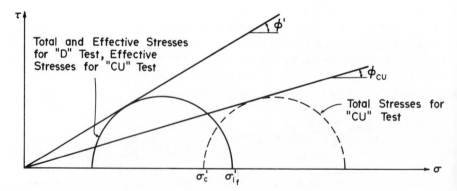

Fig. 8.7—Total and effective stress envelopes for D and CU tests on normally consolidated clay for which $A_f = 0.8$.

where

$$K_1 = \frac{\kappa \cos \phi'_e}{1 - \sin \phi'_e} \qquad (8\text{-}19a)$$

and

$$K_2 = \frac{\sin \phi'_e}{1 - \sin \phi'_e} \qquad (8\text{-}19b)$$

are both parameters of the particular soil (as well, perhaps, as the test type). Note also that

$$\frac{\sigma'_{3f}}{\sigma'_c} = \frac{\sigma_3 - u_f}{\sigma'_c} = \frac{\sigma'_c - A_f(\sigma_1 - \sigma_3)_f}{\sigma'_c} = 1 - A_f \frac{(\sigma_1 - \sigma_3)_f}{\sigma'_c} \quad (8\text{-}20)$$

Substituting Equation 8–20 into Equation 8–18 and arranging terms lead to

$$\frac{(\sigma_1 - \sigma_3)_f}{2\sigma'_c} = \frac{K_1 + K_2}{1 + 2K_2 A_f} \qquad (8\text{-}21)$$

Because A_f is constant for a given over-consolidation ratio, Equation 8–21 states that, for a normally consolidated soil,

$$\frac{(\sigma_1 - \sigma_3)_f}{2\sigma'_c} = K = \text{const} \qquad (8\text{-}22)$$

Equation 8–22 is a most important and useful result. It shows that for a *normally consolidated clay* the principal stress difference at failure is proportional to the consolidation pressure. This relationship is shown in Figure 8.8 for standard CU and D tests, where it can be seen that log $[(\sigma_1 - \sigma_3)_f/2]$ plotted as a function of e is a straight line parallel to the virgin e–log σ'_c curve. Note also that corresponding to every void ratio, or consolidation pressure *on the* virgin curve, there is a single value for the principal stress difference in a compression test, independent of the drainage conditions or means of consolidation.* This relationship was deduced from experimental results before it was generally realized that it was implied by the Hvorslev hypothesis (Rutledge, 1944).

Figure 8.8 is very useful because it is independent of u, yet expresses effective stress relationships. This is the case because, for normally consolidated soils, σ'_f in a compression test is a function of the void ratio at failure.

Effect of Anisotropic Consolidation

Equation 8–22 pertains to a specimen which has been consolidated under an all-round pressure such as the initial consolidation conditions in the triaxial compression chamber. *In situ*, however, the initial conditions

---

* Refer to footnote on p. 363.

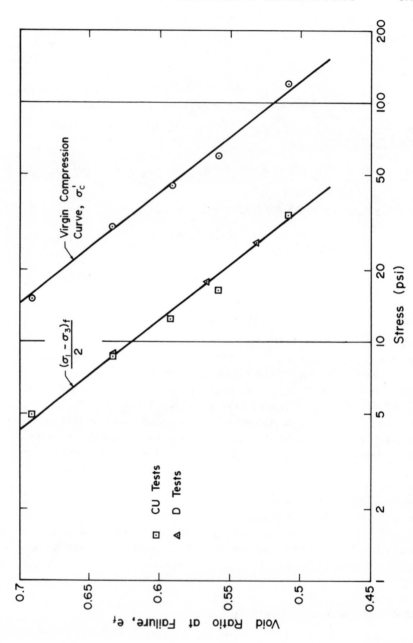

Fig. 8.8—Relationship between consolidation and strength for CU and D triaxial compression tests on remolded weald clay. (Data from Parry, 1960.)

are likely to be nearly one-dimensional, that is, K_0 consolidation. For normally consolidated soils K_0 can be related empirically to ϕ' by (Jaky, 1948)*

$$K_0 = 1 - \sin \phi' \tag{8-23}$$

in which $K_0 = \sigma_h'/\sigma_v'$, σ_h' and σ_v' are the horizontal and vertical effective stresses, respectively, for conditions of zero horizontal strain. For such cases, Equation 8–22 expressed in terms of the apparent angle of shearing resistance (using effective stresses) ϕ', becomes (Skempton and Bishop, 1954)

$$\frac{(\sigma_1 - \sigma_3)_f}{2\sigma_v'} = \frac{(1 - \sin \phi' + A_f \sin \phi') \sin \phi'}{1 + (2A_f - 1) \sin \phi'} = K' \tag{8-24}$$

For comparison, Equation 8–22 for isotropic consolidation, using the apparent angle of shearing resistance in terms of effective stresses appears as

$$\frac{(\sigma_1 - \sigma_3)_f}{2\sigma_c'} = \frac{\sin \phi'}{1 + (2A_f - 1) \sin \phi'} = K \tag{8-25}$$

Hence for normally consolidated soils consolidated under K_0 conditions, the ratio of undrained shearing resistance to the vertical consolidation pressure (this is often called in the literature the c/p ratio) is still constant, although not the same constant as for isotropic consolidation.

8.5 OVER-CONSOLIDATED CLAYS

An over-consolidated clay has been prestressed to stress levels greater than the existing *in situ* pressure. Thus it is not surprising that it is less compressible than a similar clay which is normally consolidated to the same stress level. Similarly, we would expect an over-consolidated clay to be stronger than a normally consolidated clay at the same consolidation pressure. The Hvorslev failure criterion, which is equally valid for over-consolidated and normally consolidated clays, reflects this fact.

Referring to Figure 8.5 we see that, for a specimen consolidated under some high pressure and then rebounded under a pressure equal to σ_{c1}', the void ratio, e_2, will be less than the void ratio, e_1, corresponding to this pressure on the virgin curve. Therefore, as shown by Equation 8–2, $c_{e2} > c_{e1}$. Since A at failure is less for an over-consolidated clay than for a normally consolidated clay (see Figure 8.4b), σ_f' will also be larger. Thus the over-consolidated clay will exhibit higher shear strength than a normally consolidated clay at the same consolidation *pressure*, and Equations 8–4 and 8–5 can be used to determine the effect of over-consolidation on strength.

The Mohr circles for standard consolidated-undrained (CU) tests at three degrees of over-consolidation will be considered below in order to elucidate

---

* We recall that this expression also applies for normally consolidated sand, as discussed in Chapter 3.

the form of the conventional failure envelopes in terms of total and effective
stresses.

The first case considered is that of an highly over-consolidated clay for
which $A_f < 0$. *Total* stresses at failure are shown by the left hand Mohr
circle in Figure 8.9a. Since $A_f < 0$, u_f will be negative and the *effective*
stress circle lies to the *right* of the *total* stress circle.

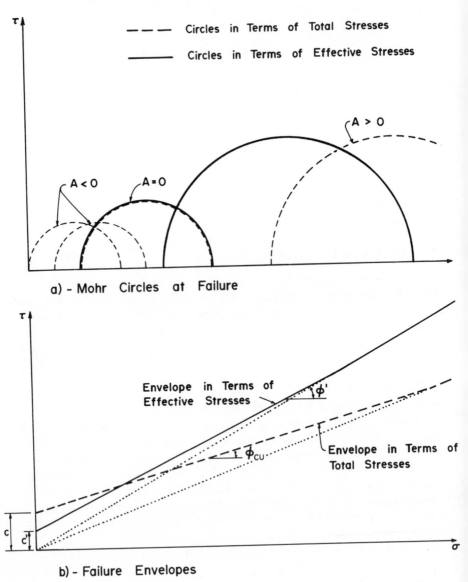

a) - Mohr Circles at Failure

b) - Failure Envelopes

Fig. 8.9—Mohr circles and failure envelopes for overconsolidated clay

The second case considered is that for which the degree of over-consolidation ($\sigma'_p/\sigma'_c \simeq 4$) leads to $A_f = 0$. In this special case, $u_f = 0$, and the Mohr circles in terms of *total* and *effective* stresses will coincide, as shown in Figure 8.9a.

The third case considered is that of a lightly over-consolidated clay for which $0 < A_f < 1$. In this case u_f is positive and the *effective* stress circle lies to the *left* of the *total* stress circle (see Figure 8.9a).

The failure envelopes for all Mohr circles in terms of total and effective stresses are shown in Figure 8.9b. The envelopes in the over-consolidated range are not straight lines, but may be approximated as such, where each envelope corresponds to a particular preconsolidation pressure. Note that, as expected, in the range where $A_f < 0$, the envelope in terms of total stresses lies above the envelope in terms of effective stresses. The envelopes cross where $A_f = 0$, and, where $A_f > 0$, the total stress envelope lies below the effective stress envelope. The implications of these differences for long and short term stability will be discussed in Chapter 12.

In the over-consolidated range the failure envelopes can be approximated by

Effective stresses: $\tau_f = c' + \sigma'_f \tan \phi'$ $\qquad\qquad$ (8–26)

Total stresses: $\tau_f = c + \sigma_f \tan \phi_{CU}$ $\qquad\qquad$ (8–27)

Again, only a single subscript f is used to indicate conditions *at* failure, but *not on* the failure surface. Equation 8–26 was first proposed by Coulomb (1776) on the basis of direct shear tests. Since the Hvorslev equation (Equation 8–1) had the same form it is frequently called the *Coulomb-Hvorslev* equation.

In order to appreciate the relationship between the effective stress failure envelopes as determined conventionally, expressed in Equation 8–26, and the true Hvorslev failure envelopes, Equation 8–1, we shall examine the way in which these two failure criteria are applied to the three effective stress Mohr circles at failure shown in Figure 8.9. This is illustrated in Figure 8.10; Figure 8.10b shows the three effective stress circles. To see how these circles might be obtained, let us consider a specimen of the clay, normally consolidated to a void ratio e_1 and loaded to failure in a standard CU triaxial compression test. Assuming that we can determine the effective stress on the failure plane at failure, we could plot the failure void ratio e_1 and the effective stress on the failure plane at failure, as shown in Figure 8.10a; Point ① indicates the effective normal stress on the failure plane. The Mohr circle at failure is shown in Figure 8.10b; the Hvorslev envelope for a void ratio e_1 is shown as a solid line tangent to this circle at point ①. The conventional failure envelope for the normally consolidated state is shown as a dashed line up to the point where it is tangent to the Mohr circle at failure. The conventional envelope is, of course, tangent to the circle at a different point from that which represents the plane of failure.

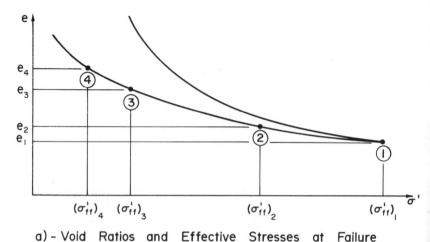

a) - Void Ratios and Effective Stresses at Failure

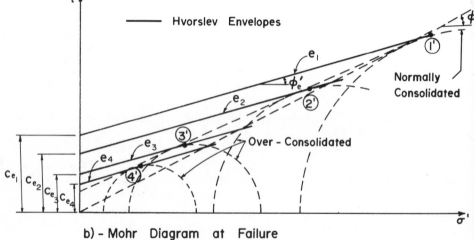

b) - Mohr Diagram at Failure

Fig. 8.10—Relationship between Hvorslev envelopes and conventional failure envelope for over-consolidated clays.

If we next consolidate an identical specimen to a void ratio e_1 and then rebound it to a larger void ratio e_2, the effective stress at failure on the failure plane in a CU test will be that shown as point ② in Figure 8.10a. The Mohr circle at failure in Figure 8.10b is tangent to the Hvorslev envelope corresponding to the void ratio e_2, at point ②'. Note that this Hvorslev envelope continues beyond the point at which it is tangent to circle ②', because that circle corresponds to an over-consolidated material. The end point of the Hvorslev envelope is the *normally consolidated* case. Circles ③' and ④' are obtained by rebounding identical samples from the void ratio e_1

to the void ratios e_3 and e_4, respectively, prior to undrained compression. Each of these circles is tangent to the appropriate Hvorslev envelope. The line drawn tangent to circles ②, ③ and ④ is the conventional envelope for an over-consolidated material. It is tangent to those circles at a different point from that corresponding to the failure plane.

It is most important to recognize that, had the void ratio at which the specimens were rebounded for variety been different, the conventional failure envelope would have been different. However, the Hvorslev envelopes *would have been the same* for all preconsolidation pressures.

The undrained shearing resistance of over-consolidated soils can also be conveniently related to the consolidation pressure using Equation 8–22. For A_f a function of the over-consolidation ratio, $(\sigma_1 - \sigma_3)_f/2\sigma'_c$ is also a function of the OCR. This is illustrated for two remolded clays in Figure 8.11. Using these and other data, Perloff and Osterberg (1964) have shown that the strength may be expressed as

$$\frac{(\sigma_1 - \sigma_3)_f}{2\sigma'_c} = r\left(\frac{\sigma'_p}{\sigma'_c}\right)^s + t \tag{8-28}$$

in which r, s, t are constants for a constant strain rate of testing.

$$\sin\phi = \frac{\sigma'_1 - \sigma'_3}{\sigma'_1 + \sigma'_3}$$

8.6 UNCONSOLIDATED-UNDRAINED TEST

One of the most easily and commonly performed tests is the *unconfined compression* test. This is a special case of the *unconsolidated-undrained* (UU) test (also called *quick* [Q] test), which is illustrated in Figure 8.12. If a saturated clay specimen is removed from the ground, without being given a source of free water, there will be an initial negative pore water pressure, u_i (tension). If this specimen is subjected to axial stress only ($\sigma_3 = 0$) and drainage is prevented, a total stress Mohr circle for failure, such as the left-hand circle in Figure 8.12, will result. The pore water pressure at failure will be

$$u_f - u_i =$$
$$\Delta u$$

$$u_f = u_i + A_f(\sigma_1 - \sigma_3)_f$$

The effective stresses at failure are shown by the dashed circle. If an all-round pressure, σ_3 is applied to the specimen before the axial stress is applied but *no drainage is permitted*, then the only effect of σ_3 will be to increase the pore water pressure at failure. The effective stress at failure will not be changed, as shown by Equation 8–3. Thus an infinite number of *total* stress Mohr circles, all with the same diameter, will occur for an infinite possible variation in σ_3. Several special cases are illustrated in Figure 8.12. However, these circles all correspond to a single *effective* stress Mohr circle. Thus, the envelope, in terms of *total stresses* for a series of UU tests will simply be an horizontal line.

The unconfined compression test ($\sigma_3 = 0$) is simple and useful in many situations. The principal stress difference at failure, $(\sigma_1 - \sigma_3)_f$, is referred

$$\Delta u = A_f \Delta \sigma_1$$

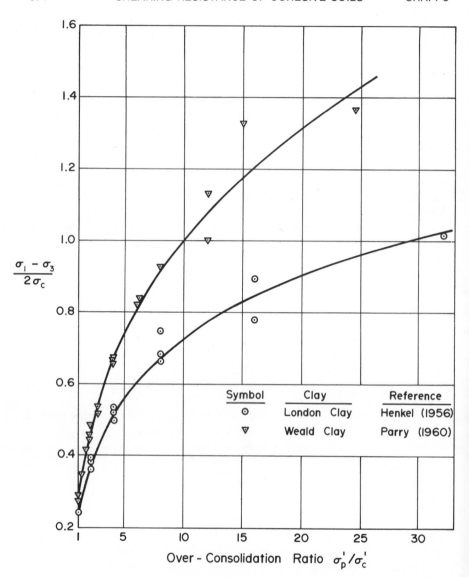

$$\frac{C}{P}$$

Fig. 8.11—Effect of over-consolidation ratio on stresses
at failure for remolded clay.

to as the "unconfined compressive strength" (q_u). This quantity represents
twice the maximum shear stress which can be exerted on the soil in its *in
situ* condition. It is an indicator of the undrained shear strength of the soil
at the time the soil sample is taken, that is, the *immediate* shear strength, but
contains no implication about the effect on strength of changes in void
ratio and/or effective stresses.

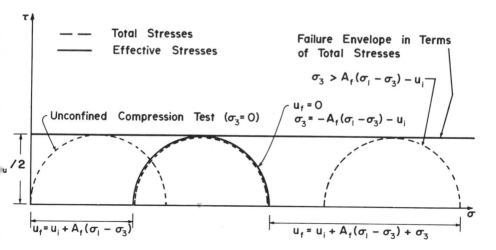

Fig. 8.12—Failure circles and envelopes for UU-test.

The most important fact to keep in mind when considering strength properties of soil, is that these properties are determined primarily by void ratio and the *effective* stresses. This principle underlies all of the points discussed above.

We shall discuss an example which illustrates the interaction between strength and consolidation, and the application of strength measurements to field conditions.

Example 8.1

The long embankment shown in Figure 8.13 is to be constructed at the surface of the profile given in the figure. An undisturbed specimen taken from the top of the normally consolidated clay stratum exhibited an unconfined compressive strength, $q_u = 0.67$ tsf. Due to the fact that the clay layer is so weak it has been estimated that the embankment must be built in two stages.

(a) How high may the first stage be built without causing failure of the clay? Assume that during construction the embankment side slopes are maintained at 30° to the horizontal.

(b) How long must the contractor wait to place the rest of the embankment?

Solution. (a) Conventional methods of stability analysis require that we consider the potential for failure everywhere along some assumed failure surface. Methods by which this is done are discussed in detail in Chapter 12. For purposes of this example we shall adopt a conservative approach and neglect the effect of the strength of the sand overlying the clay. Thus the embankment will be assumed to be on the point of failure when the shear stress at any point within the clay is equal to the shear strength of the clay at that point.

The approach which we shall take is to relate the strength available and the strength required at every point within the clay stratum at appropriate times. This is illustrated

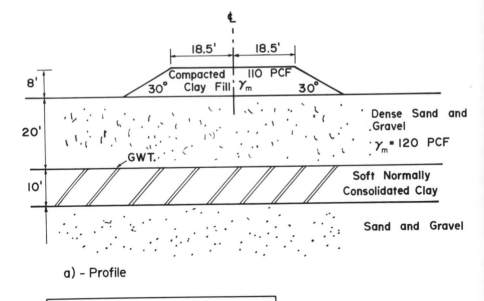

a) - Profile

γ_{sat} = 120 PCF

ϕ' = 25°

A_f = 0.8

at Top of Stratum, q_u = 0.67 TSF

c_v = 40 ft/year

Fig. 8.13—Data for Example 8.1.

b) - Data for Soft Clay Layer

in Figure 8.14 which shows conditions at a typical point in the clay layer. The initial stresses prior to placement of any fill are shown in Figure 8.14a. If the clay were brought to a state of failure by rapid placement of fill so that no drainage occurred, failure conditions would be as shown in Figure 8.14b. The initial conditions have also been indicated in this diagram for comparison. As we shall see, these failure conditions correspond to approximately 5 ft of fill placed on the surface. If the clay is permitted to consolidate under the loading from this 5 ft of fill, the state of stress will be that shown in Figure 8.14c. The initial conditions are again given for comparison purposes. If the clay is then brought to a state of failure by increasing the height of fill, the stress state shown in Figure 8.14d will result. This stress state corresponds to slightly more than the 8-ft-high completed embankment. The changes in effective stresses due to placement of the additional 3 ft of fill can be seen by comparing the two circles in Figure 8.14d.

This kind of comparison must be made for each point in the stratum because it is not immediately obvious which point will be the most likely to fail at a given time.

Because of the rapidity with which the embankment will be constructed, it is rea-

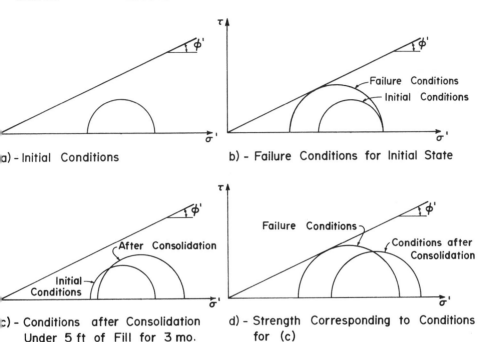

a) - Initial Conditions

b) - Failure Conditions for Initial State

c) - Conditions after Consolidation Under 5 ft of Fill for 3 mo.

d) - Strength Corresponding to Conditions for (c)

Fig. 8.14—Conditions at typical point in clay layer.

sonable to suppose that the *in situ* shearing resistance, that is, q_u, will remain constant during the construction of the first stage. Thus the principal stress difference at failure will be*

$$(\sigma_1 - \sigma_3)_f = q_u = (\sigma'_1 - \sigma'_3)_{initial} + (\Delta\sigma_1 - \Delta\sigma_3) \qquad (8–29)$$

where $(\sigma'_1 - \sigma'_3)_{initial}$ is the principal stress difference for the initial *in situ* conditions, and $\Delta\sigma_1 - \Delta\sigma_3$ is the increment in principal stress difference applied by the embankment. From the geometry of the problem, and the fact that the soil is normally consolidated, the critical point in the clay mass for the initial loading will be at the top of the stratum. The vertical effective stress at this point for the initial conditions is

$$(\sigma'_v)_{initial} = (\sigma'_1)_{initial} = \frac{(120)(20)}{2000} = 1.2 \text{ tsf} \qquad (8–30)$$

The initial horizontal effective stress can be estimated from Equation 8–23:

$$K_0 = 1 - \sin\phi' = 1 - 0.423 = 0.577 \qquad \qquad (8–31)$$

and

$$(\sigma'_h)_{initial} = (\sigma'_3)_{initial} = K_0(\sigma'_v)_{initial}$$
$$= (0.577)(1.2) = 0.693 \text{ tsf} \qquad (8–32)$$

---

* We shall see in the next section that the strength may also depend on the way in which the stresses are applied. However, for this problem, the effect is likely to be minor.

in which K_0 is the coefficient of earth pressure at rest, ϕ' is the apparent angle of shearing resistance in terms of effective stresses, and $(\sigma'_v)_{initial}$ and $(\sigma'_h)_{initial}$ are the effective vertical and horizontal stresses respectively. Thus, from Equation 8–29 the maximum increment in the principal stress difference which can be applied by the embankment before failure will occur is

$$(\Delta\sigma_1 - \Delta\sigma_3)_f = 0.67 - (1.2 - 0.693) = 0.163 \text{ tsf} \qquad (8\text{–}33)$$

The height of embankment which will produce this principal stress difference at the top of the clay stratum has been determined using the influence diagram for maximum shear stress under a linear-elastic embankment on a linear-elastic half-space (see Figure A.23). Various values of the height, H, lead to difference L/H ratios. Thus we require a number of trials as in Table 8.2. These lead to the conclusion that the maximum height, for a factor of safety of one against shear failure within the clay is approximately $4\frac{1}{2}$ ft. Because of the various conservative assumptions that we have made, we shall, for purposes of discussion, round this off to 5 ft.

TABLE 8.2
Trial Values for Maximum Height of First Stage
of Embankment for Example 8.1 ($\alpha = 30°$).

H (ft)	L (ft)	L/H	Y/H	$\dfrac{\sigma_1 - \sigma_3}{2\gamma H}$	$\sigma_1 - \sigma_3$ (tsf)
4	25.4	6.4	5.00	0.34	0.149
5	23.7	4.7	4.00	.32	.174
6	21.9	3.7	3.33	.32	.211
7	20.2	2.9	2.86	0.32	0.248

NOTE: Data of fifth column from Figure A.23; stresses are taken at top of clay stratum.

(b) We must now determine how long it is necessary to wait until the embankment can be completed. In doing so we must recognize that the degree of consolidation at various points within the stratum will be different at a given time. Thus the effective consolidation pressure, and therefore the strength will vary throughout the stratum. Consequently, it is not sufficient to permit consolidation to occur until the average strength is adequate to support the embankment. Rather we shall focus our attention upon the distribution of stresses within the clay stratum.

It is most convenient to do this by considering the effective vertical consolidation pressure required to sustain the shear stresses imposed by the completed embankment. From Equation 8–24 for a normally consolidated clay,

$$\frac{(\sigma_1 - \sigma_3)_f}{2\sigma'_v} = K' \qquad (8\text{–}34)$$

which, from the *in situ* conditions, gives

$$K' = \frac{0.67}{(2)(1.2)} = 0.28$$

Thus the vertical effective stress required to support the 8-ft high embankment can be determined for each depth in the clay stratum by computing the principal stress difference imposed by the embankment using Figure A.23 and adding to that the existing principal stress difference determined in a similar fashion to that above. The magnitude of this required σ_v' is shown as a function of depth in Figure 8.15. Shown for comparison is the initial overburden stress as well as the ultimate σ_v' when consolidation under the weight of the 5-ft fill has been completed determined from Figure A.13).

The degree of consolidation required to insure that every point in the clay stratum is subjected to σ_v' no less than that required can be determined using Figure 7.7. Although the center of the stratum may not be the critical point, it serves as a useful guide. In this case the degree of consolidation required at the center in order to produce a sufficient increment in vertical effective stress is

$$U_z = \frac{\Delta\sigma_v'}{\Delta\sigma_v} = \frac{0.106}{0.213} = 0.5$$

Referring to Figure 7.7, this corresponds approximately to a time factor of 0.37, for which the time is

$$t = \frac{H^2}{c_v}T = \frac{(5)^2}{40}(0.37) = \underline{0.23} \text{ years} = \underline{2.8 \text{ months}}$$

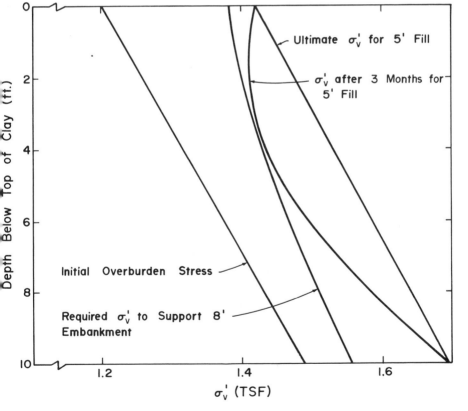

Fig. 8.15—Vertical effective stresses in clay layer, Example 8.1.

This is the time after which the last 3 ft of fill can be added. The effective vertical stress after this time is shown in Figure 8.15. We note that the critical combination of required vertical stress and degree of consolidation is not at the center of the layer but somewhat above the center.* The average degree of consolidation (from Table 7.4) is nearly 70 per cent.

8.7 RELATIONSHIP BETWEEN STRESSES IMPOSED IN THE FIELD AND THOSE APPLIED DURING LABORATORY TESTING

Figure 8.16 shows schematically some typical stress systems imposed on the soil in the field by various common types of construction activity. Clearly, only the first case (Figure 8.15a) introduces stresses which are even approximately similar to those applied during the usual triaxial compression test. If the Hvorslev failure criterion were valid irrespective of the stress system applied, and if the effective stresses imposed by the various stress systems could be determined, then it would not be important whether the stress system in the laboratory were similar to that in the field. Even when the soil is isotropic so that the Hvorslev criteria apply independent of the stress system, the pore water pressures developed by some of these stress systems lead to failure conditions which are not predicted by the conventional triaxial compression test (Ladd and Varallyay, 1965). To see how these states of stress produce pore water pressures, and therefore modify effective stresses at a given point in the ground, it is useful to consider a notion called the *stress path*.

The Stress Path

Let us visualize a point in the soil beneath the centerline of a circular loaded area, such as point A in Figure 8.16a. Assume, for the moment, that before the area is loaded, the point is subject to an isotropic state of stress. The Mohr circle corresponding to this state of stress is point ① in Figure 8.17a. If we neglect the possible effect of ground water, then the total and effective stresses will be the same at equilibrium before the application of additional load. The application of some load on the area causes the Mohr circle for total stress to increase in size, say to circle ②, shown in Figure 8.17a. Assuming that the load is applied sufficiently rapidly that no drainage occurs and that the soil is normally consolidated (with $A = 1$ at all strains, for convenience in plotting results) then the effective stress circle will be that shown as circle ②'. As the load is increased still further, the circle will increase in size until we observe circle ③ for the total stresses, and ③' describing the effective stress state. If the load is increased sufficiently,

---

* We shall see a similar situation arise in our discussion of *preloading* of foundations to reduce post-construction settlements, in Chapter 11.

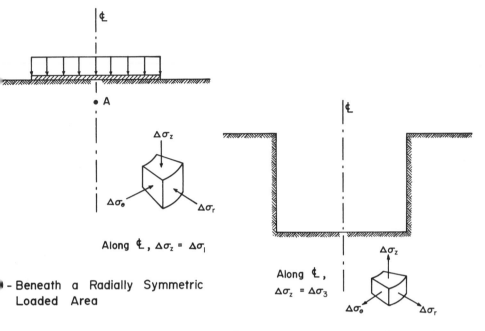

Along ₵, $\Delta\sigma_z = \Delta\sigma_1$

- Beneath a Radially Symmetric
  Loaded Area

Along ₵,
$\Delta\sigma_z = \Delta\sigma_3$

b) - Beneath a Radially Symmetric
   Excavation

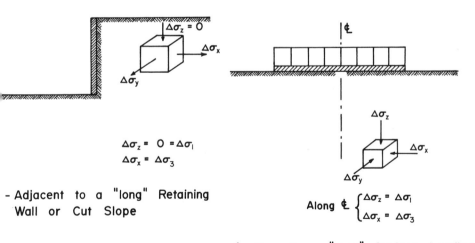

$\Delta\sigma_z = 0 = \Delta\sigma_1$
$\Delta\sigma_x = \Delta\sigma_3$

- Adjacent to a "long" Retaining
  Wall or Cut Slope

Along ₵ $\begin{cases} \Delta\sigma_z = \Delta\sigma_1 \\ \Delta\sigma_x = \Delta\sigma_3 \end{cases}$

d) - Beneath a "long" Surface Loading

Fig. 8.16—Some typical stress systems in
the field. (Modified from Ladd and Varallyay, 1965.)

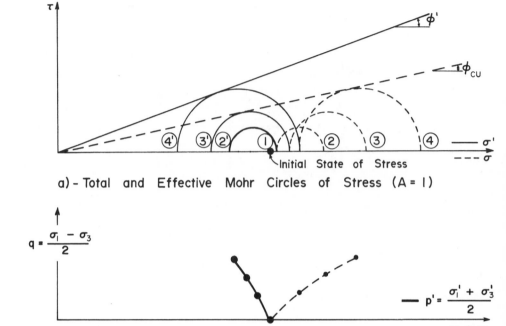

a) - Total and Effective Mohr Circles of Stress (A = I)

b) - Total and Effective Stress Paths

Fig. 8.17—Stresses due to circular loading.

the soil can be brought to a state of failure, in which case circles ④ for total stresses, and ④' for effective stresses, will result.

The difference between any two corresponding points on equivalent circles for total and effective stress is of course the pore water pressure. Thus, if we are interested in observing the changes in the state of stress (either total or effective, or both) during a change in loading, we could plot a sufficient number of the Mohr circles at different stages of loading to illustrate the path which these stresses follow. However, it soon becomes confusing to plot all of the circles shown on the page. Fortunately, it is not necessary to show each in its entirety in order to obtain the information we wish. We can completely define the circle by plotting one point on each circle, tangent to a known direction. For example, we might choose to observe the state of stress on the plane which eventually becomes the plane of failure. In that case, each circle would be represented by a single point; the radius passing through that point would define the plane on which failure occurs. The locus of all such points is called a *stress path* and defines the locations of the Mohr circles during a change in stress. As a practical matter, it is frequently more convenient to plot the peak point on the circle, rather than the point on the potential failure plane. Often it is not known which plane is the potential failure plane; whereas the peak point on the circle is always easy to identify. In fact, it has the coordinates

$$\frac{(\sigma_1 + \sigma_3)}{2}, \quad \frac{(\sigma_1 - \sigma_3)}{2}$$

These quantities are often given the symbols p and q, respectively.* A plot of the stress paths for the Mohr circles in Figure 8.17a is shown in Figure 8.17b, using the peak point on the circle.

We note that the example presented assumed initially isotropic stresses for illustrative purposes. The actual initial *in situ* stress state is, of course, likely to be anisotropic.

Stress Paths Imposed by the Triaxial Compression Test—Drained Tests

We shall follow the usual assumption that σ_r is the minor principal stress when the specimen is failed by axial compression; keeping in mind, however that as discussed in Chapter 2, this may not be the case. The stress paths for drained tests shown in Figure 8.18 are determined using this assumption. These two stress paths, which are for both total and effective stresses, represent four types of loading situations which are simulated approximately in the triaxial compression test. The line inclined to the right is a *loading curve*. That is, one stress is increased to failure while the other is held constant. In the conventional triaxial *compression* test, the axial stress is increased to failure and the lateral pressure is held constant. This approximates conditions beneath the axis of a circular footing (Figure

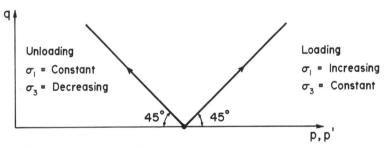

Fig. 8.18—Total and effective stress paths for drained tests.

* Some investigators reserve the symbol p for the mean principal stress.

$$p \equiv \frac{\sigma_1 + \sigma_2 + \sigma_3}{3}$$

This definition may be more generally useful than that given above, but because of the intermediate principal stress does not appear in the failure criterion we have adopted, the above definition will be useful.

8.16a). In the *extension test* (loading), the lateral pressure is increased to failure while the axial stress is held constant. This test is a radially symmetric equivalent to a passive pressure failure.

The stress path which is inclined to the left represents an *unloading* curve, in which one stress is decreased to failure while the other is held constant. Again, there are two alternative ways in which this test can be conducted. When the axial stress is reduced, the test simulates approximately the conditions under the axis of a radially symmetric excavation, Figure 8.16b. When the radial stress is reduced, the test is a radially symmetric representation of the active pressure condition on a retaining structure.

For all these cases, *when drained tests are being performed*, the shear stress at failure for an isotropic soil depends only upon the effective normal stress at failure and the void ratio at failure, and can be determined from the Hvorslev parameters.* Thus in those cases of field loading where the loading is increased sufficiently slowly for essentially drained conditions to apply, the Hvorslev hypothesis appears valid for prediction of the maximum shear stress which the soil can withstand.

Stress Paths Imposed by the Triaxial Compression Test— Undrained Tests

The stress paths for a conventional consolidated-undrained triaxial compression test on a normally consolidated clay are shown in Figure 8.19 for both total and effective stresses. In this test the axial stress is the major principal stress and is increased until failure occurs. The lateral pressure is the consolidation pressure and is held constant during the shearing portion

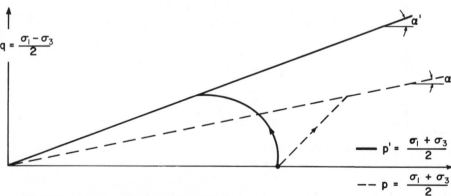

Fig. 8.19—Stress paths for conventional triaxial compression test on normally or lightly over-consolidated saturated clay.

* Assuming that suitable correction has been made for the effect of the rate of volume change with respect to strain at failure.

of the test. The envelopes of the failure points of such curves are designated by their slopes α and α', in terms of total and effective stresses respectively. These lines simply define the peak points on the Mohr circles at failure.

The increasing vertical stress is the case most commonly encountered in practice (see, for example, Figures 8.16a and 8.16d). In those cases where the horizontal stress is increased to failure, or the axial vertical stress is decreased to failure, the magnitude of the pore water pressure in the consolidated-undrained test is different from that in the conventional CU test, as predicted by Equation 8–12. The effective stress paths for such tests are of the form shown in Figure 8.20, because of the different pore water pressures developed in two tests. For the case of an extension test in which

$$\Delta\sigma_a = \Delta\sigma_3$$
$$\Delta\sigma_r = \Delta\sigma_1 = \Delta\sigma_2 \qquad (8\text{--}35)$$
$$\Delta\tau_{\text{oct}} = \sqrt{2}(\Delta\sigma_r - \Delta\sigma_a)$$

the pore water pressure is predicted by Equation 8–12b to be

$$\Delta u = \Delta\sigma_a + (\sqrt{2}\alpha + \tfrac{2}{3})(\Delta\sigma_r - \Delta\sigma_a) \qquad (8\text{--}36a)$$

or

$$\Delta u = \Delta\sigma_3 + (\sqrt{2}\alpha + \tfrac{2}{3})(\Delta\sigma_1 - \Delta\sigma_3) \qquad (8\text{--}36b)$$

which in terms of the Skempton A-factor is

$$\Delta u = \Delta\sigma_3 + (A + \tfrac{1}{3})(\Delta\sigma_1 - \Delta\sigma_3) \qquad (8\text{--}36c)$$

where A is determined in a compression test.

Prediction of stress paths for extension test conditions from compression test results is illustrated in Figure 8.21. This figure shows the results of compression and extension tests on normally consolidated remolded Boston blue clay. The compression-test stress paths are shown as solid lines, the extension-test stress paths as dashed lines. The circles are points predicted

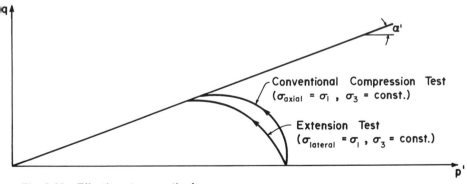

Fig. 8.20—Effective stress paths for consolidated undrained triaxial tests.

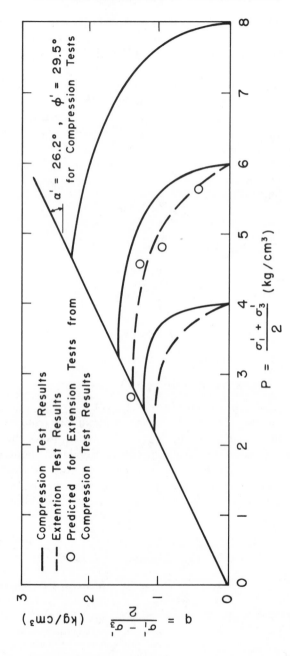

Fig. 8.21—Effective stress paths for compression and extension tests on normally consolidated Boston blue clay, consolidated from a slurry. (Modified from Ladd, 1965.)

for the extension tests determined from the compression test results using Equations 8–36. The agreement is best at failure, where the effect of different strains is less pronounced.*

If the pore water pressures developed at failure are determined appropriately, the Hvorslev parameters apply equally well for undrained conditions.

8.8 SUMMARY OF KEY POINTS

Summarized below are the key points which we have discussed in the previous sections. Reference is made to the place in the text where the original more detailed description can be found:

1. The shearing resistance of saturated uncemented cohesive soils can be interpreted usefully in terms of the Hvorslev hypothesis (Equation 8–1). Introducing the notion of an equivalent consolidation pressure (Section 8.2) leads to the description of strength in terms of the Hvorslev parameters: the effective cohesion κ, and the effective angle of shearing resistance ϕ'_e (Equations 8–4, 8–5).
2. Application of the Hvorslev hypothesis to field conditions generally requires knowledge of the pore water pressure developed in the soil for undrained failure conditions (Section 8.3).
    a. This may be estimated in terms of the changes in the octahedral stresses imposed at each point, and an empirically determined soil parameter α (Equation 8–12). This parameter can be obtained indirectly from the Skempton A factor at failure A_f which is determined from pore water pressure measurements in conventional undrained triaxial compression tests (Equation 8–14b).
    b. The magnitude of A_f in a triaxial compression test on a given clay depends primarily on the stress history experienced by the soil, as expressed by the over-consolidation ratio.
3. The complexity of applying the Hvorslev hypothesis to field conditions leads to a variety of empirical simplified approaches to evaluating shear strength of clays for specific conditions:
    a. The immediate undrained shearing resistance of a saturated clay subjected to stress increments in the field, which are of similar type to those in a compression test, can be applied directly to stability problems which are significant prior to any changes in void ratio by using equivalent strength parameters: c = the undrained shear strength, $\phi = 0$ (Section 8.6).
    b. When drainage is significant, the shearing resistance of normally consolidated clays can be expressed in either of two ways which

---

* A rational approach has been taken to the prediction of the pore water pressure under general undrained conditions (Schofield and Wroth, 1968; Roscoe and Burland, 1968); however, additional developments are necessary to permit application to field conditions.

make use of the fact that changes in void ratio and effective stress are simply related (Section 8.4):

    (1) A measure of the shear strength can be described in terms of effective stresses at failure and an apparent angle of shearing resistance (Equation 8–16).

    (2) The principal stress difference at failure is proportional to the effective consolidation pressure, in which the proportionality constant depends on the stress conditions during consolidation (Equations 8–24, 8–25).

  c. The strength of over-consolidated clays, when drainage is significant, can also be described in two alternate ways (Section 8.5):

    (1) Failure envelopes which confound changes in effective stress and void ratio, and which depend upon the magnitude of the preconsolidation pressure, can be expressed approximately in terms of an apparent cohesion and apparent angle of shearing resistance, both in terms of effective stress (Equation 8–26). Less usefully, total stresses can be used (Equation 8–27).

    (2) The principal stress difference at failure can be related directly to the effective consolidation pressure and the over-consolidation ratio (Equation 8–28).

4. The relationship between stress conditions in the field and those in a laboratory test can be visualized conveniently by considering the stress path imposed in the two cases (Section 8.6). The stress path, consequent pore water pressure developed and shearing resistance at failure depend markedly on the nature of the applied loading. For isotropic soils under idealized conditions, the effect of stress path can be accounted for by evaluating the developed pore water pressure correctly and using the Hvorslev parameters to describe the strength.

PROBLEMS

8.1. Given the following data from drained direct-shear tests (stresses in tsf):

Normally Consolidated Clay	Over-consolidated Clay (max. past pressure = 6.0)
$\sigma'_c = 2.0$ $\tau_f = 1.0$ $e_f = 1.07$	$OCR = 3.0$ $\tau_f = 1.65$ $e_f = 0.923$
$\sigma'_c = 4.0$ $\tau_f = 2.0$ $e_f = 0.935$	$OCR = 6.0$ $\tau_f = 1.15$ $e_f = 0.965$
$\sigma'_c = 6.0$ $\tau_f = 3.0$ $e_f = 0.855$	$OCR = 2.0$ $\tau_f = 2.0$ $e_f = 0.900$

Determine:

    (a) C_c for virgin compression

    (b) ϕ'_e

    (c) c_e for $e = 0.85, 0.90, 0.95$

    (d) κ

8.2. The following data are obtained from two consolidated-undrained triaxial compression tests on identical specimens of a normally consolidated clay:

Condition	Test No. 1	Test No. 2
All-round consolidation pressure	$\sigma_c' = 0.5 \text{ kg/cm}^2$	1.0 kg/cm^2
At failure	$\sigma_1 = 0.75 \text{ kg/cm}^2$	1.5 kg/cm^2
	$u = 0.25 \text{ kg/cm}^2$	0.5 kg/cm^2
	$e = 0.990$	0.880

What will be the maximum stress difference at failure for a third specimen consolidated under an all-round pressure of 2.0 kg/cm²? What will be the pore water pressure at failure? What is C_c?

8.3. Consolidation tests on a soft, normally consolidated clay indicate that $C_c = .50$. A single cylindrical specimen was consolidated to a void ratio of 1.00 under an all-round pressure of 2 tsf, and failed in an undrained triaxial compression test under an *additional* axial stress of 1.0 tsf.

(a) What will be the maximum principal stress difference for this clay in the field, under an effective overburden pressure of 0.6 tsf, if the failure stress is applied rapidly?

(b) What will be the void ratio at failure for part a? Estimate the pore water pressure at failure. State any assumptions used.

8.4. An undisturbed specimen of a natural saturated clay is excavated from a depth of 15 ft below the ground surface. The overlying material provides an overburden stress of 1200 psf, which represents the largest vertical effective stress the clay has ever withstood.

A series of triaxial compression tests are to be performed on specimens of this clay. Test No. 1 is a drained (D) test performed under a lateral pressure of 2 tsf. Failure occurs under a total vertical pressure of 5 tsf. On the basis of this information, can you estimate the *total* and *effective* vertical pressure to be expected in each of the following tests on specimens identical to the above? If so, do so. If not, why not?

(a) A drained (D) test with a lateral pressure of 4.0 tsf.

(b) A consolidated-undrained test (CU) test with a lateral (confining) pressure of 4.0 tsf.

(c) An unconsolidated-undrained test (UU) with an all-round pressure of 4.0 tsf on a specimen which had been fully consolidated under a pressure of 2.0 tsf.

(d) An unconfined compression test on a specimen taken directly from the ground.

Note: Make any reasonable assumptions you wish.

8.5. Stress at a point in the deep deposit of saturated clay is as depicted, using an *xyz*—coordinate system. Initial *in situ* stresses are:

$$\sigma_x = 1.6 \text{ tsf} \qquad \tau_{xy} = \tau_{xz} = \tau_{yz} = 0$$
$$\sigma_y = 1.6 \text{ tsf} \qquad u = 1.0 \text{ tsf}$$
$$\sigma_z = 2.0 \text{ tsf}$$

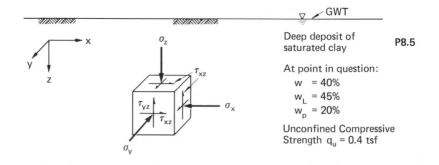

At point in question:

w = 40%

w_L = 45%

w_P = 20%

Unconfined Compressive Strength q_u = 0.4 tsf

P8.5

As a result of nearby construction, the stresses will change as indicated below:

$$\Delta\sigma_x = 0.20 \text{ tsf} \qquad \Delta\tau_{xz} = 0.10 \text{ tsf}$$
$$\Delta\sigma_y = 0.15 \text{ tsf} \qquad \Delta\tau_{xy} = \Delta\tau_{yz} = 0$$
$$\Delta\sigma_z = 0.10 \text{ tsf}$$

(a) What are the total principal stresses *before* and *after* the stress change?

(b) Estimate the change in pore water pressure, due to the change in stresses, before any dissipation of excess pore pressure has occurred. (Make whatever assumptions are necessary, but state them clearly.)

8.6. The effective overburden pressure at a depth of 20 ft below the top of a normally consolidated clay stratum is 1 tsf. The shear strength of the soil at this depth, $1/2(\sigma_1 - \sigma_3)_f$, is determined from a vane shear test to be 0.3 tsf. For the clay, $\gamma_{sat} = 112.4$ pcf. The stratum is 40 ft thick and is to be subjected to an initial increment of vertical stress, $\Delta\sigma_z = 500$ psf uniformly throughout its depth as the result of a fill of wide extent in the area. In order to investigate stage construction of the fill, determine the strength, $1/2(\sigma_1 - \sigma_3)_f$ at depths of 10 and 20 ft below the top of the stratum 1 year after the initial fill is placed if—

(a) the fill is drained at both top and bottom.

(b) a very thin sand layer at a depth of 20 ft below the top of the clay serves as a drain.

For the clay, $c_v = 20$ ft²/year.

8.7. A specimen of clay with the following properties is tested in a CU triaxial compression test:

$$c = 0.2 \text{ tsf} \qquad \phi = 10° \qquad c' = 0.15 \text{ tsf} \qquad \phi' = 20°$$

The major principal stress at failure was:

$$\sigma_{1_f} = 0.8 \text{ tsf}$$

(a) Is this specimen normally consolidated, lightly over-consolidated, or highly over-consolidated?

(b) What are σ'_{1_f} and σ'_{3_f}?

(c) Can you determine the shear stress on the failure surface at failure, τ_{ff}? If so, do so. If not, why not?

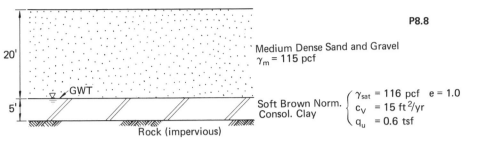

P8.8

Medium Dense Sand and Gravel
γ_m = 115 pcf

Soft Brown Norm.
Consol. Clay

γ_{sat} = 116 pcf   e = 1.0
c_v = 15 ft^2/yr
q_u = 0.6 tsf

20'

5'

GWT

Rock (impervious)

8.8. An oil tank is to be constructed at a site for which the stratigraphic profile is that shown below. The tank will impose a maximum shear stress of 0.341 tsf, approximately uniformly throughout the depth of the normally consolidated clay layer. In order to prevent a failure, the entire area is to be surcharged with a fill exerting a vertical stress of 0.7 tsf throughout the entire depth of the clay. *Ignoring any initial shear stresses which may exist in the clay,* how long must we wait before constructing the oil tank? Neglect any factor of safety, and assume that the initial strength is approximately uniform throughout the depth of the clay.

9

Structure of Cohesive Soils

9.1 INTRODUCTION

The article "Electro-Osmosis Stabilizes Earth Dam's Tricky Foundation Clay" (pages 10–16 of Chapter 1), describes a case in which continued application of an electrical potential to the cohesive foundation material for the West Branch Dam created a marked increase in strength of the clay. This process permitted completion of the dam on schedule in spite of the unanticipated failure of the subsurface material. Based upon our knowledge of the behavior of coarse-grained materials, we would not expect to see a similar stabilizing effect on the strength of such soils due to a simple electrical potential.

Many other examples of the difference between cohesionless and cohesive materials can be seen. For instance, when a cohesive soil is *remolded* at constant water content, it will, in general, lose strength. This property, called *sensitivity*, is especially pronounced in some materials. The sensitivity of Leda clay from southeastern Canada is illustrated in Figure 9.1. The cylindrical specimen in the left hand side of the picture is supporting 11 kilograms. The liquid being poured from the pitcher in the right hand side of the picture is some of the identical material which has been remolded. Disastrous earthslides are common in such highly sensitive clays which occur along coastal regions in Norway, the southeastern portion of Canada, and Japan. These slides are frequently earth "flows." That is, the remolding caused by the initiation of sliding produces liquefaction of the sliding mass and it flows as a fluid. One such slide is illustrated in Figures 9.2 and 9.3. The figures show a photograph and topographic map of the earth slide at Ullensaker, Norway, in December 1953. This area is in the southeast corner of Norway, where spontaneous earth slides are common. The report of the slide prepared by the Norwegian Geotechnical Institute

Fig. 9.1—Demonstration of undisturbed and remoulded strength of Leda clay. (From Crawford, 1963.)

(Bjerrum, 1955) stated:

> The most characteristic feature of this slide is the change of the clay in the process of the sliding. As the clay became involved in the slide, and therefore remoulded, it changed to a viscous liquid with a consistency like heavy oil: Through the opening formed by the initial slide the clay slurry moved from the cavity and descended with considerable velocity along the brook. (Figure 9.2). Flakes of the stiff upper drying crust were taken by the slurry and carried downwards. In all, about 200,000 cu. m of clay flowed away from the cavity of the slide. At a distance of 1.5 km from the slide the brook runs under a road embankment. At this place the clay slurry was dammed up, resulting in a big lake. Next morning the farmer found the demolished remains of his houses, a cow and a horse were extricated while still alive.

Although this process bears some superficial similarity to liquefaction of granular materials, it is evident that there are important differences in the observed behavior, and therefore probably in the cause.

In Chapter 8 we learned that the strength of cohesive soils was, in part, independent of the effective normal stress on the failure plane at failure. Again, we see an apparently fundamental distinction between cohesive and cohesionless materials. We know that for sand the forces between individual grains govern the mechanical behavior. These forces arise

Fig. 9.2—Flow slide at Ullensaker, December 1953; air photo. (From Bjerrum, 1955.)

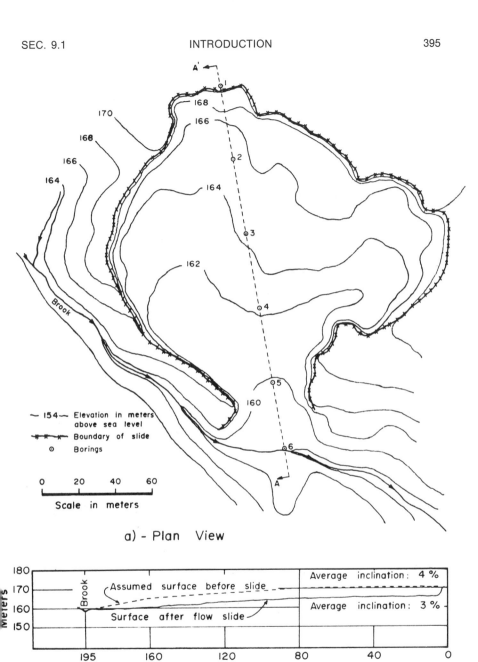

a) - Plan   View

b) - Section   A - A'

Fig. 9.3—Flow slide at Ullensaker, December 1953; plan showing boundary of slide, position of boring, cross-section through flow slide. (From Bjerrum, 1955.)

primarily from externally applied loads, gravity, and occasionally other body forces such as seepage. Interparticle forces are also fundamentally important in the behavior of clay soils. But in contrast to sands, these forces involve electrochemical as well as mechanical effects. In order to understand this key point better, we shall examine clays from a microscopic and submicroscopic point of view.

9.2 CLAY PARTICLES—PRELIMINARY CONSIDERATIONS

Clay Mineral Concept

Cohesive soils are composed of particles of varying grain size. However, the distinctive behavior of these soils is determined by the presence of certain very small crystalline particles. These particles, called *clay minerals*, are hydrous aluminum silicates containing sometimes iron, magnesium, and other metals. It is the electrochemical properties of these minerals which produce many of the properties we associate with cohesive soils. Consequently let us consider the nature of the electrochemical forces which can act between clay particles.

Electrochemical Forces Between "Small" Particles

Primary valence bonds are extremely strong forces which act over very small distances of the order of 1 to 2Å, and bind atoms together within molecules. These bonds are thought to be too strong to be broken in any of the soil engineering applications with which we will be concerned, and will not be discussed further.*

Secondary valence forces, also called *van der Waals forces*, act between particles of molecular size and larger. They are approximately a hundredth as strong as primary valence bonds. They are due to electrical moments which result from the interaction between nonsymmetrical (*dipole*) molecules and other dipoles or electrical fields. A dipole is a charged particle in which the centers of action of the negative and positive charges are not coincident. This is illustrated in Figure 9.4a. An example of a dipolar molecule is the water molecule, shown schematically in Figure 9.4b.

Secondary valence forces arise from three effects:

1. Forces between *permanent dipoles* (whether attractive or repulsive) depend on the relative orientation of the dipoles. This part of the attraction is called the *orientation effect*.
2. Temporary polarity, and attendant dipole moments, may be induced in a nonpolar molecule by an electrical field, either external or

---

* It has been postulated (Mitchell et al., 1969) that primary valence bonds exist between all soil particles in contact. Although some experimental results are consistent with this view, the supporting evidence is still inconclusive.

Symmetrical (Non - Polar)   Non - Symmetrical   (Dipole)

a) - Molecular   Symmetry

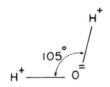

Fig. 9.4—Polar and non-polar molecules.

b) - Polar Water   Molecule

from a nearby permanent dipole. This is termed the *induction effect*.

3. In all molecules the nuclei and electrons are vibrating. Therefore, at any given instant of time, even symmetrical molecules appear to have "statistical" dipoles. These temporary dipoles induce in nearby molecules dipoles in phase with them. Therefore, even though the average dipole moment is zero, there is a net attraction. The thermal oscillations producing this effect are also called the *dispersion effect*.

Van der Waals forces act over distances of 5Å and more. They are inversely proportional to r^3 up to r^7, depending on the types of molecules involved.

The *hydrogen bond* occurs when a proton (H^+) is equally attracted to two other ions and forms a bridge between them. An example of this is water, illustrated in Figure 9.5. If it were not for the hydrogen bond, water would be a gas at normal temperatures, like H_2S. The hydrogen bond acts over distances of 2–3Å, and is approximately 10 times as strong as van der Waals forces.

A *cation* bond occurs when a cation is equally attracted to two negatively charged molecules. This is similar to the hydrogen bond except that it is

Fig. 9.5—Hydrogen bond in water.

very much weaker because all other ions are so much larger than a proton.

Coulombic (electrostatic) attraction and repulsion exist between all charged particles. The force between two particles is inversely proportional to r^2.

The way in which these forces act between clay particles is discussed in the following sections.

9.3 CLAY PARTICLES AS COLLOIDS

Specific Surface

The electrochemical forces described above are important only when they become large compared to the mass forces. Because these forces act over such small distances, this will be true only when a large proportion of the mass is concentrated near the surface of the particle. The degree to which this occurs is measured by a quantity called *specific surface*, which is defined as the ratio of the surface area to the volume of the particle. Thus, a large specific surface indicates that most molecules are at or near the surface, and that if electrochemical forces are present, they will be important. The specific surface is a function of particle size and shape. For a sphere:

$$\text{Specific surface} = \frac{\pi d^2}{\pi d^{3/6}} = \frac{6}{d} \qquad (9-1)$$

Assuming spherical particles, the specific surface for different soils is shown in Table 9.1.

TABLE 9.1
Specific Surface for Spherical Particles

Material	d (cm)	Specific Surface (cm²/cm³)
Gravel	1	6
Medium sand	0.1	60
Silt	0.001	6000
Clay	0.00001 ($= 0.1\ \mu$)	6×10^5 or approximately 20 m²/g

Clay particles are not spherical, but rather distinctly plate or lath shaped, as shown in Figures 9.6–9.9. These are electron microphotographs of four common clay mineral types. The particles have been dispersed to permit clearer viewing. Because plates have a larger specific surface than spheres of the same average diameter, the effect of small-size clay particles is enhanced. Thus, approximately 70 grams of clay has a surface larger than a football field!

However, size is not sufficient to make electrochemical forces of significance. For instance, finely ground quartz will not exhibit clay-like properties. The additional attribute which clays exhibit is that the clay particles have a

Fig. 9.6—An electron micrograph showing the particle shape of the 1–2 micron fraction of a well-crystallized, low-viscosity Georgia kaolin. (From Woodward, 1955.)

Fig. 9.7—Electron micrograph of a dispersed sample of Wendover, Utah, halloysite. (From Bates and Comer, 1955.)

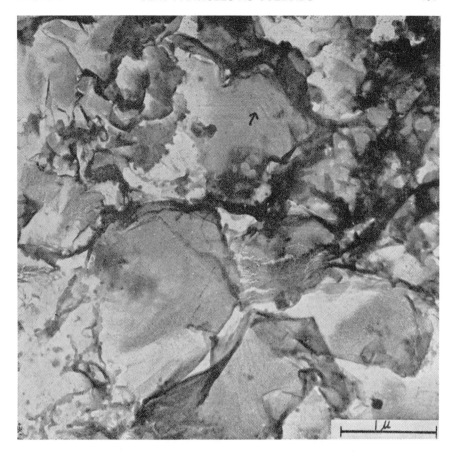

Fig. 9.8—Electron micrograph of montmorillonite from Montmorillon, France. (From Jeffries and Bates, 1960.)

net negative charge. This derives from a variety of factors, which will not be discussed in detail herein.* Nonetheless, it is well recognized that the result of this fact is that forces of the type discussed above are important to the behavior of clays.

Interaction Between Single Particle and Aqueous Ionic Suspensions

Because clays are negatively charged and of the appropriate size, the particles act as colloids when placed in aqueous suspensions. When a clay particle is in water, two primary effects occur:

1. Exchangeable cations are attracted to the vicinity of the particle surface (some anions may be attracted to the positively charged

---

* For more detailed consideration of the makeup of clay particles, see Grim (1968).

Fig. 9.9—Electron micrograph of illite from Oswego formation, State College, Pennsylvania. (From Jeffries and Bates, 1960.)

edges of the particle as well). Our current method of analyzing this effect is based upon work done independently by Gouy (1910) and Chapman (1913) in the early part of this century, and called the Gouy-Chapman double-layer theory for dilute colloidal suspensions. Assuming that the clay particle is a flat plate of unlimited extent with a uniform distribution of charges on its surface, and an ionic solution which is sufficiently dilute so that there is no interference between the ions, the theoretical ion concentration as a function of distance from the particle surface is shown in Figure 9.10 for two equilibrium ion concentrations. Note that for the assumption of no interference by other particles, the effect of the charged particle surface is observed at considerable distance from the surface. Note also that the concentration has an important effect on this distance. This point will be considered in more detail below.

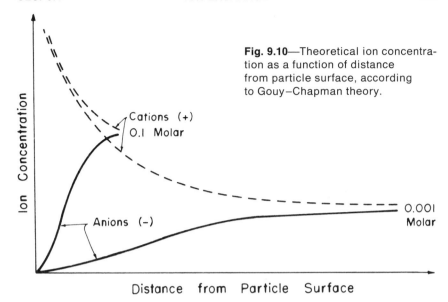

Fig. 9.10—Theoretical ion concentration as a function of distance from particle surface, according to Gouy–Chapman theory.

Distance from Particle Surface

2. Within the zone near the particle surface that is strongly influenced by particle charges, the water molecules are also affected. Because of the dipolar nature of water (Figure 9.4b), the water molecules tend to be oriented in the electrical field created by the particle. This effect is illustrated schematically in Figure 9.11. Because the attractive force on the water is so strongly a function of distance, the water molecules in the near vicinity of the particle are very tightly held to the particle. The water in this zone, which is approximately 10 to 20Å thick, is called *adsorbed* or *bound water*. It is very difficult to remove from the particle, except at temperatures in excess of 100°C. Outside of this adsorbed water layer, the water is less strongly attracted to the particle out to a point at which the orientation of the water is essentially random, as in the bulk solution. These combined layers are called the *double layer*. Because the layer boundaries are not really fixed, but rather diffuse, this zone around the particle is often called the *diffuse double layer*.

The adsorbed and double layer water is different from ordinary water, because of its relatively more ordered structure. The details of this structure are not complete, although the evidence for some ordering is strong. The double layer water has a higher boiling temperature, a lower freezing point, and probably a higher viscosity. Thickness of the double layer depends upon mineral type, ion type, and concentration, and other factors which will be discussed below.

9.4　ION EXCHANGE

The ability of a particle to attract, that is, adsorb, cations to its surface is determined by the net negative charge on the particle. This is measured

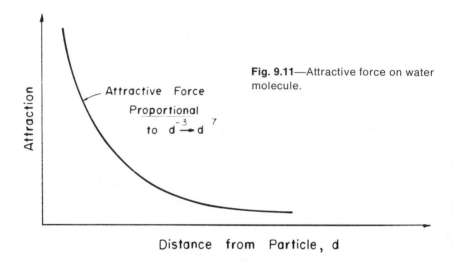

Fig. 9.11—Attractive force on water molecule.

by the number of milliequivalents of the combining compound which produces the cation, per 100 grams of clay. One equivalent is the gram-molecular weight of the compound producing one positive charge. For example, one equivalent of sodium expressed as Na_2O would be a combining weight of 31 grams, and 1 meq./100 g would be equal to 0.031% Na_2O. Typical ranges for cation exchange capacity of the clay minerals shown in Figures 9.6–9.9 are given in Table 9.2.

The cations from the pore water which are attracted to the surface of clay particles are often termed *exchangeable* ions because other ions can be exchanged for them. The relative ease with which one ion replaces another of the same concentration depends primarily upon the valence and size of the hydrated ion as shown below:

$$Li^+ < Na^+ < H^+ < K^+ < NH_4^+ < Mg^{++} < Ca^{++} < Al^{+++}$$

That is, the replacing power of a given ion is less than that of the higher valence or smaller hydrated ion to the right of it. The exception to this is potassium which, once it has been adsorbed on to the clay, is exceedingly

TABLE 9.2
Cation Exchange of
Clay Minerals in
Milliequivalents/100 g

Kaolinite	3–15
Halloysite	5–40
Illite	10–40
Montmorillonite	80–150

Modified from Grim (1962).

difficult to replace because of the way it fits cavities in the surface of the particle where it is very tightly held.

The exchanging power is also a function of ion concentration. If a calcium clay is leached with a sodium solution of high concentration, the sodium will replace some or all of the calcium. However, it requires a significantly higher concentration for a monovalent ion to replace a divalent one.

9.5 STRUCTURAL ARRANGEMENT OF CLAY PARTICLES

Interaction of Factors

The engineering behavior of clay soils is influenced more by the structural arrangement of the particles than other single factors. This arrangement is determined by interaction of the forces discussed above, acting between clay particles. The relative magnitude of these forces is determined by the type of clay mineral, and the size, type, and concentration of exchangeable cations, as well as other characteristics of the pore fluid. When two particles approach each other, they are attracted to each other by van der Waals forces which become very strong as the particles approach closely. They are further attracted to each other by cation bonds between the particles, and by edge-to-face electrostatic (Coulombic) attraction. However, when the particles approach sufficiently closely that the double layers overlap, there is electrostatic repulsion. The structural arrangement of the particles is the result of the interplay between these attractive and repulsive forces, which depends to a major extent upon the size of the double layer. When the double layer is small in size so that particles can approach sufficiently closely for van der Waals forces to act before the double layers overlap, the particles tend to stick together, or *flocculate*. This type of flocculation occurs most readily under conditions of high salt concentration, such as during deposition in sea water. Under such circumstances, the particles tend to stick together in whatever way they touch first. This is illustrated schematically in Figure 9.12a and is called *salt flocculation*. This type of flocculation can be promoted by increasing the ion concentration, increasing the valence of the adsorbed ion, or decreasing the size of the hydrated exchangeable ion.

Even without high-salt conditions, if the fluid is acidic (low pH), there will be a tendency for H^+ ions to attach themselves to the edges of the particle, thereby increasing the tendency for edge-to-face electrostatic attraction. Thus, particles will tend to stick if they come into edge-to-face contact leading to a *non-salt* flocculation.

Those factors which tend to increase the double layer will tend to promote repulsion between the particles, and a *disperse* structure, as illustrated in Figure 9.12c. The difference between these is dramatically illustrated at the outlet of a large muddy river such as the Mississippi, where the fresh

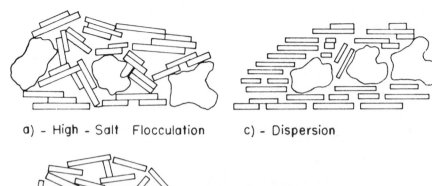

a) - High - Salt   Flocculation   c) - Dispersion

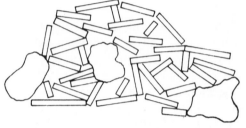

Fig. 9.12—Schematic representation of clay structure.
(Modified from Lambe, 1958.)

b) - Low - Salt   (acid)   Flocculation

water carrying sediments runs into a strong salt solution. A major portion of the solids carried in suspension settle almost immediately. This is in part due to the reduction in flow velocity, but even more importantly, is due to the strong tendency for flocculation created by the salt concentration.

The theoretical effect of these system characteristics on soil structure is illustrated in Figure 9.13, which represents, schematically, a series of small-scale experiments in which a fixed quantity of powdered clay is placed in a test tube full of different solutions, shaken vigorously and permitted to settle. The sediment volume is a measure of the tendency toward flocculation; the larger the volume, the stronger the tendency toward flocculation and therefore the larger volume occupied by the soil mass. The effects of cation concentration, temperature, ion valence, size of hydrated ion, dielectric constant, and pH are illustrated in the figure.

Fabric of Natural Pure Clays

Determination of the structure of natural pure clays is difficult. However, by making a "replica" of a freshly broken clay surface, it is possible to obtain electron micrographs which indicate this natural arrangement. Figure 9.14 is a replica micrograph of a kaolinite which has been broken along a natural cleavage surface, more or less parallel to the majority of the particles. This picture may be somewhat disconcerting, because it suggests that the concept of kaolinite particles as distinct entities may be oversimplified.

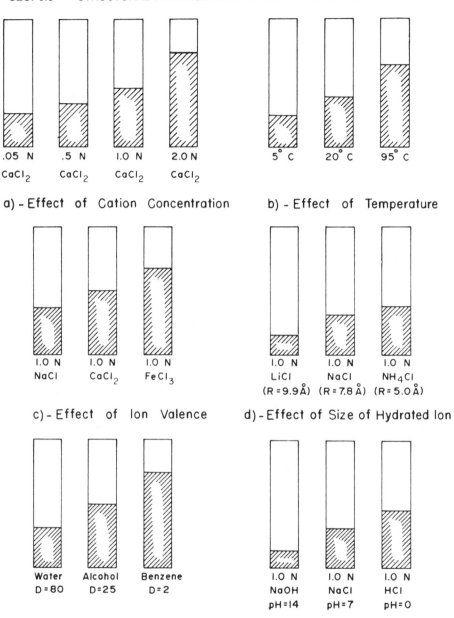

Fig. 9.13—Effect of system characteristics on soil structures. (From Lambe, 1958.)

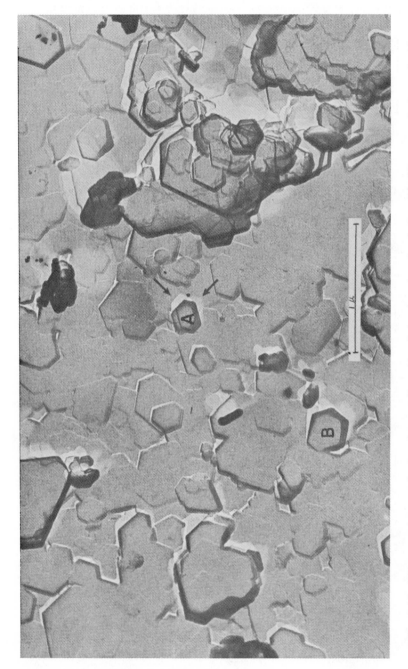

Fig. 9.14—Electron micrograph of surface replica of kaolinite, Langley, South Carolina; A, arrows indicate double shadow from particle; B, cleavage fragment partly broken away from underlying crystal. (From Bates and Comer, 1955.)

Notice that many of the particles merge into other particles: some of them consist of layers of different-size fragments. In fact, this picture and many other results, indicate that the concept of a unique particle size distribution may be fallacious, and the results obtained depend upon the energy applied in dispersing the clay. Figure 9.15 is an electron micrograph of a replica

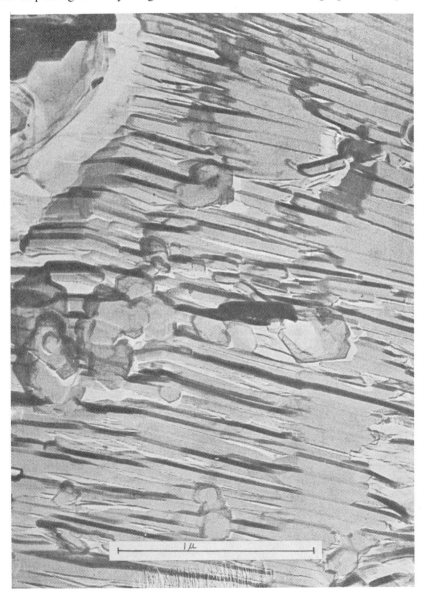

Fig. 9.15—Electron micrograph of surface replica of kaolinite, Langley, South Carolina, edge view of kaolinite book. From Bates and Comer, 1955.)

of a broken section normal to the plane shown in Figure 9.14. This figure shows a mass of kaolinite in which the particles form leaves of a "book" which are not entirely distinct. It is apparent that sufficient energy would produce cleavage along surfaces. However, these surfaces are not, in this photograph, planes of separation.

An electron micrograph of a surface replica of halloysite is shown in Figure 9.16. Here the individual particles are more distinct, but still appear to be combined into a large "mass."

Fig. 9.16—Electron micrograph of surface replica of halloysite from Wendover, Utah, showing tubular morphology of crystals.

Thus, it appears that the idealized concepts of structure which we have developed, based primarily on consideration of double layer theory are oversimplified. Nonetheless they provide considerable intuition for us when we attempt to understand certain phenomena that are observed in clay soils. These are discussed in the following sections.

9.6 OBSERVED FEATURES OF CLAY BEHAVIOR

Sensitivity

The loss in strength of a cohesive soil when its structure is disturbed at constant water content is called *sensitivity*. This property is described quantitatively as the ratio of undisturbed shear strength to the remolded shear strength at the same water content. Sensitivity of clays ranges from approximately 1.5 to almost infinity. The sensitivity of ordinary clays is less than 4; for sensitive clays it ranges from 4 to 8. When the sensitivity is more than 8, the clay is considered to be extra-sensitive or *quick*.

The sudden landslides associated with quick clays are illustrated in Section 9.1. On the basis of the previous discussion, we now can explain the reason for this behavior of the Norwegian quick clays. A combination of two factors is involved:

1. The quick clays were deposited as marine clays with a salt concentration in the pore water of approximately 36 grams per liter. As a result of centuries of uplifting of the land and leaching of the clay by rain water, the salt concentration has reduced to less than 5 grams per liter in these Norwegian clays. This has resulted in a significant increase in the size of the double layer and an increased tendency for repulsion between the particles. As a consequence, the strength of these clays has decreased with time, to the point that earth slides start spontaneously.

2. Remolding tends to cause particle reorientation. If an initially flocculent structure, idealized in Figure 9.17a, is sheared until the structure becomes more dispersed, even with the same average particle spacing, as in Figure 9.17b, the net attraction between the particles will be reduced. This is so because the van der Waals

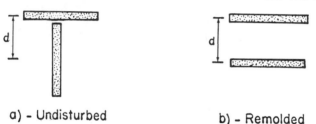

a) - Undisturbed   b) - Remolded

Fig. 9.17—Particle reorientation due to remolding.

forces are inversely related to a high power of the spacing. Thus, although the average spacing may be the same, the points of closest approach are much further apart. This leads to a reduced attraction, or perhaps even a net repulsion. When remolding due to initial sliding causes the particles to reorient and reduces the van der Waals attractions between them, the combination of reduced attraction and increased repulsion causes a drastic reduction in the strength.*

Based on this reasoning, one of the most common ways to treat highly unstable areas of limited extent in Norway is simply to spread salt on the ground. The salt is carried into the soil by rainfall and increases the ion concentration of the pore water, with a concomitant reduction in the double layer thickness and interparticle repulsion.

Thixotropy

The property of *thixotropy* was first observed in dilute colloidal suspensions and was defined as a reversible sol-gel transformation without a change in water content. For natural soils at water contents much below that of a dilute suspension, thixotropy is considered to be recoverable sensitivity at *constant* water content. Imagine a number of identical specimens of an undisturbed clay. One of them is sheared to failure and the undisturbed shear strength is noted. The remainder are thoroughly remolded, and the strength of a second specimen is immediately tested. This is called the *remolded strength*. A third specimen is tested one day later, a fourth one week later, and so on. If the measured strength of a thixotropic clay is plotted as a function of time after remolding, a curve similar to that shown in Figure 9.18 results. The figure illustrates the thixotropic increase in shear strength, measured by the unconfined compression test, for a highly sensitive Laurentian clay from Canada. As indicated in Figure 9.18 for the nearly two year period considered, the full undisturbed strength generally will not be recovered because of cementation and other long-term effects. However, a major portion of the initial strength may be regained.

It has been suggested that thixotropic strength regain results from a tendency for the particles to seek an orientation of minimum energy (Mitchell, 1960). In many clays this is a flocculent arrangement. Remolding the clay changes the flocculent structure toward a dispersed fabric. The net interparticle attractions are reduced producing a reduced remolded strength. With time, the particles gradually reorient toward the more stable flocculent structure, leading to a stronger material. Thixotropic effects are most pronounced in montmorillonites and least noticeable in kaolinites.

* The explanation of these formerly enigmatic failures was given by Rosenqvist (1946).

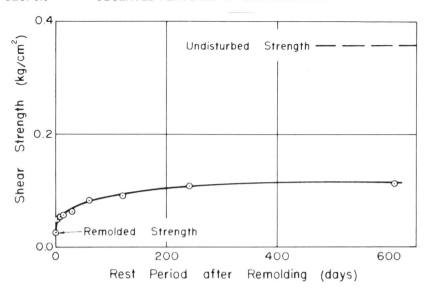

Fig. 9.18—Thixotropic strength regain for laurentian clay from beauharnois, Canada. (Data from Moretto, 1948.)

Electroosmosis

In the early nineteenth century Reuss (1809) found that water would flow through a porous material in response to an electrical potential. Helmholtz (1879) developed a mathematical description of this process based upon a highly idealized picture of the pore space within a soil. He visualized the pore to be analogous to a capillary tube in which the tube walls, corresponding to the soil particles, were negatively charged. Helmholtz (1879) considered a double layer within the tube, as shown in Figure 9.19, consisting of a tightly held adsorbed layer, and a less strongly held outer portion. He envisaged that the application of a potential difference across the length of the tube would cause the cations in the outer part of the double layer to flow toward the cathode. The free water trapped between the moving double layers would be carried along with it. It was assumed that the most strongly held adsorbed layer would remain fixed to the mineral surface and offer resistance to flow. The velocity distribution resulting from such a process is shown in Figure 9.19a.

By contrast, laminar flow due to an hydraulic gradient would occur only in the free water, with a velocity distribution across the tube of the characteristic parabolic shape, as in Figure 9.19b (see Chapter 6).

The result of this investigation, slightly modified (Casagrande, 1952), was an expression describing the quantity of electrolyte flowing per unit

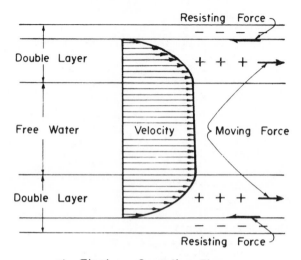

a) - Electro - Osmotic   Flow

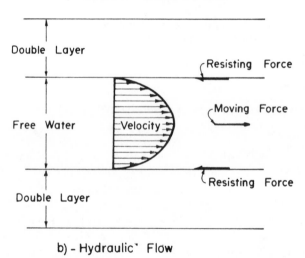

b) - Hydraulic· Flow

Fig. 9.19—Comparison of electro osmotic flow with hydraulic flow in a single capillary. (From Casagrande, 1952.)

of time through a single capillary due to electroosmosis:

$$q_e = \frac{EDr^2\zeta}{4\eta L} \qquad (9\text{--}2)$$

in which E is the electrical potential, D is the dielectric constant of the liquid, r is the radius of the capillary, ζ is the potential existing between the rigid and movable parts of the double layer, η is the viscosity of the liquid, and L is the length of the capillary between the electrodes.

In order to compare this result with hydraulic flow, we set $E/L = i_e$, the electrical potential gradient, $D\zeta/4\pi\eta = c_1$ (constant) and $\pi r^2 = a$, the cross-sectional area of the capillary. Thus for electroosmotic flow, we obtain

$$q_e = c_1 i_e a \qquad (9\text{-}3)$$

This result can be compared to Poiseuille's law for laminar flow in a tube (see Chapter 6):

$$q_h = c_2 i_h a^2 \qquad (9\text{-}4)$$

where q_h is the quantity per unit time of flow due to an hydraulic gradient, c_2 is a constant, i_h is the hydraulic gradient, and a is the area of the capillary tube.

If the soil mass is visualized as a bundle of such capillary tubes with a total cross-sectional area A, and with a void ratio e, the total quantity of *hydraulic* flow will be

$$q_h = \left(a\,\frac{e}{1+e}\,c_2\right) i_h A = k_h i_h A \qquad (9\text{-}5)$$

Thus we see that for this simplified model the hydraulic coefficient of permeability will depend upon the pore size. The rate of *electroosmotic* flow in the same bundle of capillaries is

$$q_e = \left(\frac{e}{1+e}\,c_1\right) i_e A = k_e i_e A \qquad (9\text{-}6)$$

which suggests that the electroosmotic coefficient of permeability is *independent* of the size of the capillary.

Many tests conducted by Casagrande (1952) demonstrated that in fact the electroosmotic permeability was approximately constant for a wide range of soils from sands to clays, with a value of 0.5×10^{-4} cm/sec for a gradient of 1 V/cm. More recent work (Gray and Mitchell, 1967) indicates that k_e does vary for a given material, and is a function of water content among other factors. Nonetheless, the electroosmotic coefficient of permeability is of the order of 10^{-4} cm/sec with a gradient of 1 V/cm. Consequently if the hydraulic permeability of a soil is sufficiently less than this figure, electroosmosis may be a valuable aid in promoting flow of water out of the soil with concomitant reduction in pore water pressure and increase in strength due to consolidation. Many examples of the practical application of this method in stabilizing excavations and increasing the strength of cohesive materials have been described (Casagrande, 1952).

Application to West Branch Dam

In the case of the West Branch Dam, the hydraulic permeability of the clay foundation material in which high pore water pressure was developed ranged from 1.2 to 6.5×10^{-8} cm/sec. By direct experimentation, the electroosmotic permeability was found to range from 0.30 to 0.62 $\times$ 10^{-4} cm/sec for a gradient of 1 V/cm. These figures suggest that electroosmosis would be beneficial in the case of the West Branch Dam, as in fact it was.

9.7 EFFECT OF WATER ON CLAYS

Consistency and Atterberg Limits

Water has a tremendous effect on clay soils. The water content influences the size of the double layer, and thereby the physicochemical forces. Thus, a clay soil with a high water content is likely to be much weaker and more easily deformed than the same soil at a lower water content.

The relative ease with which a soil can be deformed is called the *consistency*. In 1911, A. Atterberg, a Swedish soil scientist, classified the ranges of consistency of clay soils as a function of changes in water content. His classification, which is very useful for engineering purposes, is shown in Figure 9.20.

Because the "plastic" range of consistency is the state in which cohesive soils below the groundwater table are most commonly found, it is especially useful to know, for a particular soil, the water contents which define the upper and lower limits for this range. The water content which is the boundary between the "plastic" and "semiliquid" states is called the *liquid limit*. It is determined by an arbitrary test developed by A. Casagrande (1932). The test is performed on the fraction of the soil which passes a No. 40 (0.42 mm) sieve. This fine fraction of soil is mixed with sufficient water to produce a consistency similar to soft gelatin. It is then spread into the bottom of a shallow metal cup. A groove of standard dimensions is cut into the soil paste as shown in Figure 9.21. The cup is then dropped a standard distance (approximately 1 cm), striking a base of standardized material. When this is repeated a sufficient number of times, the groove in the clay will close in response to the impact. The number of blows required to cause the groove to close over a distance of $\frac{1}{2}$ in. is determined for successive water contents. If the number of blows required to close the groove is plotted to a logarithmic scale as a function of the water content, a straight line results. One such typical line, called a *flow curve*, is shown in Figure 9.22. The water content at which 25 blows will cause the groove to close over a distance of $\frac{1}{2}$ in. is defined as the *liquid limit*, w_L. It has been found that the number of blows required to close the groove is a measure of the shear strength; and at the liquid limit, the shear strength is approximately 20 to 25 gm/cm^2 (Seed et al., 1964). Thus, all soils have the *same strength* (but *not* the same water content) at the liquid limit. Typical values of liquid limit for some clay minerals are listed in Table 9.3.

It has been found empirically that the compression index, C_c, of a normally consolidated clay can be related approximately to the liquid limit by the

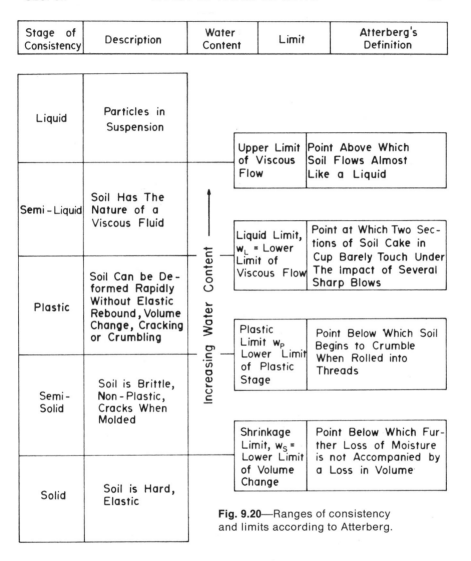

Stage of Consistency	Description	Water Content	Limit	Atterberg's Definition
Liquid	Particles in Suspension			
			Upper Limit of Viscous Flow	Point Above Which Soil Flows Almost Like a Liquid
Semi-Liquid	Soil Has The Nature of a Viscous Fluid			
			Liquid Limit, w_L = Lower Limit of Viscous Flow	Point at Which Two Sections of Soil Cake in Cup Barely Touch Under The Impact of Several Sharp Blows
Plastic	Soil Can be Deformed Rapidly Without Elastic Rebound, Volume Change, Cracking or Crumbling			
			Plastic Limit w_P Lower Limit of Plastic Stage	Point Below Which Soil Begins to Crumble When Rolled into Threads
Semi-Solid	Soil is Brittle, Non-Plastic, Cracks When Molded			
			Shrinkage Limit, w_S = Lower Limit of Volume Change	Point Below Which Further Loss of Moisture is not Accompanied by a Loss in Volume
Solid	Soil is Hard, Elastic			

(Center vertical label: Increasing Water Content)

Fig. 9.20—Ranges of consistency and limits according to Atterberg.

expression (Terzaghi and Peck, 1967)

$$C_c = 0.009(w_L - 10\%) \tag{9-7}$$

This relation permits a rough estimate of settlements of normally consolidated clays on the basis of a simple classification test.

The *plastic limit* defines the water content between the "plastic" and "semisolid" ranges. The plastic limit is determined by remolding a small amount of soil and allowing it to air-dry. At periodic intervals during the drying process, the soil is rolled into a long thread. The water content at

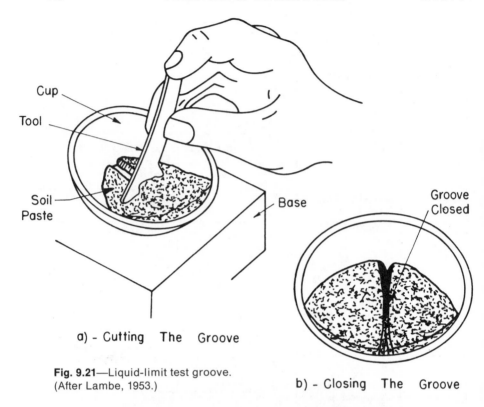

Cup

Tool

Soil
Paste

Base

Groove
Closed

a) - Cutting   The   Groove

Fig. 9.21—Liquid-limit test groove.
(After Lambe, 1953.)

b) - Closing   The   Groove

which this thread will no longer deform plastically, but cracks at a thickness of $\frac{1}{8}$ in. is defined as the plastic limit. The apparently artificial test used to determine this water content leads to surprisingly consistent results, even from inexperienced technicians. Values for several clay minerals are given in Table 9.3.

When the water content is within the plastic range, the soil exhibits *plasticity*. This is defined as that property of a soil which enables it to withstand rapid deformation without elastic rebound, volume change, cracking or crumbling.

The range in water content over which the soil is in this plastic state is called the *plasticity index*, $I_p = w_L - w_P$.

The liquid and plastic limit and the plasticity index, along with others that will be described below, are usually called the *Atterberg limits* and *indices*. They are very helpful in classifying clay soils in terms of engineering behavior. The plasticity index for clay soils varies over a wide range depending upon the type of clay and the adsorbed cations. For example, in Table 9.3 the plasticity index for montmorillonite varies from approximately 40 to approximately 600, whereas that for kaolinite ranges from 8 to approximately 40. The reason for the high water-holding capacity

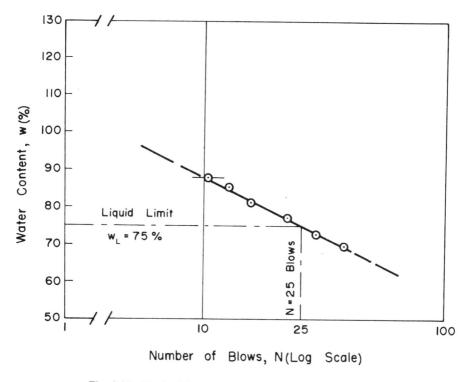

Fig. 9.22—Typical flow curve for liquid-limit test.

of thoroughly dispersed montmorillonite is primarily the large specific
surface. Note that these measurements are made on the soil in a *remolded*
condition.

Various artifacts influence the measured values of liquid limit, and
thereby the plasticity index. Among these is the effect of drying of the clay
before determination of the liquid limit. Oven drying of clay soils appears
to lower the measured liquid limit, by removing, irreversibly, tightly bound
and interlayer water. Partial air drying may or may not reduce the liquid
limit depending upon whether the removal of irreplaceable water is more
important than the possibility of particle breakdown which occurs during
the drying and rewetting process. The intensity of mixing also affects
results. If a soil is partially dried before preparation, an increase in the
intensity of mixing will tend to produce an increase in the liquid limit.
If the soil is at its natural water content, the intensity of mixing has little
apparent effect.

It has been found that a plot of the liquid limit and plasticity index (termed
a *plasticity chart*) can be used in the process of classification of clay soils.
The results of many Atterberg limits determinations have been plotted

TABLE 9.3
Atterberg Limit Values for Various Clay Minerals

Clay	Ca^{++}			Mg^{++}			K^+			NH_4^+			Na^+			Li^+		
	w_P	w_L	A_c	w_P	w_L	A_c	w_P	w_L	A_c	w_P	w_L	A_c	w_P	w_L	A_c	w_P	w_L	A_c
Montmorillonite (1)	65	166	1.26	59	158	1.24	57	161	1.30	75	214	1.74	93	344	3.14	80	638	6.98
(2)	65	155	1.20	51	199	1.97	57	125	0.91	75	114	0.52	89	443	4.72	59	565	6.75
(3)	63	177	1.34	53	162	1.24	60	297	2.79	60	323	3.09	97	700	7.09	60	600	6.35
(4)	79	123	0.44	73	138	0.65	76	108	0.32	74	140	0.66	86	280	1.12	82	292	2.10
Attapulgite	124	232	1.08	109	179	0.70	104	161	0.57	97	158	0.61	100	212	1.12	103	226	1.23
Illite (1)	40	90	0.50	39	83	0.44	43	81	0.38	42	82	0.40	34	61	0.27	41	68	0.27
(2)	36	69	0.33	35	71	0.36	40	72	0.32	37	60	0.23	34	59	0.25	38	63	0.25
(3)	42	100	0.58	43	98	0.55	41	72	0.31	39	76	0.37	41	75	0.34	40	89	0.49
Kaolinite (1)	36	73	0.37	30	60	0.30	38	69	0.31	34	75	0.41	26	52	0.26	33	67	0.34
(2)	26	34	0.08	28	39	0.11	28	35	0.07	28	35	0.07	28	29	0.01	28	37	0.09
Halloysite (2H₂O)	38	54	0.16	47	54	0.07	35	39	0.04	32	43	0.11	29	36	0.07	37	49	0.12
(4H₂O)	58	65	0.07	60	65	0.05	55	57	0.02	56	61	0.05	54	56	0.02	47	49	0.02

KEY: Montmorillonite: (1) Pontotoc, Miss.; (2) Cheto, Ariz.; (3) Belle Fourche, S. Dak.; (4) Olmsted, Ill. (contains about 25 per cent illite in mixed layers).
Attapulgite: Quincy, Fla.
Illite: (1) Fithian, Ill.; (2) Jackson County, Ohio; (3) Grundy County, Ill. (contains about 5% montmorillonite in mixed layers).
Kaolinite: (1) Anna, Ill.; (2) Dry Branch, Ga.
Halloysite: (2H₂O) Eureka, Utah; (4H₂O) Bedford, Ind.
Activity (A_c) values of montmorillonites (1), (2), and (3) are greater than $I_P/100$ because they contain nonclay minerals.
SOURCE: After Grim (1962).

this way, and have led to some general conclusions about the character of the materials in relation to their position on the plasticity chart. A simplified plasticity chart is shown in Figure 9.23. It has been found that the large portion of inorganic clays plot within the shaded range. The lower limit of this range, the "A line," has the formula shown on the chart. Clays with liquid limits less than 30 are considered to be of "low" plasticity. Those with liquid limits between 30 and 50 exhibit "medium" plasticity, and those with liquid limits above 50 exhibit "high" plasticity. The regions in which certain other materials are found to lie are shown on the chart.

The plasticity index is also useful in estimating the strength of normally consolidated clays. It has been found (Skempton, 1957) that the ratio of undrained strength to consolidation pressure for normally consolidated clays $(\sigma_1 - \sigma_3)_f/2\sigma_c'$, can be described as a function of I_p. Figure 9.24 shows results for many soil types presented by Skempton (1957) to illustrate this effect.*

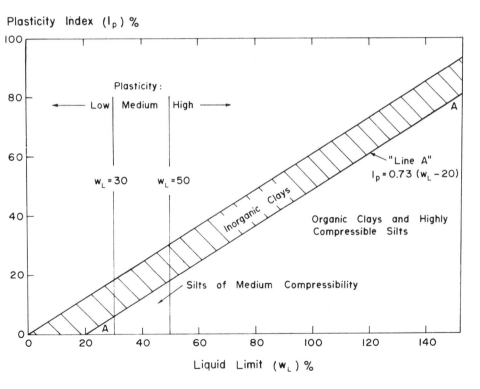

Fig. 9.23—Simplified plasticity chart.

* This ratio is often termed the "c/p" ratio, in which c is the undrained shear strength and p is the overburden pressure.

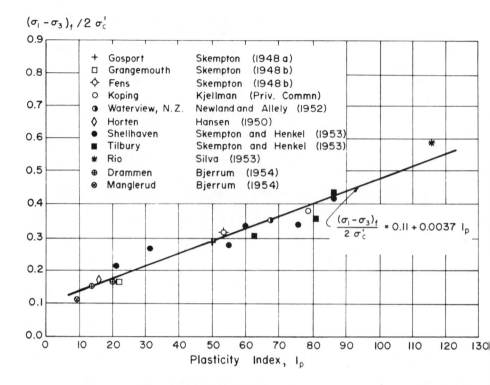

$(\sigma_1 - \sigma_3)_f / 2 \, \sigma_c'$

Plasticity Index, I_p

Fig. 9.24—Relation between strength of normally consolidated clays and plasticity index. (After Skempton, 1957.)

The Atterberg limits provide a rough measure of sensitivity of a soil. We define the *liquidity index*, I_L, as:

$$I_L = \frac{w_{\text{nat}} - w_P}{w_L - w_P} = \frac{w_{\text{nat}} - w_P}{I_p} \qquad (9\text{-}8)$$

where w_{nat} is the natural water content of the soil at which the liquidity index is being determined. When the liquidity index is 1.0, the soil is at the liquid limit and obviously possesses very little strength in the remolded condition. Thus, a liquidity index of 1 for a natural soil suggests high sensitivity. Conversely, a liquidity index of 0 indicates that the soil is at the plastic limit and is probably insensitive. The relation between the liquidity index and sensitivity for some two dozen different clays from all over the world is shown in Figure 9.25 from Skempton and Northey (1952). This figure indicates that a measure of the sensitivity of the soil can be obtained simply by performing Atterberg limit determinations.

The liquidity index is also useful in estimating the order of magnitude of the shear strength to be expected in a remolded soil. Figure 9.26 shows

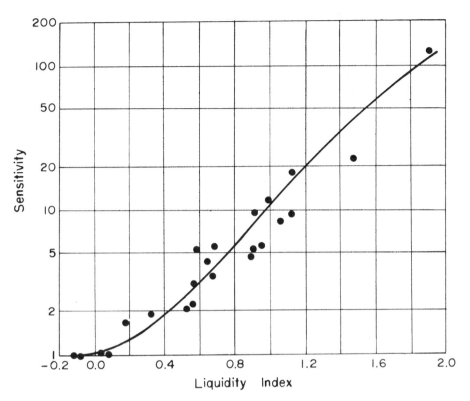

Fig. 9.25—Relation between sensitivity and liquidity index. (From Skempton and Northey, 1952.)

the relationship between remolded shear strength and the liquidity index for four clays of different character. Note that the shear strength at a liquidity index of 1, that is, the liquid limit, is somewhat different from that previously cited. This results from the fact that the liquid limit device used by the British is slightly different from the standard American model, leading to small differences in determination of the Atterberg limits.

Activity

In almost every natural clay, the actual clay-size material (smaller than 2 microns) constitutes only a fraction of the total. The degree to which the soil acts like a clay depends then upon the character of the clay-size material and its relative amount in the soil. If we take a number of samples of a naturally occurring clay and determine the plasticity index of the soil, and the proportion by weight of clay-size material, these are found to be linearly related as seen in Figure 9.27, from Skempton (1953). This figure

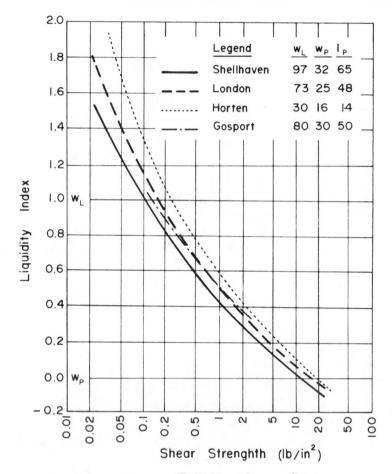

Fig. 9.26—Relation between liquidity index and shear strength of remolded clays. (From Skempton and Northey, 1952.)

shows the relationship between plasticity index and the clay fraction for four clays from Great Britain. Note that straight lines through the origin are reasonable approximations of these sets of data. The slope of each of these lines is then characteristic of the material, and in fact, is the plasticity index divided by 100 for the clay-size material. This slope has been termed the *activity*, A_c:

$$A_c = \frac{I_p}{\% < 2\mu} \tag{9-9}$$

The activity of a clay, then, is a measure of the likely degree to which it will exhibit colloidal behavior. Thus, this property is often termed the *colloidal*

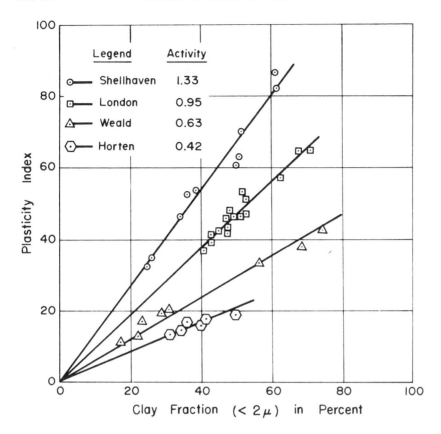

Fig. 9.27—Relation between plasticity index and clay fraction. (From Skempton, 1953.)

activity. The activities for several almost-pure clays are given in Table 9.3. Clays with an activity less than 0.75 are termed *inactive clays. Normally active* clays have activities between 0.75 and 1.25. Those clays with activity greater than 1.25 are termed *active clays.* The data given in Table 9.3 indicate that there is only a general correlation between the clay mineral type and the colloidal activity. This occurs because factors such as particle size and cation type and concentration influence A_c as well. However, the kaolinites and halloysites generally have the lowest activity, while the montmorillonites have the highest activity. Illite is ordinarily intermediate between these. It is apparent that a high activity indicates a soil which would be expected to cause problems for the engineer due to its generally high water holding capacity, cation exchange capacity, thixotropic and sensitivity properties. Conversely, there is some evidence that sensitive clay of low activity may be very difficult to sample in the field without causing so much disturbance that they are not useful for engineering tests (Skempton, 1953). Nonetheless, it is generally the high activity materials which produce potential problems.

9.8 SOIL CLASSIFICATION

Classification has been called "the first handmaiden of science", and it is generally true that an appropriate classification system is one of the first developments in any scientific discipline. The object of classification is to organize knowledge into some *useful* logical arrangement. Thus, different classification systems are determined by the end use of the system. For example, wood may be classified as to—

1. species
2. hardness
3. coniferous or deciduous
4. selling price
5. workability

The classification system applied, will reflect the viewpoint and interest of the classifier. This notion applies equally well to soil classification.

Examples of Soil Classification Systems

Geologic: A geologic classification system is based upon—

1. mineralogy
2. origin and grain size
   a. parent material
   b. mode of deposition

Thus, geologic classifications, such as "residual silty clay" or "quartz sand and gravel outwash," provide inferences about the geological processes which created these materials. Although a knowledge of geologic factors is very useful to the engineer, the system does not give direct information about engineering properties of soils.

Agricultural: Agricultural classification systems are based upon the origin and grain size of the material. These systems are primarily concerned with textural features of the soil which relate to its suitability as an agricultural material.

Engineering: Several engineering classification systems have been developed:

1. *Classification based upon grain size.* Such a classification is shown in Table 2.4. This classification system may be a fair indicator of some properties of coarse-grained materials, but we recognize that it is inadequate to describe the engineering behavior of fine-grained materials.
2. *American Association of State Highway and Transportation Officials* (*AASHTO*) classification system is based primarily on the suitability

of the material for use as a highway subgrade. Thus, this system does not indicate general engineering properties and is recognized as being of limited application.

3. *The Unified Soil Classification System (USCS)* has been jointly adopted by the U.S. Army Corps of Engineers and the U.S. Bureau of Reclamation as their standard system of classification, and is generally considered to be the most useful engineering classification system available. It is based upon the general engineering behavior of the soil, and uses grain size and grain-size distribution to classify coarse-grained soils and the Atterberg limits to classify fine-grained soils. A description of the system and identification procedures is given in Table 9.4. Figure 9.28 illustrates the step-by-step process by which a soil sample is classified in the laboratory.

The USCS is the most useful of available systems in delineating classification groups which are distinguished by differences in engineering properties. Some comparisons of relative merit of USCS classification groups with respect to engineering properties and construction uses are given in Table 9.5. The numbers shown in the table correspond to relative desirability with 1 as the greatest desirability and 14 as the least.

9.9 SUMMARY OF KEY POINTS

Summarized below are the key points which we have discussed in the preceding sections:

1. Most of the properties peculiar to cohesive soils result from the electrochemical properties of clay minerals (Section 9.2). The several types of electrochemical forces which act within and between the particles of a clay soil are:

Type of Force	Source	Relative Strength	Acts over distance of
a. Primary valence bond	Electron Lending	Very strong	$1-2\text{Å}$
b. Hydrogen bond⎱	Cation	Stronger than c.	$2-3\text{Å}$
c. Cation bond ⎰	Attraction	Stronger than d.	
d. van der Waals	Dipole moments	Weak	$>5\text{Å}$
e. Electrostatic	Electrical	Very weak	$\gg 5\text{Å}$

2. a. Clay particles exhibit colloidal action because they are both negatively charged and have large specific surface (Section 9.3). When in an aqueous environment, clay particles tend to:
   (1) Attract and orient the dipolar water molecules near their surface.
   (2) Attract cations to the particle surface. The concentration of such adsorbed cations is inversely related to the distance from the particle surface.

TABLE 9.

Major Division			Group Symbols	Typical Name	Field Identification Procedures (Excluding particles larger than 3 in. and basing fractions on estimated weights)
1	2		3	4	5

Coarse-grained Soils More than half of material is *larger* than No. 200 sieve size. (The No. 200 sieve size is about the smallest particle visible to the naked eye.)	Gravels More than half of coarse fraction is larger than No. 4 sieve size. (For visual classification, the 1/4-in. size may be used as equivalent to the No. 4 sieve size.)	Clean Gravels (Little or no fines)	GW	Well-graded gravels, gravel-sand mixtures, little or no fines.	Wide range in grain sizes and substantial amounts of all intermediate particle sizes.
			GP	Poorly graded gravels or gravel-sand mixtures, little or no fines.	Predominantly one size or a range of sizes with some intermediate sizes missing.
		Gravels with Fines (Appreciable amount of fines)	GM	Silty gravels, gravel-sand-silt mixture.	Nonplastic fines or fines with low plasticity (for identification procedures see ML below).
			GC	Clayey gravels, gravel-sand-clay mixtures.	Plastic fines (for identification procedures see CL below).
	Sands More than half of coarse fraction is smaller than No. 4 sieve size.	Clean Sands (Little or no fines)	SW	Well-graded sands, gravelly sands, little or no fines.	Wide range in grain size and substantial amounts of all intermediate particle sizes.
			SP	Poorly graded sands or gravelly sands, little or no fines.	Predominantly one size or a range of sizes with some intermediate sizes missing.
		Sands with Fines (Appreciable amount of fines)	SM	Silty sands, sand-silt mixtures.	Nonplastic fines or fines with low plasticity (for identification procedures see ML below).
			SC	Clayey sands, sand-clay mixtures.	Plastic fines (for identification procedures see CL below).

			Group Symbols	Typical Name	Identification Procedures on Fraction Smaller than No. 40 Sieve Size		
					Dry Strength (Crushing characteristics)	Dilatancy (Reaction to shaking)	Toughness (Consistency near PL)
Fine-grained Soils More than half of material is *smaller* than No. 200 sieve size.	Silts and Clays Liquid limit is less than 50		ML	Inorganic silts and very fine sands, rock flour, silty or clayey fine sands or clayey silts with slight plasticity.	None to slight	Quick to slow	None
			CL	Inorganic clays of low to medium plasticity, gravelly clays, sandy clays, silty clays, lean clays.	Medium to high	None to very slow	Medium
			OL	Organic silts and organic silty clays of low plasticity.	Slight to medium	Slow	Slight
	Silts and Clays Liquid limit is greater than 50		MH	Inorganic silts, micaceous or diatomaceous fine sandy or silty soils, elastic silts.	Slight to medium	Slow to none	Slight to medium
			CH	Inorganic clays of high plasticity, fat clays.	High to very high	None	High
			OH	Organic clays of medium to high plasticity, organic silts.	Medium to high	None to very slow	Slight to medium
Highly Organic Soils			Pt	Peat and other highly organic soils.	Readily identified by color, odor, spongy feel and frequently by fibrous texture.		

SOURCE: After USCE Report (1953).

(1) Boundary classifications: Soils possessing characteristics of two groups are designated by combinations of group symbols. For example GW-GC, well-graded gravel-sand mixture with clay binder. (2) All sieve sizes on this chart are U. S. standard.

FIELD IDENTIFICATION PROCEDURES FOR FINE-GRAINED SOILS OR FRACTIONS

These procedures are to be performed on the minus No. 40 sieve size particles, approximately 1/64 in. For field classification purposes, screening is not intended, simply remove by hand the coarse particles that interfere with the tests.

Dilatancy (reaction to shaking)

After removing particles larger than No. 40 sieve size, prepare a pat of moist soil with a volume of about one-half cubic inch. Add enough water if necessary to make the soil soft but not sticky.

Place the pat in the open palm of one hand and shake horizontally, striking vigorously against the other hand several times. A positive reaction consists of the appearance of water on the surface of the pat which changes to a livery consistency and becomes glossy. When the sample is squeezed between the fingers, the water and gloss disappear from the surface, the pat stiffens, and finally it cracks or crumbles. The rapidity of appearance of water during shaking and of its disappearance during squeezing assist in identifying the character of the fines in a soil.

Very fine clean sands give the quickest and most distinct reaction whereas a plastic clay has no reaction. Inorganic silts, such as a typical rock flour, show a moderately quick reaction.

Unified Soil Classification System

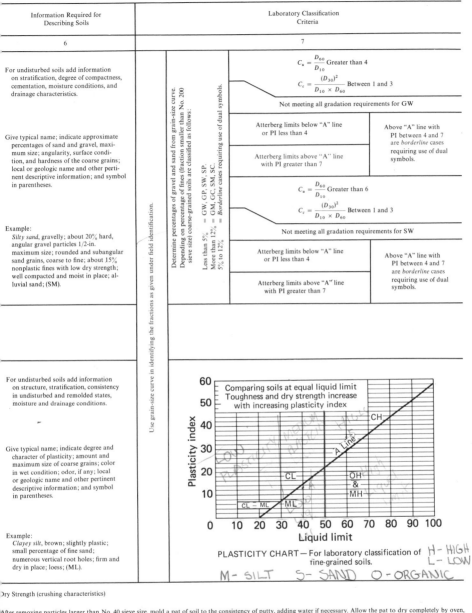

Information Required for Describing Soils	Laboratory Classification Criteria
6	7

For undisturbed soils add information on stratification, degree of compactness, cementation, moisture conditions, and drainage characteristics.

$$C_u = \frac{D_{60}}{D_{10}} \text{ Greater than 4}$$

$$C_c = \frac{(D_{30})^2}{D_{10} \times D_{60}} \text{ Between 1 and 3}$$

Not meeting all gradation requirements for GW

Give typical name; indicate approximate percentages of sand and gravel, maximum size; angularity, surface condition, and hardness of the coarse grains; local or geologic name and other pertinent descriptive information; and symbol in parentheses.

Atterberg limits below "A" line or PI less than 4

Atterberg limits above "A" line with PI greater than 7

Above "A" line with PI between 4 and 7 are *borderline* cases requiring use of dual symbols.

$$C_u = \frac{D_{60}}{D_{10}} \text{ Greater than 6}$$

$$C_c = \frac{(D_{30})^2}{D_{10} \times D_{60}} \text{ Between 1 and 3}$$

Not meeting all gradation requirements for SW

Example: *Silty sand, gravelly; about 20% hard, angular gravel particles 1/2-in. maximum size; rounded and subangular sand grains, coarse to fine; about 15% nonplastic fines with low dry strength; well compacted and moist in place; alluvial sand; (SM).*

Atterberg limits below "A" line or PI less than 4

Atterberg limits above "A" line with PI greater than 7

Above "A" line with PI between 4 and 7 are *borderline* cases requiring use of dual symbols.

Determine percentages of gravel and sand from grain-size curve. Depending on percentage of fines (fraction smaller than No. 200 sieve size) coarse-grained soils are classified as follows:
Less than 5% = GW, GP, SW, SP.
More than 12% = GM, GC, SM, SC.
5% to 12% = *Borderline* cases requiring use of dual symbols.

Use grain-size curve in identifying the fractions as given under field identification.

For undisturbed soils add information on structure, stratification, consistency in undisturbed and remolded states, moisture and drainage conditions.

Give typical name; indicate degree and character of plasticity; amount and maximum size of coarse grains; color in wet condition; odor, if any; local or geologic name and other pertinent descriptive information; and symbol in parentheses.

Comparing soils at equal liquid limit
Toughness and dry strength increase with increasing plasticity index

Example: *Clayey silt, brown; slightly plastic; small percentage of fine sand; numerous vertical root holes; firm and dry in place; loess; (ML).*

PLASTICITY CHART — For laboratory classification of fine-grained soils.

H - HIGH
L - LOW
M - SILT   S - SAND   O - ORGANIC

Dry Strength (crushing characteristics)

After removing particles larger than No. 40 sieve size, mold a pat of soil to the consistency of putty, adding water if necessary. Allow the pat to dry completely by oven, sun, or air-drying, and then test its strength by breaking and crumbling between the fingers. This strength is a measure of the character and quantity of the colloidal fraction contained in the soil. The dry strength increases with increasing plasticity.

High dry strength is characteristic for clays of the CH group. A typical inorganic silt possesses only very slight dry strength. Silty fine sands and silts have about the same slight dry strength, but can be distinguished by the feel when powdering the dried specimen. Fine sand feels gritty whereas a typical silt has the smooth feel of flour.

Toughness (consistency near plastic limit)

After particles larger than the No. 40 sieve size are removed, a specimen of soil about one-half inch cube in size, is molded to the consistency of putty. If too dry, water must be added and if sticky, the specimen should be spread out in a thin layer and allowed to lose some moisture by evaporation. Then the specimen is rolled out by hand on a smooth surface or between the palms into a thread about one-eighth inch in diameter. The thread is then folded and rerolled repeatedly. During this manipulation the moisture content is gradually reduced and the specimen stiffens, finally loses its plasticity, and crumbles when the plastic limit is reached.

After the thread crumbles, the pieces should be lumped together and a slight kneading action continued until the lump crumbles.

The tougher the thread near the plastic limit and the stiffer the lump when it finally crumbles, the more potent is the colloidal clay fraction in the soil. Weakness of the thread at the plastic limit and quick loss of coherence of the lump below the plastic limit indicate either inorganic clay of low plasticity, or materials such as kaolin-type clays and organic clays which occur below the A-line.

Highly organic clays have a very weak and spongy feel at the plastic limit.

429

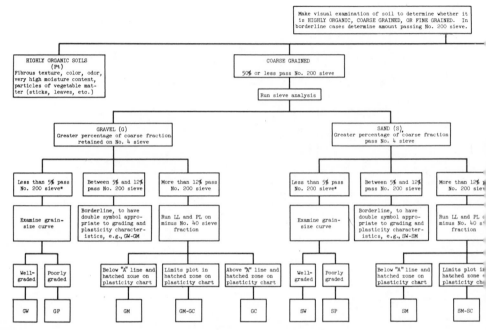

Note: Sieve sizes are U. S. Standard.

* If fines interfere with free-draining properties use double symbol such as GW-GM, etc.

The zone of oriented water and adsorbed cations surrounding the clay particles is termed the diffuse double layer.

b. The ability to adsorb cations is related to the net negative charge on the particle surface, and is measured by the cation exchange capacity. The ease with which one cation can be exchanged for another depends upon the size and valence of the hydrated ion, and relative concentrations of the ions. (Section 9.4).

3. The structural arrangement of clay particles is a function of the interaction between attractive and repulsive forces between the particles, which depends upon the type of clay mineral and the characteristics of the pore fluid (Section 9.5). The structure is characterized as:

a. Flocculent. A flocculent structure occurs when either:

(1) The double layer is small enough so that particle attractions occur at almost all relative orientations (high salt flocculation), or

(2) The presence of a positive charge concentration at the edges of the particle produces a tendency for edge-to-face attraction (non-salt flocculation).

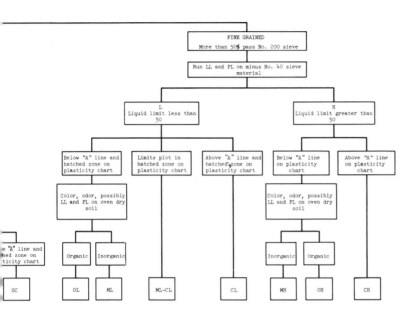

Fig. 9.28—Procedure for laboratory determination of USCS classification. (After USCE Report, 1953.)

     b. Disperse. The dispersed structure occurs when the double layer is large enough that repulsive forces tend to keep particles separate.

     Flocculation is enhanced by increasing ion concentration, increasing adsorbed cation valence, decreasing size of exchangeable cations or decreasing pH of the fluid.

  4. Electrochemical forces are important in a number of observed phenomena characteristic of cohesive soils (Section 9.6). These include:

     a. Sensitivity, S_t.

     b. Thixotropy.

     c. Electroosmotic flow.

  5. The effect of water on clay soils is very important for engineering purposes and is often described in terms of simple index properties (Section 9.7):

     a. The ranges of consistency are delineated by the Atterberg limits: the liquid limit, w_l, and the plastic limit, w_p, are of most use to us. The range of water contents over which plastic behavior occurs is defined by the plasticity index, I_p.

     b. Useful indicators of the engineering behavior of clay soils

TABLE 9.5

Typical Names of Soil Groups	Group Symbols	Important Properties			
		Permeability When Compacted	Shearing Strength When Compacted and Saturated	Compressibility When Compacted and Saturated	Workability as a Construction Material
Well-graded gravels, gravel-sand mixtures, little or no fines	GW	pervious	excellent	negligible	excellent
Poorly graded gravels, gravel-sand mixtures, little or no fines	GP	very pervious	good	negligible	good
Silty gravels, poorly graded gravel-sand-silt mixtures	GM	semi-pervious to impervious	good	negligible	good
Clayey gravels, poorly graded gravel-sand-clay mixtures	GC	impervious	good to fair	very low	good
Well-graded sands, gravelly sands, little or no fines	SW	pervious	excellent	negligible	excellent
Poorly graded sands, gravelly sands, little or no fines	SP	pervious	good	very low	fair
Silty sands, poorly graded sand-silt mixtures	SM	semi-pervious to impervious	good	low	fair
Clayey sands, poorly graded sand-clay mixtures	SC	impervious	good to fair	low	good
Inorganic silts and very fine sands, rock flour, silty or clayey fine sands with slight plasticity	ML	semi-pervious to impervious	fair	medium	fair
Inorganic clays of low to medium plasticity, gravelly clays, sandy clays, silty clays, lean clays	CL	impervious	fair	medium	good to fair
Organic silts and organic silt-clays of low plasticity	OL	semi-pervious to impervious	poor	medium	fair
Inorganic silts, micaceous or diatomaceous fine sandy or silty soils, elastic silts	MH	semi-pervious to impervious	fair to poor	high	poor
Inorganic clays of high plasticity fat clays	CH	impervious	poor	high	poor
Organic clays of medium to high plasticity	OH	impervious	poor	high	poor
Peat and other highly organic soils	Pt	—	—	—	—

SOURCE: After Wagner (1957).

ngineering Application Use Chart for USCS Classification Groups

Relative Desirability for Various Uses									
Rolled Earth Dams			Canal Sections		Foundations		Roadways		
							Fills		
Homo-generous Embank-ment	Core	Shell	Erosion Resist-ance	Com-pacted Earth Lining	Seepage Im-portant	Seepage not Im-portant	Frost Heave not Possible	Frost Heave Possible	Sur-facing
—	—	1	1	—	—	1	1	1	3
—	—	2	2	—	—	3	3	3	—
2	4	—	4	4	1	4	4	9	5
1	1	—	3	1	2	6	5	5	1
—	—	3 if gravelly	6	—	—	2	2	2	4
—	—	4 if gravelly	7 if gravelly	—	—	5	6	4	—
4	5	—	8 if gravelly	5 erosion critical	3	7	8	10	6
3	2	—	5	2	4	8	7	6	2
6	6	—	—	6 erosion critical	6	9	10	11	—
5	3	—	9	3	5	10	9	7	7
8	8	—	—	7 erosion critical	7	11	11	12	—
9	9	—	—	—	8	12	12	13	—
7	7	—	10	8 volume change critical	9	13	13	8	—
10	10	—	—	—	10	14	14	14	—

derived from the Atterberg limits are:

(1) Liquid limit, from which the compressibility of normally consolidated clays can be estimated.
(2) Plasticity index, from which the undrained shearing resistance of normally consolidated clays can be estimated.
(3) Liquidity index, I_L, which is a measure of sensitivity.
(4) Activity, A_C, which indicates the relative importance of the clay fractions in determining the plastic behavior of the material. Implications include:

Activity	Clay Is	Comments
<0.75	Inactive	May be difficult to sample if sensitive
0.75–1.25	Normal	
>1.25	Active	Generally high w and sensitive

6. Three classification systems for soils are in common use for engineering purposes (Section 9.8). They are:

System	Comment
Grain size	May be fair indicator of properties for coarse materials but is inadequate for fine-grained materials.
AASHTO	Based on suitability of materials for highway subgrades—limited application
USCS	Based on general engineering behavior of the soil—uses grain size and distribution to classify coarse materials, Atterberg limits to classify fine-grained soils.

PROBLEMS

9.1. What would be the effect of each of the following:
(a) Leaching a sodium (Na^+) clay with a calcium (Ca^{++}) salt? (The ionic radii are about equal.)
(b) Leaching a marine clay with fresh water?
(c) Leaching a sodium clay (Na^+ radius = 0.95 Å, with ammonium (NH_4^+, radius = 1.42 Å) salts?
Note: Include in your discussion the changes you would expect in the soil micelle, the liquid limit, the plasticity index, the undistorted strength, and the sensitivity. Would the resulting soils be better or worse foundation materials than the original soils?

9.2. For the effect of soils on the load-carrying capacity of piles:
(a) Consider a situation where foundation piles for a building are driven into a medium stiff layer of montmorillonite rich clay. A load test, to determine the ultimate load which these piles can withstand, is performed immediately after driving. Would you expect the results of this test to be indicative of the load carrying capacity of these piles after the building is constructed? Explain your answer. *Note*: Driving piles in clay remolds the soil in the vicinity of the pile.
(b) Suppose the piles were driven into sand instead of clay. How would you answer the above question? Explain.

9.3. Regarding the composition of clay minerals:
(a) Why is bentonite (a very pure Na^+ montmorillonite clay) used to make "drilling mud"? Why not use a pure kaolinite?
(b) In what order would you expect the activity of kaolinite, montmorillonite, and illite to rank. Why?

9.4. In order to stabilize a weak soil for use in a compacted fill, powdered slaked lime, $Ca(OH_2)$, is mixed with the soil before compaction.
(a) Why should such a procedure be efficacious?
(b) Would you expect the resultant permeability to be higher or lower than that of the unstabilized soil when compacted?
(c) Is this a coarse grained or fine-grained soil? Why?

9.5. Two soils, one a Laurentian clay, the other a Philadelphia clay, are compared by means of their Atterberg limits. The following data are obtained in the laboratory:

	Laurentian Clay		Philadelphia Clay	
Liquid Limit	N (blows)	w (%)	N (blows)	w (%)
	10	50	10	66
	12	48	16	63
	30	38	34	58
	40	35	40	57
Plastic Limit	$w_p = 25\%$		$w_p = 25\%$	

The natural water contents in the field are 44.0% for the Laurentian clay and 51.0% for the Philadelphia clay.
(a) Which clay has a higher degree of plasticity? Explain.
(b) Which clay would probably be a better foundation material if remolded? Why?
(c) For which material would the strength be more dependent on changes in water content? Explain.
(d) Do you expect these soils to contain organic material? Explain.

9.6. Knowing only the plasticity indices for several soils, could you estimate which one has the highest liquid limit? Discuss the various possibilities posed by this question.

9.7. A saturated stiff clay specimen removed from the ground below the groundwater table is found to have a natural water content of 40%, $G = 2.75$, and a liquidity index of 0.9. You are anticipating excavating a basement for a large structure and will cut into this material at the bottom of the excavation. What major construction difficulty would you expect to encounter? Why?

10

Shallow Foundations, Bearing Capacity

10.1 OVERVIEW OF PROBLEMS FACING THE FOUNDATION ENGINEER

A *foundation* is that part of a structure which has the primary function of transmitting loads from the structure to the natural ground. Although it is usually hidden from sight when the structure is complete, the successful performance of a foundation has a pronounced effect on the overall performance of the structure. On this subject, Terzaghi (1951) has said:

> On account of the fact that there is no glory attached to the foundations, and that the sources of success or failure are hidden deep in the ground, building foundations have always been treated as step-children; and their acts of revenge for the lack of attention can be very embarrassing.

There are three basic requirements of a foundation:

1. It must be safe from complete, or ultimate, failure of either the earth materials on which the foundation rests, or the materials of which the foundation elements are made.
2. It must not experience displacements which are sufficient to cause undesirable settlement or distortion of the superstructure.
3. It must be economically feasible in terms of the overall structure.

The foundation design must meet these requirements considering a variety of adverse environmental factors, some of which may act only occasionally, such as:

1. *Frost action.* Frost heave and rapid settlement following thaw may be a problem for certain types of foundations in those areas where freezing of the ground occurs during the winter months. Figure 10.1 shows, in a generalized way, the maximum depth of frost penetration in the United States. It is evident that frost penetration of several

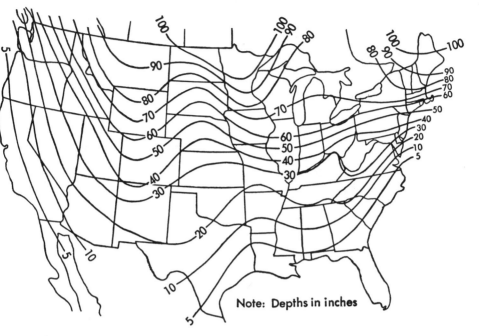

Fig. 10.1—Maximum depth (in inches) of frost penetration in the United States. (After Sowers and Sowers, 1967.)

feet will lead to potentially significant magnitudes of frost heave if the foundation is supported within the zone of frost action.

2. *Shrinking or swelling soils.* In many parts of the United States, particularly the Southwest, surface and subsurface soils undergo large volume changes in response to changes in moisture resulting from rainfall and/or variations in the groundwater level. Such volume change phenomena can cause significant distress in a structure if the potential volume change is not considered in the foundation design.

3. *Adjacent structures, excavations, and property lines.* An owner may be liable for the damage done to an adjacent structure by the stresses created by his own foundation, even if no physical contact between the two structures has occurred. Thus, consideration must be given in the design to the influence of the foundation upon adjacent or nearby structures. A similar cautionary note applies to the effect of existing nearby excavations on the stability of a foundation. Property lines represent a legal barrier to what might otherwise be an optimum location for a foundation, and cannot be overlooked.

4. *Groundwater.* Groundwater can be a treacherous enemy if neglected. The fact that groundwater may not be located near the foundation depth at a given point in time, does not necessarily insure its permanent absence. Potential changes in the groundwater level, and

the possibility of artesian pressures must be evaluated. The presence of groundwater and/or artesian pressures can significantly influence both *stability* and *settlements*.

5. *Underground defects.* The existence of cavities in apparently sound bedrock, rock faults which may be filled with unsuitable material, rock ledges of limited thickness, and other "nonstandard" underground features are highly dangerous and must not be overlooked. This is particularly important in those areas where geologic evidence indicates a significant probability of their occurrence.

6. *Earthquakes.* Earthquakes influence not only the magnitude of the structural loads transmitted to the foundation, but also the capability of the subsurface material to support those loads (see Chapter 2). Thus, foundation design is affected in at least two ways by potential earthquake activity. The degree to which this factor may be important depends upon the nature of the foundation soils as well as frequency and severity of earthquake activity in a given area. Figure 10.2 shows the relative risk of earthquake damage in various parts of the United States. It is interesting to note that even in the Midwest, earthquakes may present a significant hazard.

7. *Scour and wave action.* In the case of foundations for structures impinged upon by surface water, for example bridge piers, scour occurring at times of rapid flow often removes soil around the foundation to a depth below the normal surface equal to the rise in the water surface; in some cases the scour may be several times the rise in the water surface. In addition, wave action may induce horizontal forces against the foundation.

The role played by many of these factors in the design process will be discussed in subsequent sections.

10.2   METHOD OF APPROACH OF FOUNDATION DESIGNER

Suppose that we are faced with the following question: For a particular desired structure, what will be the most economical foundation that meets the requirements described above? In order to answer this question, the designer must carry out an iterative process which ordinarily involves the following steps:

1. Establish the scope of the problem. That is, he must decide on the overall size and type of problem he is facing. For example, an earth dam which is $\frac{1}{2}$ mi wide and 2 mi long clearly requires considerations at a different scale than a 20-story structure, 200-ft square in plan. This structure, in turn, is of much larger scale than a single pier for a highway bridge. Thus, the broad boundaries of the problem must be delineated.

2. Investigate conditions at the site. This step may actually consist of a number of individual investigations ranging from a library study of the geologic history of the area to direct observation of subsurface conditions

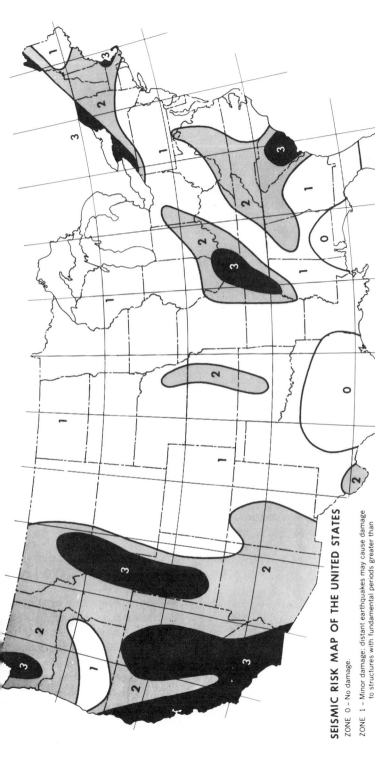

SEISMIC RISK MAP OF THE UNITED STATES

ZONE 0 – No damage.

ZONE 1 – Minor damage: distant earthquakes may cause damage to structures with fundamental periods greater than 1.0 seconds; corresponds to intensities V and VI of the M.M.* Scale.

ZONE 2 – Moderate damage: corresponds to intensity VII of the M.M.* Scale.

ZONE 3 – Major damage: corresponds to intensity VIII and higher of the M.M.* Scale.

This map is based on the known distribution of damaging earthquakes and the M.M.* intensities associated with these earthquakes; evidence of strain release; and consideration of major geologic structures and provinces believed to be associated with earthquake activity. The probable frequency of occurrence of damaging earthquakes in each zone was not considered in assigning ratings to the various zones.

*Modified Mercalli Intensity Scale of 1931.

439

from the side of a test pit at the site. Ordinarily the preliminary stages of the investigation are sufficient to lead to steps 3 and 4, from which the requirements for more detailed investigation may arise.

3. Formulate trial design. It is at this point that the experience of the designer is especially valuable in assisting him in selecting a trial design which subsequent steps show to be both functional and economical.

4. Establish a model to be analyzed. This is a key aspect of the design process, because this step may have more influence on the success or failure of the design than any other phase of the design process. The need for this step arises because some objective means must be established by which the trial design can be evaluated and compared to other trial designs. In this step, the designer must first decide what features of the foundation behavior he wishes to analyze, and second, determine the model by which he will represent the foundation and the subsurface materials. Any model is invariably a simplification of the actual *in situ* conditions, and represents a compromise between the need for a model which *can* be analyzed and our best estimate of the *true* conditions. The degree to which these simplifications distort the predictions of the analysis from the results which would occur *in situ*, is a major factor in determining the ability of the designer to make a rational decision. This is illustrated in the following example.

Example 10.1

Figure 10.3 shows schematically an elevation view of the foundation for a multistory structure, and the stratigraphic profile beneath it, determined from test borings. The profile in the figure shows the foundation located within a layer of medium dense sand and gravel. This layer is underlain by a stiff desiccated silty clay which is in turn underlain by a soft silty clay containing fine sand lenses. It is probably reasonable to infer that the stiff clay layer is a dried crust of the softer underlying material.

When the rate of settlement due to consolidation is being considered, the model assumed by the Terzaghi one-dimensional consolidation theory might be accepted as a reasonable approximation to the real situation. In this case, a drainage length of $\frac{31}{2} = 15\frac{1}{2}$ ft. might be used in the calculations.

However, it must be remembered that the profile shown is inferred from the logs and samples obtained from a few test borings. If the thin sand lenses are interconnected at various places, and, as often happens, are drained, the drainage length may vary radically in both vertical and horizontal directions. Because the time for one-dimensional consolidation is inversely proportional to the square of the drainage length, this inhomogeneity may lead to very rapid rates of settlement which vary in a horizontal direction. In this case, the usual simplified model may lead to a very unconservative result.

The assumption of the Terzaghi one-dimensional consolidation model may be unreasonable for the materials, geometric conditions and loads imposed in the problem.

Another potential source of error is the assumption of an isotropic, homogeneous elastic medium which was made in order to compute the stresses. The distributing effect of the sand layer and stiff crust may be more than is recognized. In addition, the

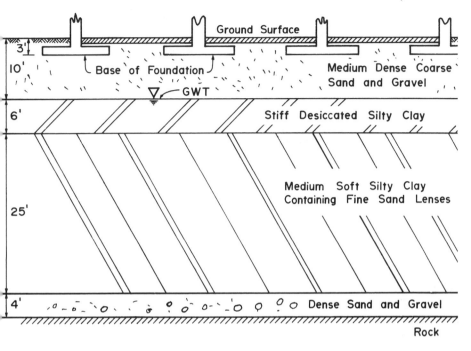

Fig. 10.3—Schematic diagram of foundation
and subsurface profile for Example 10.1.

inelastic behavior of the materials may produce important discrepancies between computed and measured stresses.

Finally, the assumed initial load distribution and changes in loading with time may not be correct. The "time-average" loads are usually based on estimates which are difficult to make accurately. Furthermore, if the structure has significant rigidity, the distribution of loads on the foundation will change as the differential movements of the foundation occur.

The separate and combined effects of these discrepancies between the model assumed for this problem and the real problem may be of decisive importance in the design.

---

5.   Determine parameters of the model. In this step it is necessary to assign numerical values to the properties of the materials involved, the magnitude of the loads to be imposed, etc. For instance, in the case of a frame structure placed above a layer of cohesive material, the designer would need to evaluate the magnitude of the loads distributed to the individual columns and the degree to which the columns will interact, the compressibility and consolidation characteristics of the soil, the shear strength and changes in shear strength with time, and other factors of this sort. This may require additional investigation at the site as in item 2 above, laboratory tests, and/or the exercise of considerable judgment.

6.   Carry out the analysis.   In this step, the designer actually applies the appropriate available analytical tools to the model which he has established for analysis, incorporating the determined parameters. Thus, this step ordinarily leads to numerical results. These are most useful if they are actually predictions of the performance of the designed system. Often, however, they are simply estimates of the degree to which specific characteristics of the design exceed some critical minimum.

7.   Compare results.   The results obtained from a trial design must be compared with those of other trial designs as well as with the experience of the designer. In fact, the experienced designer may bypass steps 3 through 6 for many designs and simply eliminate them on the basis of his experience in a specific area or with soils of a particular nature. Decisions about such matters as the significance to be attributed to a given "degree of safety" or the reliability of predicted performance of a particular kind must be made at this point.

8.   Modify the design.   On the basis of the results obtained in the previous steps, the designer is now in the position to refine, modify, or change completely his design, as the circumstances warrant. Having done so, he then returns to step 4 and completes the subsequent parts of the cycle. This iterative procedure if applied with skill and judgment, will lead to the optimum design consistent with the available methods for analysis.

9.   Observe construction.   On large projects, especially those in which significant uncertainties about the actual subsurface conditions exist, it is often necessary to monitor the construction. Such observations may lead to design changes where it becomes clear that *in situ* conditions are at variance with the design model in an important manner. The usefulness of this step is emphasized by Peck (1969).

It is now evident that in earlier chapters, our attention was focused on parts of items 4, 5, and 6 above. That is, in connection with a number of problems, we studied the appropriate models, the significant parameters of these models and how they might be determined, and some of the analyses applied to the models. In this and subsequent chapters we shall extend our consideration of items 4 and 6 to include certain other aspects of foundations and earth slope behavior. We hope that this will lead to development of the capacity to carry out steps 3 and 7. Other features of the design process are beyond the scope of this text.

10.3   TYPES OF FOUNDATIONS

There are two basic types of foundations: *shallow* and *deep*. A foundation is ordinarily considered *shallow* if it transmits the load to the soil at a depth close to that of the base of the structure which it is supporting. Thus, a spread footing foundation located immediately beneath a sub basement floor which is 30 ft beneath the ground surface, is still considered a shallow foundation. A *deep* foundation, by contrast, transmits all, or a portion of, the load to some significant depth below the base of the structure. It is ordinarily the function of a deep foundation to bypass weak or undesirable materials and cause the structure to be supported on some stratum below the unsuitable soils.

The choice of foundation type depends upon a variety of factors. However, it is most common to select that foundation type which is the least expensive to construct, and which meets the requirements of all foundations described above. Because shallow foundations are usually less expensive than deep foundations, they are ordinarily considered first in the design process. Herein, we shall restrict our attention to shallow foundations, which provide important applications of the principles examined in earlier chapters.

There are a number of types of shallow foundations illustrated in Figure 10.4. The simplest of these, the *wall* or *strip footing* and the *spread footing,* are shown in Figures 10.4a and b, respectively. These simple structural

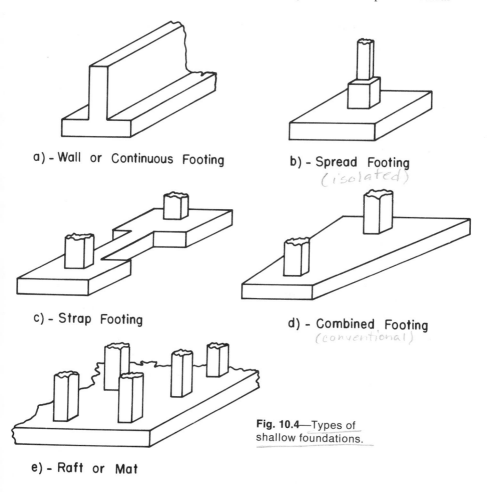

a) - Wall or Continuous Footing

b) - Spread Footing
(isolated)

c) - Strap Footing

d) - Combined Footing
(conventional)

e) - Raft or Mat

Fig. 10.4—Types of shallow foundations.

members serve to enlarge the base of a load bearing wall or column so that the average pressure across the cross section is reduced when it is transmitted to the soil. They are ordinarily constructed of plain or reinforced concrete. When suitable, they are usually the most economical foundation types.

The *strap* or *pump-handle footing*, shown in Figure 10.4c consists of two footings joined by a tie beam which has the purpose of reducing the magnitude of the stresses under one of the footings resulting from applied couples (or eccentric loads) by transferring some of the load to the other footing. The *combined footing* shown in Figure 10.4d will fulfill the same function as well as bridging over soft pockets between the two columns. If columns are closely spaced, it may be more economical to combine two footings into one combined footing. *Mat* or *raft* foundations as shown in Figure 10.4e consist of a single large continuous footing which supports the entire structure. Mat foundations are especially useful in cases where the subsoils are soft and highly erratic, or where resistance against hydrostatic uplift and watertightness are especially important. They become economical when the required area of individual spread footings exceeds approximately one-half of the total plan area of the structure.

There are several aspects of shallow foundation behavior which must be considered in the design process. It is customary to use different models when investigating these problems. The first of these which we shall investigate is the ultimate failure, or *bearing capacity* of the soil-foundation system.

10.4 BEARING CAPACITY OF SHALLOW FOUNDATIONS

The Transcona Grain Elevator Failure

One of the best-known foundation failures occurred in October 1913 at North Transcona, Manitoba, Canada. The accompanying article describes (pages 445–447) the circumstances surrounding the construction of a million-bushel grain elevator at the site, and its subsequent failure. It was fortunate that there were eye witnesses to the failure, and the account of one of them is also included (pages 448–449).

We must ask ourselves the following question: What knowledge might have forestalled this failure? Clearly, our state of knowledge has progressed in the intervening years because such failures are rare today. If we consider the steps involved in a foundation design, it is apparent that they were not followed in the design of this particular foundation. Therefore, let us go back and carry out those steps, at least in a limited way, to find out where the original design was faulty.

Because we are dealing with an already constructed problem, Step 1 is taken care of, and we recognize that our attention may be focused on an area of relatively limited extent. However, Step 2, which is crucial to all that follows, was not carried out. At the time, it was not recognized that the

Failure of the Transcona Grain-Elevator

The remarkable foundation failure of a Canadian Pacific Ry. elevator at Transcona, near Winnipeg, briefly reported in our last issue, is made clear by the four photographic views herewith reproduced, for which we are indebted to the courtesy of J. G. Sullivan, Chief Engineer of the Canadian Pacific Ry., Winnipeg, Man. The reinforced-concrete binhouse settled into the ground by the crushing out of the foundation subsoil, and at the same time tilted to the position shown. The movement was slow, taking 24 hr. in all. The displacement of the underlying strata of earth is well shown by the upheaved ridges of earth around the elevator. The detail view, Fig. 4, gives some idea of the surface disturbance. This shows the ridge along the east of the binhouse, seen sharply outlined in Fig. 1.

The elevator was built in the summer of 1912. It consists of a working-house, seen in the views standing undisturbed just south of the displaced structure, and a reinforced-concrete binhouse. The latter consisted of 65 circular bins of 14 ft. 4 in. inside diameter and 90 ft. deep, arranged in five rows of 13 bins each (5 along the working-house, 13 in the direction at right angles to it). The 18 interspaces between the circular bins were also used for grain storage. The binhouse was 10 ft. north of the working-house, and structurally distinct from it, being connected with it only at the conveyor level, just above the top of the bins, where the conveyor which operates in the low cupola over the binhouse is carried across to the working-house. The principal machinery floors of the working house are above the conveyor level.

The soil at the site is soft clay ground. Rock is 45 to 50 ft. below the surface. The binhouse was founded on a reinforced-concrete slab, 77x195 ft., about 12 ft. below the surface of the ground. In excavating for the construction of this foundation, the upper five or six of ground were found rather soft. Below this, however, a fairly stiff clay was struck. Load tests were made at the time, and they indicated that the soil was well able to bear a load of 4 to 5 tons per sq. ft. The maximum loaded weight of the binhouse averaged a little over $2\frac{1}{2}$ tons per sq. ft. The foundation was put in during July, 1912, and the soil was perfectly dry, with no indication of water. Barnet-McQueen Co., Ltd., of Minneapolis, designed and built the elevator.

Rafts or floating foundations of the kind used here are commonly used for elevator foundations in the Winnipeg territory. The working-house of the present elevator rests on a similar slab foundation.

The occurrence of the accident is described as follows: On Saturday, Oct. 18, between 11 and 12 a.m., the elevator then being practically full, movement was noticed on the bridges between the working-house and the binhouse. By 1 p.m. the binhouse had settled about 1 ft. The ground for a distance of 25 to 30 ft. on the north, west and east sides of the binhouse heaved up 4 or 5 ft. In the afternoon the settlement became a little more rapid on the west side, producing a list to the west. The building continued to settle and list for about 24 hours, until noon, Sunday, Oct. 19. Its final position is at an angle of about 28° from the vertical. The east side of the house is raised about 5 ft. above normal level, while the west side is some 30 ft. below normal. The conveyor cupola and roof structure slid off the binhouse when the tilt became too great, which occurred about midnight Saturday.

The structure of the binhouse shows practically no damage in its visible part, except for a few small hair cracks.

The question of saving the grain in the elevator occupied first attention. The binhouse held nearly its full capacity of a million bushels. The bins were tapped through the

Settlement of the Transcona
1,000,000-Bu. Grain Elevator

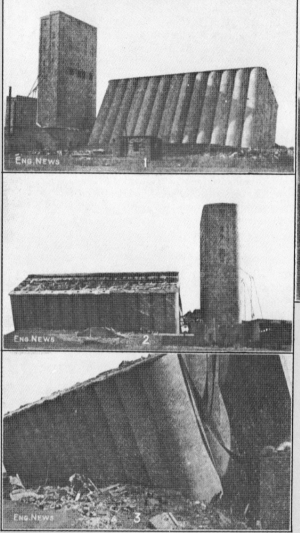

Fig. 1. View looking southwest—Working-house undisturbed though only 10 ft. from the bin-house. The openings halfway up were connected to the binhouse cupola by conveyor bridges. The ridge of earth in front of the house is the result of the subsurface flow of soil.

Fig. 2. West side of the elevator; bin tops after cupola slid off. The view shows the 30-ft. subsidence on this side; the whole length of the binhouse settled practically uniformly.

Fig. 3. Southwest corner of bin-house, showing side of greatest subsidence. The wreckage of the cupola and roof lies at the foot of the house.

Fig. 4. Detail view along east side, with adjacent soil upheaved (see Fig. 1).

sides and a temporary conveyor erected alongside for taking the grain to railway cars, and the removal when started progressed at the rate of about 40 carloads a day. It was feared at first that ground water would enter the bins at the bottom and would wet the lower part of the grain. This was found not to be the case, at least during the earlier stages of the transfer. The absence of water is held to indicate that the accident is not chargeable to wet condition of the subsoil. The above statement of facts is based on information furnished by Mr. Sullivan.

Attention is called to the very similar settlement of two reinforced-concrete warehouses in Tunis some years ago, described in ENGINEERING NEWS of Apr. 25, 1907, page 458. The two buildings there which settled were also on reinforced-concrete slab foundation. The maximum subsidence on one side in that case amounted to ten or twelve feet. It proved possible to right the buildings by loading the high side and doing some excavating underneath. In final restored position the buildings were about 15 ft. below their original level.

conditions at depth below the foundation level might be significant. However, *we* know that they may be, and are immediately faced with the question of what sort of investigation we should carry out. A useful place to start is to learn as much as one can about the geology at the site. This can provide tremendous insight into the development of a more elaborate investigation program. There are numerous geologic records for many places throughout the world. In the United States, U.S. Geological Survey maps and reports are available, among many other sources.

Such a library study would reveal that the bedrock is known to be limestone of Ordovician age. Above the limestone, one would expect to find sediments deposited in the waters of glacial Lake Agassiz, which covered extensive parts of the North American continent when the Wisconsin ice sheet blocked the region's northern outlet. This information is very helpful because immediately we suspect that we will find fine-grained soils, possibly soft, which may present problems with regard to either strength or compressibility. Therefore, we will probably wish to make some direct determination of the character of the subsurface materials. Recognizing that samples of the pertinent materials will probably be required in order to determine the parameters of our model (Step 5), the acquisition of some test *borings* seems useful at this point.

It is fortunate, for our purposes, that two continuous borings were made at the site in 1951 (see again the excerpt on pages 448 and 449). The *log* of one of these borings is shown in Figure 10.5. The subsurface profile revealed by this log confirms our suspicion that soft clays are present at the site.

It is not necessary for us to formulate a trial design (Step 3), because we are considering a design that already exists. However, at this point we must establish the model to be analyzed (Step 4). In point of fact, if we were designing this structure, we would investigate several models to insure that we met all the requirements described previously. In this case, however, we shall focus upon that model related to the observed failure of the foundation. It seems likely from the relatively rapid nature of the failure, as well

447

Transcona Elevator Failure: Eye-Witness Account

L. Scott White, O.B.E.

"The Grain Elevator referred to in the following article was one of the most important structures built by the Canadian Pacific Railway Company in their extensive railroad yards at North Transcona. The yards covered several square miles and were built on partly farmed prairie land. The ground was relatively flat for miles around.

"Excavation for the foundations of the elevator was in open cut about 12 feet in depth ... Bearing tests were made by loading a plate laid upon a prepared smooth surface; the test loading was applied from a wooden gantry erected for the purpose. To the best of the Writer's memory no borings were taken.

"The tests appeared to satisfy the requirements of the engineers, who assumed that the "blue gumbo" had similar characteristics and depth to that on which many heavy structures had been founded in the vicinity of Winnipeg, 7 miles to the west.

"The construction of the elevator proceed at a rapid rate during the autumn and winter of 1912, the circular bins being raised at the rate of 3 feet per day ...

"In the spring of 1913, when the thaw set in after heavy winter snows, a great deal of trouble was caused throughout the yards by the movement of the top surface of the clay under the pressure of deep ballast-fills which carried the tracks ... In one case high-level tracks, which were laid on a 30-foot fill, subsided several feet, throwing up great rolls of ground to the sides and breaking up massive drainage culverts constructed of 12-inch by 12-inch timbers. This subsidence was remedied by driving hundreds of 60-foot timber piles through the ballast to form a staging on which the tracks were relaid.

"This was not the end of the troubles, for on the 18th of October, 1913, when the engineering staff of the C.P.R. were having their lunch at a camp about a mile distant from the elevator, there were excited cries that the elevator was collapsing. Everyone rushed out to witness the phenomenon not believing it possible, but, sure enough, the bin structure had already taken a considerable tilt to the west and was still moving ... As the structure tilted to the west, the earth on that side bulged up forming a cushion which slowed down the movement ... On the east side a wide gap was left in the ground to the depth of the raft foundation.

"The movement of the bin structure throughout was gradual and barely susceptible to the eye, but a considerable amount of commotion was caused by the connecting bridges carrying the conveyor belts breaking adrift and crashing to the ground.

"The main concern of everyone at the time was whether the work-house could stand the disturbance. Check levels were therefore taken and it was found with considerable relief to be standing firm.

"Movement continued at a diminishing rate for the rest of the day. In the night the cupola structure over the bins, which housed the conveyor belting, suddenly collapsed and fell to the ground. This reduced the load and there was subsequently little further movement.

"Naturally a great strain was put on the comparatively thin walls of the storage bins but they were so well constructed that hardly any cracks appeared in them.

Reprinted from R. B. Peck and F. G. Bryant, "The Bearing Capacity Failure of the Transcona Elevator," *Geotechnique*, Vol. 3, No. 5, 1952.

as the presence of soft clay deposits underneath the foundation, that the strength of the soil was exceeded. Therefore, we shall concentrate upon the development of a model which will incorporate this effect.

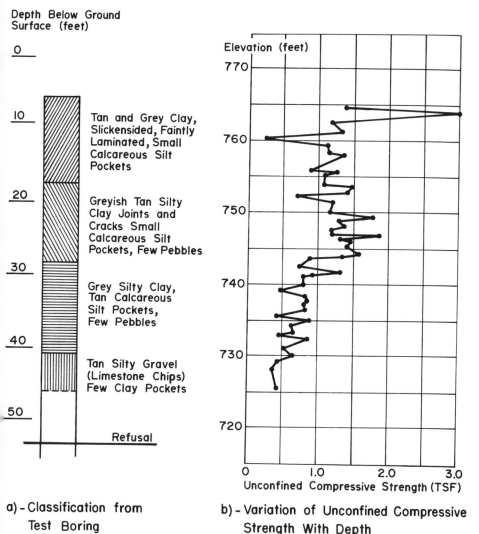

a) - Classification from
Test Boring

b) - Variation of Unconfined Compressive
Strength With Depth

Fig. 10.5—Results of test boring at site of Transcona grain elevator. (From Peck and Bryant, 1953.)

449

Establishment of Model

From the boring log in Figure 10.5, it appears that, for predicting the behavior of a rigid foundation subjected to loadings which are significant relative to those required to cause failure, even a fairly realistic model must incorporate a description of the constitutive relationship (that is, the stress-strain-time relationship) for a system consisting of several layers. Unfortunately, to attempt this highly nonlinear problem at the present time would be overly ambitious, and would lead to a formulation which would be difficult to solve.

However, we can learn a great deal by adopting a different point of view. Let us model one aspect of the behavior of the system which may be of particular value, that is, assume that the system of interest is a *rigid–perfectly plastic* material which is at the point of failure. Using this assumption we can determine the magnitude of the loads required to bring failure about, and compare them to the loads which we estimate will actually be imposed by the structure. This approach, often referred to as *limit design* or *plastic design*, is also used in Chapter 3. In order to formulate this problem, let us begin by making some very simplifying assumptions, and later examining the implication of these assumptions to our analysis of the real structure.

We shall first assume that, instead of being a rectangular foundation on an earth mass, the whole system which we are investigating is "very long," that is, we shall examine what happens in a particular vertical plane which passes through the foundation, and assume that all other planes parallel to this plane are exactly the same. This sort of *plane strain* problem was considered in Chapter 3. As before, the equations of equilibrium in two dimensions (Equations 3–2) must be satisfied. They are repeated here for ready reference:

$$\frac{\partial \sigma_x}{\partial x} + \frac{\partial \tau_{xz}}{\partial z} = 0$$

$$\frac{\partial \tau_{xz}}{\partial x} + \frac{\partial \sigma_z}{\partial z} - \gamma = 0 \tag{10–1}$$

where x and z are the horizontal and vertical coordinate directions, respectively, and γ is the unit weight of the soil.

We certainly require more information than this because there are two equations and three unknowns. Because we have assumed our soil mass to be at the point of failure, we can use a yield (strength) criterion as an additional relationship, as we did in Chapter 3. Again, we shall adopt the Mohr-Coulomb failure law for this purpose. However, we can now use one or another of the more general forms discussed in Chapter 8. For example,

$$\tau_f = c + \sigma_f \tan \phi \tag{10–2}$$

where τ_f is the shear stress on some plane at failure, σ_f is the normal stress on the same plane at failure, and c and ϕ are strength parameters, the significance of which depends upon whether the total or effective stresses are being considered, and whether the stresses described are acting upon the failure plane at failure or some other plane. Whatever the significance of the strength parameters chosen for the particular problem, the form of the equation describes the straight line relationship shown in Figure 10.6. From this figure, it is evident that the yield criterion can be expressed in terms of σ_x, σ_z, and τ_{xz} as:

$$\sqrt{\frac{(\sigma_z - \sigma_x)^2}{4} + \tau_{xz}^2} = \left(\frac{\sigma_x + \sigma_z}{2} + c \cot \phi\right) \sin \phi \qquad (10\text{--}3)$$

Equations 10–1 and 10–3 now form a set of three simultaneous equations with three unknowns, and, in principle, are soluble when the boundary conditions are specified.

These equations have been solved for two special cases which may not seem to be related to our problem at first glance, but will be found to be quite helpful. The first is the simple Rankine problem, which we solved in Chapter 3 for a cohesionless medium. The boundary conditions assumed

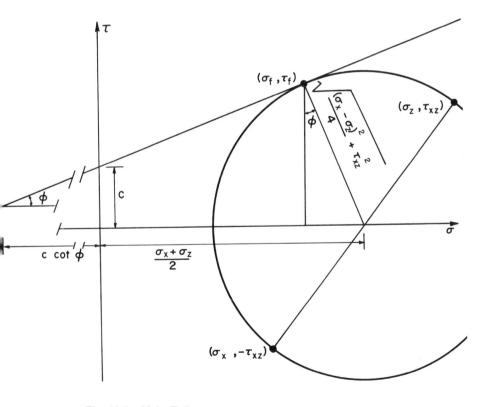

Fig. 10.6—Mohr Failure diagram for cohesive soil.

in this problem are shown in Figure 10.7: the homogeneous, isotropic medium extends indefinitely in all horizontal directions and the downward vertical direction, and the stress on the surface acts normal to it. The assumption of a rigid material before yield insures that the displacements which occur before yielding are sufficiently small that the geometry of the problem is unaffected by them.

In this case, we can integrate Equations 10–1 directly, as in Chapter 3, which leads to

$$\sigma_z = \gamma z + q \qquad (10\text{--}4)$$

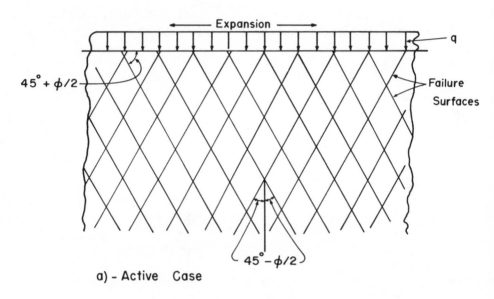

a) - Active   Case

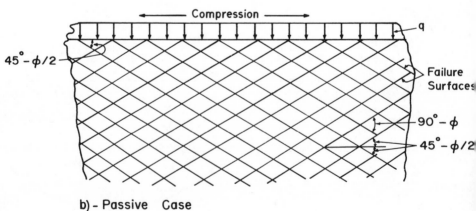

b) - Passive   Case

Fig. 10.7—Failure conditions from Rankine analysis.

Substituting this result into Equation 10–3 gives

$$\frac{\gamma z + q}{2} - \frac{\sigma_x}{2} = \left(\frac{\sigma_x + \gamma z + q}{2} + c \cot \phi\right) \sin \phi \tag{10-5a}$$

Rearranging

$$\sigma_x = (\gamma z + q)\frac{1 - \sin \phi}{1 + \sin \phi} - 2c\frac{\cos \phi}{1 + \sin \phi} \tag{10-5b}$$

or

$$\sigma_x = (\gamma z + q) \tan^2 (45° - \phi/2) - 2c \tan (45° - \phi/2) \tag{10-5c}$$

Equations 10–5 determine the active pressure (horizontal extension) case illustrated in Figure 10.7a. This result occurs because it was assumed in Equation 10–3 that σ_x was less than σ_z. Had we chosen σ_z to be less than σ_x and reversed their order in the first term on the left side of the equation, the passive pressure case (Figure 10.7b) would have resulted, that is,

$$\sigma_x = (\gamma z + q) \tan^2 (45° + \phi/2) + 2c \tan (45° + \phi/2) \tag{10-6}$$

In the Rankine problem described, the surfaces along which failure occurs, called *slip surfaces*, are two families of planes which appear as straight lines in Figure 10.7.

The second case which is of interest in our development is the plane strain problem illustrated in Figure 10.8. This figure shows a weightless wedge-shaped body, the boundaries of which are defined by two straight rays emanating from an apex and extending indefinitely. It is assumed that the body is in a state of incipient yielding along each ray. Thus the major principal stress is inclined to the rays at the angle $45° - \phi/2$. It can be shown (Nadai, 1963) that the stresses in such a problem can be described in terms of the partial derivatives of a special type of potential function,[*] F:

$$F = \frac{C_0}{2} \cos^2 \phi r^2 e^{-2\theta \tan \phi} - \frac{r^2}{2} c \cot \phi \tag{10-7}$$

where C_0 is a constant determining the intensity of the state of stress, r is the radial distance to a point, θ is the angle measured as shown in Figure 10.8, and c and ϕ are from Equation 10–2. By definition, the stresses are then

$$\sigma_r = \frac{1}{r}\frac{\partial F}{\partial r} + \frac{1}{r^2}\frac{\partial^2 F}{\partial \theta^2} = C_0(1 + \sin^2 \phi)e^{-2\theta \tan \phi} - c \cot \phi \tag{10-8a}$$

$$\sigma_t = \frac{\partial^2 F}{\partial r^2} = C_0 \cos^2 \phi e^{-2\theta \tan \phi} - c \cot \phi \tag{10-8b}$$

$$\tau_{rt} = -\frac{\partial}{\partial r}\left(\frac{1}{r}\frac{\partial F}{\partial \theta}\right) = C_0 \sin \phi \cos \phi\, e^{-2\theta \tan \phi} \tag{10-8c}$$

---

[*] This type of potential function is called an *Airy stress function* after the nineteenth century elastician G. B. Airy who first proposed such functions for plane elastic problems (Airy, 1862).

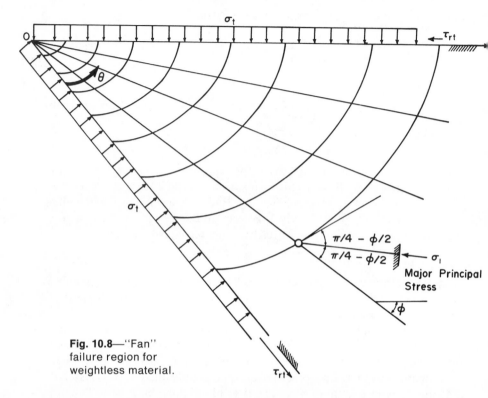

Fig. 10.8—"Fan"
failure region for
weightless material.

The rays are of course one family of slip surfaces. The other family of slip surfaces (appearing as lines) must maintain a constant angle with the rays of $90° - \phi$. This is illustrated in Figure 10.9. From the figure, the tangent of the angle ϕ, which is a constant, equals:

$$\tan \phi = \text{constant} = \frac{dr}{rd\theta} \tag{10–9}$$

Integrating this directly shows that the second family of slip lines are logarithmic spirals:

$$r = R_0 e^{\theta \tan \phi} \tag{10–10}$$

where R_0 is the magnitude of r at $\theta = 0$.

The two solutions described above were combined by Prandtl (1920, 1921) to solve a problem similar to the one in which we are interested. Prandtl was concerned with the punching of metals, and he investigated the situation illustrated in Figure 10.10a. This figure is a special case of his general strip loading of high intensity, q, required to produce failure of a weightless, rigid–perfectly plastic, incompressible material. He decided that, at failure, the material beneath the load could be divided into five regions consisting of "Rankine" zones and "fans" of the type discussed

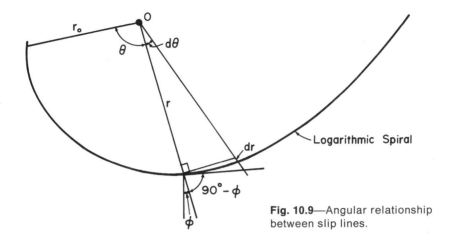

Fig. 10.9—Angular relationship between slip lines.

above. The failure zones were combined in such a fashion that the boundaries of the fans coincided with the slip lines in the two Rankine zones on either side of them, as shown in Figures 10.10b and c. Prandtl concluded that under the failure load q, the Rankine zones would be in an "active" state, that is, the largest stress would be vertical and the smallest stress horizontal. Conversely, under the surcharge load, the "passive" state would obtain, that is, the largest stress would be horizontal and the smallest stress would be vertical. The fans provided the transition. If the magnitude of q_s is given, then the horizontal stresses are known in the outer passive Rankine zones. Thus, σ_t and τ_{rt} are readily determined at the outer parts of the fan. This condition permits evaluation of the quantity C_0 in Equations 10–8 for the ray at $\theta = \pi/2$, where the fan and the passive Rankine zone coincide. Equations 10–8 can then be solved for $\theta = 0$, from which the stress q can be determined. Carrying out these operations leads to the following expression for q:

$$q = c \cot \phi \left(\frac{1 + \sin \phi}{1 - \sin \phi} e^{\pi \tan \phi} - 1 \right) + q_s \frac{1 + \sin \phi}{1 - \sin \phi} e^{\pi \tan \phi} \qquad (10\text{–}11a)$$

which, for convenience, is usually written:

$$q = cN_c + q_s N_q \qquad (10\text{–}11b)$$

in which

$$N_c = \cot \phi \left(\frac{1 + \sin \phi}{1 - \sin \phi} e^{\pi \tan \phi} - 1 \right), \qquad N_q = \frac{1 + \sin \phi}{1 - \sin \phi} e^{\pi \tan \phi} \qquad (10\text{–}12)$$

Note that N_c and N_q are functions of ϕ only. This point will be discussed further below.

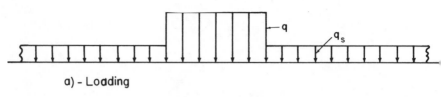

a) - Loading

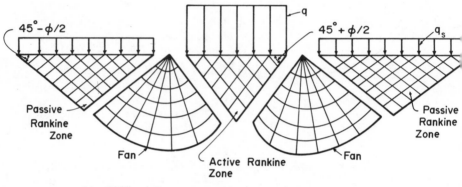

b) - Failure Zones Used

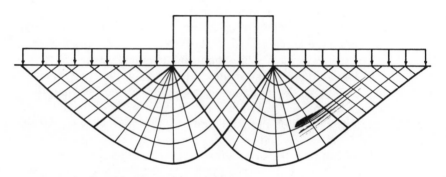

c) - Combined Failure Pattern

Fig. 10.10—Prandtl problem.

For a purely cohesive material (for which $\phi = 0$) and no surcharge, Equation 10–11a reduces, after the application of L'Hospital's rule, to:

$$q = (2 + \pi)c = 5.14c \qquad (10-13)$$

Application of Results to Soils

Because the strength of metals is very great compared to their weight, assumption of weightlessness is approximately valid in the case of such materials. Terzaghi (1943) realized, however, that such was not the case for soils. Thus, when he applied the Prandtl (1920, 1921) analysis to the determination of the *bearing capacity* of "long" shallow foundations on soils, he superimposed upon the bearing capacity of a weightless material an additional component due only to the weight of the soil and its frictional resistance. Terzaghi (1943) originally assumed that the failure surfaces for the material with weight were the same as those adopted by Prandtl (1920, 1921) for the weightless case. That this assumption is in error is illustrated in Figure 10.11, which shows the theoretically determined slip surfaces and required load distribution for failure with weight and no surcharge. These results were obtained by combining Equations 10–1 and 10–3, and integrating the resulting equation numerically. Comparison of the slip surfaces shown in Figure 10.11 with those in Figure 10.10 illustrates the difference. We note also that the stress distribution required to produce failure in the case shown in Figure 10.11 is clearly not linear. Nonetheless, the mode of failure observed in tests is qualitatively similar to that assumed by Prandtl. This is indicated graphically in Figure 10.12, which presents a photograph of a model test of a long footing on sand.

Terzaghi (1943) also assumed that the soil above the base of the footing acted solely as a surcharge weight, but did not contribute its own shearing resistance. Consequently he restricted his attention to "shallow" foundations with a depth of embedment less than their width.

More recent analyses (Meyerhof, 1955; Terzaghi and Peck, 1967) have computed the component of the bearing capacity due to weight and friction based upon an assumed failure surface, illustrated in Figure 10.13a, which is different from that used by Prandtl (1920, 1921). The angle ψ is not fixed at $45° + \phi/2$, but is permitted to assume whatever value will lead to the lowest bearing capacity. The weight effect is determined by finding the force P' required to fail the curved wedge shown in Figure 10.13b. This is

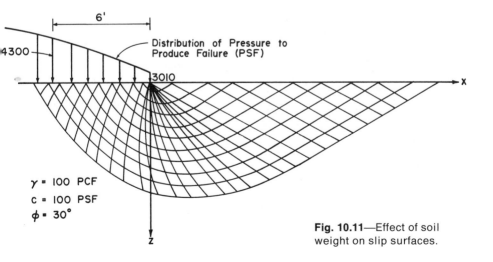

Fig. 10.11—Effect of soil weight on slip surfaces.

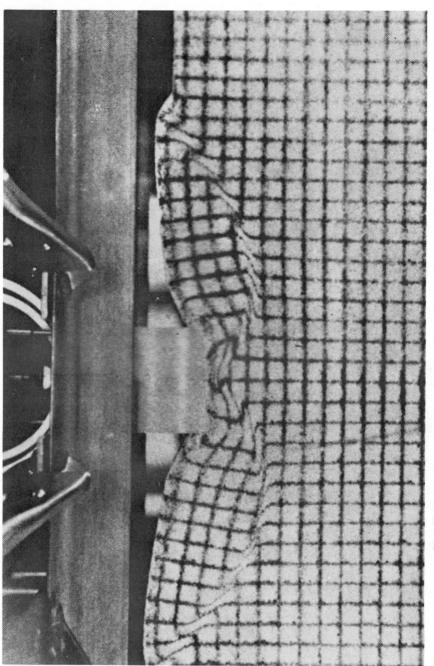

Fig. 10.12—Static test of 3-in. footing after failure. (After Selig and McKee, 1961.)

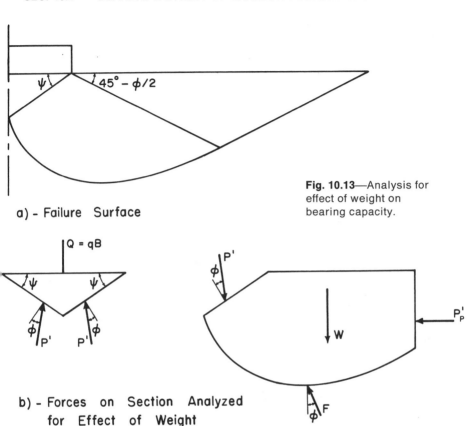

a) - Failure Surface

Fig. 10.13—Analysis for effect of weight on bearing capacity.

b) - Forces on Section Analyzed
    for Effect of Weight

a passive pressure problem which can most easily be solved graphically. Appropriate methods for doing this are discussed in detail in Chapter 13. At this point it suffices to recognize the solution will depend only upon ϕ and ψ. But of course the ψ which produces the smallest bearing capacity is selected.

We can estimate then the component of bearing capacity due to weight of the failure mass and friction from vertical equilibrium of the triangle of soil beneath the footing (Figure 10.13b). Neglecting the weight of this small triangle,

$$Q = qB = 2P' \cos (\psi - \phi) \qquad (10\text{–}14)$$

which can be rewritten as

$$q = \tfrac{1}{2}\gamma B N_\gamma \qquad (10\text{–}15)$$

where

$$N_\gamma = \frac{4P'}{\gamma B^2} \cos (\psi - \phi) \qquad (10\text{–}16)$$

Thus, the total bearing capacity for a long *strip* footing is given by (Terzaghi and Peck, 1967)

$$q = cN_c + \gamma D_f N_q + \tfrac{1}{2}\gamma B N_\gamma \qquad\qquad (10\text{-}17)$$

in which N_c, N_q, and N_γ, defined by Equations 10–12 and 10–16 are called the *bearing capacity factors*, and D_f is the depth of the base of the footing so that γD_f is the surcharge q_s. Note that the bearing capacity factors are functions of ϕ only and can be plotted for ready reference. This has been done for N_c and N_q, shown as solid lines in Figure 10.14. The parameter N_γ, defined by Equation 10–16, gives values close to those obtained by more complicated methods (Meyerhof, 1955), which are also shown in Figure 10.14 as a solid line.

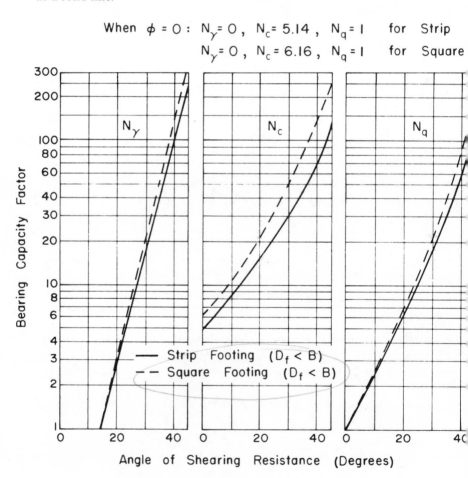

When $\phi = 0$: $\ N_\gamma = 0$, $\ N_c = 5.14$, $\ N_q = 1$ for   Strip
$N_\gamma = 0$, $\ N_c = 6.16$, $\ N_q = 1$ for   Square

—— Strip Footing ($D_f < B$)
- - - Square Footing ($D_f < B$)

Fig. 10.14—Bearing capacity factors for shallow foundations.

Geometry and Loading Effects

The results discussed thus far apply to a "strip" foundation subjected to a vertical concentric load. Meyerhof (1963) has suggested a number of semi-empirical modifications to the general bearing capacity equation to include the influence on bearing capacity of foundation shape, eccentricity and inclination of the load, and shear strength of the soil above the foundation. The modified equation takes the form

$$q = \lambda_c d_c i_c c N_{c_{\text{strip}}} + \lambda_q d_q i_q \gamma D_f N_{q_{\text{strip}}} + \lambda_\gamma d_\gamma i_\gamma \frac{B'}{2} \gamma N_{\gamma_{\text{strip}}} \qquad (10-18)$$

in which q is the *vertical* component of ultimate unit bearing capacity of the foundation, $N_{c_{\text{strip}}}$, $N_{q_{\text{strip}}}$, $N_{\gamma_{\text{strip}}}$ are the bearing capacity factors *for the strip load* given in Figure 10.14, c and γ are as previously defined. The other factors in the equation are discussed below.

Based on theoretical and experimental results, Meyerhof (1951, 1955) obtained the bearing capacity factors for square and circular footings given as the dashed lines in Figure 10.14. The λ symbols provide for linear interpolation between the strip and square bearing-capacity factors in the case of rectangular footings:

$$\lambda_c = 1 + \frac{B}{L}\left(\frac{N_{c_{\text{square}}}}{N_{c_{\text{strip}}}} - 1\right) \qquad (10-19a)$$

$$\lambda_q = 1 + \frac{B}{L}\left(\frac{N_{q_{\text{square}}}}{N_{q_{\text{strip}}}} - 1\right) \qquad (10-19b)$$

$$\lambda_\gamma = 1 + \frac{B}{L}\left(\frac{N_{\gamma_{\text{square}}}}{N_{\gamma_{\text{strip}}}} - 1\right) \qquad (10-19c)$$

In these equations B is the width and L is the length of the footing.

Note that Equations 10–19 indicate that the factor N_γ is larger for square footings ($B/L = 1$) than for strip footings ($B/L = 0$). For $\phi' > 30°$, that is, granular materials in a moderately dense condition, this is contrary to experimental results. Meyerhof (1961a) attributed this difference to the effect of the intermediate principal stress on ϕ' in the case of a long foundation. Based upon the fact that the measured ϕ' angles in laboratory plane strain tests on cohesionless soils are approximately 10 per cent higher than determined in the triaxial compression test, he recommended that the ϕ' angle for a rectangular foundation on cohesionless soil be given by:

$$\phi_r' = \left(1.1 - 0.1\frac{B}{L}\right)\phi_t' \qquad (10-20)$$

where ϕ_r' is the angle of shearing resistance to be used in determining the bearing capacity factors for a rectangular foundation, ϕ'_t is the angle determined in the triaxial compression test. For circular foundations, $B/L = 1$.

The symbols d in Equation 10–18 account for the increase in bearing capacity resulting from the strength of the soil above the foundation level. They are:

$$d_c = 1 + 0.2 \frac{D_f}{B} \tan (45° + \phi/2) \tag{10–21a}$$

$$d_q = d_\gamma = 1 \quad \text{for } \phi = 0 \tag{10–21b}$$

$$d_q = d_\gamma = 1 + 0.1 \frac{D_f}{B} \tan (45° + \phi/2) \quad \text{for } \phi > 10° \tag{10–21c}$$

These factors apply for $D_f < B$. That is, if $D_f > B$, then D_f/B in Equations 10–21 is to be set equal to one.

The factor i accounts for the effect of inclined loads for concentrically loaded "rough" foundations. With loads inclined at an angle δ from the vertical, Meyerhof (1963) recommended that:

$$i_c = i_q = \left(1 - \frac{\delta°}{90°} \right)^2 \tag{10–22a}$$

$$i_\gamma = \left(1 - \frac{\delta}{\phi} \right)^2 \tag{10–22b}$$

The influence of load inclination is illustrated graphically in Figure 10.15 for a strip footing at the ground surface, for a purely cohesive material ($\phi = 0$) and a cohesionless soil ($c = 0$, $\phi = 35°$). Recalling that the bearing capacity calculated from Equation 10–18 is the vertical component of the inclined stress, the ordinate in Figure 10.15 is determined from

$$\frac{Q_{\text{inclined}}}{Q_{\text{vertical}}} = \frac{q_{\text{inclined}}}{q_{\text{vertical}}} = \frac{q}{\cos \delta q_{\text{vertical}}} \tag{10–23}$$

in which Q_{inclined} is the total inclined bearing capacity applied at the angle δ from the vertical, and Q_{vertical} is the total bearing capacity for a *vertical concentric* load. The relative importance of load inclination for the bearing capacity of footings on cohesionless soils, as shown in Figure 10.15, appears most often in the design of foundations for retaining walls, as discussed in Chapter 13.

The effect of eccentric loading on the bearing capacity is considered by reducing the foundation dimensions B and L, to effective foundation dimensions B' and L'. If the eccentricity is one-way, as shown in Figure 10.16a, then the effective foundation dimensions are (Meyerhof, 1953)

$$B' = B - 2e_x, \quad L' = L \tag{10–23a}$$

or

$$B' = B, \quad L' = L - 2e_z \tag{10–23b}$$

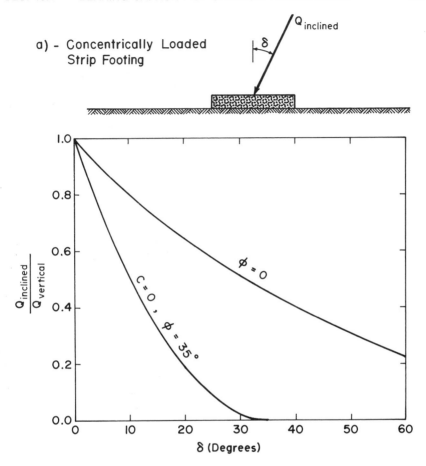

a) - Concentrically Loaded Strip Footing

b) - Effect on Bearing Capacity

Fig. 10.15—Effect of load inclination on bearing capacity of concentrically loaded strip footing, $D_f = 0$.

in which e_x and e_z are the eccentricity of the resultant load in the x and z directions, respectively. When the eccentricity is two-way, then the effective foundation width is determined so that the resultant load is located at the centroid of the "effective" area A', shown in Figure 10.16b. The *vertical component* of the total load carried by an eccentrically loaded footing is then

$$Q = qA' = qB'L' \qquad (10\text{--}24)$$

in which A' is the area of greatest length L' such that its centroid coincides with the resultant load. Thus B' is defined as A'/L', as illustrated in Figure 10.16b and c.

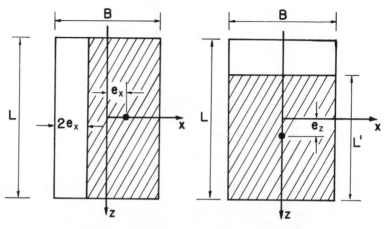

(a) - Single Eccentricity for Rectangle

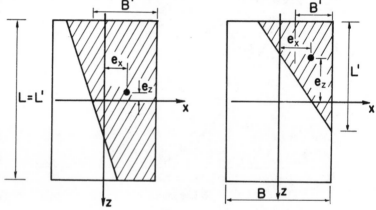

(b) - Double Eccentricity for Rectangle

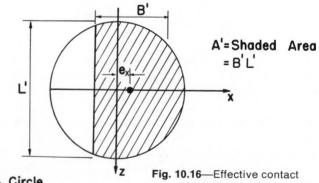

(c) - Circle

A'=Shaded Area
= B'L'

Fig. 10.16—Effective contact area of foundation with eccentric load. (After Meyerhof, 1953.)

In the case of eccentrically loaded square and rectangular foundations the shape factors λ are determined from Equations 10–19, in which the effective foundation shape B'/L' is used in place of the actual B/L.

The influence of eccentricity is illustrated quantitatively in Figure 10.17, which shows the calculated reduction in total bearing capacity as a function of eccentricity for strip footings on cohesive ($\phi = 0$) and cohesionless ($c = 0$, $\phi = 35°$) soils. We see that the effect of eccentric loading is more important in the case of cohesionless soils.

The applicability of the concept of a reduced effective foundation width is indicated clearly in Figure 10.18, which is a photograph of a model test conducted on a sand in which the foundation was loaded with a vertical eccentric load. Although the displacement of the model in this photograph is very large in order to provide a more graphic demonstration of the failure pattern, the reduced area of contact between the foundation and the soil is evident. Note that the recommendation for reduced foundation width by Meyerhof (1953) is conservative because it assumes a uniform distribution of pressure under the eccentric loading.

When the load is applied both inclined *and* eccentrically the two effects may be enhanced or reduced depending upon the arrangement of the forces. This is illustrated in Figure 10.19, which shows the results of a theoretical investigation of this question (Wack, 1961). As shown in Figures 10.18 and 10.19, the effect of a vertical eccentric load is to force the failure pattern to the side of the footing away from the eccentricity. The effect of an inclined load is to force the failure pattern toward the direction of the horizontal component of the inclined load. Hence when the horizontal component of the inclined loading points toward the side of the footing away from the eccentricity the failure pattern is reduced in size, as shown in Figure 10.19a. Conversely, when the horizontal component of the inclined loading points toward the side of the footing in which the eccentric load is placed, the two effects tend to counteract each other and the failure pattern is enlarged, Figure 10.19b. This leads to a partial compensation of one effect by the other. The degree to which this compensation occurs depends upon the relative magnitudes of the eccentricity and the inclination. Comparison with the theoretical analysis by Wack (1961) leads to these conclusions:

1. If the eccentricity and inclination are combined in the manner shown in Figure 10.19a, then Equation 10–24, where q is determined from Equations 10–18, will lead to an approximately correct and conservative result.

2. When the eccentric and inclined loads combine as indicated in Figure 10.19b, then the bearing capacity can be determined by following three steps:
    a. Compute the total bearing capacity from Equation 10–24 assuming the load to be applied *eccentrically* but *not* inclined.

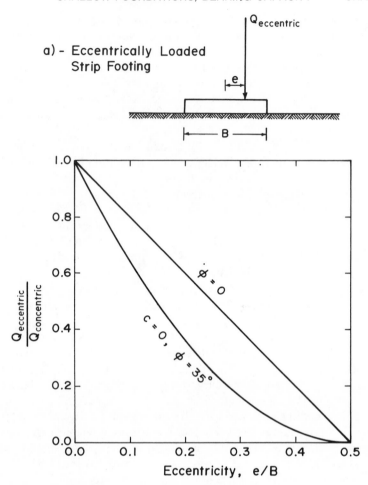

a) - Eccentrically Loaded
    Strip Footing

b) - Effect on Bearing Capacity

Fig. 10.17—Effect of load eccentricity on bearing capacity of vertically loaded strip footing, $D_f = 0$.

b. Determine the total bearing capacity using Equation 10–24 assuming that the load is *inclined* but *centrally* located.

c. Determine the total bearing capacity from Equation 10–24 assuming the load to be *vertical* and *centrally* located and multiply this value by the ratio of the smaller load computed above to the larger load. That is, if the bearing capacity due to an eccentric vertical load determined in (a) above is less than that for the inclined central load determined in (b) above, then the first of these is divided by the second and conversely.

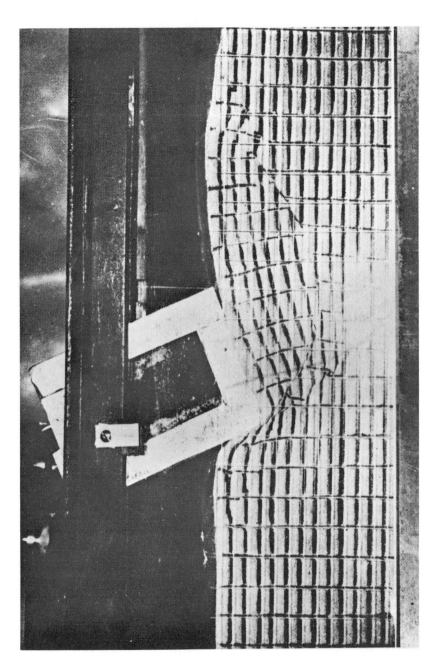

Fig. 10.18—One-sided rupture surface from a vertical, eccentrical load. (After Jumikis, 1956.)

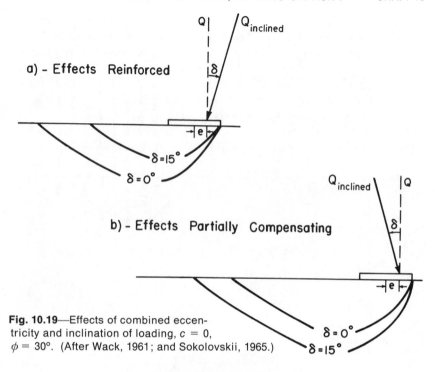

Fig. 10.19—Effects of combined eccentricity and inclination of loading, $c = 0$, $\phi = 30°$. (After Wack, 1961; and Sokolovskii, 1965.)

Determination of ϕ' for Cohesionless Soils

The direct *in situ* determination of ϕ' for cohesionless materials is infrequently carried out. It is more common to evaluate the *in situ* relative density using an empirical correlation with the *standard penetration test** performed at the site. The resistance to penetration of the soil has been found to be a function of the relative density and the confining pressure. Figure 10.20 shows an empirical correlation between these parameters. This correlation appears to be independent of grain size distribution and grain shape. Note however that it *does* depend upon the position of the groundwater table. The reason for this is evident from our discussion of shearing resistance in Chapter 2. Because the penetration test occurs very rapidly, it corresponds to a kind of undrained shear test of the soil. If the granular material is dense, the tendency for volume increase produces an increase in effective stress and an apparent strength which is too high. Conversely the penetration resistance of a loose sand below the groundwater table will appear less than its resistance to much more slowly applied loadings.

---

* The *standard penetration test* is conducted as a part of most test boring operations. It consists of driving a standard size tube into the soil at the bottom of a bore hole using a 140-lb hammer falling freely 30 in. The number of blows of the hammer required to drive the tube a distance of 1 ft. into the soil is termed the *standard penetration resistance, N*.

The relationship between ϕ' and relative density for the particular material in question can then be determined in the laboratory. A very approximate correlation between ϕ' and relative density, which is conservative in many cases, is also given in Figure 10.20. This table indicates a unique relationship between relative density and ϕ' although, for a variety of soils there will be a variation in ϕ' at a given relative density. Hence it is only an approximate guide and must be used with caution.

As we shall see subsequently, except for the unusual case of a narrow footing on loose cohesionless material with a groundwater table close to the

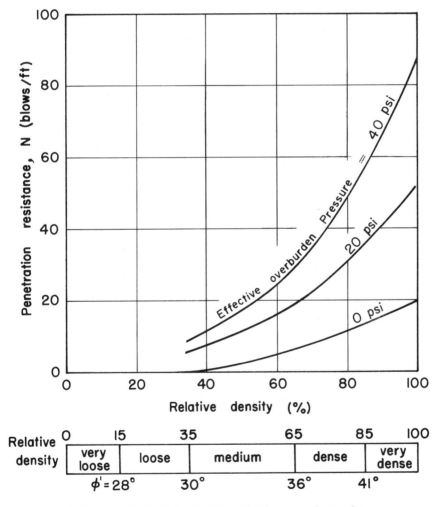

Fig. 10.20—Relationship between standard penetration resistance, relative density, and angle of shearing resistance, for sands. (After Gibbs and Holtz, Meyerhof, 1956.)

footing elevation, the controlling feature in design of shallow foundations on *cohesionless* soils is settlement, not bearing capacity. Thus, for such materials, the evaluation of the *in situ* ϕ' is often not required.

Effect of Groundwater Table

The effect of groundwater must be considered in determining the appropriate unit weights to use in Equation 10–18. In those cases where ϕ' is greater than zero the shearing resistance along the potential failure surface depends upon the weight of the material above that surface. Hence if the groundwater table is located within the potential failure zone, the effective weight of the soil must be used. Because this failure zone extends to a depth below the foundation approximately equal to the width of the foundation the effect of the groundwater table need be considered only if it is located within a depth B below the foundation. In such a case, for materials in which $\phi' > 0$, the unit weight γ in the term containing N_γ must be the submerged unit weight, γ'. Similarly, if the groundwater table is above the base of the foundation, then the term containing N_q must be modified to account for that part of the material which is submerged.

10.5   RECONSIDERATION OF TRANSCONA GRAIN ELEVATOR FAILURE

We have now developed a model, highly idealized to be sure, to predict the ultimate bearing capacity of a shallow foundation on a uniform mass of soil. It is, therefore, appropriate for us to reconsider the Transcona grain elevator failure at this point and see if we have developed sufficient information to predict the failure which was observed.

According to Peck and Bryant (1953), the average pressure on the base of the 77 by 195 ft foundation mat was 3.06 tsf. They found that the average unconfined compressive strength of the upper stratum beneath the foundation was 1.13 tsf, that of the lower stratum was 0.65 tsf, and the weighted average was 0.93 tsf. Because of the relatively rapid failure, it seems reasonable to suppose that the "immediate" strength, that is, $c = \frac{1}{2}q_u$, where q_u is the unconfined compressive strength, and $\phi = 0$, is appropriately applied in this case. Thus Equation 10–20 becomes

$$q = \lambda_c d_c c N_{c_{\text{strip}}} + \lambda_q d_q \gamma D_f N_{q_{\text{strip}}} \qquad (10\text{--}25)$$

where

$$N_{c_{\text{strip}}} = 5.14, \ N_{c_{\text{square}}} = 6.16, \ N_{q_{\text{strip}}} = N_{q_{\text{square}}} = 1$$

$$\lambda_c = 1 + \frac{77}{195}\left(\frac{6.16}{5.14} - 1\right) = 1.08$$

$$d_c = 1 + 0.2\left(\frac{12}{77}\right) = 1.03$$

$$\lambda_q = d_q = 1$$

At the time the test boring, Figure 10.5, was taken, the groundwater level was found to be approximately at the foundation level (Peck and Bryant, 1953). Thus we can assume $\gamma = \gamma_m$ above the foundation level. Assuming $\gamma = 120$ pcf $= 0.06$ tcf,

$$q = (1.08)(1.03)\left(\frac{0.93}{2}\right)(5.14) + (1)(1)(0.06)(12)(1)$$

$$= 2.66 + 0.72$$

$$= \underline{3.38} \text{ tsf}$$

This compares reasonably well with the actual failure load of 3.06 tsf.

A model which may approximate the case we are considering, even more closely than that described above, has been developed by Button (1953). He considered the case of a long foundation placed upon a stratified system, consisting of two layers of purely cohesive material, illustrated in Figure 10.21. For a purely cohesive material, for which $\phi = 0$, the curved part of the failure surface is circular. Button chose to assume that the *entire* failure surface could be approximated by the arc of a circle. He then found that circular arc which produced the minimum bearing capacity for various combinations of depth of the top layer to the foundation width, d/B, and ratios of shear strength of the bottom layer to that of the top layer, c_2/c_1. The results of his study are given in Figure 10.21. The circular arc assumption leads to $N_c = 5.4$ for a homogeneous material ($c_2/c_1 = 1$), approximately five per cent above that for the more realistic assumption of a composite surface.

In the case of the Transcona grain elevator, it seems reasonable to suppose that the top layer may be considered as those strata between the base of the foundation at a depth of 12 ft, and a depth of 28 ft (see Figure 10.5). As indicated above, the average compressive strength of this layer is approximately 1.13 tsf, and that of the lower layer below a depth of 28 ft, is 0.65 tsf. Thus, the ratio, $c_2/c_1 = 0.575$. The depth of the upper stratum beneath the foundation is 16 ft; the foundation width is 77 ft, which gives a d/B ratio of 0.21. Thus, the bearing capacity becomes

$$q = \lambda_c d_c c_1 N_c' + \lambda_q d_q \gamma D_f N_{q_{\text{strip}}} \tag{10--26}$$

in which, from Figure 10.21, $N_c' = 3.5$, and $c_1 = 0.565$ tsf, and the other parameters are defined in connection with Equation 10-25. Substituting in Equation 10-26 gives

$$q = (1.08)(1.03)(0.565)(3.5) + (1)(1)(0.06)(12)(1)$$

$$= 2.2 + 0.72$$

$$= \underline{2.92} \text{ tsf}$$

which can be compared to the actual failure load of 3.06 tsf. The difference between the actual and calculated values is due, at least in part, to the fact that the shape of actual failure surface was probably not circular, but was restricted by the bedrock.

In the case of this relatively simple subsurface condition, our ability to predict the failure of the Transcona grain elevator is quite good. For stratified soils with internal friction, or for more complex inhomogeneities, there is no theoretically valid method for determining the bearing capacity. However, in the usual case of a footing placed upon a layer of stronger soil overlying a less competent material, it is customary to divide the analysis into two separate problems:

1. The stronger material upon which the foundation rests is assumed to extend to an infinite depth. The bearing capacity of the material is computed on this basis.

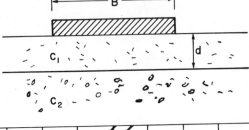

Fig. 10.21—Bearing capacity for a stratified deposit with a constant shear strength c. (After Button, 1953.)

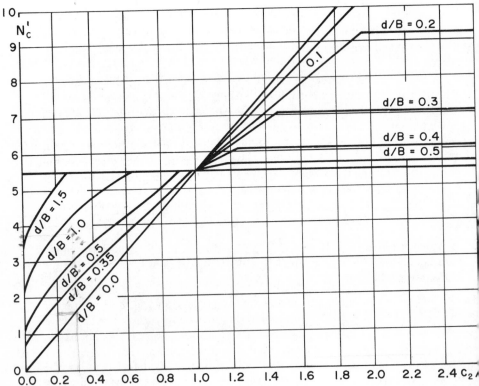

2. The average vertical stress transmitted from the foundation to the surface of the weaker material is determined by an approximate method such as that illustrated in Figure 4.10. The surface of the underlying material is then considered to be the base of an equivalent foundation with dimensions determined from Figure 4.10 and carrying a unit pressure equal to that transmitted from the foundation in addition to the weight of the overburden material above the equivalent foundation.

The smaller of the two values computed is determined to be the bearing capacity of the foundation.

10.6 FACTOR OF SAFETY

Foundations are usually designed to support a load which is a small proportion of the load required to bring about failure. The parameter which is used to compare the applied load with the bearing capacity is the *factor of safety* (FS):*

$$FS = \frac{\text{Total bearing capacity}}{\text{Applied total load}} = \frac{Q}{Q_{applied}} \qquad (10-27)$$

This parameter indicates the proportion of the failure load which is actually applied. It is perhaps inappropriately named because there is not any way presently available to relate quantitatively the factor of safety with the likelihood of failure. We know that when FS = 1 the foundation has failed. When FS = 3, the likelihood of failure will be less than if FS = 2 at the same site. However we have no way of knowing *how much* safer is a factor of safety of 3 than a factor of safety of 2. As we shall see in subsequent chapters the acceptable value of factor of safety is different in different problems. That is, acceptable factors of safety have been determined by experience. In the case of shallow foundations, a factor of safety of 3 when considering dead load plus time-averaged live load or a factor of safety of 2.5 when considering dead load plus maximum live load is used. It is customary to compute both of these values and use the more conservative in a particular case.

10.7 IMPORTANCE OF PROPER SUBSURFACE EXPLORATION

It is interesting to note that the possibility of foundation failure of the Transcona elevator was considered by the engineers concerned with construction. They conducted some plate loading tests in the excavation for the foundation. Although a detailed description of these tests is not available, it is likely that they were conducted upon 1-ft-diameter plates. These

---

* This is defined in a manner analogous to that for failure of a retaining structure with respect to sliding in Section 3.6.

tests indicated a bearing capacity of approximately 4 to 5 tsf. Because the soil in that general zone was desiccated and somewhat stiffer than the material underlying it, this was a reasonable bearing capacity for such a plate. However, the depth of the failure zone in these tests extended only a few feet into the uppermost part of the strata extending to a depth of about 17 ft below the ground surface. Thus, the bearing capacity measurement was not relevant to the determination of that required for a structure of the dimensions of the Transcona grain elevator. This difference in scale is illustrated in Figure 10.22. In this figure the failure zones corresponding to a 1-ft bearing plate and a 77-ft mat are sketched to scale. It is obvious that the failure zone beneath the smaller plate will not even contain the most significant materials with respect to failure of the larger area. This illustrates the importance of a subsurface exploration program which is consistent with the size and scope of the structure.

10.8   SUMMARY OF KEY POINTS

Summarized below are the key points which we have discussed in this chapter. The section in which they were presented is noted so that reference can be made to the original more detailed treatment:

1. A foundation is that part of a structure whose primary purpose is to transmit structural loads to the earth. Satisfactory performance of this function requires that the foundation (Section 10.1):
   a. Must be safe against catastrophic failure (bearing capacity),
   b. Must not experience excessive displacements, and
   c. Must be economically feasible.

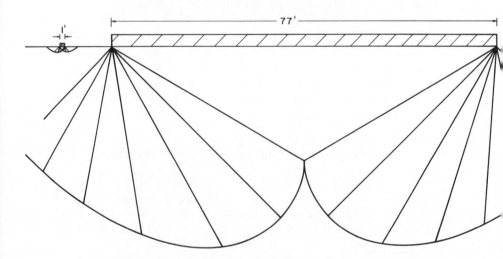

Fig. 10.22—Effect of footing size on scale of failure zone

These requirements must be met considering a variety of possible adverse environmental factors, including:

   a. Frost action,
   b. Shrinking or swelling soils,
   c. Adjacent structures, excavations, and property lines,
   d. Unfavorable groundwater conditions,
   e. Underground defects,
   f. Potential earthquake activity,
   g. Scour and wave action.

2. Design of a foundation involves the following iterative process (Section 10.2):

   a. Establish the scope of the problem to be considered,
   b. Investigate conditions at the site,
   c. Formulate trial design,
   d. Establish a model (or models) to be analyzed,
   e. Determine parameters of the model(s),
   f. Carry out the analysis,
   g. Compare results for one trial design with those of other trials,
   h. Modify the design,
   i. Observe construction.

3. There are two basic types of foundations (Section 10.3):

   a. Shallow foundations, which transmit the structural loads to the earth at a depth close to the base of the structure. Shallow foundation types include:
      (1) Wall or strip footings,
      (2) Spread footings,
      (3) Strap or pump handle footings,
      (4) Combined footings,
      (5) Mat or raft foundations.
   b. Deep foundations, which transmit at least part of the load to a significant depth below the base of the structure. The usual function of a deep foundation is to bypass weak or undesirable subsurface materials. The choice of foundation type is generally dictated by economic considerations.

4. a. Evaluation of the bearing capacity of a shallow foundation is based upon an idealized model which incorporates the following assumptions about the nature of the foundation material (Section 10.4):
      (1) The material is rigid–perfectly plastic,
      (2) The Mohr-Coulomb failure criterion applies to the material,
      (3) The material is weightless,
      (4) Plane strain conditions apply,
      (5) The boundary stress is uniform in magnitude and applied normal to the boundary surface.

   Using two special cases for which solutions, subject to the above assumptions, have been obtained, Prandtl (1920, 1921) was able to determine the magnitude of uniform surface strip loading required to produce failure of such a material (Equations 10–11).

b. Prandtl's solution was modified empirically to include the effect of soil weight (Equations 10–17).

c. Several additional empirical modifications to the bearing capacity formula have been made to account for (Equations 10–18):
(1) Foundation shape,
(2) Strength of the soil above the foundation level,
(3) Inclination of the resultant load on the foundation,
(4) Eccentricity of the resultant load.
The latter two effects may reinforce or partly compensate each other depending upon the arrangement of the resultant forces on the foundation (Section 10.4).

d. The magnitude of ϕ' to be used in the bearing capacity formula for cohesionless soils is commonly inferred from the standard penetration test (Section 10.4).

e. As in the case of all problems concerning soil strength, the influence of groundwater conditions on the result must be considered.

f. When the foundation consists of two cohesive strata of significantly different strength, the simplified approached developed by Button (1953) is useful (Figure 10.21, Section 10.5).

5. Foundations are designed with a *factor of safety* against bearing capacity failure. The significance of a given magnitude of factor of safety is not clear, but acceptable values for specific problems have been established by experience.

6. The applicability of small-scale loading tests to prediction of the bearing capacity of large foundations cannot be assumed *a priori*. It is necessary first to insure that the subsurface materials involved in the potential failure in the prototype foundation are similarly related to the failure of the load test.

PROBLEMS

10.1.

(a) For the case of the 5×5 ft footing plot-the bearing capacity as a function of depth of embedment from 0 to 5 ft for two soils ($\gamma = 100$ pcf):
1. $c = 1000$ psf, $\phi = 0$
2. $c' = 0$ $\phi' = 30°$

(b) For the same two soils, plot the effect of square-footing size on bearing capacity at the surface for $B = 1$ ft to $B = 10$ ft.

(c) For the same two soils, and $B = 3$ ft, plot the effect on bearing capacity at the surface of the footing length, from $L = 3$ ft to $L = 20$ ft.

(d) On the basis of these results, what general statements can you make concerning the bearing capacity of concentrically loaded shallow foundations?

10.2. The point of application and direction for the load applied to a 10×10 ft footing is as shown. Determine the load which can be applied to this footing. (*Note*: Ignore weight of the footing.)

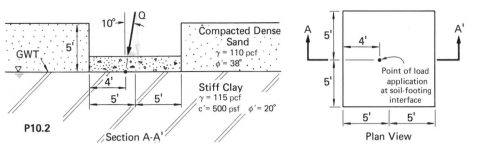

P10.2

Section A-A'

Plan View

10.3. The electric power transmission tower shown in schematic is to be constructed on a foundation consisting of four individual 10 × 10 ft concrete footings. The resultant forces on the tower are shown in the sketch. The resultant horizontal force includes wind loads and the effects of a broken

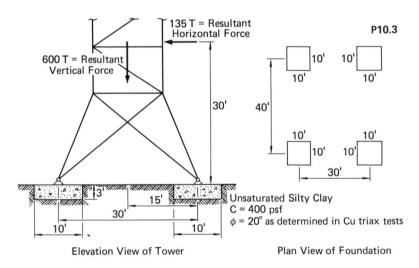

Elevation View of Tower

Plan View of Foundation

power cable on one side of the tower. The horizontal load could act in either direction (left or right in the sketch) on the tower. The resultant vertical force includes the weight of the tower (without foundation), as well as cable, ice, and snow loads.

The subsoil consists of a relatively deep deposit of unsaturated silty clay; the groundwater table is below the zone of interest.

Determine the factor of safety against a bearing-capacity failure. For this purpose, you may neglect the effect of the footing thickness on the point of application of the resultant load on the soil, and any lateral resistance on the side of the footing. Concrete weighs 150 pcf.

10.4. Plot the ultimate load Q that can be carried by the square footing shown as a function of D_f for $0 \leqslant D_f \leqslant 10$ ft.

10.5. It has been determined that the proposed oil-storage tank shown below cannot be filled completely without danger of a bearing-capacity failure.

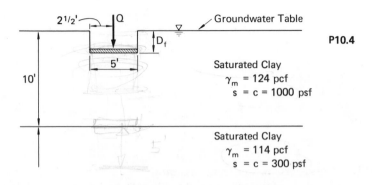

P10.4

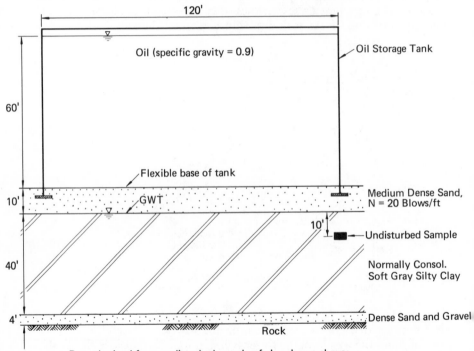

Data obtained from undisturbed sample of clay shown above:
$\gamma_m = 102.4$ pcf   e = 1.20
Unconfined Compressive Strength $q_u = 800$ psf
$C_c = 0.40$, $C_e = 0.02$, $c_v = 40$ ft²/yr

P10.5

(a) To what depth *can* it be filled without bearing-capacity failure?
(b) If the tank were constructed and *actually* filled until failure of the foundation soil was evident, how would you expect the depth of oil producing the failure to compare to that which you calculated in part a? Justify your answer briefly.

Note: Weight of tank may be neglected.

11

Shallow Foundations; Settlements

11.1 INTRODUCTION

Even though a foundation may be safe against a bearing capacity failure, we recognize that the function or value of a structure may be impaired as the result of displacement or distortion of the structure stemming from foundation settlements.

One of the most famous examples of failure (and success) resulting from settlement of a structure is the Leaning Tower of Pisa. The tower and schematic subsurface profile are shown in Figure 11.1. The tower is founded on a massive ring foundation approximately 20 m in diameter with a central hole approximately 5 m in diameter, resting in a layer of sand overlying a deep soft cohesive material. The figure shows the distribution of vertical stress (circa 1934) as estimated by Terzaghi (1934).

Figure 11.2 shows the loading history and time-differential settlement diagram for the structure. The total average settlement is approximately 2.4 m (8 ft), with the north edge having settled 3.2 m and the south edge 1.6 m. Note that the construction occurred in several stages. This was found necessary because of the large and relatively rapid differential settlements which occurred almost immediately after construction had commenced. It is evident that in the case of this structure, tilting has become so serious that the increase of stress on the subsoil below the lower part of the foundation has influenced the rate of settlement. It is probably fortunate for the town of Pisa that engineers in the twelfth century did not possess our current understanding of compressibility phenomena, because the town would now contain a little-known tower which might attract only a few tourists each year. Nonetheless, we can hardly count upon such a happy combination of circumstances in everyday design.

Fig. 11.1—Stress distribution in soil, the leaning tower of Pisa. (After Terzaghi, 1934 and Jumikis, 1962.)

(a) Vertical section

(b) Plan of base of footing

(c) Linear distribution of contact pressure on soil

(d) Curvilinear pressure distribution on top surface of clay, c–c 8.0 m below the base of footing.

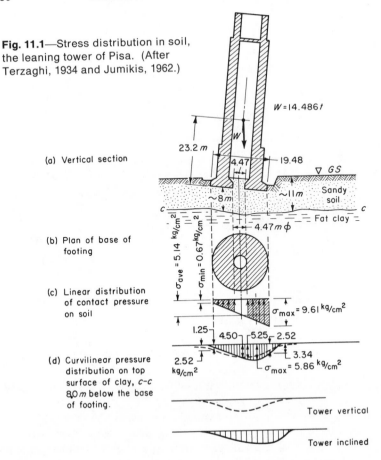

Such settlements result from the response of the underlying earth materials to the stresses induced by the structure, even when the stresses are significantly less than those required to produce failure. In Chapter 5 we saw that settlements at the surface of saturated cohesive soils are usually evaluated as the sum of a variety of factors which are considered separately and superimposed (Equation 5–1):

$$\rho = \rho_d + \rho_c + \rho_s \qquad (11-1)$$

in which ρ is the total vertical displacement (settlement) at the point in question, ρ_d is the initial *distortion settlement*, ρ_c is the time-dependent settlement due to *consolidation*, and ρ_s is the *secondary compression* effect. In this chapter we shall discuss the way in which the models developed in Chapters 5 and 7 can be applied to the prediction of settlements of foundations on cohesive soils. In addition we shall consider some further development in models for settlement prediction.

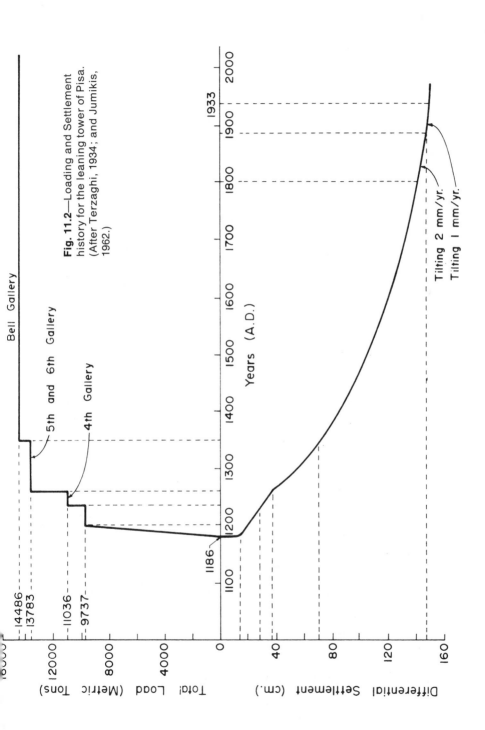

Fig. 11.2—Loading and Settlement history for the leaning tower of Pisa. (After Terzaghi, 1934; and Jumikis, 1962.)

11.2 EFFECT OF SIZE ON SETTLEMENTS OF A LOADED AREA

Elastic Distortion Settlements

We recall from Chapter 5 that the settlement of a loaded area resting on the surface of a linear-elastic half-space is (Equation 5–2)

$$\rho_d = C_s q B \left(\frac{1 - \mu^2}{E} \right) \tag{11–2}$$

in which C_s is a factor which depends upon the shape and rigidity of the loaded area, q is the magnitude of the average stress on the loaded area, B is the width of the loaded area, E is Young's modulus, and μ is Poisson's ratio. That the size of the loaded area may have a pronounced effect on the magnitude of elastic distortion settlements is illustrated in the following examples:

Example 11.1

Consider a sqaure footing, B by B, supporting an individual concentric column load Q. We shall inquire, what is the effect on the elastic distortion settlement of increasing the footing width by a factor n?

Solution. The average stress on the original footing is $q_1 = Q/B^2$. The average stress under the larger footing is $q_2 = Q/n^2 B^2$. Substituting these values into Equation 11–2 gives

$$\rho_{d_1} = C_s \frac{QB}{B^2} \left(\frac{1 - \mu^2}{E} \right), \qquad \rho_{d_2} = C_s \frac{QnB}{(nB)^2} \left(\frac{1 - \mu^2}{E} \right) \tag{11–3}$$

The ratio of the two settlements is

$$\frac{\rho_{d_2}}{\rho_{d_1}} = \frac{\dfrac{Q}{nB}}{\dfrac{Q}{B}} = \frac{1}{n} \tag{11–4}$$

Equation 11–4 shows that increasing the size of a square footing supporting a given *total* load leads to a reduction in elastic settlement in proportion to the increased linear dimension.

Example 11.2

Consider now a strip footing, with constant load Q per unit length of wall. We shall investigate the effect on elastic settlement of increasing the footing width to reduce the applied stress on the soil.

Solution. In this case $q_1 = Q/B$ and $q_2 = Q/nB$. Then,

$$\frac{\rho_{d_2}}{\rho_{d_1}} = \frac{\dfrac{Q}{nB} \cdot nB}{\dfrac{Q}{B} \cdot B} = 1 \tag{11–5}$$

Thus we observe that increasing the width of a strip footing to reduce the unit pressure under a wall which supports a constant *load* per unit length does *not* reduce the elastic settlement.

Example 11.3

We shall now investigate the elastic distortion settlement of two square footings of width B and nB respectively, which are subjected to the same *unit pressure, q*.

Solution. In this case, of course, the total load carried by each of the footings is not the same, but the average pressure is. Thus the ratio of settlements is

$$\frac{\rho_{d_2}}{\rho_{d_1}} = \frac{qnB}{qB} = n \qquad (11\text{--}6)$$

That is, the larger footing will settle n times as much as the smaller footing. This clearly illustrates the incorrectness of the common misconception that different size footings subjected to the same pressure will have equal settlements. We shall see below that a similar conclusion can be drawn with regard to settlements arising from compressibility of the soil.

Consolidation Settlement

As noted in Chapter 5, time-dependent consolidation settlements arise from the compressibility of the soil skeleton and, therefore, are likely to be significant in general only for cohesive soils. In order to appreciate the effect of footing size on such settlements we shall consider a simplified analysis of two footings of width B and nB, respectively, resting on the surface of a saturated cohesive medium, illustrated schematically in Figure 11.3. Also shown in the figure are the pressure bulbs underneath the two footings, as represented by a suitable isobar. Identified in the figure are points ① and ②, which are assumed to be located at similar points within each pressure bulb and which can be considered the "average" point; that is, it is assumed that the footing settlement can be calculated by multiplying the height of the pressure bulb by the vertical strain at this average point. If we assume that—

1. compression is approximately one-dimensional,
2. the magnitudes of the stresses involved lie wholly within either the recompression or virgin compression range,
3. the initial void ratios of the soil at both points ① and ② are the same, and
4. the groundwater table is at the surface,

then the compression settlements may be written as

$$\rho_{c_1} \propto \log\left(\frac{\Delta q_1 + \sigma'_0}{\sigma'_0}\right) H, \qquad \rho_{c_2} \propto \log\left(\frac{\Delta q_2 + n\sigma'_0}{n\sigma'_0}\right) nH \quad (11\text{--}7)$$

in which ρ_{c_1} and ρ_{c_2} are the consolidation settlements of footings of width B and nB respectively, Δq_1 and Δq_2 are the increase in vertical effective stress

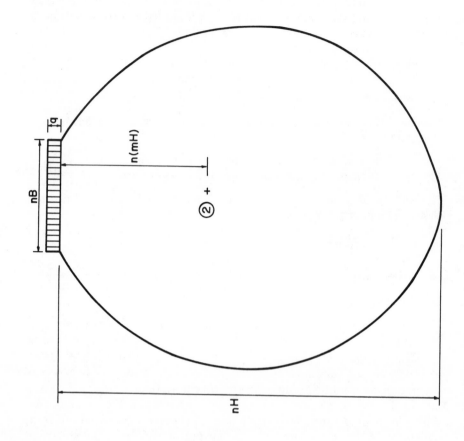

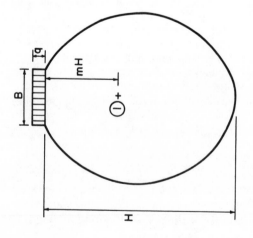

Fig. 11.3—Pressure bulbs under two loaded areas.

at points ① and ② respectively, σ'_0 and $n\sigma'_0$ are the initial overburden stress at points ① and ② respectively, and H and nH are the respective heights of the pressure bulbs corresponding to the footings of width B and nB. The ratio of the two settlements is then

$$\frac{\rho_{c2}}{\rho_{c1}} = \frac{\log\left(\dfrac{\Delta q_2 + n\sigma'_0}{n\sigma'_0}\right)}{\log\left(\dfrac{\Delta q_1 + \sigma'_0}{\sigma'_0}\right)}\, n \qquad (11\text{–}8)$$

Consolidation settlements for the three cases discussed above for elastic distortion settlement are described in the following examples:

Example 11.4

We consider first a square footing, B by B, supporting a given total load Q, and shall investigate the effect on the consolidation settlement of increasing the footing width by a factor n.

Solution. The average stress on the original footing is $q_1 = Q/B^2$. The average stress under the larger footing is $q_2 = Q/n^2B^2$. Because points ① and ② are at similar locations within their respective pressure bulbs, the ratio of the increase in stress at those points will be

$$\frac{\Delta q_2}{\Delta q_1} = \frac{q_2}{q_1} = \frac{1}{n^2} \qquad (11\text{–}9)$$

Hence from Equation 11–8 the ratio of the settlements is

$$\frac{\rho_{c2}}{\rho_{c1}} = \frac{\log\left(\dfrac{\Delta q_1}{n^3\sigma'_0} + 1\right)}{\log\left(\dfrac{\Delta q_1}{\sigma'_0} + 1\right)}\, n \qquad (11\text{–}10)$$

Equation 11–10 shows that the relative settlement of the two footings depends not only upon their relative sizes, but also upon the magnitude of the applied stress in relation to the existing overburden pressure, the *pressure-increment ratio*. Figure 11.4a shows a plot of Equation 11–10 for three values of the pressure-increment ratio. It is evident from this figure that over a wide range of pressure-increment ratios, an increase in the footing size supporting a constant total load results in a marked reduction in the consolidation settlement. The effect is somewhat more pronounced in the case of small pressure-increment ratios.

Example 11.5

We now investigate a strip footing with constant load Q per unit length of the footing, and the influence on consolidation settlements of increasing the footing width by a factor n.

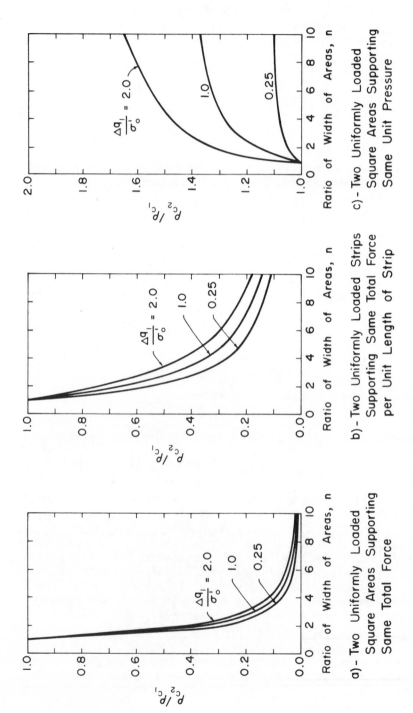

Fig. 11.4—Influence of size of loaded area on settlements due to compressibility of subsoil (approximate analysis).

Solution. In this case $q_1 = Q/B$ and $q_2 = Q/nB$. The ratio of the pressure increments at points ①and② is

$$\frac{\Delta q_2}{\Delta q_1} = \frac{1}{n} \tag{11-11}$$

from which

$$\frac{\rho_{c_2}}{\rho_{c_1}} = \frac{\log\left(\dfrac{\Delta q_1}{n^2 \sigma_0'} + 1\right)}{\log\left(\dfrac{\Delta q_1}{\sigma_0'} + 1\right)}\, n \tag{11-12}$$

Equation 11–12 is plotted in Figure 11.4b for a range of pressure-increment ratios. Again it is evident that in the case of consolidation settlements the effect of increasing footing width is to reduce the consolidation settlement, and this effect is more pronounced for smaller pressure-increment ratios.

Example 11.6

Finally, we shall consider the consolidation settlement of two square footings of width B and nB respectively, which are subjected to the *same unit pressure, q.* We note again that the total load carried by each footing in this case is not the same.

Solution. The pressure increments at points ①and② will be the same in this case and

$$\frac{\rho_{c_2}}{\rho_{c_1}} = \frac{\log\left(\dfrac{\Delta q_1}{n \sigma_0'} + 1\right)}{\log\left(\dfrac{\Delta q_1}{\sigma_0'} + 1\right)}\, n \tag{11-13}$$

This expression is plotted in Figure 11.4c. We note here that an increase in the footing size produces an increase in the settlement due to one-dimensional compression, and that the effect is more important as the pressure-increment ratio increases. We see, therefore, that both elastic and consolidation settlements will increase as footing size increases if the footing sustains the same unit pressure.

The foregoing discussion pertains to the case in which the compressible material extends from the surface down. In the important case in which the compressible stratum is located at considerable depth below the foundation, the conclusions concerning the effect of foundation size are different. For if the seat of the settlement is sufficiently far below the foundation, then the distribution of load on the foundation has a negligible effect on the settlement; it is only the total load which is of significance. This is illustrated in Figure A.2 which shows the distribution of vertical normal stress under the center of a circular loaded area on an elastic half-space. We note that at a depth of 3 radii below the loaded area, the vertical stress is independent of the distribution of the load even if the load is assumed to be concentrated at a point! This is illustrated further in the following example.

Example 11.7 ————————————————————————————

A structure supported by nine columns spaced at a distance $2B$ on centers carry a load Q each, as shown in plan in Figure 11.5a. The foundation is to be placed at the surface of a soil profile in which a compressible layer is found, with its midheight at a depth $4B$ below the foundation. In order to estimate settlements of the structure due to the compressibility of the clay, we wish to determine the distribution of vertical stress

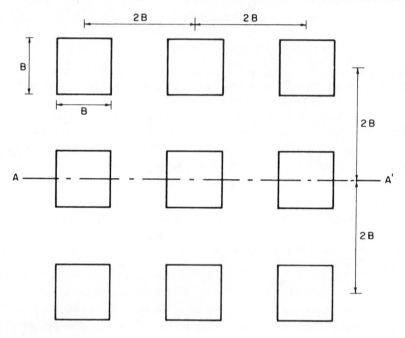

a) - Arrangement of Footings

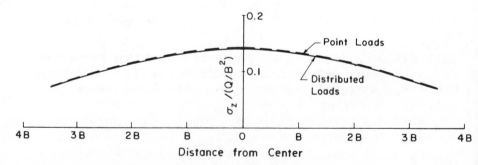

b) - Vertical Stresses Along Section A-A' at Depth of 4 B

Fig. 11.5—Effect of footing size on vertical stresses below foundation for Example 11.7.

along section $A - A'$ at the center of the clay, assuming—

    (a) Each column load is distributed uniformly over a $B \times B$ footing.
    (b) Each column load is applied as a point load.

Solution. (a) The distribution of vertical stress at a depth of $4B$ feet below the building along the line indicated was calculated for the nine distributed loads using Figure A.5. Results are shown, in Figure 11.5b, as the solid curve, in which the stress is given as the ratio of the calculated stress to Q/B^2.

    (b) The vertical stress due to the building load assumed to consist of nine point loads each of magnitude Q was determined using Figure A.1 as described in Example A.1. The resulting vertical stress distribution is shown in Figure 11.5b as the dashed line.

Thus we see that, if the point loads were located at the columns in this particular structure, the vertical stress distribution in the compressible stratum, and therefore its settlement, would have been the same irrespective of the size of the individual footings under those columns. And in general, when a compressible material is located at considerable depth below a shallow foundation, settlements in the compressible stratum will be nearly independent of the size of the individual footings.

11.3 DISTORTION SETTLEMENT OF FOOTINGS ON SAND

Factors Involved in Size Effects

Cohesionless masses cannot even be approximated as homogeneous linear elastic materials because, among other things, the stiffness varies as a function of depth. Current expressions relating the settlement of spread foundations on sand to the size of the foundation are empirical. One such expression which relates the settlement of a prototype footing to that in a plate loading test is (Sowers, 1962):

$$\rho_{d_f} = \left[\frac{B_f(B_p + 1)}{B_p(B_f + 1)} \right]^2 \rho_{d_p} \qquad (11\text{--}14)$$

in which ρ_{d_f} and ρ_{d_p} are the distortion settlements of the footing and plate of width B_f and B_p (in feet) respectively. For the case of a 1-ft-wide plate, this expression reduces to that given by Terzaghi and Peck (1967):

$$\rho_{d_f} = \left(\frac{2B_f}{B_f + 1} \right)^2 \rho_{d_p} \qquad (11\text{--}15)$$

Equation 11–15 is shown as the solid line in Figure 11.6. This relation is often used to extrapolate the settlement of small scale plates to those for prototype foundations.

    Also shown in the figure are a variety of field data, including case records from fourteen European sites assembled by Bjerrum and Eggstead (1963)

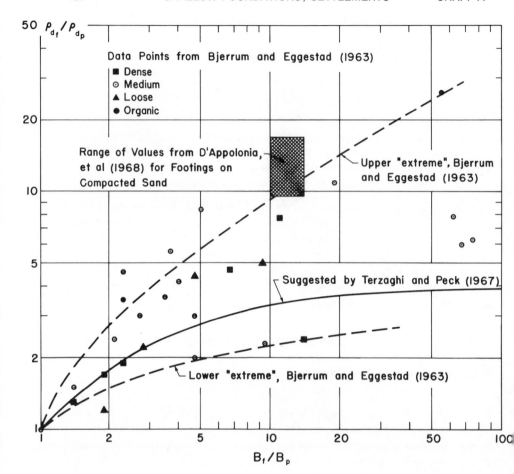

Fig. 11.6—Effect of size of loaded area on distortion settlement on sand as indicated by field data.

and measurements made on a dune sand deposit by D'Appolonia et al. (1968). The data from Bjerrum and Eggstead (1963) include collected results for foundations and load tests on sands with densities varying from loose to very dense, grain sizes ranging from fine to coarse and sandy gravel. The two dashed lines shown represent approximate "extremes" interpreted by Bjerrum and Eggstead (1963). No consistent trend is observable in the data on the basis of the relative density of the sand. That is, as many points near the upper extreme correspond to dense sand as near the lower extreme. The implication of these results is that current methods for extrapolating the results of a loading test on a small scale plate to estimate the settlement of a foundation on cohesionless material are inadequate.

In order to appreciate some of the reasons for this, and to acquire a rationale for interpreting the results of plate loading tests, we shall consider an approximate analysis of the problem. As in the case for the preceding discussion of size effects on one-dimensional consolidation settlements, the approximate analysis serves primarily as a guide to understanding of the relative significance of the phenomena involved.

We begin by considering a simplified description of two footings of width B and nB resting on the surface of a homogeneous cohesionless medium, as illustrated schematically in Figure 11.3. As in the case of the consolidation settlement analysis we assume that points ① and ② can be considered "average" points in the sense that the footing settlement can be calculated by multiplying the height of the pressure bulb by the vertical strain at this average point. Thus

$$\rho_{d_1} = \varepsilon_{z_1} H, \qquad \rho_{d_2} = \varepsilon_{z_2} nH \qquad (11\text{--}16)$$

in which H is the height of the pressure bulb under footing ① as represented by a suitable isobar. Thus we can focus our attention upon determination of the vertical strain at the average point. We then investigate the factors which influence the magnitude of the *increment* in vertical strain of a small element of a cohesionless mass in response to a small increment in the applied stresses. For purposes of discussion it is useful to decompose the state of stress into two components: the mean effective stress and the octahedral shear stress (recall Equations 8–8 and 8–9). Expressed in terms of principal stresses, the mean effective stress σ'_m is

$$\sigma'_m = \tfrac{1}{3}(\sigma'_1 + \sigma'_2 + \sigma'_3) \qquad (11\text{--}17)$$

and the octahedral shear stress τ_{oct} is

$$\tau_{oct} = \tfrac{1}{3}[(\sigma_1 - \sigma_2)^2 + (\sigma_2 - \sigma_3)^2 + (\sigma_3 - \sigma_1)^2]^{\frac{1}{2}} \qquad (11\text{--}18)$$

The vertical strain increment in response to increments in the applied stresses $d\sigma'_m$ and $d\tau_{oct}$ is determined by the magnitude of those increments and the resistance, or stiffness, of the soil mass. For a given soil, this stiffness depends upon the state of compaction (that is, the relative density), and the total magnitude of σ'_m and τ_{oct} (Domaschuk and Wade, 1969). We then make these further assumptions:

1. An infinitesimal stress increment produces a corresponding infinitesimal strain increment proportional to its magnitude.
2. The stiffness of the soil mass is proportional to the mean effective stress and independent of the magnitude of the octahedral shear stress. This is equivalent to assuming that the bulk and shear moduli are proportional to the mean stress and independent of the octahedral shear stress. The investigation by Domaschuk and Wade

(1969) indicates that the shear modulus can be considered approximately independent of the octahedral shear stress up to a stress magnitude of 40 per cent of that required to produce failure. Thus this latter assumption appears reasonable. Although they found that the bulk modulus was related to the mean stress in a nonlinear fashion, we shall make the linearized assumption as a first approximation.

3. The distribution of stresses within the soil mass, resulting from the applied footing loads at the surface of the mass, can be calculated using the theory of elasticity, assuming that the footing loads are represented by a uniformly distributed vertical stress at the surface. We recall from the discussion in Chapter 4 that, although this assumption is patently incorrect, the discrepancies between the calculated stresses and those which would be determined by a more realistic assumption may not be large.

From these assumptions, the increment in mean effective stress and octahedral shear stress produced by a small increment of vertical stress dq, applied at the surface of the mass, is proportional to the magnitude of that increment. Consequently the increment in vertical strain may be written

$$d\varepsilon_z \propto \frac{dq}{\sigma'_m} \qquad (11\text{-}19)$$

in which the constant of proportionality depends upon the relative density of the soil and the position selected for the "average" point. The mean effective stress can be written as

$$\sigma'_m = \sigma'_{mo} + \sigma'_{m_q} \qquad (11\text{-}20)$$

in which σ'_{mo} is the mean effective stress due to the *in situ* overburden pressure prior to application of the footing load and σ'_{m_q} is the mean stress due to the total magnitude of applied footing load prior to application of the small increment dq. From the foregoing assumptions, the mean effective stress due to the applied footing load is proportional to it:

$$\sigma'_{m_q} = aq \qquad (11\text{-}21)$$

in which a is a constant of porportionality. Thus expression 11–19 becomes

$$d\varepsilon_z \propto \frac{dq}{\sigma'_{mo} + aq} \qquad (11\text{-}22)$$

Integrating this expression with $\varepsilon_z = 0$ for $q = 0$ gives

$$d\varepsilon_z \propto \log\left(\frac{aq}{\sigma'_{mo}} + 1\right) \qquad (11\text{-}23)$$

To determine the ratio of vertical settlement of two footings of width B and

nB subjected to the same magnitude of unit load, we substitute expression 11–23 into Equations 11–16 to obtain

$$\frac{\rho_{d_2}}{\rho_{d_1}} = \frac{\log\left(\dfrac{aq}{\sigma'_{mo_1}} + 1\right)}{\log\left(\dfrac{aq}{\sigma'_{mo_2}} + 1\right)} \, n \tag{11-24}$$

So we see that for a homogeneous cohesionless stratum at approximately constant relative density, the effect of size of a loaded area on settlements is related to the initial stress state prior to load application and the magnitude of the applied load, as well as the footing size.

The initial mean stress can be expressed as

$$\sigma'_{mo} = \tfrac{1}{3}(\sigma'_{xo} + \sigma'_{yo} + \sigma'_{zo}) = \tfrac{1}{3}(2K_0 + 1)\sigma'_{zo} \tag{11-25}$$

in which K_0 is the coefficient of earth pressure at rest. We recall from the discussion in Section 3.3 that K_0 for normally consolidated sands is a function of the same properties which determine its angle of shearing resistance, and further depends to a marked degree upon the stress history (recall Figure 3.17).

Using our approximate analysis, we investigate the effect of loading history and footing size on the relative magnitude of settlements of footings on sand in the following example.

Example 11.8 ————————————————————————————

Consider a square footing at the surface of a homogeneous dry sand mass. Determine the effect of footing size on settlement, relative to that of a 1-ft-square plate, for applied uniform pressures of 1000 and 4000 psf. Assuming $\gamma = 100$ pcf, investigate three cases:

(a) The sand is normally consolidated;
(b) The sand was preloaded by 20 ft of overburden which had been removed at some time in the past;
(c) The sand was preloaded by 40 ft of overburden which had been removed at some time in the past.

Solution. If we consider the "average" point to be at a depth of $B/2$ below the centerline of the footing and, for purposes of the stress calculation that $\mu = 0.4$ for the sand, the proportionality coefficient a is 0.0189.* The mean stress at the average point below the 1-ft-square footing is then

$$\sigma'_{mo_1} = \tfrac{1}{3}(1 + 2K_{0_1})\gamma\,\frac{B_1}{2} = 16.7(1 + 2K_{0_1}) \tag{11-26}$$

* This was obtained using values tabulated by Ahlvin and Ulery (1962) for an equivalent circular uniformly loaded area. The equivalent circle was determined as in Example A.1.

and that below the n-ft-square footing is

$$\sigma'_{mo_2} = 16.7n(1 + 2K_{0_2}) \qquad (11-27)$$

Substituting these values into Equation 11–24, the relative settlement is

$$\frac{\rho_{d_2}}{\rho_{d_1}} = \frac{\log \dfrac{0.0189q}{(1 + 2K_{0_2})n} + 1}{\log \dfrac{0.0189q}{1 + 2K_{0_1}} + 1} \, n \qquad (11-28)$$

in which q is expressed in pounds per square foot.

The results for the six cases are shown in Figure 11.7. The figure shows the effect of footing size on the settlement relative to that of a 1-ft-square plate. The solid lines pertain to an average footing pressure of 1000 psf; the dashed lines are for an average footing pressure of 4000 psf.

(a) For the case of the normally consolidated sand, $K_{0_1} = K_{0_2} =$ constant over all depths of interest. The results shown are for $K_0 = 0.4$. They indicate that the relative settlement is larger at larger footing pressure, and that this effect becomes more pronounced as the size of the footing increases. Recall of course, it is assumed that for all pressures considered, the sand beneath the smallest footing is far from a state of failure.

(b) For the preloaded sand, K_0 is a function of depth because the over-consolidation ratio depends upon the depth. Thus, for the one-foot square plate, for which the depth of the "average" element is 0.5 feet, the over-consolidation ratio is 41 and an appropriate magnitude of $K_0 = 2.40$ has been extrapolated from Figure 3.17. As the footing increases in size, the over-consolidation ratio, and therefore K_0 reduces. Using values of K_0 determined from Figure 3.17, the influence of footing size for the two pressures considered is shown in Figure 11.7. As in the case of the normally consolidated sand, the higher average pressure on the footing produces a greater influence of footing size on the relative settlement. We note also that for the same footing pressure, the relative settlements on the over-consolidated sand are greater than for the normally consolidated sand, and the effect increases as the footing size increases.

(c) Results similar to those discussed above are shown in Figure 11.7 for the sand preloaded by 40 ft of overburden. In this case, the over-consolidation ratio at the "average" point beneath the one-foot square plate is 81, which leads to a K_0 magnitude approaching $K_p = 3.69$.

The results of this simplified analysis show that the most important factors which affect the relative settlement of two square footings on a sand mass are their relative size, the magnitude of the applied stress, and the stress history of the sand (because of its influence on the initial *in situ* mean stress). The fact that a high prestress produces a corresponding large magnitude of K_0 in the field is illustrated by measurements by D'Appolonia et al. (1969) of horizontal stresses in sand during and after field compaction. They report values of K_0 exceeding 2.5 at a depth of 2 ft below the surface of the fill. The magnitude of K_0 is undoubtedly greater than this at shallower depths. These measurements were made at the same site for which relative

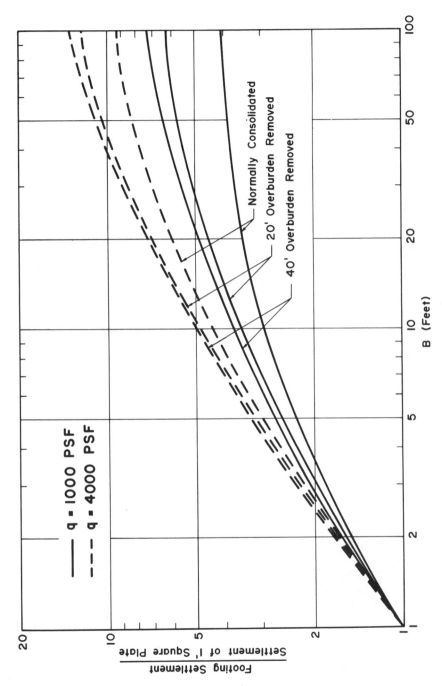

Fig. 11.7—Effect of square footing size on distortion settlement on sand (approximate analysis).

settlements reported by D'Appolonia et al. (1968) are shown in Figure 11.6. The large magnitude of relative settlement shown is reasonable in light of the apparently large prestressing resulting from the compaction process. It is thus evident that the apparent scatter of the field data in Figure 11.6 is to be expected, even if experimental artifacts could be eliminated.

We recognize that although *relative* settlement is larger for the over-consolidated sand, the settlement of the *1-ft* plate will be *less* than that for the *same* plate on the normally consolidated sand. In fact the approximate analysis applied herein indicates that the settlement of the 1-ft plate on normally consolidated sand, due to a 1000-psf applied pressure, would be more than twice the settlement when the sand has been preloaded by 40 ft of overburden.

On the basis of the above discussion, it is suggested that in extrapolating the results of load tests on small plates to predict the settlement of larger footings, consideration be given to the nature of the stress history in the area concerned. If no such information is available, then a relation such as the upper dashed curve in Figure 11.7 may be a more reasonable approximation for footings of conventional dimensions than Equation 11–15.

In those cases for which the loading history of the area is sufficiently well known that an extrapolation of load-test results appears warranted, several other disadvantages may still preclude the use of load tests:

1. Load tests on small plates cannot be extrapolated to the prediction of settlements of large footings if the *seat of the settlement* of the footing is different from that of the plate. This scale effect is shown schematically in Figures 10.22 and 11.3. The difference between large and small scale results may occur because of the presence of strata which intercept the pressure bulb under the footing and not under the plate. The groundwater table may have a similar effect.
2. Capillary tensions in moist unsaturated cohesionless soils influence the results of small scale tests much more than those of prototype footings. Capillarity produces an effect similar to the prestress which results from prior loading. Thus the relative settlement in a deposit which has not experienced external preloading will correspond to that of an over-consolidated deposit because of the prestress induced by capillarity.
3. Because of the natural variability of most granular deposits, a few elaborate plate load tests may reveal less information about average conditions than a much larger number of simpler, albeit less definitive measurements.
4. The relatively high cost of conducting plate loading tests is justified only if the results can lead to savings which are likely to exceed the cost of testing.

An alternative method for estimating the settlement of a 1-ft-square plate on a cohesionless soil is to make use of empirical correlations between the

relationship for the plate and the standard penetration resistance of the soil.*
One such correlation is given by Terzaghi and Peck (1967) and expressed
by Meyerhof (1965) in a form equivalent to

$$\rho_{d_p} = \frac{8q}{N_{40}} \tag{11-29}$$

in which ρ_{d_p} represents the settlement of the 1-ft-square plate *expressed in
inches*, q is the applied pressure *expressed in tsf*, and N_{40} is the standard
penetration resistance corrected to an overburden pressure of 40 psi using
Figure 10.20. Although not indicated by Meyerhof (1965), more recent
evidence (D'Appolonia et al., 1968) suggests that failure to correct the N
value leads to settlement predictions which are too high.

Suggested Design Procedure

It is clear from the discussion above that a practical rational solution
to the problem of predicting settlement of shallow foundations on cohesionless masses is still lacking. However, the method that was proposed by
Schmertmann (1970) leads to results which are compatible with field measurements in a variety of locales, and which provide the engineer with a more
rational basis for evaluating the relative significance of the various factors
involved in the settlement problem. The analysis is based on the following
observations:

1. The distribution of vertical strain within a linear-elastic half-space
    subjected to a uniformly distributed load over some area at the
    surface can be described by

$$\varepsilon_z = \frac{q}{E} I_z \tag{11-30}$$

   in which q is the intensity of the uniformly distributed load, E is
   Young's modulus of the elastic medium, and I_z is a strain influence
   factor depending only upon Poisson's ratio and the location of the
   point for which the strain is being evaluated. The distribution of the
   vertical strain influence factor for a uniformly loaded circular area
   at the surface of an elastic half-space is shown in Figure 11.8a for
   two values of Poisson's ratio.

2. Based upon the results of displacement measurements within sand
    masses loaded by model footings, as well as finite element analyses

---

* Because the standard penetration test is a measure of shearing resistance, it does not reflect
the important effect of prestress on the "stiffness" of the sand to the same degree as settlements.
Despite this, and other serious shortcomings, the test is relatively inexpensive and commonly
used. The best method for applying standard penetration test results to the prediction of
foundation settlements on cohesionless soil remains a subject of continuing controversy and
inquiry (Meyerhof, 1965; D'Appolonia et al., 1968; Holtz and Gibbs, 1969; Peck and Bazaraa,
1969; Bolognesi, 1969).

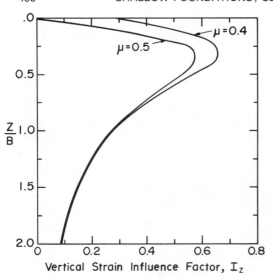

a)- Distribution in Elastic Medium

Fig. 11.8—Vertical strain distribution for settlement of cohesionless masses. (Modified from Schmertmann, 1970.)

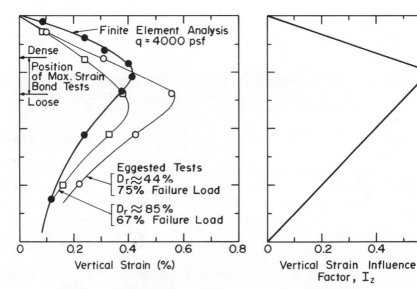

b)- Distribution in Cohesionless Mass
From Model Tests and Analysis

C)- Assumed 2B-0.6
Distribution

of deformations of a nonlinear material with assumed characteristics similar to sand, distribution of strain within such loaded cohesionless masses is very similar in form to that for a linear-elastic medium. Indicated in Figure 11.8b are some typical results of model tests and the finite element analysis reported by Schmertmann (1970). The similarity in shape of these strain distributions to that for the strain distribution in an elastic medium is evident. It appears, however, that the depth at which the maximum vertical strain occurs is somewhat lower, at a depth approximately one-half the foundation width.

On the basis of these observations, Schmertmann (1970) suggested that for practical purposes the distribution of vertical strain within a mass of cohesionless material could be expressed by Equation 11–30 in which Young's modulus might vary in value from point to point, and the strain influence factor could be approximated by a triangle, as shown in Figure 11.8c. Using the theory of elasticity as a guide, the maximum value of the strain influence factor is 0.6; on the basis of the model tests and analysis of cohesionless masses the depth of the maximum I_z is assumed to be at $z/B = 0.5$, and the magnitude of $I_z = 0$ at a depth $z/B = 2$. Schmertmann (1970) refers to this as a "2B–0.6 distribution".

The distortion settlement, ρ_d, is the integration of the strains,

$$\rho_d = \int_{z=0}^{\infty} \varepsilon_z \, dz \qquad (11\text{--}31a)$$

which may be approximated by

$$\rho_d = q \int_0^{2B} \frac{I_z}{E} \, dz \qquad (11\text{--}31b)$$

We can approximate this more usefully as a summation of settlements of convenient approximately homogeneous layers,

$$\rho_d = C_1 C_2 q \sum_{i=1}^{n} \left(\frac{I_z}{E}\right)_i \Delta z_i \qquad (11\text{--}31c)$$

in which q is the *net* load intensity at the foundation depth, I_z is the strain influence factor from Figure 11.8c, E is the appropriate Young's modulus at the middle of the ith layer of thickness Δz_i, and C_1 and C_2 are correction factors described below.

To incorporate the effect of strain relief due to embedment, and yet retain simplicity for design purposes, the method assumes that the 2B–0.6 distribution of the strain influence factor is unchanged, but its maximum value is modified. The suggested linear factor is

$$C_1 = 1 - 0.5 \left(\frac{\sigma_0'}{q}\right) \geqslant 0.5 \qquad (11\text{--}32)$$

in which σ_0' is the effective *in situ* overburden pressure at the foundation depth, and q is the net foundation pressure. In all cases, however, it is suggested that this correction factor not be less than 0.5.

A second correction factor is included to account, in part, for the time-dependent increase in settlement which appears to occur even for foundations on cohesionless soils. The proposed correction factor is

$$C_2 = 1 + 0.2 \log \ln \left(\frac{t \text{ yr}}{0.1} \right) \tag{11-33}$$

in which the time t is expressed in years. The theoretical justification for this correction is not clear. However in the case studies cited by Schmertmann (1970), the presence of time-dependent effects was apparently of such significance that the use of such a factor was warranted. The degree to which time-dependent effects arose as a result of cohesive materials mixed within the predominantly cohesionless soil or from underlying cohesive strata is not evident from the information available.

In the analysis, we have not considered the influence of foundation shape on the strain distribution. While an approximation, Schmertmann (1970) suggests that this is a refinement which is unwarranted because as the foundation shape changes from approximately axisymmetric to an approximately plane strain condition, the angle of shearing resistance increases at the same time as the stresses at a given depth also increase. These two effects tend to cancel each other, giving a strain distribution which is, perhaps, not very different for a wide range of length-to-width ratios.

Model test results, coupled with expediency, suggest that when a boundary underlain by rigid, or cohesive strata lies within the $2B-0.6$ distribution, that the distribution of strain influence factor be simply truncated at that depth.

Determination of Equivalent Young's Modulus

In order to use the method, it is necessary to estimate the stiffness in terms of an equivalent Young's modulus at various depths. Schmertmann (1970) recommends that this be done using the *Dutch cone* bearing capacity. The Dutch cone is a static-type penetrometer which has been used in Europe for more than thirty years. When provided with a sleeve to eliminate side friction it provides a convenient and rapid way to measure bearing capacity, and therefore strength, at various depths. In the case of cohesionless materials which have not been prestressed significantly to pressures above the *in situ* overburden pressure, the Dutch cone bearing capacity can be correlated with Young's modulus, (DeBeer 1965; Webb, 1970). The relationship suggested by Schmertmann (1970), and consistent with that of other investigators, is

$$E = 2q_c \tag{11-34}$$

in which q_c is the Dutch cone bearing capacity (usually expressed in kg/cm^2 or tsf:

Because prestress effects influence the *stiffness* of cohesionless materials more than the *strength*, the correlation given in Equation 11–34 is likely to underestimate the equivalent Young's modulus for over-consolidated cohesionless soils. Prestress is also likely to influence the strain distribution

(Webb, 1971). The extent to which this is true, however is not known at this time, and no means is presently available for accounting on a routine basis for the influence of prestress.

The static Dutch cone bearing capacity is much more reliable than the standard penetration test, which is a measure of the dynamic bearing capacity, the results of which are fraught with experimental difficulties and uncertainty. Nonetheless, it is recognized that standard penetration resistance data may be available for sites when Dutch cone bearing capacity data are not. In order to permit the use of standard penetration data as a temporary expedient, Schmertmann (1970) recommends the conservative correlation between Dutch cone bearing capacity and standard penetration resistance given in Table 11.1. It is, of course, much more preferable to obtain *directly* the Dutch cone bearing capacity. However, when it is necessary to use the standard penetration resistance Schmertmann (1970) suggests that as many N values as possible be obtained, to minimize by averaging, correlation errors associated with having only a few data.

TABLE 11.1
Correlation Between Dutch Cone
Bearing Capacity and Standard
Penetration Resistance

Soil Type	q_c/N
Silts, sandy silts, slightly cohesive silt-sand mixtures	2.0
Clean, fine to medium sands, and slightly silty sands	3.5
Coarse sands and sands with little gravel	5
Sandy gravel and gravel	6

NOTE: Units of q_c are kg/cm^2 or tsf; units of N are blows/ft.
SOURCE: Schmertmann (1970).

The use of the method described is illustrated in the following example, modified from one given by Schmertmann (1970):

Example 11.9

The bridge pier shown schematically in Figure 11.9a is to be constructed at the groundwater table, which is 6.6 feet below the surface of a medium-dense sand stratum of considerable depth. The Dutch cone bearing capacity profile is given in Figure 11.9b. Determine the settlement to be expected five years after construction.

Solution. We calculate the settlement in the following sequence:

1. Plot the $2B$–0.6 strain influence factor distribution as shown in Figure 11.9c. The maximum strain influence value (0.6) is at a depth $B/2$ below the foundation, and the triangle extends to a depth $2B$.

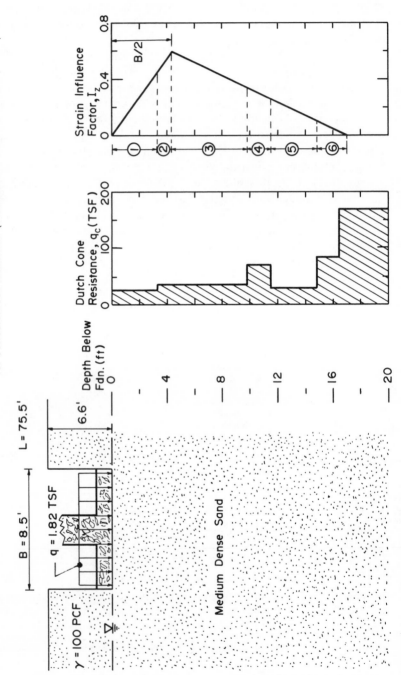

Fig. 11.9—Conditions for example settlement calculation for cohesionless material. (Modified from Schmertmann, 1970.)

2. Based on the q_c profile and the $2B$–0.6 distribution, divide the $2B$ depth into a convenient number of layers, as shown in Figure 11.9c and columns 1 and 2 of Table 11.2.
3. Determine the average q_c for each layer, and then calculate the corresponding E from Equation 11–34, as indicated in columns 3 and 4 of Table 11.2.
4. Locate the depth of the middle of each layer, column 5 in Table 11.2, and determine the I_z value for this depth from Figure 11.9c. The values are given in column 6 of Table 11.2.
5. Calculate $(I_z/E)\,\Delta z$ for each layer and sum the results as in column 7, Table 11.2.
6. Determine C_1 from Equation 11–32. The initial overburden pressure at the foundation depth, σ_0', is

$$\sigma_0' = \frac{100 \times 6.6}{2000} = 0.32 \text{ tsf}$$

The *net* foundation pressure, q, is

$$q = 1.82 - 0.32 = 1.50 \text{ tsf}$$

Thus,

$$C_1 = 1 - 0.5\frac{\sigma_0'}{\Delta p}$$

$$= 1 - \frac{0.5 \times 0.32}{1.5} = 0.89$$

7. Determine C_2 from Equation 11–33 for the time of interest,

$$C_2 = 1 + 0.2 \log\frac{t}{0.1}$$

$$= 1 + 0.2 \log\frac{5}{0.1} = 1.34$$

8. Calculate settlement from Equation 11–31c,

$$\rho_d = C_1 C_2 q \sum \left(\frac{I_z}{E}\right) \Delta z_i$$

$$= 0.89 \times 1.34 \times 1.50 \times 0.0748 = \underline{0.134 \text{ ft}} = \underline{1.6 \text{ in.}}$$

TABLE 11.2
Data for Settlement Estimate of Example 11.9

Layer No. i (1)	Δz (ft) (2)	Average q_c (tsf) (3)	Average E (tsf) (4)	Depth of Middle Below Fdn. (ft) (5)	I_z (6)	$\frac{I_z}{E}\Delta z$ (7)
1	3.3	25	50	1.65	0.23	0.0152
2	1.0	35	70	3.8	0.53	0.0076
3	5.6	35	70	7.1	0.47	0.0376
4	1.6	70	140	10.7	0.30	0.0034
5	3.3	30	60	13.1	0.185	0.0102
6	2.3	85	170	15.9	0.055	0.0008
TOTAL					$\sum \frac{I_z}{E}\Delta z =$	0.0748

SOURCE: Modified from Schmertmann (1970).

Deflected Shape of Loaded Areas and Contact Pressures

Unlike a flexible loaded area on an elastic medium, the profile of a flexible loaded area near the surface of a sand mass is that shown in Figure 11.10. The outside of the loaded area settles more than the inside because the stiffness of the sand depends upon the confining pressure, which is larger near the central zone of the footing. As a loaded area near the surface becomes wide, the zone over which the settlement is uniform increases in size, as shown in Figure 11.10. When a wide loaded area is embedded below the surface of a granular medium, the settlement may be almost uniform as indicated.

Figure 11.10 also shows the pressure distribution against the base of a rigid footing near the surface. As might be inferred from the figure, the contact pressure increases from almost zero at the edges, where there is little confinement, to some maximum value at the center of the footing.

11.4 THREE-DIMENSIONAL EFFECTS ON COMPRESSION SETTLEMENTS OF COHESIVE SOILS

Skempton-Bjerrum Correction

All of our previous discussion of settlements due to compressibility effects has been restricted to one-dimensional considerations. In those cases where the thickness of compressible strata is large relative to the loaded area, the three-dimensional nature of the problem may influence the magnitude and rate of settlement. Although numerical analysis methods offer the

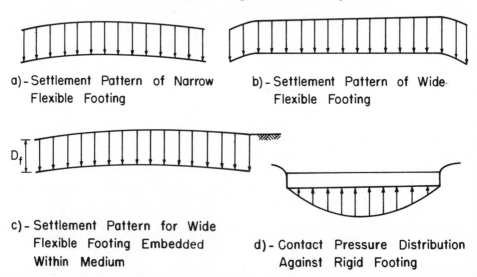

a) - Settlement Pattern of Narrow Flexible Footing

b) - Settlement Pattern of Wide Flexible Footing

c) - Settlement Pattern for Wide Flexible Footing Embedded Within Medium

d) - Contact Pressure Distribution Against Rigid Footing

Fig. 11.10—Settlement pattern and contact pressure of loaded areas on sand.

prospect of rational consideration of three-dimensional compression effects, they have not proven useful in practice to date. So, semi-empirical approaches have been used for this purpose. The most commonly applied of these methods, developed by Skempton and Bjerrum (1957), is based upon the following two assumptions:

1. Consolidation settlement occurs as a result of the dissipation of excess pore water pressure in the soil, and can be expressed by

$$\rho_c = \int_0^H \Delta u_e \frac{a_v}{1 + e_0} \, dz \qquad (11-35)$$

where ρ_c is the consolidation settlement, H is the thickness of the compressible stratum, Δu_e is the induced excess pore water pressure at each depth z due to an increment of stress applied at the surface, a_v is the coefficient of compressibility for the soil, and e_0 is the initial void ratio.

2. The induced excess pore water pressure at a point due to applied principal stress changes $\Delta \sigma_1$ and $\Delta \sigma_3$ for an axisymmetric situation (such as the triaxial compression test), in which $\Delta \sigma_2 = \Delta \sigma_3$, can be expressed by the Skempton (1953) equation for saturated soils:

$$\Delta u_e = \Delta \sigma_3 + A(\Delta \sigma_1 + \Delta \sigma_3) \qquad (11-36)$$

where A is the Skempton "A factor" (see Section 8.3) and expresses that proportion of the principal stress difference which produces pore water pressure in the material.

These assumptions imply that even though the induced excess pore water pressure results from three-dimensional effects, the settlements are one-dimensional.

If the field loading situation is approximately axisymmetric, and principal stresses are applied, then along the axis of symmetry Equation 11–36 applies, and Equation 11–35 may be rewritten

$$\rho_c = \int_0^H [A \, \Delta \sigma_1 + (1 - A) \, \Delta \sigma_3] \frac{a_v}{1 + e_0} \, dz \qquad (11-37)$$

The settlement computed from the results of a one-dimensional consolidation test is

$$\rho_c' = \int_0^H \Delta \sigma_1 \frac{a_v}{(1 + e_0)} \, dz \qquad (11-38)$$

Comparing Equations 11–37 and 11–38 leads to the conclusion that if the coefficient of compressibility a_v, and the A factor are assumed constant with depth, then the consolidation settlement, including three-dimensional effects, may be expressed in terms of the consolidation settlement predicted from a one-dimensional test as

$$\rho_c = \lambda \rho_c' \qquad (11-39)$$

where
$$\lambda = A + \beta(1 - A) \qquad (11\text{–}40)$$
and
$$\beta = \frac{\int_0^H \Delta\sigma_3 \, dz}{\int_0^H \Delta\sigma_1 \, dz} \qquad (11\text{–}41)$$

Thus it is possible to plot the factor λ as a function of the A factor for a given boundary loading. This is done in Figure 11.11 for a uniformly loaded circular area.

For a strip footing, assuming that Poisson's ratio $= 1/2$, it can be shown that (Scott, 1963)
$$\lambda = N + \beta(1 - N) \qquad (11\text{–}42)$$
where
$$N = \frac{\sqrt{3}}{2}\left(A - \frac{1}{3}\right) + \frac{1}{2} \qquad (11\text{–}43)$$

The λ factor is also plotted for strip footings in Figure 11.11.

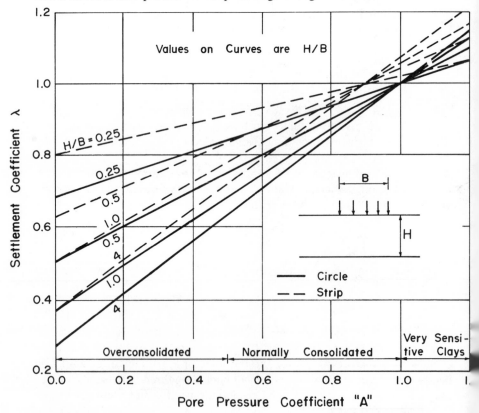

Fig. 11.11—Settlement coefficient vs. pore pressure coefficient for circular and strip footings. (After Skempton and Bjerrum, 1957; and Scott, 1963.)

This correction to the computed one-dimensional settlement should be used with caution because of the difficulty of evaluating the A factor in laboratory tests. Not applying the correction is conservative for those cases for which the A factor is less than one, which includes most normally consolidated clays and almost all over-consolidated clays. Failing to use the correction is unconservative if the A factor is greater than one, which usually corresponds to sensitive soils.

Stress Path Method

In an alternative approach described by Lambe (1964, 1967), a specimen of soil, presumed to represent the average element underneath the center line of the structure, is subjected in the laboratory to the same effective stress path which, it is anticipated, will be applied in the field (recall Section 8.6). The vertical settlements of this specimen are measured and summed over the thickness of the compressible stratum. Although its proponents suggest that this method offers significant advantages over the Skempton-Bjerrum correction (Lambe, 1964), it requires highly sophisticated laboratory techniques, which may be warranted only in the case of major projects. Furthermore, the approximations which must be made in deciding upon the average element, and the stresses to which it should be subjected may in fact introduce more error than the semi-empirical correction suggested above.

11.5   THREE-DIMENSIONAL DRAINAGE EFFECTS

Axisymmetric Flow with One-Dimensional Compression; Sand Drains

A special case of the general three-dimensional consolidation problem is of practical importance: axisymmetric fluid flow coupled, for ease of analysis, with one-dimensional compression. Consideration of this case is motivated by the use of *sand drains*. Sand drains are vertical columns of sand or other pervious material inserted through a compressible stratum at sufficiently close spacing that the longest horizontal drainage path is less than the longest vertical path. Thus the drains serve to hasten the consolidation process. Sand drains have been employed when the time delays required for consolidation to eliminate deleterious post-construction settlements, or to acquire sufficient additional shear strength, would have been excessive without their use. Sand drains are often used in conjunction with *precompression* of soft deposits, as described in Section 11.9.

The geometry of a typical sand drain installation is illustrated schematically in Figure 11.12. As indicated in the figure, the drains must be connected to a free-draining outlet in order to be effective. With the arrangement shown in Figure 11.12, each drain well has an axisymmetric zone of influence with a radius approximately equal to one-half the well spacing. The flow within this zone of influence is a combination of radial flow toward

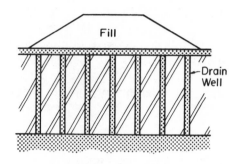

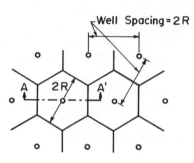

a)-Section of Sand Drain
Pattern

b)-Plan View of Sand Drain
Pattern

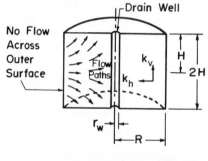

Fig. 11.12—Schematic diagram of sand drain installation.

c)-Section A-A'

the sand drain and vertical flow toward the free-draining boundary, or boundaries.

If we analyze this problem by invoking the same assumptions used to derive the one-dimensional consolidation equation in Section 7.2, except that the flow can take place in both radial and vertical directions, a linear partial differential equation results, which can be solved by the technique of separation of variables. As a consequence, the solution to the problem can be determined by superimposing the solution for the case of vertical flow only, and that for radial flow only. The resulting distribution of excess pore water pressure is

$$u_e(r, z, t) = \frac{u_{e_r}(r, t)u_{e_z}(z, t)}{u_{eo}} \qquad (11\text{-}44)$$

in which $u_e(r, z, t)$ is the distribution of excess pore water pressure as a function of position and time resulting from the combined vertical and radial drainage, $u_{e_r}(r, t)$ is the distribution of excess pore water pressure resulting from the radial flow only, $u_{e_z}(z, t)$ is the excess pore water pressure distribution due to vertical flow only, and u_{eo} is the initial distribution of excess pore water pressure. Alternatively, this can be stated in terms of the average degree of consolidation:

$$U(t) = 1 - [1 - U_r(t)][1 - U_v(t)] \qquad (11\text{–}45)$$

in which $U(t)$ is the average degree of consolidation in the compressible stratum, $U_r(t)$ is the average degree of consolidation in the stratum due to radial drainage only, and $U_v(t)$ is the average degree of consolidation due to vertical drainage only.

Thus we shall investigate the effect of sand drains by considering the vertical and radial flow problems separately. The case of one-dimensional vertical flow, described by Equation 7–10, was discussed in detail in Chapter 7. By analogy, the governing differential equation for radial flow, expressed in cyclindrical coordinates, is

$$c_{v_r}\left(\frac{\partial^2 u_e}{\partial r^2} + \frac{1}{r}\frac{\partial u_e}{\partial r}\right) = \frac{\partial u_e}{\partial t} \qquad (11\text{–}46)$$

in which u_e is the excess pore water pressure, r is the distance from the axis of symmetry (which in this case is the center of the sand drain), t is time, and c_{v_r} is the coefficient of consolidation for radial flow:

$$c_{v_r} = \frac{k_h(1 + e_0)}{a_v \gamma_w} \qquad (11\text{–}47)$$

In Equation 11–47, k_h is the coefficient of permeability in a horizontal direction, and a_v is the coefficient of compressibility as determined in a one-dimensional consolidation test.

Expressing Equation 11–46 in finite difference form, we obtain a recurrence relation for the excess pore water pressure analogous to Equation 7–21, developed for one-dimensional flow,

$$u_{e_{i,\,j+1}} = \alpha\left(1 - \frac{\Delta r}{2r_i}\right)u_{e_{i+1,\,j}} + (1 - 2\alpha)u_{e_{i,\,j}} + \alpha\left(1 + \frac{\Delta r}{r_i}\right)u_{e_{i-1,\,j}} \qquad (11\text{–}48)$$

in which $u_{e_{i,\,j+1}}$ is the excess pore water pressure a radial distance r_i from the centerline of the sand drain at a time equal to $j + 1$ increments of time after the initial excess pore pressure development, Δr and Δt are the increments in radial distance and time respectively, and

$$\alpha = \frac{c_{v_r}\,\Delta t}{(\Delta r)^2} \qquad (11\text{–}49)$$

As in the case of one-dimensional consolidation, analytical solutions have been obtained for a number of cases of practical interest. We shall consider two of these briefly:

1. "Free vertical strain." Assuming boundary conditions of no excess pore water pressure at the interface of the compressible soil and the drain well for $t > 0$, and no flow across the cylindrical outer boundary

of the zone of influence of the well, implies that settlements at the surface do not change the distribution of load to the soil mass. That is, the applied surface loading is assumed to be perfectly flexible, and the effect on the stress distribution of maintaining compatibility as the vertical strains vary in a radial direction is neglected. The analytical solution to this limiting condition is given by Barron (1948) and Richart (1957).

2. "Equal vertical strain." An alternative view of the problem considers that the load is applied by an approximately rigid system such as a stiff fill or a mat foundation, which would enforce approximately equal vertical displacements at the surface. The initial radial distribution of pore pressure corresponding to this problem is not uniform. The solution for the equal vertical strain case, however, is considerably simpler than for the free-strain case.

Richart (1957) has shown that when the radius of the zone of influence of the well is greater than approximately five times the radius of the well itself the relationships between the degree of consolidation and time are nearly the same for the two cases. Thus it is customary to use the results of the equal vertical strain analysis irrespective of the nature of the applied loading system. These results are given in Table 11.3 in terms of the average degree of consolidation corresponding to various ratios of one half the well spacing to the well diameter and *time factors*. The time factor for radial flow is defined as

$$T_r = \frac{c_{v_r} t}{R^2} \tag{11-50}$$

in which R is one-half the well spacing. Note that, because c_{v_r} may be different in magnitude from c_v, T_r may not be the same as T for one-dimensional flow corresponding to the same time.

The relative significance of well diameter and spacing on the rate of consolidation due to radial flow is illustrated in the following example.

Example 11.10 ――――――――――――――――――――――――――――――

It is desired to use sand drains to hasten the rate of consolidation of a soft compressible stratum underlying a proposed fill. We wish to determine how long it will take to achieve an average degree of consolidation in the compressible material of 80 per cent, assuming radial drainage only,* for three alternative cases:

(a)   One-ft-diameter drain wells spaced at 20 ft on centers,
(b)   One-ft-diameter drain wells spaced at 10 ft on centers,
(c)   Four-ft-diameter drain wells spaced at 20 ft on centers.

The coefficient of consolidation in a radial direction is $c_{v_r} = 20$ ft^2/yr.

―――

* We have limited our consideration in this example to radial drainage effects only to indicate the relative significance of well spacing and diameter. In general, of course, we would need to consider the combined radial and vertical drainage.

TABLE 11.3

Time Factor T_r for Radial Consolidation under Conditions of Equal Vertical Strain

Average Degree of Consolidation U_r (%)	Value of R/r_ω										
	5	10	15	20	25	30	40	50	60	80	100
0	0	0	0	0	0	0	0	0	0	0	0
5	0.006	0.010	0.013	0.014	0.016	0.017	0.019	0.020	0.021	0.023	0.025
10	0.012	0.021	0.026	0.030	0.032	0.035	0.039	0.042	0.044	0.048	0.051
15	0.019	0.032	0.040	0.046	0.050	0.054	0.060	0.064	0.068	0.074	0.079
20	0.026	0.044	0.055	0.063	0.069	0.074	0.082	0.088	0.092	0.101	0.107
25	0.034	0.057	0.071	0.081	0.089	0.096	0.106	0.114	0.120	0.131	0.139
30	0.042	0.070	0.088	0.101	0.110	0.118	0.131	0.141	0.149	0.162	0.172
35	0.050	0.085	0.106	0.121	0.133	0.143	0.158	0.170	0.180	0.196	0.208
40	0.060	0.101	0.125	0.144	0.158	0.170	0.188	0.202	0.214	0.232	0.246
45	0.070	0.118	0.147	0.169	0.185	0.198	0.220	0.236	0.250	0.291	0.288
50	0.081	0.137	0.170	0.195	0.214	0.230	0.255	0.274	0.290	0.315	0.334
55	0.094	0.157	0.197	0.225	0.247	0.265	0.294	0.316	0.334	0.363	0.385
60	0.107	0.180	0.226	0.258	0.283	0.304	0.337	0.362	0.383	0.416	0.441
65	0.123	0.207	0.259	0.296	0.325	0.348	0.386	0.415	0.439	0.477	0.506
70	0.137	0.231	0.289	0.330	0.362	0.389	0.431	0.463	0.490	0.532	0.564
75	0.162	0.273	0.342	0.391	0.429	0.460	0.510	0.548	0.579	0.629	0.668
80	0.188	0.317	0.397	0.453	0.498	0.534	0.592	0.636	0.673	0.730	0.775
85	0.222	0.373	0.467	0.534	0.587	0.629	0.697	0.750	0.793	0.861	0.914
90	0.270	0.455	0.567	0.649	0.712	0.764	0.847	0.911	0.963	1.046	1.110
95	0.351	0.590	0.738	0.844	0.926	0.994	1.102	1.185	1.253	1.360	1.444
99	0.539	0.907	1.135	1.298	1.423	1.528	1.693	1.821	1.925	2.091	2.219
100	∞	∞	∞	∞	∞	∞	∞	∞	∞	∞	∞

SOURCE: After Leonards (1962).

Solution. (a) From Table 11.3, for $R = 10$ ft, $r_w = 0.5$ ft, $T_r = 0.453$. The time required is then

$$t = \frac{TR^2}{c_v} = \frac{(0.453)(10)^2}{20} = \underline{2.26} \text{ yr} \qquad (11\text{--}51)$$

(b) Halving the well spacing without changing the well diameter, $R = 5$ ft and $r_w = 0.5$ ft, leads to $T_{80} = 0.317$, for which the time required is

$$t = \frac{(0.317)(5)^2}{20} = \underline{0.40} \text{ yr} \qquad (11\text{--}52)$$

(c) Increasing the well diameter by four times, but retaining the original well spacing so that $R = 10$ ft and $r_w = 2$ ft, leads to a time factor $T_{80} = 0.188$. Then,

$$t = \frac{(0.188)(10)^2}{20} = \underline{0.94} \text{ yr} \qquad (11\text{--}53)$$

Thus we see that halving the well spacing reduces the time required by more than a factor of five while increasing the well diameter four times reduces the waiting time to a little less than half of its original value.

As a result of the process of installing a sand drain, a zone around the periphery of the well is disturbed. This effect, called *smear*, produces an area around the outside of the well which is likely to exhibit a horizontal coefficient of permeability considerably less than that for the undisturbed material. Flow toward the well is impeded by this disturbed zone. Barron (1948) incorporated the influence of smear on radial flow in his analysis. Charts illustrating this effect were given by Richart (1957) which indicate that the effect of smear can be accounted for by using a reduced equivalent well diameter. It has been suggested (Leonards, 1962) that an equivalent well diameter equal to one-half the actual well diameter is often used to approximate the effect of smear. Recent developments in non-displacement sand drain installation methods (Landau, 1966; Genini, 1968) appear to lead to markedly reduced disturbance effects around the well and accompanying smaller influence of smear.

The procedures described above are useful in the design of an installation involving sand drains. However, caution must be exercised in interpreting the results of an analysis of the problem for the field situation because of the many factors which may lead to marked discrepancies between predicted and observed rates of excess pore water pressure dissipation (Rowe, 1968; Casagrande and Poulos, 1969). In particular, the influence of inhomogeneity of natural deposits and the difficulty in measuring appropriate average values of k_h and k_v may be cited. For this reason, when the rate of pore pressure dissipation is of importance to the construction sequence, the use of *piezometers* in the field to measure actual pore pressure dissipation rates is essential.

General Three-Dimensional Effects

There have been numerous attempts at theoretical analysis of time-rate of consolidation effects incorporating the three-dimensional nature of the general problem. These are summarized by Schiffman et al., (1969). Even the most realistic of such analyses presently available, is the result of important simplifying assumptions. Nonetheless, it is apparent that when the three-dimensional nature of the deformations is considered in light of the requirements of compatibility, the *total* stresses as well as the effective stresses are a function of time. Schiffman et al., (1969) illustrated the significance of this effect with the example of a strip load applied normal to the surface of a saturated compressible half-space and showed that the components of total stress change sufficiently that the maximum shear stress beneath the strip load increases during the consolidation process to a maximum magnitude nearly 1.8 times its initial and final values. As a

further consequence of this effect, the change in total stresses during consolidation led to an increase in excess pore pressure in a zone beneath
the strip load (Schiffman et al., 1969).

Although three-dimensional analyses have not yet reached the stage of
development to permit their application to routine field problems, they
already provide insight into factors which have heretofore not been
recognized.

11.6   SETTLEMENTS DUE TO SECONDARY COMPRESSION

Secondary compression is considered to be that portion of the time-
dependent settlement which occurs at essentially constant effective stress.
Thus, during secondary compression, the rate of volume strain is not controlled by the rate at which pore water can flow from the soil, and therefore,
should not depend upon the thickness of the layer considered. Consequently,
the field rate may be estimated directly from a laboratory test.

It has been observed in many laboratory and field measurements that the
relationship between the magnitude of secondary compression and time is
approximately a straight line on a semilogarithmic plot after the primary
consolidation has been completed, as shown in Figure 11.13. Thus the void-
ratio change may be expressed approximately as

$$\Delta e = -C_\alpha \log \frac{t_2}{t_1} \qquad (11-54)$$

where C_α is the slope of the straight line on the semilogarithmic plot, and
is referred to as the coefficient of secondary compression, t_2 is the time at
which the magnitude of secondary compression is desired, and t_1 is the time
on the extrapolated secondary compression curve corresponding to the
100 per cent primary consolidation point (see Figure 11.13).

Although the data are still limited, it appears that secondary compression
settlements in the field can be determined from Equation 11–54 (Kapp
et al., 1966). The magnitude of C_α depends primarily upon the soil type and
the magnitude of the effective pressure. The importance of the magnitude
of secondary compression, as expressed by C_α, relative to the magnitude of
primary consolidation, depends markedly upon the pressure-increment
ratio, that is, the magnitude of the newly applied pressure to the existing
in situ effective stress. This is illustrated in Figure 11.14. This figure shows
the ratio of the coefficient of secondary compression C_α to the void ratio
change at 100 per cent consolidation, Δe_{100}, as a function of the pressure-
increment ratio, $\Delta \sigma'/\sigma_0'$, for two soils. The pronounced effect of the pressure-
increment ratio on the significance of secondary compression for a particular
soil is evident. These results are typical of those for many soils.

In a great many cases of practical interest, secondary compression is a
minor effect relative to the magnitude of primary consolidation. However

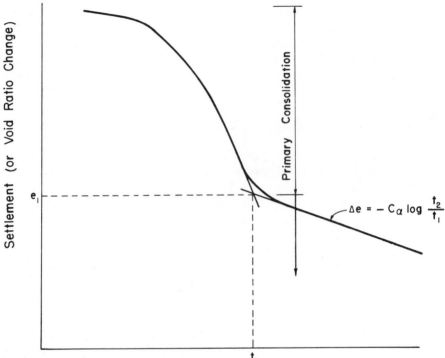

Fig. 11.13—Typical time-settlement curve for soft cohesive soil.

in some cases where very soft soils are involved, particularly those which contain some organic matter, or where deep compressible strata are subjected to small pressure-increment ratios by the imposed foundation loads, secondary compression may contribute a major component of the settlement.

11.7 EFFECT OF QUASI-PRECONSOLIDATION

The existence of a *quasi-preconsolidation* pressure for normally consolidated clays may have an important influence on the magnitude of settlements due to compressibility. This effect, which is discussed in detail in Chapter 5, may be decisive in the selection of foundation types, and should be considered in those cases where the presence of a quasi-preconsolidation pressure can be demonstrated in the laboratory. Although field evidence to support the importance of the quasi-preconsolidation effect in the settlements experienced by structures is still limited, it is impressive (Bjerrum, 1967a).

The way in which the quasi-preconsolidation pressure may be considered in settlement analysis is illustrated in Figure 11.15. Shown in the figure is a loaded area overlying a normally consolidated compressible stratum. The overburden pressure, which in this case is equal to the preconsolidation

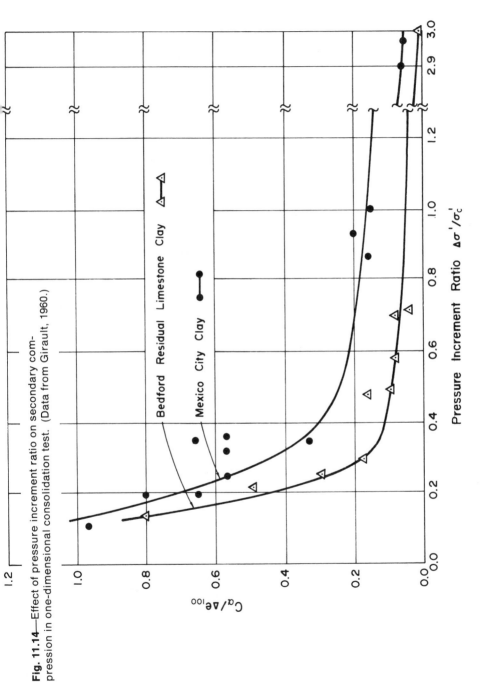

Fig. 11.14—Effect of pressure increment ratio on secondary compression in one-dimensional consolidation test. (Data from Girault, 1960.)

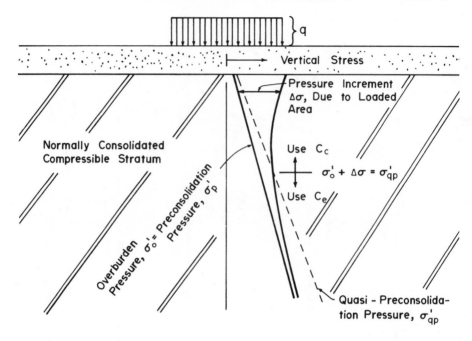

Fig. 11.15—Effect of Quasi-preconsolidation pressure on settlement analysis for normally consolidated clays.

pressure, is shown as a function of depth as the straight solid line. The quasi-preconsolidation pressure determined from laboratory tests is shown as the dashed line. Superimposed upon these results is the plot of the pressure increment transmitted to the soil along the centerline of the structure. The depth at which the sum of the overburden pressure and the pressure increment due to the loaded area equals the quasi-preconsolidation pressure has been shown. The material above this depth behaves as if it were normally consolidated, and one-dimensional settlements can be calculated using the compression index C_c. Below this point, the material exhibits little settlement, and these settlements can be calculated conservatively using the recompression index C_e (see Section 5.3).

11.8 EFFECT OF SETTLEMENT ON STRUCTURES

The settlements of representative points of a structure are computed using full dead load plus live load for cohesionless soils, and full dead load plus a time-averaged live load (e.g., 40 per cent) for consolidation settlements. The magnitude of settlement which is considered tolerable varies depending upon the type of structure, and the limiting factors involved. Table 11.4 indicates some typical values for limiting settlements, both total and differential, currently used in practice. The values in this table have been

TABLE 11.4
Limiting Settlements for Structures

Type of Movement	Limiting Factor	Maximum Allowable Settlement or Differential Movement
Total settlement	Drainage	6 to 12 in.
	Access	12 to 24 in.
	Probability of nonuniform settlement	
	Masonry-walled structure	1 to 2 in.
	Framed structures	2 to 4 in.
	Smokestacks, silos, mats	3 to 12 in.
Tilting	Stability against overturning	Depends on height and weight
	Tilting of smokestacks, towers	$0.004b$
	Rolling of trucks, etc.	$0.01L$
	Stacking of goods	$0.01L$
	Machine operation—cotton loom	$0.003L$
	Machine operation—turbogenerator	$0.0002L$
	Crane rails	$0.003L$
	Drainage of floors	$0.01L$ to $0.02L$
Differential movement	High continuous brick walls	$0.0005L$ to $0.001L$
	One-story brick mill building, wall cracking	$0.001L$ to $0.02L$
	Plaster cracking (gypsum)	$0.001L$
	Reinforced-concrete building frame	0.0025 to $0.004L$
	Reinforced-concrete building curtain walls	$0.003L$
	Steel frame, continuous	$0.002L$
	Simple steel frame	$0.005L$

NOTE: L = distance between adjacent columns that settle different amounts, or between any two points that settle differently.
Higher values of allowable settlement are for regular settlements and more tolerant structures. Lower values are for irregular settlements and critical structures.
SOURCE: After Sowers (1962).

determined as the result of studies of cracking and other forms of distress of existing structures in many locations throughout the world (Sowers, 1962).

11.9 REDUCTION OF DETRIMENTAL SETTLEMENTS

In cases where structures are founded on compressible subsoils it often happens that the settlements computed are of a magnitude which is deemed to be detrimental to the structure. In such cases a *deep foundation* may be indicated. However the additional cost of such foundations usually recommends consideration of alternative measures which would permit use of a shallow foundation. Such measures include the following:

1. Alteration of the structure. This approach is usually most promising in the case of buildings. There are four ways in which alteration

of the structure may be useful in reducing settlement or its undesirable effects:

a. The structure may be modified to a more *rigid* form. By increasing the rigidity of the structure, the tendency for differential settlement will be compensated by a tendency for redistribution of loads. This technique, however, is usually more expensive than other alternatives, and includes many uncertainties. Because the relationship of the redistributed loads and foundation settlements is difficult to predict, a certain amount of overdesign is frequently necessary. In the event that the structure is insufficiently rigid, cracking and localized distortion may result.

b. A simpler and often less expensive method is to make the structure more *flexible*, and thereby more tolerant to distortion. Simply supported members rather than rigid connections, a larger number of lighter members, and a more flexible type of wall construction will all assist in such a procedure.

c. In the event that the structure naturally divides into two or more parts, the use of *construction joints* between these parts may be of considerable value.

d. In those cases where large movements of the foundation are estimated and the cost of alternative foundation types is prohibitive, means of compensating for this movement may be included in the foundation itself. Thus jacks can be placed between the structure and the foundation permitting adjustment of the foundation at periodic intervals so that the structure does not experience distortion. This solution requires regular monitoring of foundation movements and periodic compensation of these movements.

2. Modify foundation. In some cases it may be possible to change the foundation, retaining a shallow type foundation but reducing the net load on the soil. One method to do this is to partially or completely *compensate* for the weight of the building by excavating an amount of soil which is equal to some desired proportion of the total building weight. Such a foundation is called a *compensated foundation*.* The depth to which the foundation must be carried in such cases can be estimated from the rule of thumb that one foot of excavation is approximately equivalent to one story of a typical office or apartment building. In such cases the foundation is frequently a mat, because of ground water conditions and the desirability of bridging over localized variability in the deposits under the structure. The success of this type of foundation depends upon the presence of a subsurface profile which is relatively uniform in horizontal directions.

---

* Early examples of this technique in the Boston area are described by Casagrande (1947) and Aldrich (1952). More recent cases are discussed by Leonoff and Ripley (1961), and Roberts and Darragh (1963).

The pressure-increment ratio applied to the compressible material by a compensated foundation is often low, say less than 0.5 even at shallow depths. In the case of normally consolidated deposits in which the quasi-preconsolidation pressure is not exceeded over a major portion of the depth, the benefits of partial compensation are enhanced.

3. Modify the soil. The most common method of modifying the soil to reduce the settlements is to increase the effective stress by *precompression*, also called *surcharging* or *preloading*. This method consists of increasing the load on the soil *above* that which would result from the intended construction, in order to produce more rapid consolidation of compressible materials. If the effective stress can be increased sufficiently, postconstruction settlements may be largely eliminated. An example of this is described in "Surcharging a Big Warehouse Floor Saves $1 Million," at the beginning of Chapter 4 (pages 161–166). In Chapter 7 we developed the background necessary to predict the magnitude of settlement of Building 301 during the time the floor was surcharged. We shall now extend our considerations to determining the relationship between the magnitude of surcharge load and the required waiting time.

In recent years, extensive use has been made of precompression for the foundations of oil storage tanks, light buildings, highway embankments and bridges (Aldrich, 1965). Precompression is most commonly accomplished by applying dead load in the form of an earth fill or water. Those foundation soils which are most amenable to such treatment include soft fine-grained silts and clays, organic deposits, loose silt and sandy strata and even rubbish fills. However, the time available during construction usually limits the thickness of strata subject to precompression to approximately 15 feet. For layers of greater thickness, sand drains may be used to promote horizontal drainage and reduce the effective length of the longest drainage path. For building foundations, however, the cost of sand drains is frequently a limiting factor in their use (Aldrich, 1965). In the case of earth fills, where time delays in construction may be very expensive, sand drains are used more frequently. Precompression also leads to an increase in strength of compressible materials which may, under some circumstances, be the most important benefit.

The rationale for determining the relationship between the magnitude of the preload and the time required to eliminate consolidation settlements is presented in Figure 11.16. Figure 11.16 shows a compressible stratum which is drained both top and bottom. The overburden stress prior to placement of the preload fill is shown as a solid line. The overburden stress at the level which we shall subsequently determine to be the critical plane is indicated as a point on this line with the magnitude σ_0'. Also shown is the distribution of stress due to the final load, p_f. The magnitude of the stress due to this final load at the critical plane is shown as a point on this

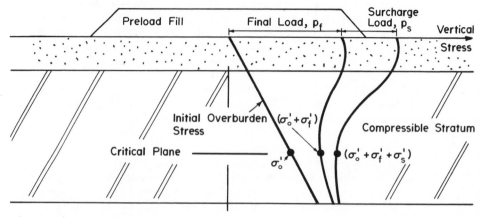

a)- Stress Conditions in Compressible Stratum Due to Preload Fill

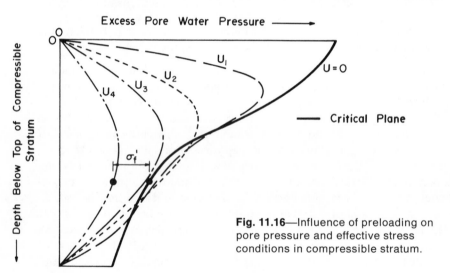

Fig. 11.16—Influence of preloading on pore pressure and effective stress conditions in compressible stratum.

b)- Excess Pore Water Pressure in Compressible Stratum at Various Times

curve with magnitude $\sigma'_0 + \sigma'_f$. The magnitude of the preload fill is equal to the final load p_f plus an additional *surcharge load* p_s. The distribution of stress due to this additional increment of load is shown in the figure. The magnitude of the stress due to the initial overburden and the total preload fill at the critical plane is shown as the point labeled $\sigma'_0 + \sigma'_f + \sigma'_s$.

The objective of applying a surcharge load in excess of the final load is to increase the effective stress at every point in the compressible stratum to a magnitude greater than or equal to the effective stress under the final loading, thereby reducing or eliminating postconstruction settlements. The basis for the analysis is illustrated in Figure 11.16, and is essentially similar to the problem of stage construction to permit development of strength in a weak stratum described in Chapter 8. Isochrones corresponding to the various degrees of consolidation under the preload fill, assuming that the fill was placed instantaneously, are shown in the figure. If consolidation were permitted to continue until it was essentially complete, the increase in vertical effective stress at every point in the compressible stratum would be equal to the stress due to the preload fill. However, in order to insure that when the preload fill is removed and the final load is applied, the settlements will be along the recompression curve, it is only necessary that the stratum be consolidated until the effective stress change is equal to σ'_f at every depth. The critical plane is that plane which is the last depth to develop the requisite effective stress. The location of the critical plane depends upon the shape of the isochrones and the distribution of stress due to the final loading, p_f. We see that the degree of consolidation at the critical plane will be, in general, considerably less than the average degree of consolidation. In the case illustrated, this corresponds to the degree of consolidation U_4. This implies that the rest of the compressible stratum will be preconsolidated to a pressure in excess of that applied by the final loading. We note also that in this example, as a result of the initial distribution of U_e ($U = 0$), some rebound occurred low in the stratum at early times. Many other features of the use of precompression are discussed by Johnson (1970).

11.10   EVALUATION OF PRELOADING AT BUILDING 301

In Chapters 4–7, we considered a number of aspects of the preloading of a compressible foundation soil at Building 301 in Port Newark. Let us now return to review the selection of the magnitude of surcharge and time of loading. In making such a comparison it is useful to write the required degree of consolidation at each point in the stratum as

$$U_z = \frac{\sigma'_f}{\sigma'_f + \sigma_s} = \frac{1}{1 + \dfrac{\sigma_s}{\sigma'_f}} \tag{11–55}$$

Thus we see that the degree of consolidation required depends upon the ratio of the stress produced by the surcharge to the stress produced by the final loading. This is, of course, a function of the *surcharge ratio*, p_s/p_f. When the loaded area is very large in extent compared to the depth of the compressible material, the increment in vertical stress throughout the stratum is equal to the applied surface load and

$$\frac{\sigma_s}{\sigma'_f} = \frac{p_s}{p_f} \tag{11–56}$$

We recall from Chapter 7 that the time factor for one-dimensional consolidation of the compressible silt when the surcharge was removed, 14 months after loading, was $T = 0.35$, and that the degree of consolidation at the center of the stratum was 46 per cent. Knowing that the total of the surcharge and final load, $p_s + p_f = 1300$ psf, we calculate the design effective vertical stress at the center to be 600 psf. So for this case the surcharge ratio used was

$$\frac{p_s}{p_f} = \frac{1300 - 600}{600} = 1.17 \tag{11-57}$$

In investigating the economics of surcharging we should compare the effect on the required waiting time of varying the surcharge ratio. In the case of a homogeneous stratum loaded uniformly with depth, the critical plane is at the center and Figure 7.7 can be combined with Equation 11-55 to find the *average* degree of consolidation required to produce the necessary effective stress at the center. The result is shown in Figure 11.17. Using this figure and Table 7.7 we can see how the time required is related to the surcharge ratio at Building 301. Results for several different surcharge ratios are shown in Table 11.5. The decision about the surcharge ratio to use depends upon a number of factors among which are the relative cost of the surcharge material and the waiting time, and especially the possibility of instability under the surcharge loading. In the case of Building 301 the material was already available and there was no stability problem so the building was filled to the maximum height possible as illustrated in the reprinted article at the beginning of Chapter 4.

11.11 SUMMARY OF KEY POINTS

Summarized below are the key points which we have discussed in this chapter. Reference is made to the place in the text where the original more detailed description can be found:

1. The effect of size of a loaded area on its settlement depends on the type of soil and the component of settlement being considered (Section 11.2):
    a. "Elastic" distortion settlement of cohesive deposits, approximated as a linear elastic half-space is
        (1) Inversely proportional to the size of a square area carrying a given total load.
        (2) Independent of the width of a strip load with constant total load per unit length along the strip.
        (3) Proportional to the size of a square area supporting a given unit pressure.
    b. One-dimensional consolidation settlement of cohesive deposits depends on the depth of the top of the compressible stratum below the loaded area.

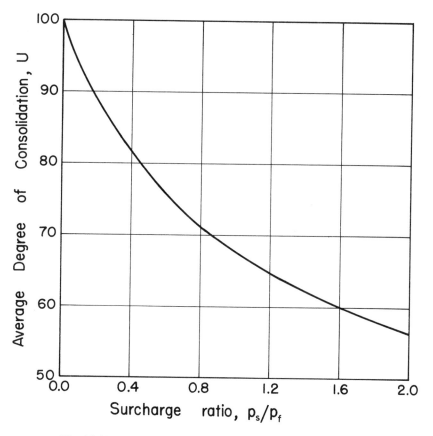

Fig. 11.17—Average degree of consolidation required to prestress uniformly loaded compressible stratum to p_f.

TABLE 11.5
Waiting Time Required for Various Surcharge Ratios at Building 301, Port Newark, N.J.

Surcharge Ratio p_s/p_f	Fill Height (ft)	Average U Required (%)	Time Required (months)
0.43	8	79	23
0.80	10	72	17
1.17	12	66	14

    (1) For a thick compressible stratum with its surface near the loaded area, consolidation settlement is

        (a) Inversely related to footing size for square areas supporting a given total load.

        (b) Inversely related to footing size for strip loads of constant total load per unit length of strip.

        (c) Directly related to size of square loaded areas subjected to a constant unit pressure.

        The settlement is also influenced by the pressure increment ratio.

    (2) For a stratum located at considerable depth below a loaded area, the size of the loaded area, that is, the distribution of load, is unimportant.

  c. Distortion settlement of footings on sand increases as the size of loaded area increases. The relative increase depends, however, on both the stress history of the deposit and the pressure increment ratio (Section 11.3).

2. For design purposes, the method proposed by Schmertmann for predicting the settlement of shallow foundations on sand is recommended (Section 11.3, Equations 11–31 to 11–34).

3. A variety of factors which are inconsistent with an assumed one-dimensional compression analysis may require additional consideration:

  a. The initial magnitude of excess pore water pressure may be significantly different from the vertical stress increment at a point when the thickness of the compressible stratum is large relative to the dimensions of the loaded area. This effect can be accounted for approximately by correcting the settlement calculated from conventional assumptions by a semi-empirical factor (Section 11.4, Equations 11–40, 11–43).

  b. Three-dimensional drainage effects may be important, especially when sand drains are used to hasten consolidation (Section 11.5). The effect of axisymmetric consolidation can be superimposed on the usual vertical one-dimensional consolidation (Equation 11–45). Radial effects can be determined from numerical solutions (Equation 11–48), or for special cases, from analytical solutions (Table 11.3).

4. Secondary compression occurs at essentially constant effective stress; the rate of volume strain does not depend on the thickness of the compressible stratum, and is inversely proportional to time (Section 11.6, Equation 11–54). The coefficient of secondary compression C_α, which is the void ratio change per log cycle of time, is influenced primarily by the pressure increment ratio.

5. The effect of the quasi-preconsolidation pressure can be incorporated in settlement predictions for normally consolidated clays by assuming normally consolidated behavior where σ'_{qp} is exceeded, and overconsolidated response where it is not exceeded (Section 11.7).

6. Time-averaged loadings should be used in predicting settlements of structures due to consolidation of cohesive strata (Section 11.8).

7. The tolerance of structures to settlement depends to a great extent on the type of construction (Table 11.4). When predicted settlements of a shallow foundation are deemed intolerable, various measures are available to avoid the use of a deep foundation. These include (Section 11.9):

   a. Alteration of the structure to either a more rigid, or more flexible form; incorporation of construction joints; provision of jacks to permit compensatory adjustments to reduce structural distortion.

   b. Use of a compensated foundation to reduce the net load on the soil.

   c. Reduction of compressibility of the critical strata by preloading. A surcharge load greater than the design load is usually applied to increase the effective stress to the desired value within an acceptable time period. Possible instability of the preload, or other factors, may limit the magnitude of surcharge which can be applied.

PROBLEMS

11.1. Given two columns with column *loads* Q_1 and Q_2, respectively, supported by large square footings on a deep-saturated cohesive deposit: What are the footing sizes, B_1 and B_2, such that—

(a) elastic settlements will be equal?

(b) consolidation settlements will be equal (assuming that unit pressures will be entirely within either the virgin compression or recompression range)?

11.2. The center of a highway bridge is to be supported on two piers, as shown. The piers carry 256 tons each, are supported by square footings, 8×8 ft, at a depth of 3 ft below the ground surface.

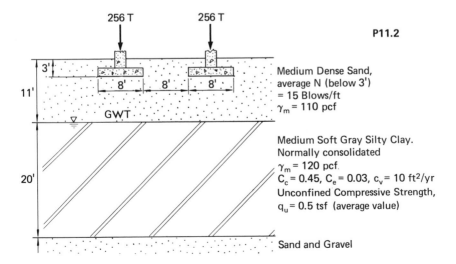

256 T   256 T

P11.2

3'
8'   8'   8'
11'
GWT

Medium Dense Sand,
average N (below 3')
= 15 Blows/ft
γ_m = 110 pcf

Medium Soft Gray Silty Clay.
Normally consolidated
γ_m = 120 pcf.
C_c = 0.45, C_e = 0.03, c_v = 10 ft²/yr
Unconfined Compressive Strength,
q_u = 0.5 tsf (average value)

20'

Sand and Gravel

The average standard penetration resistance of the sand below footing depth is $N = 15$ blows/ft. Conventional consolidation tests indicate that the underlying clay is normally consolidated. Estimate the settlement to be expected at the center of each footing *five years* after construction.

11.3. Regarding penetration resistance:

(a) Estimate the settlement of point A (see sketch) due only to consolidation settlement of the clay, 10 years after construction—
   1. for the case shown.

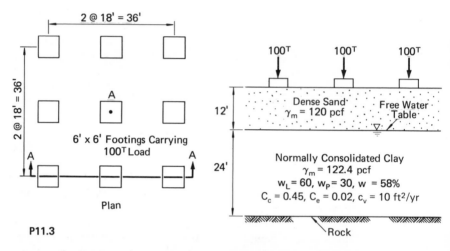

2 @ 18' = 36'

2 @ 18' = 36'

A

6' x 6' Footings Carrying 100^T Load

A   A

Plan

P11.3

100^T 100^T 100^T

12'

Dense Sand
γ_m = 120 pcf

Free Water Table

24'

Normally Consolidated Clay
γ_m = 122.4 pcf
w_L = 60, w_P = 30, w = 58%
C_c = 0.45, C_e = 0.02, c_v = 10 ft²/yr

Rock

Section A-A

2. if the building had been founded on a 42×42 ft mat at the ground surface.
3. if a 12 ft-deep excavation had been made and the building had been supported on a mat at the surface of the clay.

(b) Would bearing capacity be a problem in any of these cases?

11.4. A circular elevated water-storage tank is to be constructed at the surface of the soil profile illustrated, supported on a 40-ft-diameter circular mat foundation. In order to limit potential differential settlement, it is considered desirable to allow no more than 4 in. of settlement due to consolidation of the soft clay.

This criterion has led to the decision to preload the site. Fill is available to preload the entire site over a wide area to a depth of 10 ft (γ_{fill} = 120 pcf).

(a) How long must the preload fill be left in place before removing it and constructing the tank so that consolidation settlement of the tank will not exceed 2 in.?
(b) How could the waiting time be shortened? (Suggest at least two possibilities.)

Notes: (1) For purposes of calculation neglect the weight of the tank and mat. (2) Triaxial compression tests on undisturbed samples of the clay indicate an "A" factor of $A = 0.8$ over a wide range of strains.

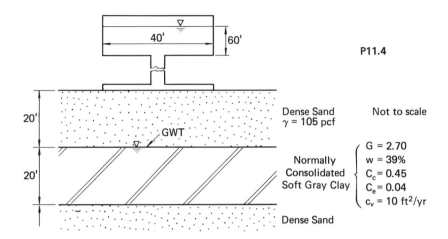

P11.4

Not to scale

11.5. The decision has been made to fill the tank described in Prob. 10.5 with oil to a depth of 25 ft and allow the clay stratum to consolidate before filling the tank further.

    (a) Determine the time-settlement history of the center of the tank during the first two years due to the 25 ft of oil, assuming the tank was constructed and filled instantaneously, and neglecting any settlement due to the sand stratum.

    (b) How would you expect the *actual* time-settlement history to compare to that you have calculated?

    (c) How long would it be necessary to wait after filling the tank to a depth of 25 ft before filling to the full 60 ft depth?

    (d) How might the waiting time be shortened?

12

Stability of Slopes

12.1 INTRODUCTION

Insuring the stability of both man-made and natural slopes affected by the works of man has proven to be one of the least successful endeavors of civil engineers. Measured by almost any standard, slope failures constitute the most important failures directly related to civil engineering. Failures of man-made fills and cuts in the earth probably occur more frequently than all other failures of civil engineering structures combined. The loss of life and property in such disasters as the Vaiont Reservoir slide in 1963 and the coal mine spoil bank failure at Aberfan, Wales, in 1966 is staggering.

Although we have developed an understanding of the major factors which lead to the failure of natural and man-made slopes, a large number of such failures occur each year. In some cases these result from our inability to predict and prevent them. In others, ignorance or negligence or both may be contributing factors. Such was the case for the slide at Aberfan.*

12.2 VARIETIES OF SLOPE FAILURES

Our discussion here will be restricted to consideration of those relatively large and relatively rapid movements associated with failure of at least some of the material comprising the slope or its foundation, as distinct from the slow, long-term slope movements commonly referred to as creep. Slope failures have been classified in a number of ways depending upon the application for which the classification system was developed. For our purposes, it is sufficient to identify three basic types of slope failures: falls, slides, and flows.

---

* The tribunal, appointed by the British Parliament to investigate the Aberfan disaster, concluded that "many other examples of the tragic result of the complete absence of any system in relation to . . . [mine spoil bank] stability could be given. When Mr. Ackner, Q. C., spoke of the National Coal Board's 'quite overwhelming responsibility for eight years of folly and neglect,' in our judgment his choice of words was justified." (Tribunal Report, 1967, p. 82.)

Falls

Falls are distinguished by a rapidly moving mass of material (rock or soil) that travels mostly through the air, with little or no interaction between one moving unit and another. A number of types of rock falls are illustrated in Figure 12.1. Such falls may or may not be preceded by minor movements. In general, falls are not subject to analysis because of the complexity of the factors involved in creating them, although recent attempts at prediction of falls (Jennings and Robertson, 1969) are encouraging.

Slides

In *slides*, the movement results from shear failure along one or more surfaces. The sliding mass may move as a relatively intact body, or may be greatly deformed. Several varieties of slides are illustrated in Figure 12.2. Note that the geologic conditions play a major role in determining the shape of the failure surface in slides. A large proportion of slope failures encountered by civil engineers are of this type. One of the best known is the Fort Peck Dam slide described in the article at the beginning of Chapter 2 (pages 28–33). Approximately 5.2 million cubic yards slid, up to a maximum distance of 1200 feet, in approximately 10 minutes. In the case of this slide the failure surface apparently passed largely through the hydraulic fill material of the dam. However the failure appears to have been initiated in the weathered shale at the base of the dam. This illustrates graphically the importance of undetected inhomogeneities or points of weakness where a failure may initiate and then subsequently propagate throughout otherwise stable material.

Another disastrous slide failure at the Vaiont Reservoir in Italy in 1963 (pages 18–24) led to the loss of almost 3000 lives.

A case of the sliding mass being carried on a thin layer of almost liquified *quick* clay is illustrated by the Vibstad slide in northern Norway (Hutchinson, 1965). As a consequence of the presence of a quick clay layer upon which sliding occurred, the slide mass moved more than 200 m, even though the mass remained relatively intact. Several similar cases occurred during the Alaskan earthquake on Good Friday, 1964 (Seed and Wilson, 1967). Much of the damage which ensued was a direct result of these slides.

Flows

Flows are characterized by movements within the displaced mass that resemble those of a viscous liquid. Although flows are most frequently associated with high water contents, dry flows of granular materials ranging in size from silt to large rock fragments can occur. There have been a number of large and catastrophic landslides of essentially dry material which still are characterized as flows (Varnes, 1958). Flows are common

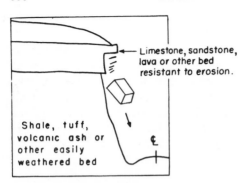

A. Differential weathering

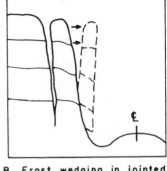

B. Frost wedging in jointed
homogeneous rock

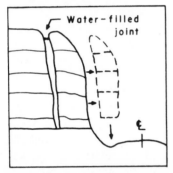

C. Jointed homogeneous rock.
Hydrostatic pressure acting
on loosened blocks.

D. Homogeneous jointed rock.
Blocks left unsupported or
loosened by overbreakage
and blast fracture.

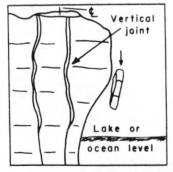

E. Either homogeneous jointed
rock or resistant bed
underlain by easily eroded
rock. Wave cut cliff.

F. Either homogeneous jointed
rock or resistant bed
underlain by easily eroded
rock. Stream cut cliff.

Fig. 12.1—Examples of rockfalls. (After Varnes, 1958.)

in the quick clays which occur along coastal regions in Norway, the south-eastern portion of Canada, and Japan, due to liquefaction of the sliding mass. A typical flow of this sort at Ullensaker, Norway, is described in Chapter 9 (pages 392–395).

Another type of flow, shown in Figure 12.3, occurred in 1948 in Greenboro, Florida. The flow occurred on a gentle slope of less than 10 degrees. Approximately 146,000 cubic yards of the clayey sand was involved. In this case, the flow apparently resulted not from changes in the physicochemical nature of the soil, but rather from a reduction in effective stress due to record 12-month rainfall which culminated in 16 inches of rain during the 30 days preceding the failure (Jordan, 1949).

The disastrous flow slide of mine tailings at Aberfan is described in the Inquiry Tribunal Report:

At about 9:15 A.M. on Friday, October 21st, 1966, many thousands of tons of colliery rubbish swept swiftly and with a jet-like roar down the side of the Merthyr Mountain which forms the western flank of the coal-mining village of Aberfan. This massive breakaway from a vast tip* overwhelmed in its course the two Hafod-Tanglwys-Uchaf farm cottages on the mountainside and killed their occupants. It crossed the disused canal and surmounted the railway embankment. It engulfed and destroyed a school and eighteen houses and damaged another school and other dwellings in the village before its onward flow substantially ceased. Then, in the words of the Attorney-General:

"With commendable speed, the work of attacking this seemingly ever-moving slimy, wet mass began as people strove to release the afflicted. Essential services were brought to the village and there began the unprecedented and Herculean task of recovery. People came in their hundreds from far and wide to lend their hands, whilst from the local collieries there hurried the officials and the sturdy experienced colliers to use their strength and skill as never before."

But despite the desperate and heroically sustained efforts of so many of all ages and occupations who rushed to Aberfan from far and wide, after 11 A.M. on that fateful day nobody buried by the slide was rescued alive. In the disaster no less than 144 men, women and children lost their lives. 116 of the victims were children, most of them between the ages of 7 and 10, 109 of them perishing inside the Pantglas Junior School. Of the 28 adults who died, 5 were teachers in that school. In addition, 29 children and 6 adults were injured, some of them seriously. 16 houses were damaged by sludge, 60 houses had to be evacuated, others were unavoidably damaged in the course of the rescue operations, and a number of motor cars were crushed by the initial fall. According to Professor Bishop, in the final slip some 140,000 cubic yards of rubbish were deposited on the lower slopes of the mountainside and in the village of Aberfan, whilst the amount actually crossing the embankment is estimated, very approximately, to have been about 50,000 cubic yards (Tribunal Report, 1967, p. 26).

An aerial view of the slide is shown in Figure 12.4. Figure 12.5 shows a plan of the area of the town which was inundated. The influence of excess

---

* Mine waste spoil banks are called *tips* in Great Britain.

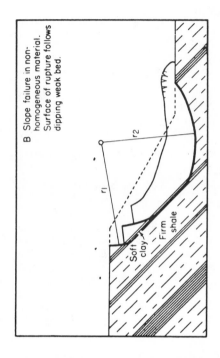

A Slope failure in homogeneous material. Circular arc.
1 Slide wholly on slope.
2 Surface of rupture intersects toe of slope.

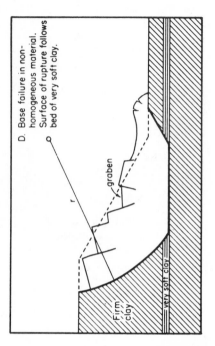

B Slope failure in non-homogeneous material. Surface of rupture follows dipping weak bed.

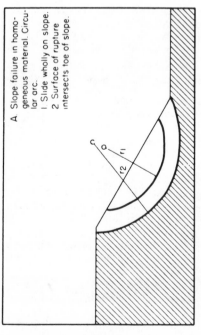

C Base failure in homogeneous clay. Slip circle tangent to firm base, center on vertical bisector of slope.

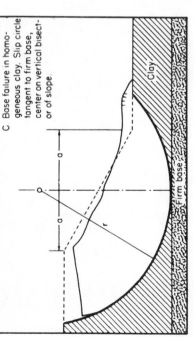

D. Base failure in non-homogeneous material. Surface of rupture follows bed of very soft clay.

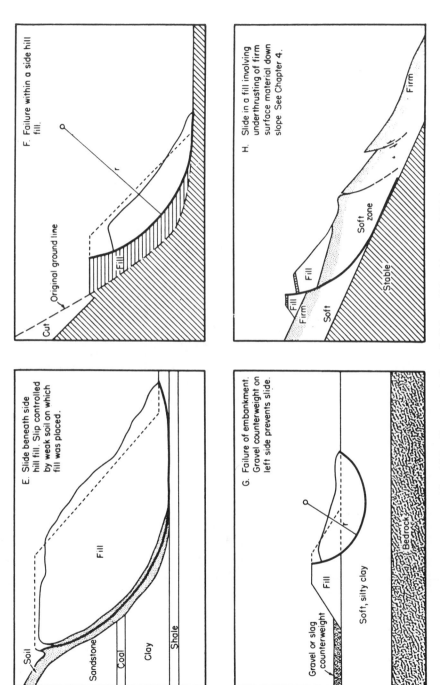

Fig. 12.2—Some varieties of slides. (After Varnes, 1958.)

E. Slide beneath side hill fill. Slip controlled by weak soil on which fill was placed.

F. Failure within a side hill fill.

G. Failure of embankment. Gravel counterweight on left side prevents slide.

H. Slide in a fill involving underthrusting of firm surface material down slope. See Chapter 4.

Fig. 12.3—Aerial view of Greenbo

ow slide. (After Jordan, 1949.)

Fig. 12.4—Aerial view of Aberfan flow taken the day of the disaster. (From Tribunal Report, 1967.)

pore water pressure in creating the flow once the slide was initiated was similar to that of the Greensboro slide (Bishop et al., 1969). We shall analyze the Aberfan slide further in Section 12.12.

Slope failures frequently exhibit more than one type of movement at different times in their development. For example, the severity of the Fort Peck slide was apparently enhanced because of the partial liquefaction of some of the sliding mass. Thus the slide was in part a flow. It is often difficult to draw a sharp distinction between these types, and in many cases it may not even be desirable to do so.

12.3 PROCESSES RESPONSIBLE FOR SLOPE FAILURES

Terzaghi (1950) divided the causes of slope failures into "internal" and "external" effects. External causes are those which produce an increase of the shearing stress at unaltered shearing resistance of the material comprising or adjoining the slope. The most common of these is a steepening or heightening of a slope by man-made excavation or river erosion. Internal causes are those which lead to a failure without any change in external conditions or earthquakes. The most common such cause is an increase in pore water pressure within the slope. A summary of processes leading to slope failures and their effects on the slope is presented in Table 12.1.

12.4 ASSUMPTIONS OF STABILITY ANALYSIS

From the foregoing comments it is obvious that the models that we have developed to investigate the stability of earth slopes do not successfully describe conditions in a significant number of circumstances. Some of the reasons for this situation will become apparent from the discussion which follows.

Current methods of analysis of slope stability are based largely upon assumptions and observations made in the eighteenth and nineteenth centuries. Although our understanding of certain aspects of the stability problem, particularly the shearing resistance of soils, has improved significantly since that time, the basic assumptions remain unchanged. These assumptions are:

1. Failure of an earth slope occurs along a particular sliding surface. That is, we assume that the failure can be represented as a two-dimensional plane problem.
2. The failure mass moves as an essentially rigid body, the deformations of which do not influence the problem.
3. The shearing resistance of the soil mass at various points along

Fig. 12.5—Plan of area in Aberfan covered by flow. (From Tribunal Report, 1967.)

TABLE 12.1

Processes Leading to Slope Failures

Name of Agent	Event or process Bringing Agent into Action	Mode of Action of Agent	Slope Materials Most Sensitive to Action	Physical Nature of Significant Actions of Agent	Effect on Equilibrium Conditions of Slope
Transporting agent	Construction operations or erosion	1. Increase of height or rise of slope	Every material	Changes state of stress in slopeforming material	Increase of shearing stresses
			Stiff, fissured clay, shale	Changes state of stress and causes opening of joints	Increases shearing stresses and initiates process 8
Tectonic stresses	Tectonic movements	2. Large-scale deformations of earth crust	Every material	Increases slope angle	Increases shearing stresses
	Earthquakes or blasting	3. High-frequency vibrations	Every material	Produces transitory change of stress	Decrease of cohesion and increase of shearing stresses
			Loess, slightly cemented sand, and gravel	Damages intergranular bonds	
			Medium or fine loose sand in saturated state	Initiates rearrangement of grains	Spontaneous liquefaction
Weight of slope-forming material	Process which created the slope	4. Progressive deformation of slope	Strongly bonded over-consolidated clay subject to weathering	Develops continuous slide surface	Progressive reduction of shearing resistance
		5. Creep on slope	Stiff, fissured clay, shale, remnants of old slides	Opens up closed joints, produces new ones	Reduce cohesion, accelerates process 9
		6. Creep in weak stratum below foot of slope	Rigid materials resting on plastic ones		
Water	Rains or melting snow	7. Displacement of air in voids	Moist sand		Decrease of frictional resistance
		8. Displacement of air in open joints	Jointed rock, shale	Increases pore-water pressure	

540

			Physical effect	Result
	...illary pressure associated with swelling	soil, assorted clay and some shales	Causes swelling	Decrease of cohesion
	10. Leaching of salts from pore fluids	Marine deposited clays	Increase interparticle electrochemical repulsion	Decrease of cohesion, spont. liquefaction
	11. Chemical weathering	Rock of any kind	Weakens intergranular bonds (chemical weathering)	Decrease of cohesion
Frost	12. Expansion of water due to freezing	Jointed rock	Widens existing joints, produces new ones	Decrease of frictional resistance
	13. Formation and subsequent melting of ice layers	Silt and silty sand	Increases water content of soil in frozen top layer	Decrease of cohesion
Dry spell	14. Shrinkage	Clay	Produces shrinkage cracks	Decrease of frictional resistance
Rapid drawdown	15. Produces seepage toward foot of slope	Fine sand, silt, previously drained	Produces excess pore-water pressure	Spontaneous liquefaction
Rapid change of elevation of water table	16. Initiates rearrangement of grains	Medium or fine loose sand in saturated state	Spontaneous increase of pore water pressure	Decrease of frictional resistance
Rise of water table in distant aquifer	17. Causes a rise of piezometric surface in slope-forming material	Silt or sand layers between or below clay layers	Increases pore water pressure	Decrease of frictional resistance
Seepage from artificial source of water (reservoir or canal)	18. Seepage toward slope	Saturated silt	Increases pore water pressure	Decrease of frictional resistance
	19. Displaces air in the voids	Moist, fine sand	Eliminates surface tension	Decrease of cohesion
	20. Removes soluble binder	Loess	Destroys intergranular bond	Decrease of cohesion
	21. Subsurface erosion	Fine sand or silt	Undermines the slope	Increase of shearing stress

SOURCE: Modified from Terzaghi (1950).

the failure surface is independent of the orientation of the failure surface; that is, the strength properties are isotropic.*

4. The factor of safety is defined in terms of the average shear stress along the potential failure surface and the average shear strength along this surface, rather than the local values at particular points. Thus the shear strength of the soil may be exceeded at some point along the failure surface whereas the computed factor of safety may be larger than 1.0.

We shall investigate the results of these assumptions for a number of special cases.

12.5 STABILITY OF INFINITE SLOPE OF COHESIONLESS SOIL

We begin by considering a very long slope bounded by a plane surface inclined to the horizontal at an angle β, Figure 12.6. The slope consists of a homogeneous cohesionless soil ($c' = 0$). We investigated the limiting equilibrium of this problem in Chapter 3, and found that such a slope would be stable for $\beta < \phi'$ (recall Figure 3.13). Furthermore, if we choose to define the *factor of safety* as the ratio of the obliquity on potential failure planes parallel to the slope angle to the maximum obliquity ϕ', we find from Equation 3–16,

$$FS = \frac{\tan \phi'}{\tau_\beta / \sigma'_\beta} = \frac{\tan \phi'}{\tan \beta} \qquad (12-1)$$

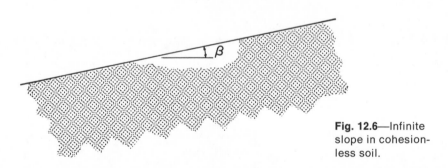

Fig. 12.6—Infinite slope in cohesionless soil.

---

* Consideration has been given to the possible effect of the anisotropic response of the soil on slope stability (Lo, 1965; Duncan and Seed, 1966b; Ranganatham et al., 1969). However, the utility of such an approach is limited at the present time by the paucity of information concerning the nature of anisotropic effects in natural deposits. Many laboratory investigations, among which are those of Casagrande and Carillo (1944), Jakobson (1955), Broms and Ratnam (1963), Ladd (1965), Duncan and Seed (1966a), Khera and Krizek (1967), and Jacobsen (1967), have provided considerable insight into the problem, but have not led to results which are applicable to current practice. Analytical considerations of anisotropic strength properties (Livneh and Shlarsky, 1964; Livneh and Komornik, 1967; Baker and Krizek, 1970) are still at an early stage of development.

in which FS is the factor of safety. Note that FS is independent of the height of the slope.

12.6 STABILITY OF SLOPES IN PURELY COHESIVE SOILS

Plane Failure Surface

Consider the simple slope in a purely cohesive soil ($\phi = 0$), shown in Figure 12.7a. Let us assume that the *plane* surface passing through point a, the toe of the slope, and inclined at an angle i to the horizontal is a potential failure surface. We can then investigate the forces which act on the wedge *abca* rather than attempt to investigate limiting equilibrium of a slope of limited extent. The force polygon in Figure 12.7b shows these forces, where

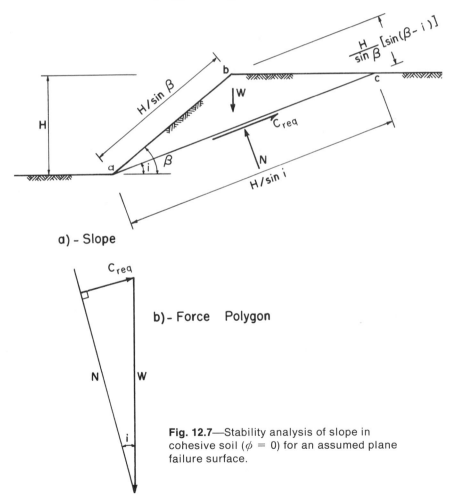

a) - Slope

b) - Force Polygon

Fig. 12.7—Stability analysis of slope in cohesive soil ($\phi = 0$) for an assumed plane failure surface.

W is the weight of the wedge $abca$, N is the normal force on the potential failure surface, and C_{req} is the magnitude of the cohesive force on the potential failure surface required to maintain equilibrium. From the force polygon

$$C_{req} = W \sin i \qquad (12\text{--}2)$$

Additionally, from the figure,

$$C_{req} = c_{req} \frac{H}{\sin i} \qquad (12\text{--}3)$$

where c_{req} is the average cohesive resistance along the assumed failure plane required to maintain equilibrium. From the diagram,

$$W = \frac{1}{2} \gamma \left(\frac{H}{\sin i} \right) \frac{H}{\sin \beta} \left[\sin (\beta - i) \right] \qquad (12\text{--}4)$$

Substituting Equations 12–2 and 12–3 into Equation 12–4 gives

$$c_{req} \frac{H}{\sin i} = \frac{1}{2} \gamma \frac{H^2}{\sin i \sin \beta} \left[\sin (\beta - i) \right] \sin i$$

or

$$\frac{c_{req}}{\gamma H} = \frac{\sin i \sin (\beta - i)}{2 \sin \beta} \qquad (12\text{--}5)$$

The parameter $c_{req}/\gamma H$ is a useful dimensionless number called the *stability number*. It simply expresses the ratio of the cohesive component of shearing resistance to γH that is required to just maintain stability, that is, for a factor of safety of 1.

Recall, however, that we have assumed an orientation i for the potential failure surface as we did for similar wedge analysis in Chapter 3. Clearly, we must find that i value which represents the most critical case, that is, the plane along which failure is most likely to occur. The critical plane will be the one for which $c_{req}/\gamma H$ is a maximum. In this simple problem we can find the minimum analytically rather than by trial and error. Therefore:

$$\frac{\partial}{\partial i} \left(\frac{c_{req}}{\gamma H} \right) = \frac{\partial}{\partial i} \left[\sin i \sin (\beta - i) \right] = 0 \qquad (12\text{--}6)$$

From which

$$i = \beta/2 \qquad (12\text{--}7)$$

Thus, for $i = \beta/2$,

$$\frac{c_{req}}{\gamma H} = \frac{\sin \beta/2 \sin \beta/2}{2 \sin \beta} = \frac{1}{4} \tan \beta/2 \qquad (12\text{--}8)$$

Example 12.1 ————————————————————————————————

Based upon an assumed plane failure surface, determine the maximum height to which a vertical unbraced slope in a purely cohesive material can stand.

Solution. For this case, from Equation 12–8,

$$\frac{c_{req}}{\gamma H} = \frac{1}{4}\tan 45° = 0.250$$

Curved Failure Surface

As early as the middle of the nineteenth century, the French engineer Collin (1846) recognized from observation of slope stability failures in the field that the failure surface was curved rather than plane. This discovery was an important step in improving stability analyses. It is evident that the natural phenomena which occur in the field are independent of whatever analysis we attempt to perform in order to predict conditions at which failure will occur. That is, if we select a plane failure surface to analyze, and the path of least resistance, which the failure surface *actually* follows is some curve, then our analysis will lead us to an unconservative result, perhaps seriously so. Thus let us look at the result of assuming a curved failure surface.

For simplicity we shall assume that the failure surface is an arc of a circle, and for the moment, that this surface passes through the toe of the slope, point *a* in Figure 12.8. Having assumed the failure surface, we can determine the magnitude of c_{req} needed to maintain equilibrium. In the case of a circular failure surface it is most convenient to evaluate the required cohesion by considering equilibrium of moments about the center of the circle. Thus it is not necessary to investigate the distribution of the normal forces around the potential failure surface. Considering that

$$\Sigma M_0 = \Sigma M_D - \Sigma M_R = 0 \tag{12–9}$$

in which ΣM_D is the sum of all of the *driving* moments, that is, those which tend to cause instability; ΣM_R is the sum of the *resisting* moments, that is, the moments which tend to resist failure. In this case the driving moment is due to the weight of the failure mass,

$$\Sigma M_D = W\bar{x} \tag{12–10}$$

where $\bar{x}$ is the horizontal distance from the center of the circle to the resultant weight vector of the sliding mass. The resisting moment is due to the cohesive forces which act on the circular arc:

$$\Sigma M_R = c_{req} R^2(2\omega) \tag{12–11}$$

It can readily be seen that both the driving and resisting moments are

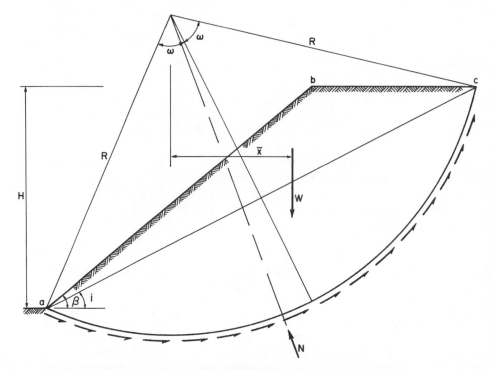

Fig. 12.8—Stability analysis of slope in cohesive soil ($\phi = 0$), for an assumed circular failure surface (shallow failure).

TABLE 12.2

Comparative Values of $c_{req}/\gamma H$ for Given Slope Angle β and Various Slide Surfaces in Cohesive Soil ($\phi = 0$)

		Circular Surface	
β	Plane Surface $c_{req}/\gamma H$	Toe Slide $c_{req}/\gamma H$	Deep Slide $c_{req}/\gamma H$
15°	0.033	0.145	0.181
30°	0.067	0.156	0.181
45°	0.104	0.170	0.181
53°	—	0.181	0.181
60°	0.145	0.191	0.181
75°	0.192	0.219	0.181
90°	0.250	0.261	0.181

NOTE: Critical values shown between dashed lines.

expressible in terms of the height of the slope H, the slope angle β, and the two angles ω and i which define the location of the assumed failure surface. Thus in this case it is necessary to determine both ω and i that will lead to the maximum $c_{req}/\gamma H$. Taylor (1937) carried out the detailed calculations involved, and presented the results shown in Table 12.2. In this table, the first column indicates the variety of slope angles for which the $c_{req}/\gamma H$ corresponding to assumed plane and circular failure surfaces passing through the toe of the slope are shown in the second and third columns respectively. We note immediately that in all cases the assumption of a circular failure surface leads to a more critical situation than that of a plane surface. Thus it is evident that, if we are going to investigate slope stability by *assuming* a specific failure surface and determining the forces acting on a sliding wedge bounded by that surface, the quality of the result obtained depends upon our choice of failure surface.

If we go one step further and do not restrict the assumed circular failure surface to passing through the toe of the slope (Figure 12.9), we find:

1. The circular failure surface will extend to the greatest depth that it can until it encounters a firm stratum.
2. Assuming that the depth to a firm stratum is great, the stability number for a purely cohesive material is 0.181. This is shown in the fourth column of Table 12.2. In this table the dashed lines

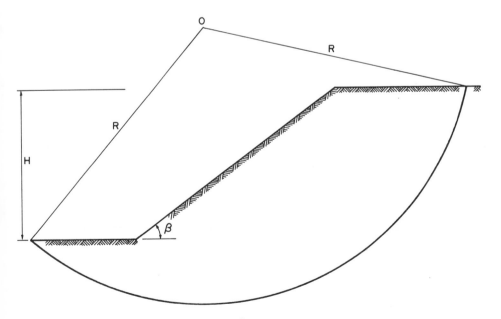

Fig. 12.9—A deep failure surface.

delineate those values of the stability factor which are critical for each specific slope angle β. We see that for $\beta > 53°$, the failure surface for a purely cohesive material will pass through the toe. For $\beta < 53°$, it will pass below the toe to as great a depth as possible.

3. When the depth to a firm stratum is limited, the results shown as dashed lines in Figure 12.10 are obtained. In this figure, the depth to a firm layer is described in relation to the slope height as nH. The beneficial effect on stability of a shallow strong layer is evident.

12.7 STABILITY OF SLOPES IN HOMOGENEOUS SOILS POSSESSING FRICTION

The ϕ-Circle Method

In the case of soils possessing both frictional and cohesive components of strength, the situation is somewhat more complicated than for purely cohesive soils. For cohesive soils the shearing resistance along the potential failure surface was independent of the normal stresses applied to that surface. Thus by taking moments about the center of the circle, it was possible to evaluate stability without considering this distribution. However, where there is a frictional component to strength, the distribution of normal stresses on the failure surface influences the distribution of shearing resistance. This is illustrated in Figure 12.11. In Figure 12.11a, a typical potential failure surface and some distribution of normal stress acting on it is shown as

$$\sigma_n = f(\theta) \qquad (12\text{--}12)$$

in which σ_n is the normal pressure and θ is the angular measure of location of the point on the failure surface. If the frictional component of strength is fully mobilized, the Mohr-Coulomb relationship must apply at that point. At such a point the resultant of the normal stress and *frictional* strength components is inclined at an angle ϕ to the normal or, in other words, will be tangent to a circle with radius $R \sin \phi$. Figure 12.11b shows two such points on the failure surface, defined by the angles θ_1 and θ_2. Note that the resultant stresses at these two points are tangent to the "ϕ circle" at different points. If each of these stresses is presumed to act over an element of differential length, then the resultant of the two forces will act through point m, and will not be tangent to the ϕ circle. Consequently in order to investigate the stability of the slope for the assumed failure surface shown in Figure 12.11 it is necessary to know the normal stress distribution on the potential failure surface. This point will be considered below.

It is most convenient to attempt to assess the stability of the slope in terms of the resultant forces which act on the potential failure mass. These forces are shown in Figure 12.12a, and are:

1. *Weight of wedge, W.* This is determined by multiplying the unit weight of the material times the area enclosed by the failure surface.

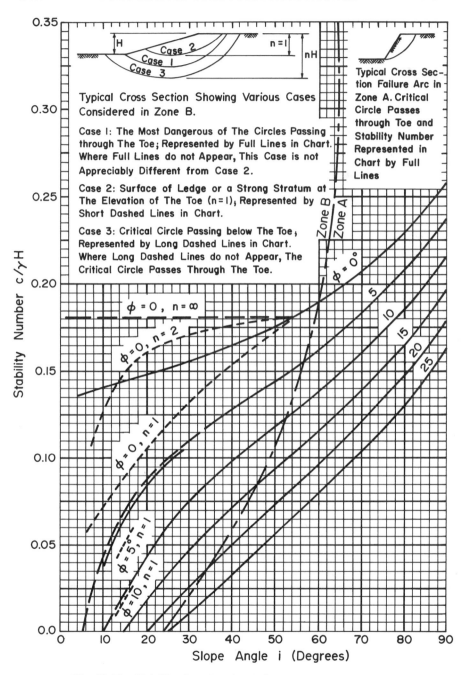

Fig. 12.10—Stability chart for slopes in homogeneous material. (After Taylor, 1948.)

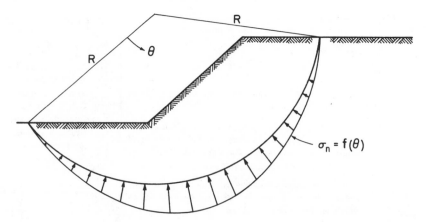

a) - Normal Pressure Distribution on Potential Failure Surface

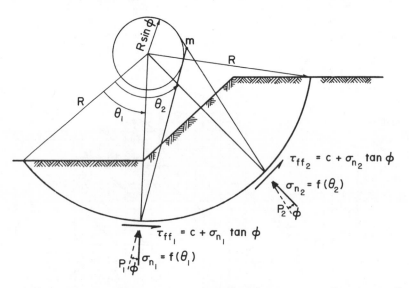

b) - Resultant Stress at a Point Tangent to "ϕ - Circle"

Fig. 12.11—Concept of "ϕ-Circle" for stresses on a failure surface.

2. *Resultant required cohesive force, C_{req}.* The magnitude of this force is reasoned to be

$$C_{req} = c_{req} \, \overline{dc} \qquad (12\text{-}13)$$

where c_{req} is the magnitude of the unit cohesive resistance required to maintain stability and $\overline{dc}$ is the length of the chord dc. The reason that the chord length $\overline{dc}$ is used rather than the arc length dc is that the components of c_{req} normal to the chord length dc will act in

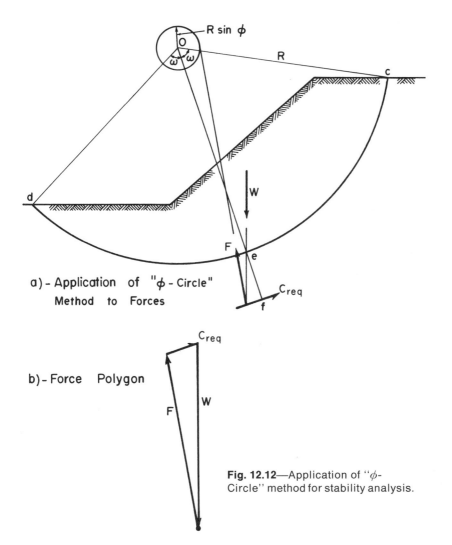

a) - Application of "ϕ - Circle"
    Method to Forces

b) - Force Polygon

Fig. 12.12—Application of "ϕ-Circle" method for stability analysis.

opposite directions on either side of the bisector Oe. Thus C_{req} will act parallel to $\overline{dc}$. The location at which the resultant cohesive force C_{req} acts is determined from the fact that the moment of that force about the center of the circle must be the same as the moment of the cohesive shear resistance acting along the arc length dc. Thus the distance Of is determined from

$$C_{req}\overline{Of} = c_{req}\,\overline{dc}R \qquad (12\text{--}14)$$

Substituting Equation 12–13 into Equation 12–14 we find that

$$\overline{Of} = R\,\frac{dc}{\overline{dc}} \qquad (12\text{--}15)$$

3. *Resultant of normal frictional forces, F.* This force must act through
    the point at which W and C_{req} intersect. If we further assume that
    the force F acts *tangent to the ϕ circle*, we then know the direction
    of this force.

Consequently we can plot the force polygon shown in Figure 12.12b.

We must, however, consider the error associated with assuming that the
resultant of the normal and frictional forces P, acts tangent to the ϕ circle,
because we know from Figure 12.11 that this assumption cannot be correct.
Unfortunately, we do not know the actual distribution of stress along the
potential failure surface when failure is imminent. Taylor (1937) investigated
the error associated with this assumption by considering (1) a uniform
normal stress distribution along the failure surface, and (2) a sinusoidal
normal stress distribution. Figure 12.13 shows the distribution of normal
stress on a circular potential failure surface within a linear elastic medium
subject to its own weight, forming a slope of height H with a slope angle
of 30°. The stress distribution is shown in terms of $\sigma_n/\gamma H$ where σ_n is the
normal pressure, and γ is the unit weight of the material. Thus it appears

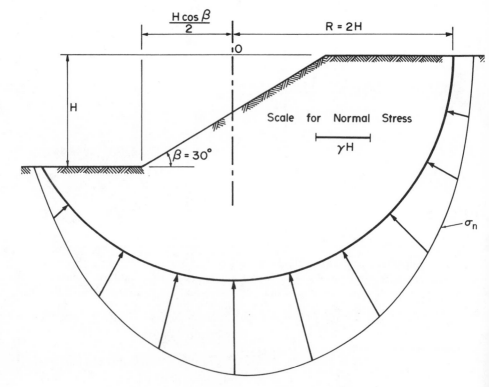

Fig. 12.13—Distribution of normal stress on circular arc
in linear elastic material subject to its own weight.

that the cases which Taylor (1937) considered represent reasonable limiting values. His results suggest that the error in the radius of the circle to which the force P is tangent is likely to be less than 15 per cent, and that the assumption shown in Figure 12.12a is reasonable.

A careful examination of Figure 12.12 reveals that it is actually possible to express the force polygon in Figure 12.12b in analytical terms, and therefore to find a direct solution to maximizing the $c_{req}/\gamma H$. Taylor (1937) carried out this calculation. Figure 12.10 shows the results of his analysis. The dashed lines considered above are for the case $\phi = 0$ with the failure surface passing below the toe. The solid lines are for slides passing through the toe. Notice that as long as ϕ is greater than zero, the critical slip surface will pass through the toe of the slope. Figure 12.10, called *Taylor's stability chart*, is very useful for providing an estimate of the stability number required for a slope in a homogeneous material. It also gives an indication of lower limits of height for slopes in layered systems in which the influence of the minimum strength material is considered to predominate.

Factor of Safety

In the foregoing discussion we have tacitly assumed that the frictional component of strength is fully mobilized, and we find the required cohesive exponent of strength for equilibrium. Thus it seems logical to define the factor of safety for such a calculation as the *factor of safety with respect to cohesion*, F_c, as

$$F_c = \frac{c}{c_{req}} \qquad (12\text{-}16)$$

where c is the cohesive component of shearing resistance.

Alternatively, we might assume that the cohesive component of strength is fully developed, and determine the required frictional component, $\tan \phi_{req}$, or the *factor of safety with respect to friction*, F_ϕ (as suggested by the analysis of infinite slopes of cohesionless soils):

$$F_\phi = \frac{\tan \phi}{\tan \phi_{req}} \qquad (12\text{-}17)$$

where ϕ is the angle of shearing resistance of the soil.

Let us inquire which of these parameters is more relevant to the evaluation of stability, keeping in mind that the relationship between the factor of safety as defined above, and the relative *degree* of safety is not known. We recognize that the shearing resistance of a soil can be mobilized only if straining of the soil occurs. Thus, given a particular level of strains in the soil, we must determine the relative degrees of mobilization of c and ϕ. Studies by Schmertmann and Osterberg (1960), Schmertmann (1962), and others indicate that c is mobilized at smaller strains than ϕ. This suggests that Equation 12–17 may be the more meaningful definition of factor of safety.

However, for soils possessing both frictional and cohesive components of strength, it is generally more convenient to think in terms of a factor of safety relative to the overall shearing resistance, that is,

$$FS = \frac{\tau_{ff}}{\tau_{f_{req}}} = \frac{c + \sigma_n \tan \phi}{c_{req} + \sigma_n \tan \phi_{req}} \qquad (12\text{--}18)$$

Alternatively, we may say

$$c_{req} + \sigma_n \tan \phi_{req} = \frac{c}{FS} + \sigma_n \frac{\tan \phi}{FS} \qquad (12\text{--}19)$$

From Equations 12–16, 12–17, and 12–19 we can see that this is equivalent to requiring that

$$F_c = F_\phi = FS \qquad (12\text{--}20)$$

If we wish to use the Taylor stability chart to determine $FS = F_c = F_\phi$ for a given slope, we must determine

$$\frac{c_{req}}{\gamma H} = \frac{c}{F_c \gamma H} \qquad (12\text{--}21)$$

in terms of a ϕ_{req} such that

$$\phi_{req} = \tan^{-1}\left(\frac{\tan \phi}{F_\phi}\right) \qquad (12\text{--}22)$$

To do this, we estimate an F_ϕ and determine ϕ_{req}. Using this value, find $c_{req}/\gamma H$ from the chart, and, from Equation 12–21, F_c. We can determine several such values for assumed F_ϕ and plot the results. This will lead us to the value of $FS = F_\phi = F_c$. The procedure is illustrated below:

Example 12.2

We wish to determine the factor of safety, FS, for the slope shown in Figure 12.14a.

Solution. Try $F_\phi = 1$, so that

$$\phi_{req} = \tan^{-1}(\tan \phi) = \phi = 5°$$

So, from Figure 12.10,

$$\frac{c_{req}}{\gamma H} = \frac{c}{F_c \gamma H} = 0.11$$

or

$$F_c = \frac{c}{0.110\gamma H} = \frac{500}{(0.110)(125)(25)} = \underline{1.45}$$

Then try $F_\phi = 1.20$:

$$\phi_{req} = \tan^{-1}\left(\frac{0.0872}{1.2}\right) = 4.1°$$

So

$$\frac{c_{req}}{\gamma H} = 0.12$$

or

$$F_c = \frac{500}{(0.12)(125)(25)} = \underline{1.33}$$

Then try $F_\phi = 1.50$:

$$\phi_{req} = \tan^{-1}\left(\frac{0.0872}{1.5}\right) = 3.3°$$

So

$$\frac{c_{req}}{\gamma H} = 0.126$$

or

$$F_c = \frac{500}{(0.126)(125)(25)} = \underline{1.28}$$

Plotting these values as in Figure 12.14b leads to FS = $F_c = F_\phi = 1.31$.

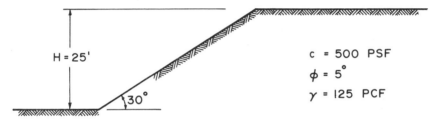

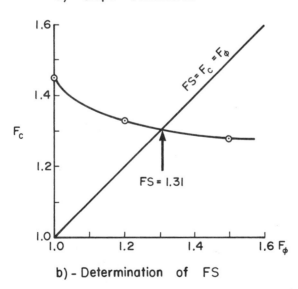

Fig. 12.14—Calculation of factor of safety with respect to shearing resistance by ϕ-circle method, Example 13.2.

12.8 APPLICATION OF METHOD OF SLICES TO STABILITY ANALYSIS

The largest number of stability analyses carried out are for cases in which the materials involved are stratified, or sufficiently inhomogeneous, that the Taylor stability chart provides only a very rough guide. In such cases, it is necessary to consider conditions within the slide mass, in order to take into account the material variability. The method by which this is done is called the *method of slices*.

Generalized Method of Slices

It is evident from Table 12.2 and Figure 12.10 that the assumption of a circular surface leads, in the cases investigated, to a more critical condition than that of a plane surface. We are immediately led to this question: Is there some other surface which would be still more critical in the general case? This question must be answered in the affirmative, especially in those cases where a layered system causes the failure surface to pass through a weaker stratum. Thus it is worthwhile to consider a generalized failure surface such as that shown in Figure 12.15a. This figure shows a slope with some loading acting on the surface, and a failure surface of a general shape. The slope may be composed of a number of materials which may vary in both the horizontal and vertical directions, with pore water pressures that vary with position. In order to investigate conditions at a point along the failure surface, the slope has been divided into a number of thin *slices*. A typical one of these, the ith slice, is shown in Figure 12.15b. The slice is of thickness Δx_i, with one of its faces at position x, the other face at position $x + \Delta x_i$. The forces which act on the slice, shown in Figure 12.15b, are the horizontal and vertical components of the resultant external loading, Q_{ih} and Q_{iv} respectively, and the weight of the slice W_i. The horizontal forces E_x and $E_{(x+\Delta x)}$ and shear forces T_x and $T_{(x+\Delta x)}$ act on the vertical faces, and represent the interaction between the slice and the slices next to it. The normal force on the bottom of the slice, N_i, and the shear force S_i, are the forces on the potential failure surface. We shall assume that Δx_i is small enough so that the segment of arc on the base can be considered as a straight line.

The situation with regard to evaluating these forces is quite discouraging. For n slices we find a total of $6n - 3$ unknowns, as shown in Table 12.3. If we assume that the slope is in equilibrium, we have at our disposal a total of $3n$ equations of equilibrium from which to determine these unknowns. Thus we must make some additional assumptions which will give us a sufficient number of conditions to make the equations available equal the number of unknowns. The assumptions we will make are:

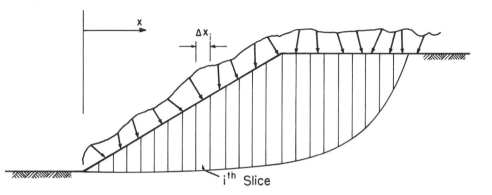

a) - Sliding   Mass

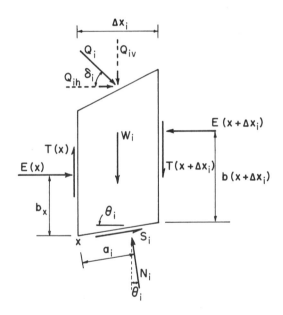

Fig. 12.15—Division of general slide mass into slices.

b) - Thin   Slice   of   Sliding   Mass

1. There is a fixed relationship between S_i and N_i, that is,

$$S_i = \frac{1}{FS}(c_i \Delta x_i \sec \theta_i + N_i \tan \phi_i) \qquad (12\text{--}23)$$

in which FS, the factor of safety, is assumed to be constant at every point along the failure surface, c_i is the cohesive component of strength, $\Delta x_i \sec \theta_i$ is the length of the failure surface within the slice.

TABLE 12.3
Factors in Equilibrium
Formulation of Slope
Stability for n Slices

Unknown	Number
E_x	$n - 1$
T_x	$n - 1$
b_x	$n - 1$
N_i	n
S_i	n
a_i	n
TOTAL UNKNOWNS	$6n - 3$

Because Equation 12–23 applies to each slice, it is equivalent to adding n additional equations so that there are now $4n$ equations available. However, the factor of safety, FS, is one additional unknown, so that the number of unknowns now becomes $6n - 2$.

2. We shall assume that at any point x, the relationship between E_x and T_x is only a function of x, that is,

$$\frac{T_x}{E_x} = \lambda f(x) \qquad (12\text{--}24)$$

in which λ is a constant, and $f(x)$ is some function of x which we shall assume. For n slices, Equation 12–24 corresponds to $n - 1$ additional relationships, increasing the number of equations to $5n - 1$. However, λ is an additional unknown bringing the total of unknowns to $6n - 1$.

3. We also require that Δx approaches zero, or at least becomes so small that a_i becomes negligible, that is $a_i \simeq 0$. This eliminates n unknowns, bringing the total to $5n - 1$ which is now the same as the number of equations.

Thus we see that for any assumed function $f(x)$, we can, in principle, determine the forces acting on each slice, and thereby evaluate the stability of the slope.

The ratio T_x/E_x really defines the orientation of the resultant force on the side of each slice. So $f(x)$ describes the way in which the direction of the resultant force on a vertical slice changes within the sliding mass. The choice of appropriate $f(x)$ is a matter of engineering judgment. The primary criteria applied to make this choice are that the location of the line of thrust, b_x be "reasonable," and that the required T_x on the sides of the slices be smaller than the maximum permitted by the shear strength of the soil (Morgenstern and Price, 1965). Because of the difficulty of choosing the appropriate $f(x)$ for a given situation, the generalized method of slices is not applicable to most problems in practice at the present time. However, it *is* the most rational method currently available and provides a standard of comparison for other simpler methods which will be discussed below.

We shall now consider two approximate methods which offer certain desirable computational simplifications.

Bishop Simplified Method of Slices

If we assume that the failure surface is an arc of a circle, we may eliminate N_i by taking moments about the center of the assumed circle. Doing this for the typical element shown in Figure 12.15b, with Δx_i suitably small, we find that

$\sum M_0 = 0$:

$$W_i R \sin \theta_i - S_i R + Q_{iv} R \sin \theta_i - Q_{ih} d_i$$
$$+ E_{x+\Delta x}(R \cos \theta_i - b_{x+\Delta x}) - E_x(R \cos \theta_i - b_x)$$
$$+ (T_{x+\Delta x} - T_x)R \sin \theta_i = 0 \quad (12\text{-}25)$$

where d_i is the moment arm of Q_{ih} about the center of the circle. Substituting Equation 12-23 into Equation 12-25 and dividing through by R, we can express the forces on the element in terms of the defined factor of safety:

$$W_i \sin \theta_i - \frac{1}{FS}(c_i \Delta x_i \sec \theta_i + N_i \tan \phi_i)$$
$$+ Q_{iv} \sin \theta_i - Q_{ih}\frac{d_i}{R} + E_{x+\Delta x}\left(\cos \theta_i - \frac{b_{x+\Delta x}}{R}\right)$$
$$- E_x\left(\cos \theta_i - \frac{b_x}{R}\right) + (T_{x+\Delta x} - T_x) \sin \theta_i = 0 \quad (12\text{-}26)$$

Our definition of factor of safety will not change if we now sum Equation 12-26 over the n slices. In this case the moment of the forces on the sides of the slice, E and T, will be cancelled by the moment of the same force acting in the opposite direction on the adjacent slices. Thus the remaining expression for the factor of safety is

$$FS = \frac{\sum\limits_{i=1}^{n}(c_i \Delta x_i \sec \theta_i + N_i \tan \phi_i)}{\sum\limits_{i=1}^{n}\left[(W_i + Q_{iv}) \sin \theta_i - Q_{ih}\frac{d_i}{R}\right]} \quad (12\text{-}27)$$

In order to determine N_i, we sum the forces acting in a vertical direction:

$\sum F_v = 0$:

$$W_i + Q_{iv} - N_i \cos \theta_i - S_i \sin \theta_i + T_{x+\Delta x_i} - T_x = 0 \quad (12\text{-}28)$$

Substituting Equation 12-23 into Equation 12-28 and rearranging terms, we solve for the normal force:

$$N_i = \frac{W_i + Q_{iv} - \dfrac{c_i\,\Delta x_i\,\tan\theta_i}{FS} + T_x - T_{x+\Delta x_i}}{\cos\theta_i + \dfrac{\sin\theta_i\,\tan\phi_i}{FS}} \tag{12-29}$$

We can eliminate N_i from Equation 12–27 by substituting Equation 12–29 into the expression. After some simplification Equation 12–27 becomes

$$FS = \frac{\sum\left[\dfrac{c_i\,\Delta x_i + [(W_i + Q_{iv}) + (T_x - T_{x+\Delta x_i})]\tan\phi_i}{\cos\theta_i + \dfrac{\sin\theta_i\,\tan\phi_i}{FS}}\right]}{\sum\left[(W_i + Q_{iv})\sin\theta_i - Q_{ih}\dfrac{d_i}{R}\right]} \tag{12-30}$$

This expression gives the factor of safety in an *implicit* form in which all the quantities in the equation are known except FS, T_x, and $T_{x+\Delta x_i}$. The result is as general as that for the "accurate" method except that a *circular* failure surface has been assumed.

Recognizing that equilibrium requires

$$\sum(T_x - T_{x+\Delta x_i}) = 0 \tag{12-31}$$

if we further assume that

$$\sum\left[\frac{(T_x - T_{x+\Delta x_i})\tan\phi_i}{\cos\theta_i + \dfrac{\sin\theta_i\,\tan\phi_i}{FS}}\right] = 0 \tag{12-32}$$

Equation 12–30 becomes a simple implicit expression in a single unknown, the factor of safety, FS:

$$FS = \frac{\sum\left[\dfrac{c_i\,\Delta x_i + (W_i + Q_{iv})\tan\phi_i}{\cos\theta_i + \dfrac{\sin\theta_i\,\tan\phi_i}{FS}}\right]}{\sum\left[(W_i + Q_{iv})\sin\theta_i - Q_{ih}\dfrac{d_i}{R}\right]} \tag{12-33}$$

We can solve this expression readily by an iterative approach, which is particularly effective when using a digital computer. Equation 12–33 can be reformulated conveniently for this purpose as described by Little and Price (1956). If the equation is solved manually, Figure 12.16 is of assistance. This figure gives the quantity

$$\cos\theta_i + \frac{\sin\theta_i\,\tan\phi_i'}{FS} \;=\; \frac{1}{M_i}$$

for various values of θ_i and $\tan\phi_i/FS$.

Note: θ_i is + When Slope of Failure arc is in
Same Quadrant as Slope of Ground Surface

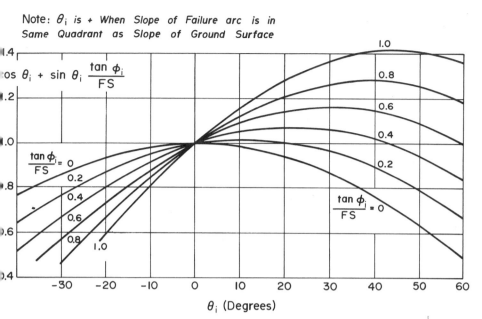

Fig. 12.16—Values of $\cos \theta_i + \sin \theta_i (\tan \phi_i/FS)$ for use in simplified method of slices. (After Janbu, et al., 1956.)

$\frac{1}{M_i}$

Bishop (1955) investigated many cases and found that generally less than a 2 per cent error was introduced by the assumption stated in Equation 12–32. This approach is called the *Bishop simplified method of slices.*

Note that Equation 12–33 assumes that the surface load Q_i affects the failure surface stresses only beneath the *i*th element. This neglects the distributional effect of the slope material, and may lead to significant error if the external loads are concentrated. Whether the error is on the safe side or unsafe side depends upon the location of Q_i. In the case of relatively uniformly distributed external loads, the assumption is reasonable.

Effect of Pore Water Pressure

In the preceding discussion we tacitly assumed the use of *total* stresses in regard to both stress and strength parameters. There are cases where the total stresses and total strength parameters will lead to accurate answers. However, in many other important circumstances, failure to consider effective stresses will lead to serious errors. Incorporating pore water pressures into Equation 12–23:

$$S_i = \frac{1}{FS} \left[c_i' \, \Delta x_i \sec \theta_i + (N_i - u_i \, \Delta x_i \sec \theta_i) \tan \phi_i' \right] \qquad (12\text{–}34)$$

where c_i' is the cohesive component of strength in terms of effective stress, ϕ_i' is the apparent angle of shearing resistance in terms of effective stresses and u_i is the pore water pressure on the base of the slice. Following a procedure analogous to that in the development of Equation 12–26 through 12–33 will lead us to a computed factor of safety in terms of effective stresses:

$$FS = \frac{\sum \left[\dfrac{c_i' \Delta x_i + (W_i + Q_{iv} - u_i \Delta x_i) \tan \phi_i'}{\cos \theta_i + \dfrac{\sin \theta_i \tan \phi_i'}{FS}} \right]}{\sum \left[(W_i + Q_{iv}) \sin \theta_i - Q_{ih} \dfrac{d_i}{R} \right]} \qquad (12\text{–}35)$$

Recall that Equations 12–33 or 12–35 permit determination of the factor of safety for a specific assumed circular sliding surface. The factor of safety for the slope will be that factor of safety corresponding to the surface along which failure is most likely to occur. Thus it is necessary to investigate a number of potential sliding surfaces, and select from among them the one which produces the lowest factor of safety.

Fellenius Method of Slices

Fellenius (1927) developed a method of slices similar to that described above except for a different assumption about the resultant forces on the sides of the slice. If it is assumed that the resultant of the forces E_x and T_x is collinear and equal in magnitude to the resultant of $E_{x+\Delta x}$ and $T_{x+\Delta x}$, then we can conveniently eliminate the effect of the side forces by summing forces normal to the base of the slice to determine N_i, rather than summing forces in a vertical direction as we did in Equation 12–28. This leads to

$$FS = \frac{\sum [c_i' \Delta x_i + [W_i \cos \theta_i + Q_i \sin (\delta_i + \theta_i)] \tan \phi_i']}{\sum \left[(W_i + Q_{iv}) \sin \theta_i - Q_{ih} \dfrac{d_i}{R} \right]} \qquad (12\text{–}36)$$

where δ_i is the angle between the force Q_i and the horizontal, measured clockwise from the horizontal.

Whitman and Bailey (1967) have shown that the Fellenius method gives results which are conservative relative to the more "accurate" method. Unfortunately the error ranges from approximately 5 to 40 per cent depending upon the factor of safety, the central angle of the circle chosen, and the magnitude of the pore water pressure. Even in terms of total stress analysis, the error is a function of factor of safety and the central angle of the trial circle, as illustrated in Figure 12.17. Furthermore, the location of the critical circle using the Bishop (1955) method is generally closer to that observed in failures in the field. Thus although the Fellenius method is simpler, the Bishop (1955) method is preferable because of its more consistent results.

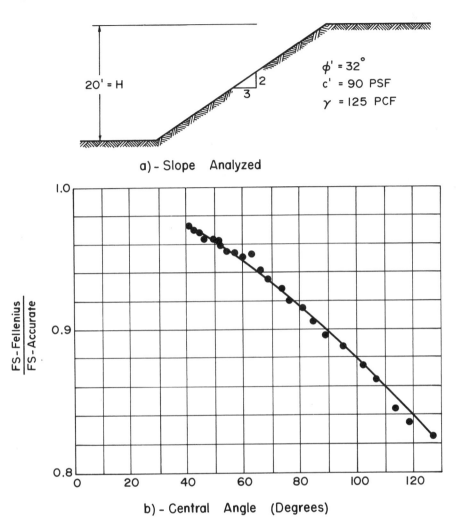

a) - Slope   Analyzed

b) - Central   Angle   (Degrees)

Fig. 12.17—Error in Fellenius method with
dry soil. (From Whitman and Bailey, 1967.)

12.9   COMPOSITE SLIP SURFACES

If a slope is underlain by one or more strata of very soft or loose material
the failure surface may not be even approximately circular. This is illustrated
in Figure 12.2d and 12.18. In Figure 12.18 is shown a *composite* surface
of sliding in which a maximum amount of the surface passes along the soft
material. This type of failure is commonly called "failure by spreading."
In such a case, the failure surface is frequently much longer than an assumed
circular failure surface. Failures of this sort often occur in the case of fills

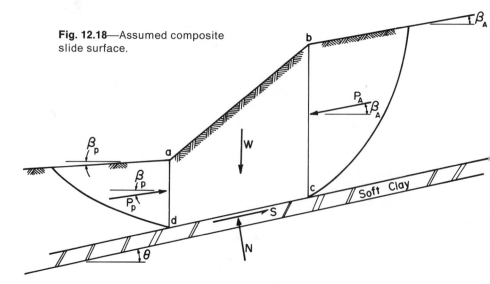

Fig. 12.18—Assumed composite slide surface.

placed upon stratified subsurface soils containing relatively thin strata of soft clay. In these circumstances, the rapidity, and to some extent the severity of movement depend upon whether the clay is relatively homogeneous, or whether it contains thin partings of more pervious materials. In the latter case, the distribution of hydrostatic pressure created by construction of the fill, through these pervious materials in the clay may reduce the strength of the clay and lead to a dramatic failure.

Composite slip surfaces may also occur in natural slopes underlain by an especially weak stratum, or one which is subject to large variations in hydrostatic pressure leading to reduction in strength. Again in this case, the failure surface tends to follow such a stratum and deviates considerably from an assumed circular surface. The slide into the Vaiont Dam Reservoir was of this type.

The most commonly applied solution for a composite failure surface is illustrated in Figure 12.18. In this case the stability of the wedge *abcd* is analyzed. It is assumed that active earth pressure acts on the face $\overline{bc}$ and passive earth pressure acts on the face $\overline{ab}$. The magnitude of the required shear force along the base can then be compared to the maximum available shear force along the base. Assuming P_A and P_P to act parallel to the slope which creates them, the required shear force S_{req} is

$$S_{req} = P_A \cos(\beta_A - \theta) - P_P \cos(\beta_P - \theta) + W \sin \theta \quad (12\text{--}37)$$

where the angles θ, β_A, β_P are shown in Figure 12.18. The available shear force along the surface $\overline{bc}$, in terms of effective stresses, is

$$S_{max} = c'L + [W\cos\theta + P_A \sin(\beta_A - \theta) - P_P \sin(\beta_P - \theta)]\tan\phi' \quad (12\text{--}38)$$

Then the factor of safety is defined as

$$FS = \frac{S_{max}}{S_{req}}$$

$$= \frac{c'L + [W\cos\theta + P_A\sin(\beta_A - \theta) - P_P\sin(\beta_P - \theta)]\tan\phi'}{P_A\cos(\beta_A - \theta) - P_P\cos(\beta_P - \theta) + W\sin\theta} \quad (12\text{-}39)$$

Note, however, that because *active* and *passive* pressures are assumed to act on the sides of the wedge, it is implied that the strains required to mobilize active and passive earth pressures in the upper material are very much less than those required to mobilize the shear strength of the softer underlying materials. Thus the factor of safety is defined with regard to the shear resistance on the bottom of the center wedge, assuming fully mobilized active and passive pressures. It is then necessary to determine the locations of $\overline{ad}$ and $\overline{bc}$ to yield the minimum FS.

Tension Cracks

If a cohesive soil at the surface is subject to drying stresses, tension cracks may form to a maximum depth z_t,*

$$z_t = \frac{2c}{\gamma} \quad (12\text{-}40)$$

in which c is the cohesive components of shearing resistance and γ is the unit weight of the soil. Such cracks have two effects:

1. The crack constitutes a separation along which no shear stress is transmitted. Thus the failure mass is delineated by circular surface up to the point at which the circle intersects the crack, above which the failure surface is the crack itself.
2. If rainfall fills an open crack with water, the hydrostatic pressure acting on the side of the crack adds to the overturning moment, perhaps to a significant extent.

12.10 MAXIMUM SHEAR STRESS METHOD

An alternative approach to the assumption of slip surfaces for stability analysis is to insure that the shear stress imposed by the slope does not exceed the shear strength within or under the slope at any point. This approach has the advantage of simplicity but it suffers from three important disadvantages:

1. Currently available solutions for determining the magnitude of the maximum shear stress within and under an earth slope, such as the

---

* This point is discussed further in Section 13.3.

results shown in Figures A.22–A.26 are based upon the assumption of linear elasticity. These solutions do not account for the redistribution of stress which occurs as a result of the nonlinear stress-strain curve at stress levels approaching failure.

2. Because this approach involves the comparison of the maximum shear *stress* at a point, with the shear *strength* at that point, it is really only helpful in the case where failure will be initiated in a cohesive material for which the immediate shear strength will be independent of the normal pressure, and therefore less dependent upon orientation than that of a granular material.

3. This method considers only conditions at the most critical point in the mass. The method in which a slip surface is assumed, considers *average* conditions along the potential slip surfaces. Hence it is to be expected that the maximum stress approach would lead to more conservative values than an assumed slip surface approach. This is illustrated in the following example.

Example 12.3 ───

Consider the simple homogeneous slope shown in Figure 12.19. From the Taylor stability chart, Figure 12.10, the stability number, $c_{req}/\gamma H = 0.181$. Thus the factor of safety for this slope, assuming a circular sliding surface is 1.85.

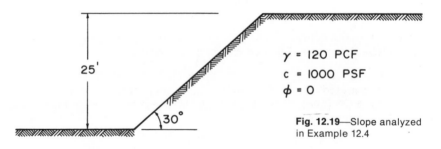

$\gamma = 120$ PCF
$c = 1000$ PSF
$\phi = 0$

Fig. 12.19—Slope analyzed in Example 12.4

By contrast, the maximum shear stress from Figure A.23, is

$$\tau_{max} = 0.32\gamma H = 0.32(120)(25) = 960 \text{ psf}$$

and occurs at a depth of approximately 20 feet below the base of the slope. Thus by this method the factor of safety against failure is only 1.02.

───

The maximum shear stress method may be useful in cases where progressive failure would prevent simultaneous development of peak shear strength at each point of the potential failure surface. This subject is discussed more fully in the following section.

12.11   SELECTION OF STRENGTH PARAMETERS FOR STABILITY ANALYSIS

Effect of Drainage Conditions

The appropriate strength parameters to be substituted into Equation 12–33, 12–35, or 12–39 depend upon a number of factors. Among the most important of these are the drainage conditions corresponding to the critical condition of the slope, that is, the minimum FS. In order to visualize this let us examine three examples described by Bishop and Bjerrum (1960) and Wu (1966).

Example 12.4 ———————————————————————

We shall first investigate the case of an embankment constructed on a deposit of clay as shown in Figure 12.20a. Let us suppose that the deposit is in equilibrium, with the pore water pressure at a typical point such as point P equal to the hydrostatic pressure due to the groundwater table. Assuming that the embankment is constructed at a uniform rate, the increase in load at the surface is shown schematically in Figure 12.20b. Similarly, the shear stress imposed by the embankment at point P is shown in this figure as a function of time, increasing through construction, and constant thereafter. The effect of the increased stress on the pore water pressure, assuming that no drainage of the clay occurs during the construction period, is shown in Figure 12.20c for two values of the Skempton A-factor: $A = 1$ corresponding to a normally consolidated soil, and $A = 0$ corresponding to a moderately over-consolidated soil. Note that the pore water pressure increases uniformly with the construction load until the end of the construction period at which time the soil consolidates and the pore water pressure dissipates, approaching its initial value. Because there is no change in water content during the construction period, the strength of the soil remains approximately constant, as indicated in Figure 12.20d. As the soil consolidates following construction, the strength of the clay at point P increases. If one were to evaluate the progress of consolidation at many points on the potential failure surface shown in Figure 12.20a, it would be possible to determine the factor of safety as a function of time. Because the undrained shear strength of the soil does not change during construction, while the shear stresses increase, the factor of safety decreases to a minimum at the end of construction as shown in Figure 12.20e. After construction the shear stress remains constant at each point and the shear strength increases, so the factor of safety increases correspondingly. Thus, in this case, the *critical* condition is clearly *at the end of construction.* Because the shearing resistance of the soil is approximately constant during construction, it is possible to estimate the stability of the embankment assuming that the total shearing resistance is equal to the undrained shearing resistance c, and $\phi = 0.*$ This approach is termed a "$\phi = 0$ analysis." It is especially convenient

---

* Recall from Section 8.6 that the undrained shearing resistance is a function of the stress path, and that the undrained strength under approximately plane strain conditions *in situ* is not quite the same as that for an axisymmetric laboratory test.

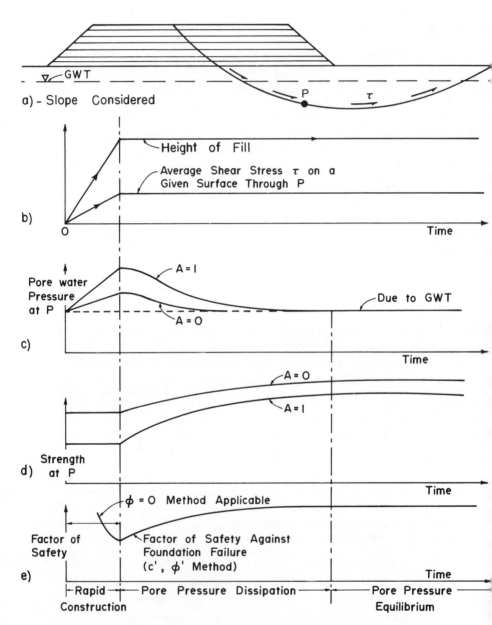

a) - Slope Considered

b)

Height of Fill

Average Shear Stress τ on a
Given Surface Through P

O   Time

c)

Pore water
Pressure
at P

A = I

A = O

Due to GWT

Time

d)

Strength
at P

A = O

A = I

Time

e)

Factor of
Safety

ϕ = O Method Applicable

Factor of Safety Against
Foundation Failure
(c', ϕ' Method)

Time

Rapid | Pore Pressure Dissipation | Pore Pressure
Construction   Equilibrium

Fig. 12.20—Variation with time of shear stress, pore water pressure, strength and factor of safety for a saturated clay foundation beneath a fill. (From Bishop and Bjerrum, 1960.)

because total stresses are used, pore water pressures do not need to be measured in the laboratory, and if the embankment is safe under the total stress conditions at the end of construction, the factor of safety will increase with time.

Example 12.5 ————————————————————————————————

We shall now investigate the degree of instability arising from a stress relief such as occurs in connection with a cut in a cohesive soil, illustrated in Figure 12.21a. The change in the height of material above a representative point P on a potential failure surface is shown in Figure 12.21b along with the average shear stress on a plane passing through point P. As in the previous example, we shall assume that the loading changes linearly with time, and that no change in water content occurs during the construction process. Figure 12.21c shows the pore water pressure at point P as a function of time, for two values of the A factor: $A = 1$ and $A = 0$. The initial and final values are of course the same, and are determined by the level of the original and final groundwater table at that point. The strengths corresponding to these normally consolidated and over-consolidated materials are shown in Figure 12.21d. Again, in the case of a fully saturated cohesive soil, the strength remains approximately constant during the construction period. As a result of the reduced load, however, the soil will swell as the negative pore water pressure is relieved. The strength will *decrease* with time.

So we see that during the period of rapid excavation the shear stress will increase while the shear strength remains constant leading to a continuously reduced factor of safety as shown in Figure 12.21e. Furthermore, at the end of the construction process, even though the shear stress is then constant, the reduction in strength will lead to a continued deterioration of the factor of safety. Thus for a cut, the factor of safety will be a minimum after the pore water pressure has increased to the hydrostatic value, that is, in the long term. Hence, the use of the *existing* undrained shearing resistance to evaluate the stability of a cut slope of the sort described, would be unsafe. In this case it is necessary to determine the *effective stress parameters* for use in the stability analysis.

In order to evaluate the long-term stability of such a slope incorporating the effective stress parameters, it is necessary to know the magnitude of the pore water pressure after equilibrium has been attained. This may require evaluation of steady-state groundwater flow. A discussion of the techniques involved will be deferred to Chapter 14. Also considered there is the dangerous case of "rapid drawdown" of a reservoir.

Example 12.6 ————————————————————————————————

A third important, and potentially dangerous category of problem, is the case in which the total stresses do not change, but a change in pore water pressure produces a concomitant change in effective stress, and thereby strength (Wu, 1966). Such a case is illustrated in Figure 12.22. A stable slope is shown in Figure 12.22a. At a time subsequent to establishment of the slope, additional construction such as driving of piles illustrated in the figure, may induce large pore water pressures. This will cause flow of pore water away from the piles as the soil in the vicinity of the loaded area consolidates. Pore water pressure changes with time at a typical point (point b) near the piles, are shown in Figure 12.22c. As the pore water flows laterally away from point b and toward point P, the pore water pressure at point P is temporarily increased

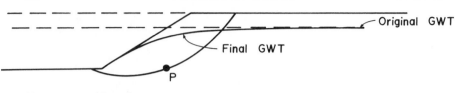

a) - Slope Considered

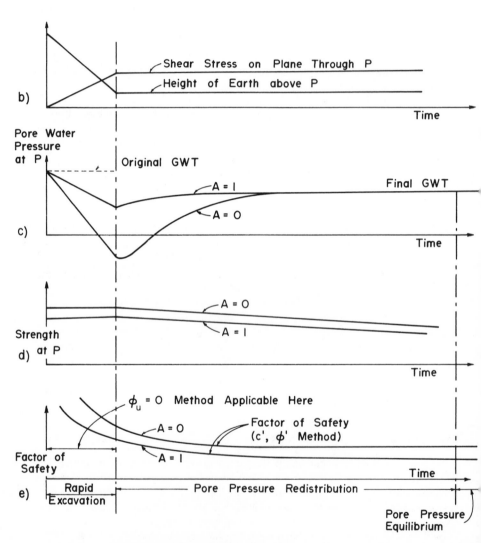

Fig. 12.21—The changes in pore pressure and factor of safety during and after the excavation of a cut in clay. (From Bishop and Bjerrum, 1960.)

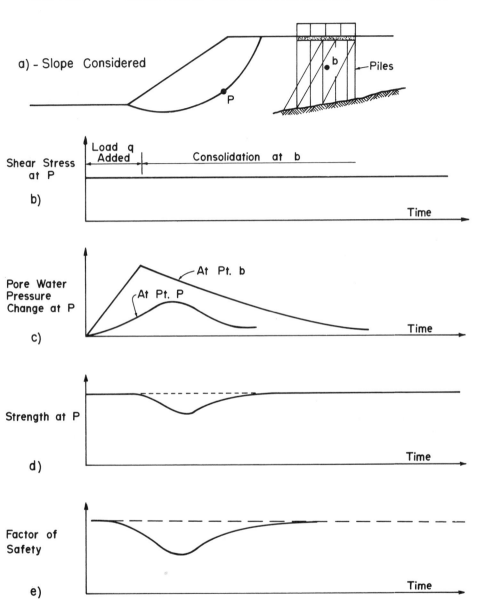

a) - Slope Considered

Shear Stress at P

b)

Pore Water Pressure Change at P

c)

Strength at P

d)

Factor of Safety

e)

Fig. 12.22—Stability conditions of a slope with load. (From Wu, 1966.)

(Figure 12.22c). In the absence of an appropriate increase in total stress, the strength at point P is reduced (Figure 12.22d), and the factor of safety of the slope will also be reduced (Figure 12.22e). Thus during some intermediate time period the factor of safety for this slope will be less than it is either *before* the additional construction, or *after* the excess pore water pressure has dissipated. Failure to consider this potentially critical situation may lead to unfortunate results. Clearly in this case, effective stress analysis must also be used.

It is evident then that the strength parameters selected for incorporation into the analysis must be chosen on the basis of the conditions which are anticipated to provide a critical situation.

Relationship of Predictions from Analysis to Field Observations

The number of assumptions introduced into the analysis makes it very difficult to decide *a priori* the degree of confidence one should assign to the results obtained from such an analysis. Among the uncertainties associated with this problem are:

1. Slope failures are generally three-dimensional in nature.
2. The results of strength tests in the laboratory (usually triaxial compression tests) bear a complex relation to the strength under states of stress in the field. Sample disturbance also influences these test results.
3. Anisotropy of strength properties in the field may cause a variation in strength parameters along the curved failure surface.

In spite of these uncertainties, the data in Tables 12.4 and 12.5 suggest that with properly obtained shear strength information, the appropriate application of $\phi_u = 0$ and effective stress (c', ϕ') analyses lead to reasonable estimates for factor of safety. In those cited cases which have actually failed, the computed factor of safety, using the appropriate strength parameters, is close to unity. For those slopes which are still stable, the computed factor of safety is significantly greater than one.

TABLE 12.4

Immediate Failures in Normally and Slightly Over-consolidated Clay: Undrained Analysis ($\phi_u = 0$)

Locality	w	w_L	w_p	Computed FS	Reference
Foundations					
1. Kippen	50	70	28	0.95	Skempton (1942)
2. Transcona silo	50	110	30	1.09	Peck and Bryant (1953)
3. Fredrikstad oil tank	45	55	25	1.08	Bjerrum and Overland (1957)
Embankments					
4. Chingford	90	145	36	1.05	Skempton and Golder (1948)
5. Newport	50	60	26	1.08	Skempton and Golder (1948)
6. Gosport	56	80	30	0.93	Skempton (1948)
Cut slopes					
7. Congress St.	24	33	18	1.10	Ireland (1954)

SOURCE: Wu (1966).

TABLE 12.5
Long-Term Stability in Clay: Effective-Stress Analysis (c', ϕ')

Locality	Clay Type	w	w_L	w_p	Computed FS	Reference
Slope failures						
1. Londalen	Overconsolidated, intact	31	36	18	1.05	Sevaldson (1956)
2. Drammen	Normally consolidated	31	30	19	1.15	Bjerrum and Kjaernsli (1957)
Stable slopes						
3. Drammen	Normally consolidated	31	30	19	1.25	Bjerrum and Kjaernsli (1957)
4. Bakklandet A.	Normally consolidated	28	25	18	1.85	Bjerrum and Kjaernsli (1957)
5. Borregaard	Normally consolidated	18	18	11	1.25	Bjerrum and Kjaernsli (1957)

SOURCE: Wu (1966); see that work for references.

Effect of Different Materials

When a slope or embankment consists of several materials with significantly different stress-strain curves it is incorrect to determine the factor of safety for the failure surface corresponding to the peak strengths from both curves. The reason for this is illustrated in Figure 12.23. A schematic cross-section of an earth dam consisting of two materials is shown in Figure 12.23a. Material ① is a flexible clay core with a stress-strain curve shown in Figure 12.23b. Material ② is a stiff, compacted slightly cohesive granular material with a stress-strain curve indicated in Figure 12.23b. Because the failure mass is presumed to move as a rigid body, displacements will be the same at each point along the potential failure surface. Noting that approximately 2/3 of the failure surface passes through material ① and 1/3 passes through material ② an average stress-strain curve can be constructed as in Figure 12.23c. It is apparent from this curve that the factor of safety computed for any value of strain would be significantly less than that determined by assuming that the peak strength is mobilized simultaneously in both materials. Thus it is essential to consider the stress-strain characteristics of dissimilar materials in deciding what strength parameters to use in a stability analysis.

Effect of Progressive Failure

In our previous discussion of slope stability it was assumed that the ratio of the shearing stress at each point on a potential failure surface to the shearing strength at that point is a constant for the entire surface. The discussion above illustrates the fact that this assumption is not always justified. Another case in which incorrect application of this assumption

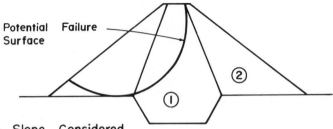

a) - Slope   Considered

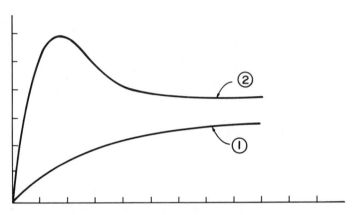

b) - Stress - Strain   Relationships   for   Materials   in   Slope

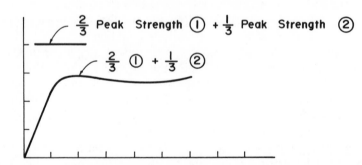

c) - Average   Stress - Strain   Relationship   Along   Failure   Surface

Fig. 12.23—Effect of differing stress-strain curves for materials in slope. (After Leonards, 1962.)

may lead to a serious overestimate of the long-term stability of a slope involves *progressive failure.* In the classic description of this problem, Skempton (1964) described cases of failures which occurred in cuts in London clay as much as fifty years after they had been constructed. Progressive failure is a problem on slopes formed by over-consolidated clays, clay-shales, or stiff compacted cohesive materials. An essential prerequisite to such failures is a stress-strain curve, characteristic of brittle materials, of the type illustrated in Figure 12.24a. This is the result of a direct shear test conducted on weathered London clay (Skempton, 1964). The key feature of this diagram is the discrepancy between the maximum shear stress sustained by the material, the *peak strength*, and the stress sustained by the material after extended straining along the failure surface, the *residual strength*. Note that the peak strength is developed at very small displacement

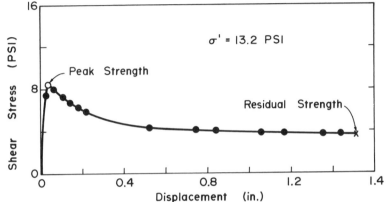

a) - Development of Peak and Residual Strengths in Direct Shear Test

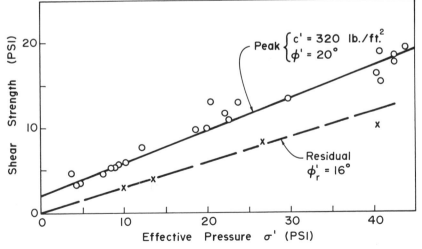

b) - Mohr Diagram for Peak and Residual Conditions

Fig. 12.24—Relationship between residual and peak strengths for London clay. (Modified from Skempton, 1964.)

or strain, whereas considerable displacement is required in order for the strength to reduce to the residual value.* Figure 12.24b shows the results of a number of such tests conducted on the London clay at different confining pressures. The strength parameters deduced from the results of such tests when the peak strength is considered are typical of many cohesive materials of this type. However, continued straining after the peak strength is reached appears to destroy any cohesive component of shearing resistance, as well as to reduce the frictional component of shearing resistance (note the residual strength curve in this figure). These data are typical of over-consolidated clays, and indicate that the residual strength must be considered a solely frictional effect in which the angle ϕ'_r is usually much less than the angle ϕ' corresponding to the peak strength.

There is considerable evidence that the residual shear strength parameter ϕ'_r is independent of the original strength of the clay and such factors as water content and liquidity index. It appears to be a parameter which depends only upon the basic constituents of the clay material (Skempton, 1964).

The difference between the peak and residual strengths becomes especially important under those circumstances in which strains occur to different degrees at different points within a slope. We can see how this might occur by referring to Figure 12.25. This figure shows the distribution of shearing stress on the base of a slope in a linear elastic material. The shearing stress is a maximum in the vicinity of the toe and decreases in a direction away from the toe. This distribution might be considered an approximation of that under a slope in a stiff clay.† In the event that the maximum shear stress developed exceeded the peak strength of the material, straining in the vicinity of the toe of the slope would occur, a failure plane would develop in that vicinity, and the strength would reduce to some lower value approaching the residual value. It is evident that in such a case the stress distribution would have to change, with the concentration of shear stress moving back into the slope. Were this process to continue, a continuous failure surface would develop progressively along some preferred path until the strength along that failure surface everywhere had been reduced. If sufficient straining took place, the strength on the surface would approximate the residual value. The evidence cited by Skempton (1964) and Bjerrum (1967b) indicates that this process may take a very long time, and that failure may not occur for many years after the construction which in fact produces the failure.

In a detailed consideration of this problem, Bjerrum (1967b) elucidated the combination of events which appears to lead to progressive failure in slopes. Highly over-consolidated clays and clay-shales tend to exhibit a

---

* The exceedingly large displacement required to reduce the shearing resistance to the true residual value may require a torsional ring shear apparatus, or less conveniently, reversal of shear direction.

† Duncan and Dunlop (1969) illustrated this point in a detailed analysis.

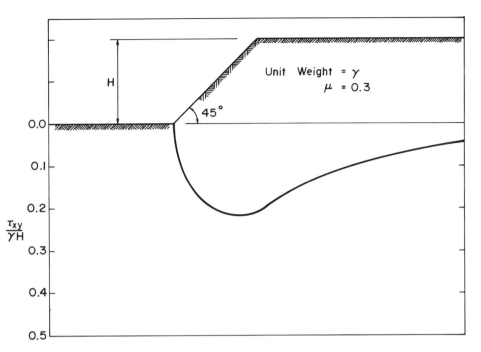

Fig. 12.25—Distribution of shear stress on base of slope in linear elastic material, $\mu = 0.3$.

greater discrepancy between the peak and residual strength values than normally, or lightly over-consolidated materials. In addition, the large horizontal stresses which remain in many over-consolidated materials lead to much more strain energy stored in the material than in a normally consolidated material under the same conditions. This large magnitude of stored internal energy produces an increased tendency for lateral expansion when a slope has been cut into such a material, either by man or nature. This tendency in turn promotes additional concentration of shear stress beneath the toe of the slope.

In those materials which are already weathered at the time they are unloaded in a horizontal direction, or in which the resistance to lateral expansion upon unloading is small, the formation of a slope will produce a relatively rapid horizontal movement. Progressive failure may accompany this movement but it will tend to occur relatively quickly. Conversely, in those materials which have a strong resistance to lateral expansion the concentration of shear stress beneath the toe may be insufficient to produce the localized failure required to initiate the process. However, subsequent weathering may reduce the peak strength sufficiently to permit this to happen after an extended period of time. Thus in considering the potential for progressive failure we must consider first the possibility that weathering will cause a gradual reduction in strength leading to an eventual progressive failure. This appears to be the basis for the long-term failures in the London clay.

In such cases, a stability analysis considering the potential of long-term failure must be conducted using the residual strength parameter rather than peak strength parameters. Table 12.6, from Bjerrum (1967b), illustrates a number of case histories in which progressive failure occurred although stability analyses conducted using *peak* strength parameters predicted an acceptable factor of safety. For those cases in which a residual strength parameter was available the table indicates a close correspondence between ϕ'_r and the ϕ' back-calculated from the actual failure. Thus, when the potential for progressive failure appears significant, the stability analysis should be conducted using the *residual* strength rather than *peak* strength.

In those materials in which the tendency for lateral expansion is small, particularly nonplastic materials, the danger of progressive failure is much less than for the more plastic clays and clay-shales (Bjerrum, 1967b).

TABLE 12.6
Case Histories of Progressive Failures

Name of Slide and Reference	Natural Slope (= NS) Cut (= C)	Data of Clay			Shear-Strength Data			Computed from Slide (c' = 0) (deg)
		w	w	w_p	c' (tons per square meter)	ϕ' (deg)	ϕ'_r (deg)	
Culebra Panama (N.A.S. 1924)	C	12	80	35	—	—	10	—
California coast (Gould, 1960)	NS	19	68	28	—	25	—	12
Waco Dam (von Auken, 1963)	—	17	80	22	4	17	6	7–8
Saskatchewan (Ringheim, 1964)	C	32	115	23	4	20	6	7–9
Dunvegan Hill (Hardy et al., 1962)	NS	22	50	24	3.8	20	—	9
Little Smoky (Hardy et al., 1962)	NS	24	42	18	2.0	22	—	12
Seattle Freeway (Peck, private communication)	C	30	55	23	—	30	—	13
Balgheim (Einsele, 1961)	C	27	61	25	1.5	18	17	14–17
Sandnes (NGI, 1964)	C	36	60	30	1.3	22	12–18	15–17
Jackfield (Henkel and Skempton, 1954)	NS	21	44	22	1.1	25	19	17
Walton's Wood (Skempton, 1964)	NS	—	53	28	1.6	21	13	13
London clay slopes (Skempton, 1964)	NS	33	80	29	1.6	20	16	16
Sudbury Hill (Skempton, 1964)	C	31	82	28	0.3	17	16	15

SOURCE: After Bjerrum (1967); for information on references, see Bjerrum (1967b).

12.12 STABILITY OF TIP NO. 7 AT ABERFAN, WALES

The spoil banks produced by wasted material that is excavated to reach underground coal seams are large and potentially dangerous structures, whose safety and adequate performance must be insured by appropriate engineering design and evaluation. The Inquiry Tribunal concluded that the failure of the National Coal Board to recognize this fact and establish a policy concerning spoil banks was the basic cause of the disastrous slide at Tip No. 7 at Aberfan, Wales (Tribunal Report, 1967, p. 131). In arriving at its conclusion, the Tribunal was assisted by the results of a number of detailed technical investigations of the conditions at Aberfan which led to the failure. Several of these have been published.* We shall review briefly the situation there, and use information from these reports (Bishop et al., 1969; Woodland, 1969) to carry out an example slip circle stability analysis.

A simplified schematic cross-section through Tip No. 7 is shown in Figure 12.26. The main body of the tip forms a slope approximately 200 ft high with an average surface slope of about 36°. The tip consists of excavated mine waste, which is predominantly broken pieces of shale rock, along with finer

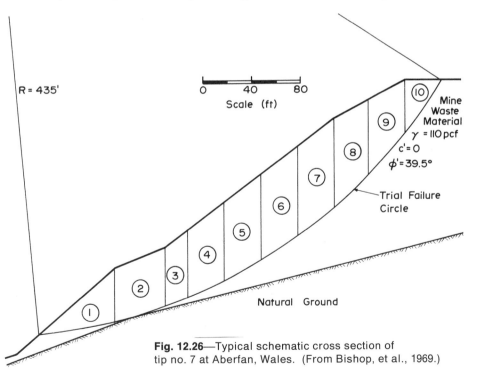

Fig. 12.26—Typical schematic cross section of tip no. 7 at Aberfan, Wales. (From Bishop, et al., 1969.)

* *A Selection of Technical Reports Submitted to the Aberfan Tribunal*, H. M. Stationery Office, London, 1969.

tailings from washing operations, and other assorted debris. Extensive laboratory strength testing of undisturbed samples of the bulk of the waste material indicated drained effective stress strength parameters $c' = 0$, $\phi' = 39.5°$ (Bishop et al., 1969).

The tip was located on the western slope of the Taff River valley. The natural ground slopes at approximately 14° at the tip. Surficial deposits vary from a few feet to more than 70 ft thick. They consist primarily of a dense glacial drift "boulder clay" with an upper mantle of detritus accumulated due to postglacial movement of materials downslope from higher elevations. Underlying the drift is a water bearing sandstone interbedded with mudstone and coal layers. The water pressure in the sandstone produced a "spring" which emanated from the natural ground beneath the toe of the tip and caused periodic erosion and undercutting of the toe. The magnitude of the water pressure producing the spring varied in relation to the rainfall at a given time.

Considerable evidence was assembled by Bishop et al., (1969) that Tip No. 7 had been unstable and sliding for a number of years prior to 1966. These slips appear to have been initiated by removal of material from the toe by the "spring," and to have been driven by the continued dumping of waste materials at the top of the tip. The slips had occurred *mainly along the same deep seated slip surface within the tip.* The trial failure circle shown in Figure 12.26 is a close approximation to this failure surface determined after the 1966 slide.

The sequence of events at the time of the disaster appeared to be the following (modified from Bishop et al., 1969):

1. As a result of heavy but not unusual rains, the water pressure in the sandstone beneath the tip was increased. This led to an increase in water pressure in the toe of the tip sufficient to reactivate sliding movements along the preexisting slip surface. The evidence suggests that the pore water pressure in the tip itself was probably very small.

2. Once commenced, sliding continued slowly for several hours leading to outward movement of the toe and a settlement of approximately 20 ft at the crest of the tip.

3. The large shearing strains in the nearly saturated loose material at the toe induced a flow slide of this material (recall the flow slide illustrated in Figure 12.3). The flow carried additional saturated surface soil lying downslope from the tip. This flow, which is depicted in Figure 12.4, appears to have been the main cause of damage to the town.

4. As the flowing mass was removed from the toe, a secondary slide in the less wet material behind it occurred. This can be seen in Figure 12.4. The secondary slide cut into the natural drift and released the water impounded in the sandstone.

5. The water gushing from the jointed sandstone mixed with tip material to form a fluid "mud-run." The "mud-run" scoured a deep channel

in the flow slide and carried the additional material further into the town. This is illustrated in Figure 12.4.

Slip Circle Analysis

The conditions which led to the initial slide are amenable to analysis, particularly because there does not appear to have been significant pore water pressure in the tip material itself prior to the slide. We shall, therefore, investigate the factor of safety for sliding along the circular surface shown in Figure 12.26, using Equation 12–35.

The trial failure mass has been divided into ten slices. Soil parameters are given in the figure. Calculations are tabulated in a convenient form in Table 12.7. Three trial factors of safety were assumed; the corresponding calculated FS are shown in the table. These were then plotted, as shown in in Figure 12.27, to yield the calculated result FS = 1.45.*

This apparently anomalous result is explained by the fact that the shearing resistance along the preformed slip surface differed from that in the bulk of the tip. Shear tests conducted on material taken from the slip zone revealed an average *residual* angle of shearing resistance $\phi_r' = 18°$! (Bishop et al., 1969). This difference between the material in the shear zone and that in the main mass appeared to result partly from physical breakdown of particles due to shear, and partly due to chemical changes in the shale clay minerals.

In fact, a reduction in the operative ϕ' to $29°$ is all that is necessary to lead to failure along the circular surface shown in Figure 12.26. Thus it appears that progressive failure effects, resulting from the large periodic movements along the initial slip surface were important factors in the disaster.

12.13 SUMMARY OF KEY POINTS

The key points discussed in the previous sections are summarized below. Reference is made to the place in the text where the original more detailed description can be found:

1. Instability of both natural and man-made slopes constitutes a major share of civil engineering failures. Slope failures can be classified usefully into three types (Section 12.2):
    a. Falls are distinguished by a rapidly moving failure mass which travels mostly through the air. Falls are not generally subject to analysis.
    b. Slides result from shear failure along one or more surfaces. The sliding mass remains coherent although it may deform greatly.
    c. Flows occur when the displaced mass exhibits movements which resemble those of a viscous liquid.

---

* The result for a noncircular surface which approximates the actual surface somewhat more closely is essentially similar.

TABLE 12.7
Calculations for Trial Failure Circle Shown in Fig. 12.27 for Tip No. 7 at Aberfan, Wales

Slice No. (A)	Δx (ft) (B)	Δz (ft) (C)	W (kips) (D)	$W \tan \phi'$ (kips) (E)	θ (deg) (F)	$\sin \theta$ (G)	$W \sin \theta$ (kips) (H)	Trial FS = 1.30 $(I_1)^a$	Trial FS = 1.30 (E)/(I₁) (J_1) (kips)	Trial FS = 1.40 $(I_2)^a$	Trial FS = 1.40 (E)/(I₂) (J_2) (kips)	Trial FS = 1.50 $(I_3)^a$	Trial FS = 1.50 (E)/(I₃) (J_3) (kips)
1	62	21	143	118	9	0.156	22	1.09	108	1.08	109	1.07	110
2	42	44	203	167	16	0.276	56	1.14	147	1.12	149	1.11	150
3	18	50	99	82	21	0.358	35	1.16	70	1.14	72	1.13	73
4	30	59	195	161	24	0.406	79	1.17	137	1.15	140	1.14	142
5	30	66	218	180	28	0.469	102	1.18	152	1.16	155	1.14	158
6	30	71	234	193	33	0.545	128	1.18	163	1.16	166	1.14	170
7	30	72	238	196	38	0.616	147	1.18	166	1.15	170	1.13	174
8	28	66	203	167	43	0.682	138	1.16	144	1.13	147	1.11	151
9	30	52	172	141	48	0.743	128	1.14	124	1.11	127	1.08	131
10	30	21	69	57	55	0.819	57	1.09	52	1.06	54	1.02	56
							$\Sigma = 892$		$\Sigma = 1263$		$\Sigma = 1289$		$\Sigma = 1315$

$$FS_1 = \frac{\Sigma (J_1)}{\Sigma (H)} = 1.42$$

$$FS_2 = \frac{\Sigma (J_2)}{\Sigma (H)} = 1.44$$

$$FS_3 = \frac{\Sigma (J_3)}{\Sigma (H)} = 1.47$$

$^a \cos \theta + \sin \theta \dfrac{\tan \phi'}{FS}$ obtained from Figure 12.16.

$\gamma = 110 \, \text{pcf}; c' = 0; \phi' = 39.5°.$

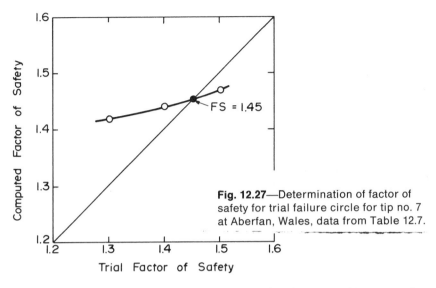

Fig. 12.27—Determination of factor of safety for trial failure circle for tip no. 7 at Aberfan, Wales, data from Table 12.7.

2. A variety of factors, both "internal" and "external" can produce slope failures (Section 12.3, Table 12.1). The most common is an increase in the steepness or height.

3. Current methods of slope stability analysis employ certain common assumptions (Section 12.4):
   a. Failure occurs along a specific cylindrical sliding surface.
   b. The failure mass moves as a rigid body.
   c. Strength properties along the failure surface are independent of orientation.
   d. The factor of safety is defined in terms of the average shear stress and average shear strength along the failure surface.

4. Stability analysis of certain special cases provides guidance for more complex problems:
   a. The limiting inclination of an infinite slope in cohesionless soil is the angle ϕ' (Section 12.5).
   b. A circular failure surface is more critical than a plane for a finite slope in cohesive soil. The critical cricle passes through the toe of the slope for all cases except for shallow slope angles in nearly purely cohesive soils (Sections 12.6, 12.7).

5. The method of slices is useful for stability analyses, involving stratified or inhomogeneous materials, or complex boundary conditions, (Section 12.8).
   a. The Bishop simplified method assumes a circular failure surface, and neglects the effect of shear forces on the sides of the slices as well as the distribution of stresses due to surface loads. The approach is amenable to manual or computer calculations, and can be used for either total stress (Equation 12–33) or effective stress (Equation 12–35) analysis.

b. The Fellenius method of slices assumes that the forces on either side of a slice are collinear. This leads to a simpler explicit result, but involves errors which may be large.

6. When a weak stratum dictates a composite failure surface, a sliding wedge analysis may be most applicable (Section 12.9). The method assumes that the earth pressure forces in front of, and behind the sliding block result from a state of limiting equilibrium.

7. The maximum shear stress approach is likely to be more conservative than the method of slices, and is useful only in the case of undrained cohesive materials.

8. The drainage conditions corresponding to the most critical condition of a slope largely determine the most appropriate strength parameters to use for a stability analysis (Section 12.11):

   a. For a embankment rapidly constructed over a soft foundation, the end-of-construction condition will be critical. In this case, a $\phi = 0$ analysis using the undrained shearing resistance for c is appropriate.

   b. The critical condition for a cut in cohesive soil occurs after hydrodynamic equilibrium has been reached. Thus effective stress strength parameters should be used.

   c. When an increase in pore water pressure occurs without a corresponding change in total stress, the stability of an existing slope is likely to be adversely affected. The stability analysis requires effective stress strength parameters.

9. When the peak strength of materials in a slope or embankment is reached at significantly different strain magnitudes, a composite stress-strain curve should be constructed. This accounts for the fact that the peak strengths of the different materials are not likely to be mobilized simultaneously (Section 12.11).

10. Progressive failure may occur in slopes composed of materials with a brittle-type stress-strain curve. Conditions appear to be most favorable for progressive failure in cuts in highly over-consolidated clays and clay-shales. In such cases the residual shear strength parameters, $c' = 0$ and $\phi' = \phi_r'$, should be used in the stability analysis for long-term conditions (Section 12.11).

PROBLEMS

12.1. A highway cut is to be made in the deposit of medium stiff over-consolidated clay shown schematically in cross-section. Results of two consolidated-undrained (CU) triaxial compression tests on undisturbed samples, believed to be representative of the region of interest in the clay, are:

Test No.	$\sigma'_c = \sigma_{3f}$ (psf)	$(\sigma_1 - \sigma_{3f})$ (psf)	u_f (psf)
1	250	744	−58
2	1000	984	+454

(*Note:* $u = 0$ after consolidation and before shearing.)

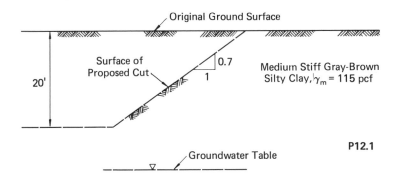

P12.1

(a) What strength parameters are most appropriate to investigate the stability of this cut? *Why*? What are the magnitudes of the parameters in question?

(b) What is the factor of safety against failure of the cut slope?

12.2. A cut is to be made in an approximately homogeneous clay deposit. Typical cross-sections through the cut expected during the dry summer

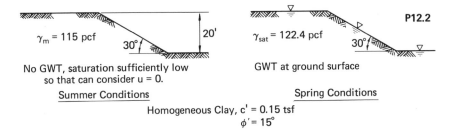

P12.2

Homogeneous Clay, $c' = 0.15$ tsf
$\phi' = 15°$

months and wet spring months, respectively, are shown in the accompanying sketch.

(a) What is the factor of safety against a slope failure in the summer months?

(b) In which case is FS smaller? Why? Read Section 14.7; then determine FS for the spring condition.

12.3. A land-reclamation project in a tidal-swamp area requires adding 20 ft of fill, some of it below water level, as shown schematically in cross-section. Fill is to be placed so that a slope angle of 30° is to be maintained during construction. This is illustrated by the cross-section (heavily stippled) at some typical time during construction. Because of the presence of the soft clay layer, it will be impossible to place the entire fill at one time.

(a) What height of fill *can* be placed initially without producing a failure?

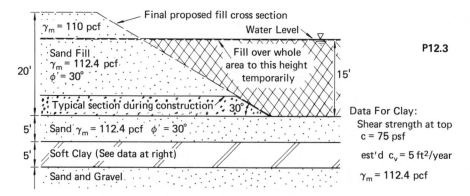

P12.3

(b)   To allow complete construction, a fill 15 ft high is to be constructed, and extended beyond the area of concern, as shown cross-hatched above. How long must the contractor wait before removing the excess material to the right side of the finished cross-section and completing the fill?

13

Elements of Design
of Rigid Retaining
Structures

13.1 INTRODUCTION

Changes in the surface elevation of earth masses can be provided by an unbraced transitional slope or a retaining structure. For most soil material, unbraced slopes require gradual changes in elevation, usually of the order of 1.5 horizontal to 1 vertical. In many situations, however, space limitations dictate abrupt elevation changes, often requiring a retaining structure to provide a vertical, or near vertical transition.

Retaining structures are to be found in harbors and along rivers as quay walls, locks, levees, breakwaters, and cellular docks and platforms. During all types of construction, the sides of open excavations are supported by retaining structures. Retaining walls are used on highways, railroads, and landscaping and in erosion control. Many foundation elements such as basement walls, bridge abutments, and the sides of box culverts, serve not only as an integral part of the structure, but also retain surrounding soil. The diverse applications requiring retaining structures make these among the most commonly constructed civil engineering works.

Design of a retaining structure involves considering the complex interaction between the structure itself and the earth mass with which it is in contact. Thus the design process usually includes the following steps:

1. Make a preliminary choice of structure type and overall dimensions.
2. Determine the magnitude and location of the resultant forces acting on the structure.
3. Investigate the stability of the structure.
4. Evaluate the *distribution* of forces on the structure.
5. Dimension the structural elements.

6. Estimate displacements of the various components of the structure, and the acceptability of these displacements in relation to the overall design problem.

In this chapter we shall devote our attention principally to items 2 to 4 above. Selection and dimensioning of the invidual structural elements is beyond the scope of this text. Settlements of the structure have already been considered in several earlier chapters and are discussed only briefly herein.

13.2 TYPES OF RETAINING STRUCTURES

We find it useful to subdivide structures which retain earth into two broad categories: *rigid* and *flexible*. "Rigid" retaining structures are those for which the deformations within the structure are sufficiently small that they do not influence the magnitude of the earth pressure transmitted to the structure. We recognize of course, that rigid-body displacements of the entire structure are of major significance in establishing the forces between the soil and the structure.

Rigid structures are commonly of concrete or masonry construction. Various types occur as illustrated in Figure 13.1a. The *gravity* wall is an unreinforced mass of concrete, brick, rubble, etc., designed so that no tensile stresses act at any point in the wall. It is a massive structure, normally stable due to its own weight. It is usually economical only for small heights.

A *semigravity* wall is less massive and requires the use of a small amount of reinforcement to support tensile stresses within the wall.

A *cantilever* wall is usually constructed of reinforced concrete in the form of a thin wall cantilevered from a base slab. The stability of the wall is enhanced by the weight of that part of the backfill which acts on the base.

Counterfort and *buttress* type walls are constructed of reinforced concrete and are similar to cantilever walls. The counterfort or buttress is used to reduce the bending moment and shear in the wall, thus making them practical for great heights, or when large surface loads on the backfill must be transmitted to the wall. The difference between the two is that counterforts are placed at the rear side of the wall and are thus in tension while buttresses are located on the front of the wall and are in compression. Buttress type walls are rarely used because the buttresses are exposed, creating a need for greater clearances and a less finished appearance.

Crib walls consist of a continuous series of hollow, bottomless cells made of prefabricated interlocking structural units of timber, concrete or steel. The cells are filled with soil, preferably cohesionless, to provide the weight required to make the wall stable. Crib walls are normally not found over approximately 20 feet in height, although a higher wall was described in Chapter 3.

By contrast, "flexible" retaining structures (Figure 13.1b) experience deformations within the structure sufficient to influence the magnitude and

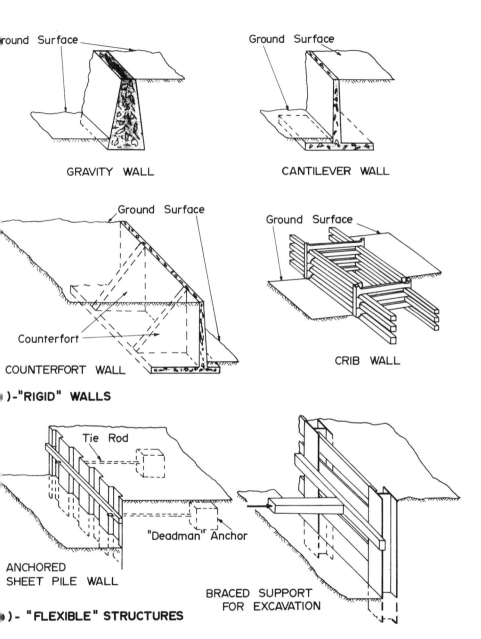

Fig. 13.1—Schematic illustration of common types of retaining structures.

distribution of the forces acting upon it. Flexible retaining walls may be *unbraced* as in the case of interlocking steel or wood sheet piling. Alternatively, they may be *braced* as in the case of anchored sheet pile walls or strutted or tied walls for excavations. The magnitude and distribution of earth pressures on flexible retaining structures depend to a marked extent upon the deformations which are permitted. Thus, the presence of an anchor or strut which prevents displacement of the wall at specific points produces localized increases in applied pressure.

In the following sections we shall focus upon certain design considerations for rigid retaining walls. The additional complexity introduced by flexible structures is discussed in detail by Teng (1962) and Bowles (1968).

13.3   MAGNITUDE AND DISTRIBUTION OF EARTH PRESSURE FORCES ON RIGID RETAINING STRUCTURES

Effect of Strains Within the Backfill

The magnitude of the forces transmitted from the soil to a retaining structure depends to a great extent upon the nature of the strains and corresponding mobilized shearing resistance in the backfill. The significance of the strain, described in terms of the displacement of the wall relative to the soil mass was discussed in detail for cohesionless soil in Chapter 3. The principles elucidated there are also applicable to cohesive soils. These are summarized briefly below:

1.   When the wall moves away from the backfill, lateral strains are *extensive*, and the lateral earth pressure on the wall reduces concomitant with mobilization of the shearing resistance of the soil. When the strength of the backfill is fully developed, the lateral pressure assumes its *minimum* value, and is termed the *active earth pressure*.

2.   If the wall is *forced* into the backfill, lateral strains are *compressive*, and the corresponding increase in mobilized shearing resistance is accompanied by an increase in the lateral earth pressure. When limiting equilibrium is reached in this case, the lateral earth pressure is a *maximum*, and is termed *passive earth pressure*. The magnitude of the passive earth pressure force can be calculated using the appropriate Mohr-Coulomb failure criterion.

3.   The lateral pressure prior to construction, corresponding to zero lateral strain, is called *earth pressure at rest*. The magnitude of the earth pressure at rest can be determined only by empirical means. It has been found, for both cohesionless and cohesive soils to be a function of the stress history as characterized by the over-consolidation ratio (OCR). For normally consolidated soil (both sand and clay) the coefficient of earth pressure at rest K_0, defined as the ratio of the

effective lateral and vertical stresses is approximately (Brooker and Ireland, 1965; this is the same as Eq. 3–30):

$$K_0 = 1 - \sin \phi' \qquad (13-1)$$

As the over-consolidation ratio increases, the magnitude of the earth pressure at rest increases up to a maximum value corresponding to the passive pressure condition. That is, for a soil which had been prestressed to a sufficiently great extent, the lateral prestress retained may be sufficient to induce a state of limiting equilibrium in the soil mass. This has been shown to be the case in the field by Skempton (1961b) for the highly over-consolidated London clay.

Methods for calculating the magnitude of active and passive earth pressures arising from cohesive backfills are discussed below.

Active Earth Pressure for Cohesive Soil

In Chapter 3 we considered two general approaches to determining the earth pressure imposed by cohesionless soils in a state of limiting equilibrium. These were:

1. The Rankine analysis, which was a special case of the generalized limiting equilibrium analysis.
2. The Coulomb analysis, which was a special case of the general wedge method of analysis.

We required in both approaches, that the soil mass, or at least certain parts of the mass, were in a state of *limiting equilibrium*, that is, on the verge of failure. Failure, for our purposes, was defined to be that stress state which satisfied the Mohr-Coulomb relationship. The Rankine approach assumed that the entire soil mass was in this state of limiting equilibrium and that the families of failure surfaces were determined from simultaneous solution of the equilibrium and strength equations everywhere in the medium. By contrast, the Coulomb method assumed a single planar failure surface. The earth pressure was calculated from the requirement of equilibrium of the soil wedge resting on the assumed failure surface. For cohesive soils, the same two fundamental approaches are employed.

Rankine Analysis. Although the assumptions in the Rankine analysis are restrictive, there are situations in which the assumptions are nearly satisfied. Such is the case for many cantilever retaining walls. Even for more complicated structures, the Rankine analysis may offer a useful "first approximation." The approach is the same as that we used in Section 3.2 for cohesionless soil: we combine the two-dimensional equations of equilibrium with the Mohr-Coulomb failure criterion. For cohesive soil, this may be written

$$\tau_f = c' + \sigma'_f \tan \phi' \qquad (13-2)$$

in which τ_f is the shearing resistance on some plane of interest at failure, c' is the cohesive component of strength in terms of effective stresses, σ'_f is the effective normal stress on the plane of interest at failure, and ϕ' is the apparent angle of shearing resistance in terms of effective stresses.

This equation is shown in Figure 13.2 along with a typical Mohr's circle at failure. From the Mohr's circle, the relationship between effective principal stresses at failure is

$$\sigma'_3 = \sigma'_1 \left(\frac{1 - \sin \phi'}{1 + \sin \phi'}\right) - 2c' \left(\frac{1 - \sin \phi'}{1 + \sin \phi'}\right) \tag{13-3}$$

which may also be written (recall Equations 10–5) as

$$\sigma'_3 = \sigma'_1 \tan^2 (45° - \phi'/2) - 2c' \tan (45° - \phi'/2) \tag{13-4}$$

or

$$\sigma'_3 = \sigma'_1 K_A - 2c' \sqrt{K_A} \tag{13-5}$$

in which $K_A = \tan^2 (45° - \phi'/2)$ is the *coefficient of active pressure.*

It is useful to consider the special case of a semi-infinite medium. In this case, horizontal and vertical planes are principal planes, and the horizontal effective stress is

$$\sigma_x = \gamma z K_A - 2c' \sqrt{K_A} \tag{13-6}$$

as illustrated in Figure 13.3a. We note that this result implies that, for the active condition, there will be tensile stresses formed near the top of the medium and diminishing to zero at a depth z_t,

$$z_t = \frac{2c'}{\gamma \sqrt{K_A}} \tag{13-7}$$

where γ is the unit weight of the soil.

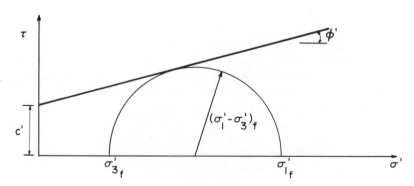

Fig. 13.2—Idealized Mohr failure diagram for a cohesive soil.

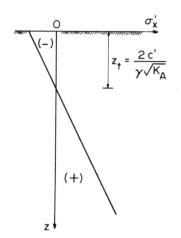

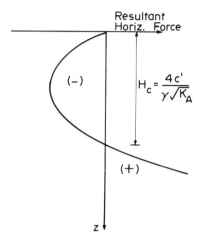

a) - Horizontal Earth Pressure

b) - Resultant Horizontal Force

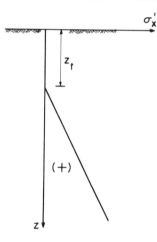

Fig. 13.3—Horizontal pressure distribution and resultant horizontal force as a function of depth for active Rankine conditions in cohesive soils.

c) - Horizontal Earth Pressure Considering Potential for Cracking

The resultant active earth pressure force on a vertical plane at any depth in the soil may be calculated by integrating Equation 13–6, and is (Figure 13.3b)

$$P'_A = \frac{\gamma H^2}{2} K_A - 2c'H\sqrt{K_A} \qquad (13\text{--}8)$$

in which H is the depth of interest. For a height

$$H_c = \frac{4c'}{\gamma\sqrt{K_A}} \qquad (13\text{--}9)$$

the resultant active earth pressure is zero, in which H_c is the height to which a *vertical unbraced* slope in a cohesive soil could stand, at least temporarily.

The results discussed above are expressed in terms of effective stresses. It is often convenient in the case of saturated, or nearly saturated clays to describe conditions in terms of the immediate undrained shearing resistance $c = q_u/2$, and $\phi = 0$. For this case, the equivalent $K_A = 1$, and Equation 13–9 becomes the same result which was obtained in Example 12.1 for wedge analysis with an assumed planar failure surface.

The zone of tensile stress predicted by the Rankine analysis is probably not found in real field situations. There are at least two reasons for this:

1. The adhesion between a retaining wall and the backfill is likely to be less than $2c$, even if no cohesionless free-draining soil is used directly behind the wall.
2. Even within the cohesive soil mass itself, the cracking which accompanies surficial drying will likely penetrate to the depth z_t and prevent tensile stresses from acting for an extended period.

Thus, in applying the Rankine analysis to the estimation of active earth pressure, it is conventional to ignore the effect of the zone above a depth z_t, and calculate the active pressure as shown in Figure 13.3c.

Coulomb Analysis. The Coulomb analysis permits consideration of more complicated backfill geometry and surface loadings, as well as shear forces between the wall and soil different from those which would obtain under the Rankine assumptions. As in the case of cohesionless soils, we assume a plane failure surface. However, the effect of the cohesive component of shear strength along both the failure surface and the interface between the wall and the backfill is considered: the Mohr-Coulomb criterion, Equation 13–2, is assumed to apply along the failure plane; along the interface between the back of the wall and the backfill the shear stress developed is assumed to be of the same form,

$$\tau_\alpha = c'_a + \sigma_\alpha \tan \delta \qquad (13\text{–}10)$$

in which τ_α is the shear stress developed between the soil and the wall, c'_a is the adhesion developed between the wall and the soil, and δ is the equivalent angle of shear resistance developed between the wall and the soil. The magnitude of c'_a is often assumed to be

$$c'_a = \frac{2}{3}c'$$

The procedure to determine the active earth pressure force is essentially similar to that described in Section 3.5, and is illustrated in Figure 13.4. A failure plane is assumed, and a free-body diagram of the potential sliding wedge is drawn, as in Figure 13.4a. From this, the force polygon shown in

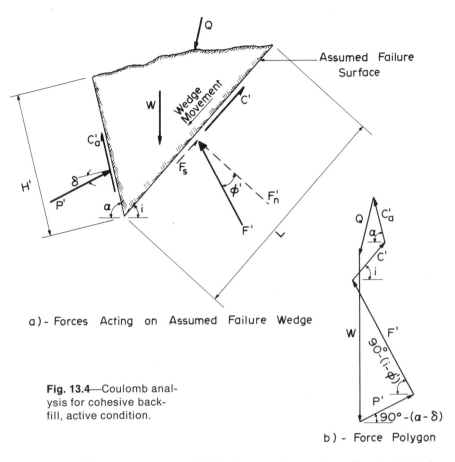

a) - Forces Acting on Assumed Failure Wedge

Fig. 13.4—Coulomb analysis for cohesive backfill, active condition.

b) - Force Polygon

Figure 13.4b can be constructed. The forces acting on the soil wedge include:

1. *Weight of soil within wedge, $W = \gamma A$*, where A is the area of the wedge shown in Figure 13.4a. The weight acts vertically so its magnitude and direction are both known.

2. *Cohesive force on failure surface, $C' = c'L$*, where L is the length of the failure surface (Figure 13.4a). The cohesive force acts parallel to the assumed failure surface in a direction opposing the motion of the wedge.

3. *Adhesive force at soil-wall interface, $C'_a = c'_a H'$*, in which H' is the length of contact surface between the wall and soil (Figure 13.4a). The adhesive shear force acts parallel to the wall, also in a direction opposing the relative motion of the wedge with respect to the wall.

4. *Effective normal and frictional shear components of force on failure surface.* These forces, shown in Figure 13.4a, are related by

$$F_s = F'_n \tan \phi' \qquad (13\text{–}11)$$

and may be replaced by a single force F', inclined at an angle ϕ' to the normal to the failure surface, to resist the movement of the sliding wedge. Thus, the direction of F' is known although its magnitude is not.

5. *Effective normal and frictional shear components of force on soil-wall interface.* These two forces, shown in Figure 13.4a are related by

$$P_s = P'_n \tan \delta \qquad (13\text{--}12)$$

and can also be combined into a single force P' acting at an angle δ to the normal to the back of the wall with an orientation opposing the motion of the wedge relative to the wall. Thus, the magnitude of P' is also unknown but its direction is known, and can be used to complete the force polygon shown in Figure 13.4b.

From the force polygon we can determine the magnitude of the force P' for the given trial wedge. By assuming a number of trial failure surfaces and determining the value of P' for each, we can determine the P' corresponding to active earth pressure conditions, for the Coulomb analysis assumptions, as the maximum value of P' obtained. The active earth pressure force between the wall and the soil is then equal to the vectorial sum of P' and C'_a. The active earth pressure force acting on the wall is, of course, equal in magnitude and opposite in direction to that acting on the soil wedge.

Because of the potential formation of tension cracks as discussed above, it is advisable to incorporate this effect into the wedge analysis. As shown schematically in Figure 13.5, the cracks reduce the lengths of both the failure surface and the contact surface between the wall and the soil, thereby reducing C' and C'_a. To a much smaller extent, the volume and therefore the weight of the wedge are also reduced. The net result is an increase in the active pressure force on the wall.

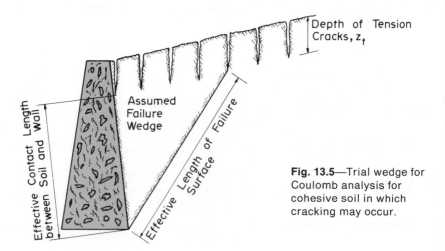

Fig. 13.5—Trial wedge for Coulomb analysis for cohesive soil in which cracking may occur.

If a static groundwater table exists behind the wall, we can account for it by using the submerged unit weight of the soil below the groundwater table to determine the active earth pressure force on the wall as we did in Section 3.7 for cohesionless soils. To this force we must add a water pressure force acting normal to the wall. As we observed in Chapter 3, if groundwater is present immediately behind a retaining wall it will significantly increase the total force on the wall.

Incorporation of the various effects discussed in determining the forces acting on a retaining wall supporting a cohesive backfill is illustrated in the following example problem.

Example 13.1

A 15-ft-high rigid retaining wall, shown schematically in Figure 13.6a, supports a cohesive backfill. The back of the wall is inclined at an angle of 85° to the horizontal. During the spring of the year the groundwater table is at a depth of 3 ft below the ground surface, as indicated in the figure. The partially saturated backfill above the groundwater table exhibits a unit weight of 115 pcf, and the approximately saturated material below the groundwater table weighs 120 pcf. Effective stress strength parameters are $c' = 100$ psf, $\phi' = 20°$. We shall assume that, between the wall and the soil, two thirds of the strength has been mobilized so that $c'_a = 67$ psf, $\delta = 7°$. Using the Coulomb analysis, estimate the force on the wall under active earth pressure conditions.

Solution. We first calculate the depth of potential cracking:

$$\sqrt{K_A} = \tan (45° - \phi'/2) = 0.7$$

and from Equation 13–7,

$$z_t = \frac{2c'}{\gamma \sqrt{K_A}} = \frac{(2)(100)}{(120)(0.7)} = 2.4 \text{ ft}$$

Trial failure wedges are then constructed, as in Figure 13.6b. The forces acting on the trial failure surfaces are shown schematically in the figure. The adhesive force between the wall and the soil, C'_a, is shown acting at the interface, and is of magnitude

$$C'_a = 67 \times \frac{12}{\sin 85°} = 807 \text{ pounds/linear foot (plf)}$$

The resultant of the normal and frictional shear forces at the wall-soil interface, P', is assumed to act at the third point (Figure 13.6b) at an angle of 14° to the normal to the wall. Calculated forces for the trial wedges are:

Wedge	Effective Weight (plf)	C′ (plf)
1	3679	1230
2	5036	1350
3	6773	1480
4	8370	1840

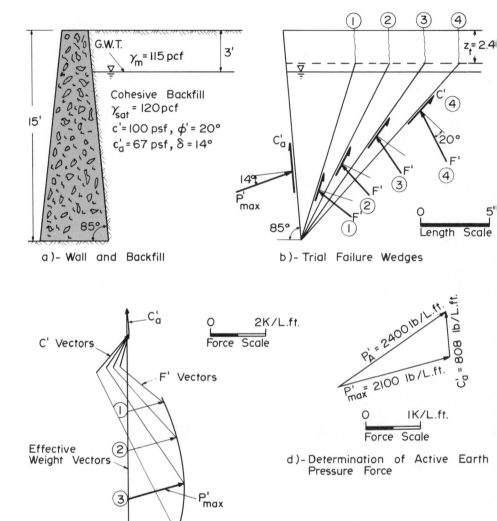

a)- Wall and Backfill

b)- Trial Failure Wedges

c)- Force Polygons for
Trial Wedges

d)- Determination of Active Earth
Pressure Force

Fig. 13.6—Determination of active
earth pressure force on retaining
wall with cohesive backfill,
Example 13.1. (continued)

From these data the force polygons shown in Figure 13.6c can be drawn. Connecting
the P' vectors gives the largest, and therefore critical value, P'_{max}. Determined graphically,

$$P'_{max} = 2100 \text{ plf}$$

Combining this force with the adhesion, as in Figure 13.6d we find the *active earth*

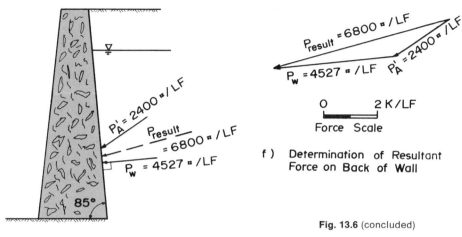

f) Determination of Resultant
    Force on Back of Wall

Fig. 13.6 (concluded)

e) Forces Acting on
     Back of Wall

pressure force to be

$$P'_A = 2400 \text{ plf}$$

To determine the resultant force acting on the back of the wall, as shown in Figure 13.6e, we must include that due to the groundwater, P_w,

$$P_w = \frac{\gamma_w H'^2}{2} = \frac{62.4}{2} \left(\frac{12}{\sin 85°} \right)^2 = 4527 \text{ plf}$$

The water force acts normal to the wall.

The resultant force can then be found using a force polygon as in Figure 13.6f, or analytically. It is determined to be

$$P_{result} = \underline{6800} \text{ plf}$$

We note that the magnitude of the water force is more than twice the active earth pressure force. This emphasizes, once again, the importance of adequate drainage behind the wall. Several types of drainage systems are described by Terzaghi and Peck (1967).

In practice, cohesionless backfill material is much more desirable than cohesive material, at least within the zone affecting the magnitude of pressure on the retaining wall. Among other reasons, this is the case because of the time-dependent characteristics of most cohesive soils. The initial yield of a retaining wall supporting either cohesionless or cohesive backfill will generally be sufficient to permit the active state of limiting equilibrium to develop. In the absence of changes in the environment, the pressure exerted by a relatively clean cohesionless backfill will remain unchanged. However, a cohesive fill will *creep* with time inducing continued movement of the wall. If wall movement is restricted, as in the case of the bridge abut-ment, then the force on the wall will increase, perhaps to a magnitude approximating the at-rest condition.

In addition to time-dependent effects inherent in the behavior of the soil, cohesive soils are typically subject to swelling and shrinking due to seasonal wetting and drying cycles. If tensile cracks do develop, the low permeability of cohesive soils may result in temporary impounding of water within the crack creating additional pressure on the wall. Further-more, expansive pressures due to freezing of soil water may be significant. Thus, unless the opportunity for continued wall movement is present, at-rest conditions may provide a more realistic estimate of earth pressures arising from cohesive backfills.

Earth Pressure Distribution and Location of Resultant

In evaluating a proposed retaining wall design, we must determine the location of the resultant earth pressure force on the wall. This requires that we obtain at least an approximate estimate of the distribution of earth pressure. However, the wedge method of analysis does not yield direct information concerning the location of the earth pressure force. When the surface of the backfill is plane and there are no loads superimposed on the surface, it is customary to assume that the resultant earth pressure force acts at one-third the height of the wall above the base. This corresponds to a pressure distribution which is linear with depth, as suggested by the Rankine analysis for a plane unloaded cohesionless backfill. If, on the other hand, there is a break in the backfill near the wall, or surface loads are located near the wall, the resulting earth pressure force is likely to act above the one-third point. As we shall see below, such a situation has an adverse effect on the stability of the wall.

The distribution of lateral pressures on retaining walls due to surface loads on the backfill has been investigated in both small- and large-scale experiments (Spangler and Mickle, 1956). The conclusions from these observations is that the *form* of the pressure distribution on a retaining wall due to surface loads on the backfill is similar to that predicted by elastic theory. However, the magnitudes are approximately twice the theoretical value. Thus, although empirical correction factors have been suggested to permit the use of elastic theory to determine the effects of surface loads on the earth pressure distribution (Spangler and Mickle, 1956; Bowles, 1968), the approximate approach described below is probably sufficiently accurate, and has the advantage of relative simplicity.

If we assume that, once the potential failure plane has been determined, all other parallel planes through the wedge are also potential planes of failure, we can easily determine the earth pressure force due to a portion of the wedge. When sufficiently small portions of the wedge are used, an approximation to the pressure distribution can be obtained. The steps involved in this process are illustrated in Figure 13.7 for a representative problem, and described below (Huntington, 1957):

1. The problem is illustrated schematically in Figure 13.7a. Using the Coulomb analysis, determine the orientation of the potential failure surface, and the magnitude of the earth pressure force P', as shown in Figure 13.7b. The force P' acts at a distance $\bar{z}$ above the base of the wall. The magnitude of $\bar{z}$ remains to be determined. The force C'_a due to the adhesion between the wall and the backfill is omitted from this consideration, and is incorporated only after the point of application of the earth pressure force has been found.
2. The failure wedge is divided into a number of smaller wedges by lines parallel to the potential failure surface. This is conveniently

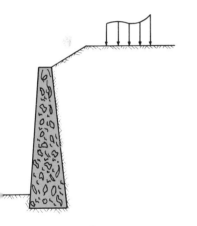

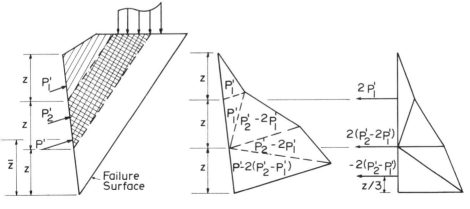

Fig. 13.7—Approximate method for determining distribution and location of resultant earth pressure.

a)- Sketch of Problem

b)- Failure Wedge and Partial Wedges

c)- Approximate Pressure Distribution

d)- Partial Earth Pressure Forces for Locating Resultant P'

done by dividing the height of wall into a number of equal increments, denoted by z in Figure 13.7b, and constructing lines from these parallel to the failure surface. In the sketch, the wall has been divided into three equal thicknesses for illustrative purposes, although any number can be used.

3. The resultant force on the back of the wall due to the uppermost wedge, P_1' is then found from a force polygon, just as if the line bounding that wedge were a failure surface. This upper wedge is shown shaded in Figure 13.7b. Next, the partial earth pressure force P_2', due to the wedge formed by the upper *two* portions of the failure wedge, is found. This area is shown in Figure 13.7b as the sum of the shaded and crosshatched areas. Had the wall been divided into more segments, the partial earth pressure due to increasingly larger portions of the failure wedge would be found in a similar manner.

4. If it is assumed that the earth pressure on the wall is distributed linearly within each segment of thickness z, then the pressure distribution can be sketched as in Figure 13.7c. Note that the area beneath each of the triangular elements shown will be of the magnitude indicated. That is, in the upper triangle, due to the earth pressure from the uppermost shaded segment, the earth pressure force is P_1'. Similarly, the earth pressure force due to the partial wedge consisting of the shaded *and* crosshatched areas is P_2'. Since the area of the uppermost triangle is P_1', the area of the triangle immediately below it is also P_1'. Thus the remaining portion of the force due to the two partial wedges is $P_2' - 2P_1'$. This same reasoning can be used to determine the area of the other triangles shown. By this means, we can determine the approximate pressure distribution.

5. The location of the resultant earth pressure force is found by noting that the moment of the resultant force, which is the moment of the total area shown in Figure 13.7c must equal the sum of the moments of each of the individual parts. The diagram can be rotated to the vertical for convenience, as illustrated in Figure 13.7d. Taking advantage of symmetry,

$$P'\overline{z} = 2P_1'(2z) + 2(P_2' - 2P_1')z + [P' - 2(P_2' - P_1')]\left(\frac{z}{3}\right) \quad (13\text{–}13)$$

from which

$$\overline{z} = \frac{z}{3P'}(2P_1' + 4P_2' + P') \quad (13\text{–}14)$$

Equation 13–14 applies when the height of the wall has been divided into three equal parts. It can be generalized for any number n equal parts as follows. If n is *odd*:

$$\overline{z} = \frac{z}{3P'}(2P_1' + 4P_2' + \cdots + 2P_{n-2}' + 4P_{n-1}' + P') \quad (13\text{–}15)$$

or, if n is *even*,

$$\bar{z} = \frac{z}{3P'} (4P'_1 + 2P'_2 + \cdots + 2P'_{n-2} + 4P'_{n-1} + P') \qquad (13\text{–}16)$$

Effect of Base Projection on Active Earth Pressure Force Calculation

The base of cantilever type retaining walls often projects behind the wall into the backfill as shown schematically in Figure 13.8. As the wall moves, permitting the soil to yield laterally and mobilize its shearing resistance, the failure surface, line $\overline{ad}$, develops. Because the soil to the left of line $\overline{ab}$ is restrained from deformation by contact with the wall, it is only within the region $abcda$ that a state of limiting equilibrium will be approached. If the angle θ is sufficiently large that a slip surface emanating from point a can form without interference by the wall, then the Rankine conditions will apply within the region $abcda$. In the case of *cohesionless* soils this will occur when (Teng, 1962)

$$\theta \geq \tfrac{1}{2}\left[90° + \beta - \phi' - \sin^{-1}\left(\frac{\sin \beta}{\sin \phi'}\right) \right] \qquad (13\text{–}17)$$

in which the parameters are defined in Figure 13.8.

When this criterion is satisfied, it is most convenient to evaluate the active earth pressure on the vertical plane shown as line $\overline{ac}$, using the Rankine analysis. The active earth pressure force is then located one-third of the distance from a to c, inclined parallel to the ground surface, and of magnitude

$$P'_A = \frac{\gamma(\overline{ac})^2}{2} \cos \beta \, \frac{\cos \beta - \sqrt{\cos^2 \beta - \cos^2 \phi'}}{\cos \beta + \sqrt{\cos^2 \beta - \cos^2 \phi'}} \qquad (13\text{–}18)$$

The soil to the left of line $\overline{ac}$ is simply included as part of the wall in calculations involving the stability of the wall. If the criterion, Equation 13–17,

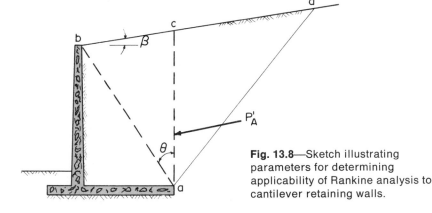

Fig. 13.8—Sketch illustrating parameters for determining applicability of Rankine analysis to cantilever retaining walls.

is not satisfied, the wall interferes with development of the slip surface corresponding to a semi-infinite medium. Then the active earth pressure on the surface $\overline{ab}$ is calculated using the Coulomb method, and the weight of the soil to the left of the line $\overline{ab}$ is included with that of the wall for subsequent calculations.

For *cohesive* backfills, the criterion for interference of the wall with the formation of the failure surfaces predicted by the Rankine analysis, Equation 13–17, is only approximately correct. However, it is usually most convenient to calculate active earth pressure force using the Coulomb method and we can use Equation 13–17 to decide whether to calculate the active earth pressure force on plane $\overline{ac}$ or on plane $\overline{ab}$.

Passive Pressure

The passive pressure condition occurs as the result of an *external agent* pushing a wall into a mass of soil with sufficient force to induce failure. The passive case may occur when a wall is used to brace some other structure (this is illustrated in Example 13.2 and Problem 13–3). If we visualize a footing as a retaining wall lying on its side, we see that the bearing capacity problem is another example of passive pressure.

When there is significant friction and/or adhesion between the wall and backfill, the Rankine or Coulomb analyses lead to overestimates of the passive pressure which can be developed against the wall. This occurs because the orientation of the principal planes changes with depth, and the failure surface departs markedly from the planar surface assumed in the Coulomb wedge analysis. The situation is illustrated for cohesionless soils in Figure 13.9. The figure provides a comparison between the *plane* potential failure surface determined by the Coulomb analysis and the "actual" failure surface determined by a more generalized limiting equilibrium analysis (Sokolovski, 1954) for both active and passive conditions in which $\delta = \phi' = 30°$. We see that in the active case (Figure 13.9a) the failure surface determined by the Coulomb method appears to be a reasonable approximation to that determined by more exact means. Consequently it is not surprising that the active earth pressure estimated by the Coulomb method, even when $\delta = \phi'$, corresponds closely to the "actual" values (Figure 13.9c).

By contrast, the "actual" failure surface for the passive case shown in Figure 13.9b is markedly different from the most critical *planar* surface. Thus it is reasonable that the Coulomb method overestimates the magnitude of the passive earth pressure coefficient as indicated in Figure 13.9c. The magnitude of the discrepancy increases as δ increases. For cohesive soils, the results are similar and lead to the same conclusion.

We should not be surprised by this observation since the passive pressure problem is one which we described briefly in Chapter 10, when we considered the analysis for the effect of weight of the soil mass on bearing capacity. This

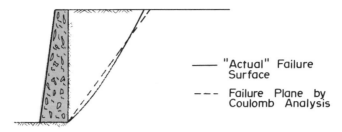

—— "Actual" Failure Surface

――― Failure Plane by Coulomb Analysis

a) - Active Case ($\phi' = 30°$, $\delta = 30°$)

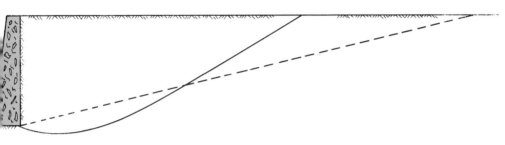

b) - Passive Case ($\phi' = 30°$, $\delta = 30°$)

		ϕ'	30°		40°	
		δ	15°	30°	20°	40°
K_A	"Actual"		0.29	0.27	0.19	0.17
	Coulomb		0.29	0.26	0.18	0.16
K_P	"Actual"		4.46	5.67	9.10	14.0
	Coulomb		4.81	8.74	11.06	70.9

c) - Earth Pressure Coefficients for Representative Values of ϕ' and δ

Fig. 13.9—Failure surfaces and earth pressure coefficients determined by "exact" and Coulomb analysis. (From Sokolovski, 1954.)

was illustrated in Figure 10.13, in which a *composite* failure surface, consisting of a logarithmic spiral and a plane, was used in the analysis. Thus we shall investigate the use of a composite sliding surface which would approximate the actual failure surface more satisfactorily than a plane. Such a composite surface is illustrated in Figure 13.10a. The region of failure behind the wall has been divided into three zones in the figure. Zone I has a curved boundary, the shape of which is a function of δ and the adhesion between the wall and the soil c'_a. The upper two zones are Rankine zones in which the limiting equilibrium stresses are not influenced by the shear stresses at the wall-soil interface.

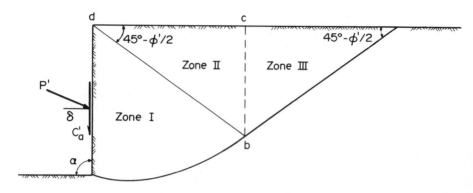

a) - Composite Sliding Surface

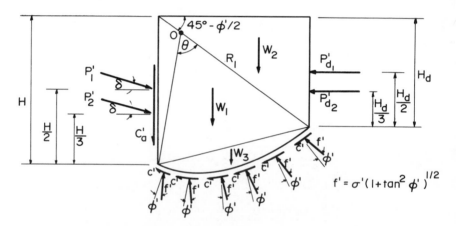

b) - Region Analyzed when Curved Surface is Assumed to Be a
Logarithmic Spiral.

Fig. 13.10—Determination of passive pressure using composite sliding surface.

A variety of choices for the shape of the curved portion of the failure
surface are possible. However, we are motivated to select a logarithmic
spiral, both on theoretical grounds (see Section 10.4) as well as by the useful
property of the log spiral that the resultant of the normal and frictional forces
on the failure surface passes through the center of the spiral (recall Figure
10.9). A trial-value passive-pressure force is then determined by analyzing
a free-body diagram of the region containing zones I and II, as shown in
Figure 13.10.

The log spiral portion of the trial failure surface is determined by ϕ' of
the soil in the failure region and the boundary geometry. We recall the

formula for the spiral (Equation 10–10) is

$$r = R_0 e^{\theta \tan \phi'} \tag{13–19}$$

in which r is the radius of the log spiral, θ is the angle to the radius measured from a reference line (usually the line from the center to the base of the wall), R_0 is the radius of the reference line, and e is the base of natural logarithms. These parameters are shown in Figure 13.10b. In order for the boundary of Zone I (Figure 13.10a) to meet the boundary of Zone III with a common tangent, the center of the spiral must lie along the line db. The center may lie within the soil mass, or on the extension of the line outside the soil mass depending on the problem geometry.

We assume for the analysis that the passive pressure forces acting on the sides of Zones I and II can be separated into two components due, respectively, to the cohesive component of strength, and the frictional component of strength combined with the weight of the pertinent materials. We assume further that the effects of these two groups of forces can be superimposed so that the passive pressure due to the cohesive effects can be determined separately from that due to frictional effects. The forces involved are illustrated in Figure 13.10b and described below:

1. *The force on the right side of Zone II due to the cohesive component of passive earth pressure in the Rankine zone, Zone III*, P'_{d_1},

    $$P'_{d_1} = 2c' H_d \tan (45° + \phi'/2) \tag{13–20}$$

    in which H_d is shown in Figure 13.10b. This force is assumed to act at a point $H_d/2$ above the bottom of Zone II, that is, at the mid-point of line $\overline{bc}$, in Figure 13.10a.

2. *The force on the right side of Zone II due to the frictional component of the passive pressure in the Rankine zone, Zone III*, P'_{d_2},

    $$P'_{d_2} = \frac{\gamma H_d^2}{2} \tan^2 (45° + \phi'/2) \tag{13–21}$$

    This force acts at a point $H_d/3$ above the base of Zone II, as shown in the figure.

3. *The weight of the region analyzed* is usually subdivided into components convenient for computation, such as W_1, W_2, W_3 shown in Figure 13.10b.

4. *The cohesive component of shearing resistance along the curved portion of the failure surface* is depicted schematically in Figure 13.10b. The moment of these cohesive forces M_c, about the center of the logarithmic spiral, denoted as point O in Figure 13.10b, is

    $$M_c = (R_1^2 - R_0^2) \frac{c'}{2 \tan \phi'} \tag{13–22}$$

in which R_0 is the radius of the logarithmic spiral where it intersects the base of the wall and R_1 is the radius of the spiral at its intersection with the boundary of Zone II.

5. *The resultant normal and frictional forces acting on the curved portion of the failure surface* are shown schematically in Figure 13.10b. Because they pass through the center of the log spiral, they do not enter into the expression for moment equilibrium and can be neglected for this purpose.

6. *The adhesive force between the wall and the soil,* C_a', is determined as in the preceding discussion of active earth pressure, except that its direction is changed to reflect the different relative movement of wall and soil under passive conditions.

7. *The unknown trial passive pressure component due to the cohesive resistance of the soil,* P_1' is assumed to act at the mid-height of the wall inclined at an angle δ to the wall, Figure 13.10b. Its magnitude is determined by summing moment equilibrium about the center of the log spiral, point O, for the forces P_{d_1}', C_a' and the cohesive forces on failure surface (Equation 13–22).

8. *The unknown component of the trial passive pressure on the wall due to frictional and weight effects,* P_2', acts at a point $\frac{1}{3}$ the height of the wall above its base, at an inclination δ to the wall. This force is calculated by summing moments about point O due to P_{d_2}' and the weight components W_1, W_2, W_3.

The resultant trial passive pressure is then determined as the resultant of the three forces acting on the wall-soil interface, P_1', P_2', and C_a'. The "actual" passive pressure P_p' is found by the same type of trial-and-error procedure used in the plane wedge (Coulomb) method. For the passive case, it is of course the minimum of all trial values.

It often occurs that the passive pressure force applied to the soil is imposed in such a way that its direction and point of action are known *a priori*. When this happens, the magnitude of P_p' can be determined directly from the moment equilibrium equation including all forces acting on the wedge. This is illustrated in the following example problem.

Example 13.2

A low inclined concrete wall has been cast in the base of a large excavation to provide the reaction force for *rakers* which brace the structural elements supporting the vertical sides of the excavation. The rakers transmit the load to the wall at its midheight inclined 10° to the normal to the wall, as shown schematically in Figure 13.11a. The soil parameters are indicated in the figure. Determine the maximum force per lineal foot of wall, P_p', which the rakers can apply without producing a passive pressure failure of the wall.

Solution. The free-body diagram to be analyzed is shown in Figure 13.11b. The free body includes both Rankine Zone II and Zone I bounded by the curved failure surface, as illustrated in Figure 13.10. The trial logarithmic spiral failure surfaces were

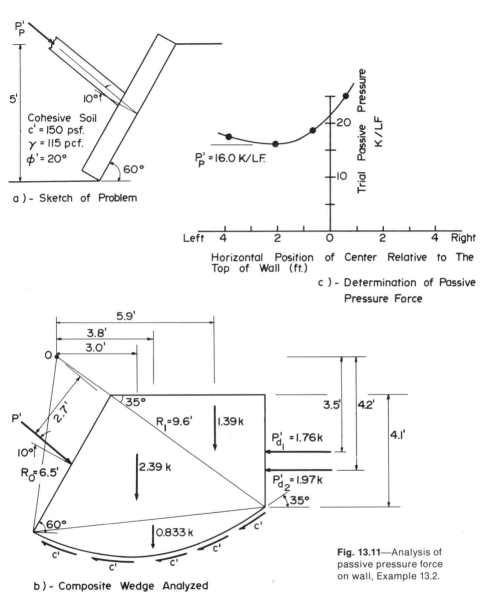

a) - Sketch of Problem

c) - Determination of Passive Pressure Force

b) - Composite Wedge Analyzed

Fig. 13.11—Analysis of passive pressure force on wall, Example 13.2.

obtained by constructing a log spiral on transparent paper, for a wide range of θ, according to the relation

$$r = e^{\theta \tan 20°}$$

The portion of the log spiral corresponding to the failure surface was then found by trial and error by moving the spiral so that its center was located on the extension of the boundary of Zone II with the curve passing through the base of the wall. The curved portion of the failure surface shown in Figure 13.11b is one such trial.

Weights and moment arms about the center of the spiral for convenient portions of the wedge are shown in the figure. The weight of the segment with the curved boundary was obtained by approximating the segment as a triangle.

Forces on the right side of Rankine Zone II were calculated using Equation 13–20,

$$P'_{d_1} = 2c'H_d \tan (45° + \phi'/2) = (2)(150)(4.1)(1.43) = 1.76 \text{ kips/linear foot}$$

and Equation 13–21,

$$P'_{d_2} = \frac{\gamma H_d^2}{2} \tan^2 (45° + \phi'/2) = \frac{(115)(4.1)^2}{2} (1.43)^2 = 1.97 \text{ kips/linear foot}$$

The moment of the cohesive forces along the spiral failure surface is determined from Equation 13–22, with R_0 and R_1 as shown in the figure,

$$M_c = (R_1^2 - R_0^2) \frac{c'}{2 \tan \phi'} = (9.6^2 - 6.5^2) \frac{150}{(2)(.364)} = 10.3 \text{ kips/linear foot}$$

From moment equilibrium the trial passive pressure force is then,

$$P' = \frac{(1.97)(4.2) + (1.76)(3.5) + (1.39)(5.9) + (2.39)(3.0) + (0.833)(3.8) + 10.3}{2.7}$$

$$= \underline{16.0} \text{ kips/linear foot}$$

Recall that, because the direction of the applied force is specified in this case, the portion of the adhesion between the wall and soil which is mobilized is automatically included in the magnitude of P'.

Having found the passive pressure force corresponding to one assumed failure surface, we must compare it to those determined for other locations of the surface, that is, other positions of the center of the spiral. We can then plot the calculated passive pressure as a function of the position of the center to determine the "actual" passive pressure force, which is the minimum value so obtained. This has been done in Figure 13.11c, in which the position of the center is indicated by its horizontal position relative to the top of the wall. We see from this figure that the trial wedge illustrated in Figure 13.11b is very close to the critical one. The "actual" passive pressure force, assuming the validity of the composite failure surface, is

$$P'_p = \underline{16.0} \text{ kips/linear foot}$$

It is interesting to compare this result with that which we would have obtained using the Coulomb plane wedge analysis. The Coulomb method leads to a calculated passive pressure of 22.3 kips/linear foot. Thus, even for the relatively small δ in this case, the plane-wedge assumption gave a nearly 40 per cent overestimate of the passive pressure relative to the more realistic assumption.

13.4 STABILITY OF RETAINING WALLS

The satisfactory performance of a retaining wall founded on its base necessitates that the base meet the requirements we have already established for a shallow foundation. This is especially important in the case of retaining structures because the generally large horizontal component of earth pressure imposed on the wall leads to both inclination and eccentricity of the resultant force on the base. In fact, in a study of retaining wall failures all over the United States, Peck et al. (1948) found the major cause of such failures to be inadequate performance of the *foundation* of the wall.

The forces on the wall foundation are illustrated in Figure 13.12. The resultant force acting on the base of the wall is determined from the requirement of equilibrium for the wall. In general, the horizontal component of the resultant, R_h, will lead to inclination of the resultant force R as shown in the figure. Furthermore, the moment of the active earth pressure force about the base will usually be sufficient to cause the resultant force to be located eccentrically on the base, nearer the toe. Thus we must investigate the performance of the base of the wall as a foundation subject to loads of the sort shown in Figure 13.12b. One of the primary considerations is, of course, the foundation stability.

In evaluating the stability of a wall, there are four potential modes of failure which must be considered:

1. Bearing capacity of the base of the wall.
2. Sliding along the base.

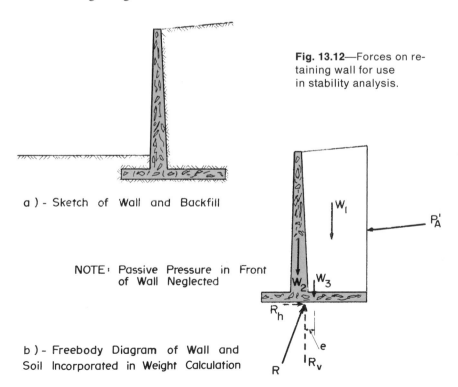

Fig. 13.12—Forces on retaining wall for use in stability analysis.

a) - Sketch of Wall and Backfill

NOTE: Passive Pressure in Front of Wall Neglected

b) - Freebody Diagram of Wall and Soil Incorporated in Weight Calculation

3. Overturning of the wall.
4. Stability of the slope created by the wall.

These items are discussed below.

Bearing Capacity

The bearing capacity of the base of the wall can be evaluated by considering the base to be a shallow foundation subjected to an inclined eccentric
force, R. The approach described in Section 10.4, in which the effects of
eccentricity and inclination are partially compensating, applies in this
situation. In the event that the wall is founded on a relatively shallow
stratum which is underlain by significantly stronger, or weaker material,
we must recognize the effect of this layered system on the potential failure
surface and the bearing capacity. In this case, the method of Section 10.5
may be applicable.

Sliding Along the Base

The component of the earth pressure and/or water pressure acting on
the back of the wall may induce *sliding* of the wall. Usually, sliding occurs
with respect to the underlying soil. We recall however, the crib wall failure
discussed in Chapter 3, in which sliding occured along an *internal* surface.
The analysis procedure used to evaluate the potential for sliding is similar
to that for failure by "spreading" of a slope, as discussed in Section 12.9.
That is, we need to identify the surface along which a block containing
the wall will slide with respect to the underlying soil, and the forces on this
block which act to assist or resist sliding. If the base of the wall is approximately horizontal, the most critical potential sliding surface will frequently
be the interface between the base of the wall and the soil on which it rests.
The factor of safety against sliding along the base is then defined as

$$\text{FS}_{\text{sliding}} = \frac{R_{h_{\max}}}{R_h} \tag{13-23}$$

in which R_h is the horizontal component of the *required* resultant force
on the base of the wall (Figure 13.12b) and $R_{h_{\max}}$ is the *maximum available*
horizontal resisting force. When the wall is underlain by essentially
cohesionless soil, this force can be written

$$R_{h_{\max}} = \mu R_v \tag{13-24}$$

in which R_v is the vertical component of the resultant force and μ is the
developed coefficient of friction between the wall and the soil. Appropriate
values for various soil are given in Table 13.1. When the underlying soil
is cohesive, the maximum horizontal force can be taken as

$$R_{h_{\max}} = sB, \qquad s \leqslant 1000 \text{ psf} \tag{13-25}$$

TABLE 13.1

Coefficient of Friction μ, Between Poured Concrete Base of Retaining Wall and Underlying Soil

Soil Type	μ
Coarse-grained sands and gravels without silt	0.55
Sands and gravels containing a significant proportion of silt	0.45
Cohesionless silts	0.35
Sound rock, roughened surface	0.60

in which B is the width of the base of the wall and s is the undrained shearing resistance, that is, one half the unconfined compressive strength, provided that s is always taken to be $\leqslant 1,000$ psf.

Note that in calculating $R_{h_{max}}$ the potential contribution of passive pressure in front of the wall has been ignored. This is desirable, except in the case of deeply imbedded walls, because of a variety of factors which make it difficult to assess the probable long-term contribution of the passive pressure. Such factors include softening of soil due to drainage through weep holes in the wall or surface drainage along the front, subsequent excavation of material from in front of the wall, and other environmental and imponderable effects.

It is customary to require the factor of safety against sliding to be no less than 1.5. In the case of cohesive soils, where long-term creep deformation may be important, a factor of safety of 2 is desirable.

When the resistance against sliding is calculated to be inadequate, the use of a *key*, cut into the underlying supporting soil as shown in Figure 13.13, is sometimes considered. However, unless the key penetrates a significantly stronger stratum, only a very nominal benefit is likely to result. This occurs because the effect of the key is simply to move the potential sliding surface lower, and lengthen it somewhat as illustrated in Figure 13.13.

If a layer of weak material occurs below the soil immediately underlying the base, we should consider the possibility of a block slide along the surface of these materials.

Overturning

The term "*overturning*" embodies two considerations: the first is the possibility that the wall could rotate about its toe and actually tip forward; the second is the possibility that the tendency for rotation about the toe would be strong enough to lead to a zone of zero contact pressure between the wall and the soil near the heel. It is, of course, desirable to maintain a positive contact pressure over the entire base in order to insure that sloughing of soil due to flowing groundwater cannot occur beneath the base of the wall. We

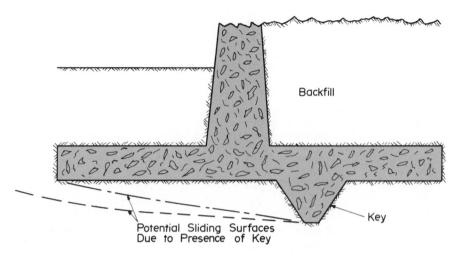

Fig. 13.13—Schematic effect of key on potential sliding surface for uniform soil beneath base of wall.

can eliminate the potential overturning problem by insuring that the resultant
force on the base of the wall lies within the middle third of the heel-toe width,
as shown in Figure 13.12. That is, the eccentricity e must be

$$e \leqslant \frac{B}{6} \qquad (13-26)$$

in which B is the width of the base of the wall.

Slope Stability

The stability considerations we have discussed above all relate to the performance of the base of the wall as a foundation, and the capacity of the
subsoil in the vicinity of the base to support the load transmitted from it. If
the underlying soils are uniform or generally increasing in strength with
depth, the potential for failure is likely to be highest near the base of the wall
where the imposed stresses are the largest. If, however, a soft or weak layer
exists at some depth below the base of the wall, a deep slide of either the
circular or wedge failure mode should be considered. We can recognize the
potential for this type of failure if we view the wall simply as a means of
increasing the slope angle of the transition in elevation between the ground
surface near the toe of the wall and the backfill. Thus the potential for a
deep seated slide must not be overlooked. This possibility can be analyzed
by methods discussed in Chapter 12 in which the slope angle is simply that
provided by the presence of the wall.

13.5 OTHER DESIGN CONSIDERATIONS

Environmental Factors

The most significant environmental effect on the performance of a retaining wall usually arises from changes in groundwater and/or frost conditions. We have seen the major influence which a rise in the groundwater table behind the back of the wall has on the lateral force on the wall. The presence of free water in the soil below the base and near the toe of the wall may also reduce the stability of the wall due to a reduced bearing capacity and horizontal resistance at the base. Proper drainage provisions will insure that the groundwater level behind the wall does not materially influence the forces imposed on it by the backfill. Details of useful drainage systems are illustrated in Terzaghi and Peck (1967). Care when removing water from the drainage systems so that it does not flow to the zone in front of the wall will assist in maintaining proper drainage in this region.

Adequate drainage provisions are also helpful in reducing frost effects. Furthermore, it is necessary to locate the base of the wall at such a depth that frost heave, and subsequent softening due to thawing do not lead to seasonal reduction in foundation stability.

Settlement

Even though our design of the wall and backfill satisfies the various requirements of stability, we must insure that wall movements are within acceptable limits. As for any foundation, settlement of the base of the wall will lead to displacements of the superstructure. In the case of retaining walls, the magnitude of wall movement is often of minor consequence, so long as stability is assured. When the wall is in proximity to, or part of another structure, however, even small movements may affect the performance of the overall system. For example, bridge abutments often serve a dual purpose, both to support the bridge and to retain the fill material of the bridge approach. In such a case, large wall movements may be disastrous. This is illustrated schematically in Figure 13.14. When the wall movement arises primarily from deformation within the foundation soil just beneath the base of the wall, the eccentric and inclined load produces a rotation of the wall away from the backfill (Figure 13.14a). In the case of the bridge abutment, this leads to reduced clearances, and if the movements are sufficient, imposition of compressive forces on the bridge structure. If, on the other hand, settlements arise primarily from compression of a deeper underlying compressible stratum, as depicted in Figure 13.14b, the major cause of the settlement will be the pressures imposed by the backfill. Thus the wall will tend to rotate *toward* the fill because settlement at the heel will be larger than that at the toe. The potential consequences of such movement are obvious.

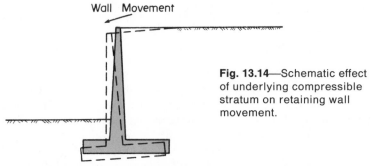

Fig. 13.14—Schematic effect of underlying compressible stratum on retaining wall movement.

a) - Wall Settlement Due Primarily to Compression of Foundation Soil Beneath Base.

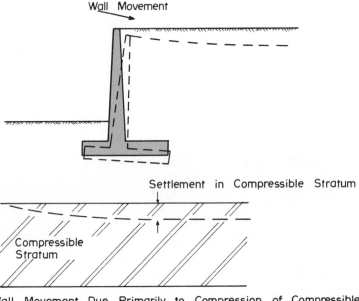

b) - Wall Movement Due Primarily to Compression of Compressible Stratum at Depth Below Backfill.

13.6 SUMMARY OF KEY POINTS

Summarized below are key points which we have discussed in the previous sections. Reference is made to the place in the text where the original more detailed description can be found:

1. Retaining structures provide an abrupt transition in elevation within an earth mass. Design of such structures requires consideration of the interaction between the structure itself and the soil with which it is in contact. The design process is, by necessity, iterative (Section 13.1).

2. Retaining walls are of two general types: "rigid" and "flexible." The distinction is made on the basis of the significance of the deformations occurring *within* the structure. Flexible walls may be *braced* or *unbraced* (Section 13.2).
3. Active earth pressure forces on rigid walls due to a cohesive backfill can be determined by (Section 13.3):
   a. The Rankine (limiting equilibrium) method if the development of potential failure surfaces can occur as in a semi-infinite medium (Equation 13–17). The effect of potential tension cracks must be included (Equation 13–7).
   b. The Coulomb (plane wedge) analysis, if shear stresses between the wall and backfill, or geometry and loading conditions render the Rankine analysis inapplicable.
4. The distribution of forces on a wall due both to earth pressures and the effect of forces on the backfill surface can be estimated using an approximate method involving dividing the potential failure wedge into a series of smaller wedges bounded by planes parallel to the potential failure surface (Equations 13–15 and 13–16).
5. Passive pressure forces cannot generally be determined by Rankine or Coulomb methods because the friction and/or adhesion between the wall and backfill produce a marked curvature in the potential failure surface (Section 13.3, Figure 13.9). To calculate passive pressure forces, a *composite* failure surface is assumed, consisting of a logarithmic spiral curved portion continuous with a straight line portion. The passive pressure force is obtained by satisfying moment equilibrium of a free-body of part of the potential failure zone (Figure 13.10).
6. Once the forces acting on a wall have been determined, the stability of the wall must be assured. Specific potential modes of failure requiring investigation are (Section 13.4):
   a. Bearing capacity of the base of the wall. This is analyzed by considering the base as a shallow foundation subject to an eccentric inclined load.
   b. Sliding along the base (Equations 13–23 and 13–25).
   c. Overturning the wall. This is circumvented by insuring that the resultant force on the base lies within the middle third.
   d. Stability of the slope created by the wall. The methods of Chapter 12 apply here.
7. Other design considerations include
   a. The effects of water and/or frost in the backfill (Sections 13.3 and 13.5).
   b. Evaluation of the magnitude, direction, and significance of wall movements due to settlement (Section 13.5).

PROBLEMS

13.1. Estimate the magnitude and location of the active earth pressure force on the wall shown.

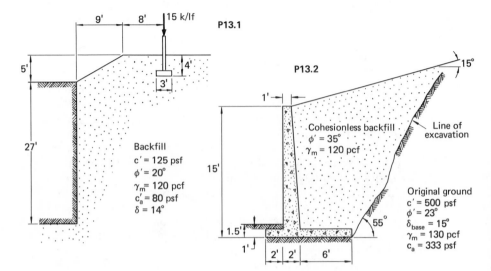

13.2. Consider the cantilever retaining wall shown.
  (a) From the point of view of foundation stability, is the design adequate as shown?
  (b) What changes would improve the design?
  13.3. It is necessary to support the side of a building to permit excavation adjacent to it below the foundation level, as shown. The load to be applied by the building is $F = 16$ kips/ft. How deep must the trench be dug (i.e., $H = ?$) to support the structure with FS = 2 against a failure of the soil?

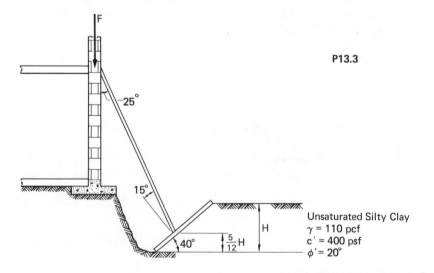

14

Steady-State Seepage
and Its Effects

14.1 INTRODUCTION

In nearly every chapter, we have emphasized the profound effect of water on soil behavior, and therefore on the performance of structures composed of, or in contact, with soil. We have seen that the presence of a *static* water table generally leads to, or implies, increased deformation and reduced stability relative to those cases where free water is absent. We also observed in Chapter 6 that the effect of *flowing* water may be far more serious, especially in relation to stability.* "Sand boils" at the toe of a levee or other water-retaining structure will, if unattended, lead to a collapse of the works. When an excavation penetrates to a depth below the groundwater table, flow of water into the base of the cut may endanger the stability of the entire excavation. Seepage out of a cut slope or earth dam enhances the potential for sliding. The seepage forces increase the driving moments, and the pore water pressures on the failure surface reduce the shear resistance. Flowing water beneath a dam indicates the presence of pore water pressure at the base of the structure which diminishes the resistance to sliding along its base. This effect may be a major factor in the design of the structure.

All the foregoing examples illustrate the often crucial effect which *seepage* of water may have on the overall performance of structures on, in, or of the earth. In order to predict the effect of flowing water upon our designs, we are usually concerned with determining one or more of the following features of the seepage problem:

1. Definition of the region in which flow is actually taking place. This may be most difficult when one surface of the *flow regime* is a *free surface*, that is, a surface at which the water pressure is atmospheric.

---

* Studies of the Vaiont Reservoir disaster and other slides into reservoirs have implicated the combination of seepage due to rainfall and the change in groundwater table due to the rise in reservoir surface as a major factor in the failure (Breth, 1967; Kenney, 1968; Taniguchi, 1967).

2. Water pressure within the flow region.
3. Changes in effective stress due to the flow, that is, *seepage forces.*
4. The quantity of seepage.

The fundamental relationships governing the flow of water in a saturated porous medium were discussed in Chapter 6. We considered the *transient* flow problem in one dimension in Chapter 7. In this chapter we shall consider two-dimensional seepage, but limit our attention to the steady-state case. That is, situations in which no flow occurs in one dimension, and in which neither the region nor the nature of the flow change with time. Some examples of two-dimensional seepage are shown in Figure 14.1. In these examples, the flow is essentially two-dimensional because one of the dimensions is much greater than the others (Figures 14.1a and b), or because of axial symmetry (Figure 14.1c). Thus many problems of practical interest fall within this category.

14.2   REVIEW OF ASSUMPTIONS AND CONCEPTS

In Chapter 6 we developed a mathematical model describing flow of an incompressible fluid through a saturated porous medium such as soil. The principles which govern fluid flow are the requirements of conservation of mass, energy, and linear momentum. We expressed conservation of mass in the form of the continuity equation for steady-state conditions in Equation 6–10. That is,

$$\frac{\partial v_x}{\partial x} + \frac{\partial v_y}{\partial y} + \frac{\partial v_z}{\partial z} = 0 \tag{14–1}$$

in which v_x, v_y, v_z, are the *discharge* velocities in the x, y, and z directions respectively.

Conservation of energy and linear momentum are embodied in the deceptively simple statement of Darcy's law,

$$\mathbf{v} = K\mathbf{i} \tag{14–2}$$

in which $\mathbf{v}$ is the discharge velocity vector, K is a symmetric matrix of directional components of the coefficient of permeability (illustrated in detail in Equation 6–81), and $\mathbf{i}$ is the hydraulic gradient, defined in general terms in Equations 6–48 and 6–49:

$$\mathbf{i} = -\frac{\partial h}{\partial \mathbf{s}} \tag{14–3}$$

Conservation of energy is introduced via *head, h,* in Equation 14–3. Head is free energy per unit volume of fluid, and from Bernoulli's equation for slow flow (Equation 6–12) is

$$h = \frac{u}{\gamma_w} + z + \text{const} \qquad (14\text{-}4)$$

in which u is the water pressure, γ_w is the unit weight of water, and z is the elevation of the point of interest.

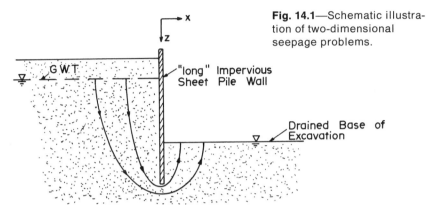

Fig. 14.1—Schematic illustration of two-dimensional seepage problems.

a)- Flow into "Long" Excavation

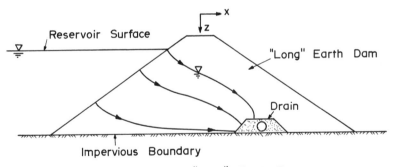

b)- Flow through "Long" Earth Dam

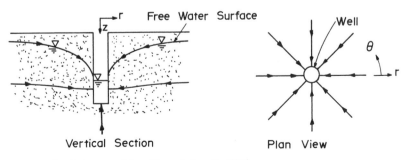

c)- Axisymmetric Flow into a Well

For the usual case, in which the coordinate axes are assumed to be principal directions of permeability, Darcy's law is written in terms of its components (Equation 6–83) as

$$v_x = -k_x \frac{\partial h}{\partial x}$$

$$v_y = -k_y \frac{\partial h}{\partial y}$$ (14–5)

$$v_z = -k_z \frac{\partial h}{\partial z}$$

When the soil is assumed homogeneous, Equations 14–5 and 14–1 may be combined to produce Equation 6–86:

$$k_x \frac{\partial^2 h}{\partial x^2} + k_y \frac{\partial^2 h}{\partial y^2} + k_z \frac{\partial^2 h}{\partial z^2} = 0$$ (14–6)

The further assumption of isotropy, so that

$$k_x = k_y = k_z = k$$ (14–7)

leads to Laplace's equation:

$$\frac{\partial^2 h}{\partial x^2} + \frac{\partial^2 h}{\partial y^2} + \frac{\partial^2 h}{\partial z^2} = 0$$ (14–8)

14.3 LAPLACE'S EQUATION

If we restrict our attention to two-dimensional problems in which $\partial h/\partial y = 0$, Equation 14–8 reduces to*

$$\frac{\partial^2 h}{\partial x^2} + \frac{\partial^2 h}{\partial z^2} = 0$$ (14–9)

A solution of this equation which also satisfies the boundary conditions of the flow region of interest is a *unique solution*. Before considering solution procedures, however, we shall find it useful to examine certain characteristics of the equation and its solution.

---

* In many references (for example, Harr, 1962, and Wu, 1966) a potential function $\phi(x, z)$ is defined as

$$\phi(x, z) = -kh$$

Equation 14–9 may then be expressed in terms of this potential function as

$$\frac{\partial^2 \phi}{\partial x^2} + \frac{\partial^2 \phi}{\partial z^2} = 0$$

This notation is used principally because of its historical significance in other areas of mathematics, science, and engineering, in which the Laplace equation has found application. For our purposes, there is no advantage to be gained by introducing the additional notation.

If a solution to Laplace's equation $h(x, z)$, exists, then it may be shown (Churchill, 1960) that another potential function $\Psi(x, z)$, must also exist, such that

$$\frac{\partial \Psi}{\partial x} = k \frac{\partial h}{\partial z} \qquad (14\text{--}10a)$$

and

$$\frac{\partial \Psi}{\partial z} = -k \frac{\partial h}{\partial x} \qquad (14\text{--}10b)$$

For future reference we note that, from Equation 14–5, this implies that

$$\frac{\partial \Psi}{\partial x} = -v_z \qquad (14\text{--}11)$$

Differentiating Equation 14–10a with respect to x and 14–10b with respect to z, and adding the results, assuming Ψ is a continuous function,

$$\frac{\partial^2 \Psi}{\partial x^2} + \frac{\partial^2 \Psi}{\partial z^2} \qquad (14\text{--}12)$$

Thus we see that the Ψ function also satisfies Laplace's equation. In fact, h and Ψ are both parts of the solution to our problem. We can see this more clearly by examining the physical significance of these parameters.

Streamlines and Equipotential Lines

We have already defined the function $h(x, z)$ as the free energy per unit volume of fluid, that is, the capacity to do work. In order to establish the physical significance of the Ψ function we consider a curve in the xz plane along which $\Psi(x, z)$ = constant. There will, of course, be an infinite number of such curves, each corresponding to a specific arbitrary value of the constant. In Figure 14.2a we see one such curve, along which

$$d\Psi = \frac{\partial \Psi}{\partial x} dx + \frac{\partial \Psi}{\partial z} dz = 0 \qquad (14\text{--}13)$$

Thus at any point on such a curve the tangent assumes an orientation defined by

$$\frac{dz}{dx} = -\frac{\dfrac{\partial \Psi}{\partial x}}{\dfrac{\partial \Psi}{\partial z}} \qquad (14\text{--}14)$$

as shown schematically in Figure 14.2a. Comparing this result with equations 14–10, we see that the slope of this tangent is

$$\frac{\partial z}{\partial x} = \frac{v_z}{v_x} \qquad (14\text{--}15)$$

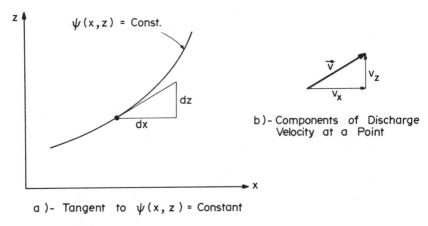

a)- Tangent to $\psi(x, z)$ = Constant

Fig. 14.2—Relation between stream function $\psi(x, z)$ and flow velocity.

As indicated schematically in Figure 14.2b, this is simply the slope of the resultant flow-velocity vector. In other words, the tangent at each point of the curve of $\Psi(x, z)$ = constant has the same orientation as the discharge velocity vector at that point. Consequently, this curve is simply the path along which a drop of water will flow (neglecting the localized deviations from this path due to the tortuosity of the continuous voids). For this reason, the Ψ function is called the *stream function* and curves of $\Psi(x, z)$ = constant are called *streamlines* or *flowlines*.

In order to see the relationship between $h(x, z)$ and $\Psi(x, z)$ we shall consider the line in the xz plane along which $h(x, z)$ = constant, called an *equipotential line*. The slope of such a curve is determined from:

$$dh = \frac{\partial h}{\partial x} dx + \frac{\partial h}{\partial z} dz = 0 \qquad (14\text{–}16)$$

which gives a slope

$$\frac{dz}{dx} = -\frac{\dfrac{dh}{dx}}{\dfrac{\partial h}{\partial z}} \qquad (14\text{–}17)$$

Substituting for the partial derivatives in this expression from Equation 14–5 leads to

$$\frac{\partial z}{\partial x} = -\frac{v_x}{v_z} \qquad (14\text{–}18)$$

Comparing Equation 14–18 with the expression for the slope of the streamlines (Equation 14–15), we see that the two families of curves are *orthogonal.* Thus all equipotential lines which are solutions to Laplace's equation will intersect the streamlines, which are also solutions, at right angles as illustrated in Figure 14.3. This result will be useful when we develop a procedure for generating a solution graphically.

Quantity of Seepage

Figure 14.4 shows two streamlines, defined by

$$\Psi(x, z) = \Psi_1$$
$$\Psi(x, z) = \Psi_2 \tag{14-19}$$

Because the resultant velocity vector at each point along a streamline is tangent to the streamline, there can be no flow across a streamline. Therefore, the quantity of flow between two specific streamlines such as those shown in the figure, remains constant, irrespective of the change in distance between the streamlines.

In order to determine this quantity of flow, we consider a point on a typical equipotential line such as that shown in Figure 14.4. The flow velocity at that point is normal to a differential area about that point, defined by a differential distance along the equipotential line ds, multiplied by a unit distance into the page:

$$d\mathbf{A} = (1)\, d\mathbf{s} \tag{14-20}$$

Thus the differential flow rate through this area is

$$dq = \mathbf{v} \cdot d\mathbf{s} \tag{14-21}$$

If we express the magnitude of the velocity vector in terms of its components in the x, z coordinate system,

$$v = v_x \cos \theta + v_z \sin \theta \tag{14-22}$$

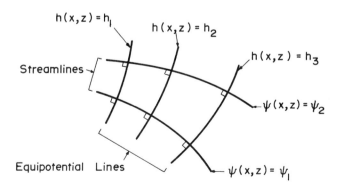

Fig. 14.3—Orthogonal relationship between equipotential lines and streamlines.

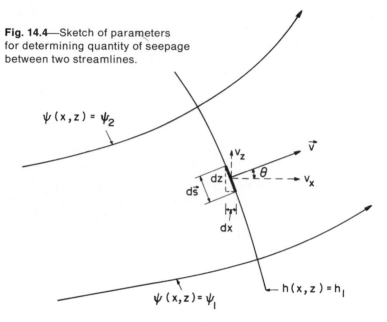

Fig. 14.4—Sketch of parameters for determining quantity of seepage between two streamlines.

the differential quantity of flow becomes

$$dq = v_x\, ds \cos \theta + v_z\, ds \sin \theta \qquad (14\text{--}23)$$

We can also describe the small distance along the equipotential line in terms of the coordinate system,

$$ds \cos \theta = dz$$
$$ds \sin \theta = -dx \qquad (14\text{--}24)$$

in which the minus sign appears because, for the orientations shown in the sketch, a positive change in s and z corresponds to a negative change in x. By substituting Equations 14–24 into Equation 14–23, we have

$$dq = v_x\, dz - v_z\, dx \qquad (14\text{--}25)$$

From Equations 14–11 we see that this is simply the total differential of Ψ:

$$dq = \frac{\partial \Psi}{\partial z}\, dz + \frac{\partial \Psi}{\partial x}\, dx = d\Psi \qquad (14\text{--}26)$$

If we now integrate Equation 14–26 from Ψ_1 to Ψ_2, the increment in flow between the two streamlines, Δq, becomes

$$\Delta q = \Psi_2 - \Psi_1 \qquad (14\text{--}27)$$

Thus we find the useful result that the quantity of flow between any two streamlines (along which $\Psi(x, z)$ is constant) is simply the difference between the two values of Ψ.

14.4   BOUNDARY CONDITIONS

In order to obtain the solution to a seepage problem, we must consider the influence of the various possible boundary conditions. These conditions lead to two general categories of problems:

1. *Confined flow.* In this case all boundaries of the flow domain are *initially* defined, and the flow must occur within these boundaries. That is, the limits of the flow region can be established before the solution is obtained. This is illustrated in Figure 14.5a in which the flow domain is limited by the dam and sheet-pile above and the impervious material below.

2. *Unconfined flow.* In this case, the position of at least one boundary of the flow domain is *initially* undefined. The location of that

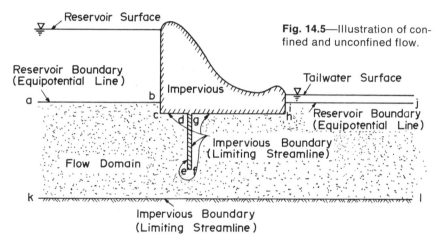

Fig. 14.5—Illustration of confined and unconfined flow.

a) - Confined Flow

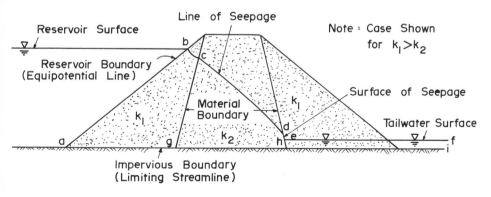

b) - Unconfined Flow

boundary is determined by the solution itself. This is illustrated schematically in Figure 14.5b, where the free-water surface, or *line of seepage*, at the top of the flow region through an earth dam is not known *a priori*. The free-water surface flowing into a drainage well, as indicated in Figure 14.1c, is initially undetermined, making this an unconfined flow problem too. In both of these cases the position of the free-water surface is determined by the geometry of problem, the flow characteristics of the soil, and other factors which influence the solution. Methods for determining the position of the free-water surface are discussed below.

There are five types of boundary conditions which we need to consider:

1.   Reservoir boundaries,
2.   Impervious boundaries,
3.   Boundaries between materials of different permeabilities,
4.   Line of seepage,
5.   Surface of seepage.

These boundary types are illustrated in Figure 14.5. The first three types pertain to both confined and unconfined flow problems; the last two types apply only to unconfined flow problems. Reservoir and impervious boundaries are discussed below. Description of the latter three types is deferred to another section.

Reservoir Boundaries

The flow into and out of a reservoir produces energy losses which are generally small compared to losses within the soil mass in contact with the reservoir. Thus, the reservoir may be assumed to be a standing body of water within which the total head is constant. The reservoir boundaries are then equipotential lines. In the cases shown in Figure 14.5, surfaces $\overline{ab}$ and $\overline{ij}$ in Figure 14.5a, and surfaces $\overline{ab}$ and $\overline{hi}$ in Figure 14.5b are equipotential lines by virtue of being reservoir boundaries.

Because reservoir boundaries are equipotential lines, the streamlines must intersect them at right angles.

Impervious Boundaries

One or more boundaries of a flow problem are generally defined by a soil or rock stratum whose permeability is negligible compared to that for the soil within the flow region. Such boundaries are considered *impervious*. If a boundary surface is impervious, then by definition there can be no flow normal to it. Assuming no gaps within the flow region, and recognizing that streamlines cannot intersect one another, we see that the impervious boundaries must also be streamlines. In fact, they are the bounding streamlines of the flow region. Such limiting streamlines are illustrated by lines $\overline{kl}$ in Figure 14.5a, and line $\overline{agh}$ in Figure 14.5b.

In the case of confined flow, there is also an upper impervious boundary which is frequently the contour of the water impounding structure. Thus in Figure 14.5a the base of the dam and the surface of the sheet-pile cutoff wall, line $\overline{bcdefghi}$, is the upper limiting streamline.

14.5 SOLUTION PROCEDURES

There are a variety of procedures available for the solution of Laplace's equation, (Equations 14–9 and 14–12), subject to the boundary conditions of a specific problem. These methods may be classified as follows:

1. Closed-form solutions
2. Physical models and analogue solutions
3. Approximate solutions

These approaches are discussed in detail below.

Closed-Form solution

A closed-form solution is a mathematical expression which satisfies the governing equations of the problem as well as all of the boundary conditions. Such closed-form solutions cannot be obtained for every seepage problem. But when one is available, or can be obtained, it is the most desirable solution. This is the case because the iterative design process requires considering alternative trial designs; alternatives which frequently involve changes in the geometric parameters of the problem. The effect of varying geometric design features is often determined most easily when a closed-form solution is available.

One useful method for obtaining a closed-form solution is to *guess* at the form of the solution, and then determine the constants so that the function satisfies Laplace's equation and the boundary conditions. We shall illustrate this procedure with a simple, albeit instructive example:

Example 14.1 ———————————————————————

Two reservoirs are connected by a tabular pervious aquifer of approximately uniform thickness. This is illustrated schematically in Figure 14.6a. Determine the solution, $h(x, z)$, for flow through the aquifer, and the rate of seepage.

Solution. If the reservoir and aquifer are long in a direction normal to the plane of the sketch, then we might expect that the problem reduces to a simple one-dimensional case. We have already deduced the results of this case without proof in Section 6.6. Now, we can examine the basis for the assumptions made then.

Based on our intuition about this problem we shall model the aquifer analytically as a saturated homogeneous medium of uniform thickness with isotropic permeability k. We shall guess that the solution form is linear,

$$h(x, z) = rx + sz + t \qquad (14\text{–}28)$$

in which x is the distance along the flow path from line $\overline{ab}$ to $\overline{cd}$, as shown in Figure 14.6a, z is the elevation measured from an arbitrary elevation datum, r, s, t are constants which must be evaluated from the boundary conditions. In this simple case, our "solution" satisfies Laplace's equation identically.

At the inlet end of the aquifer, surface $\overline{ab}$, the head is constant because the reservoir can be considered a standing body of water. Assuming the midheight of the aquifer to be the elevation datum, $z = 0$, the depth to the midline is equal to the head at the inlet end,

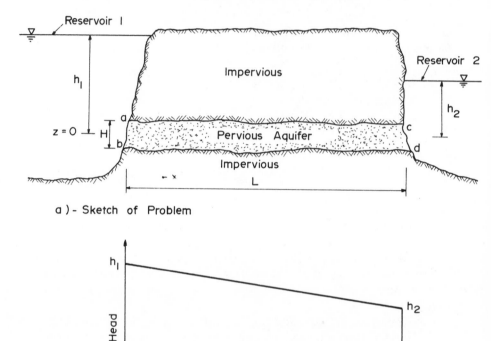

a) - Sketch of Problem

b) - Distribution of Head along Flow Path Determined in Solution

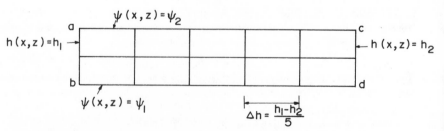

c) - Typical Flow Lines, and Equipotential Lines Corresponding to Five Equal Increments of Head Loss.

Fig. 14.6—Illustrative one-dimensional seepage problem.

h_1. That is, for $x = 0$,

$$h_{ab} = h_1 = sz + t \tag{14-29}$$

From this expression we see that

$$s = 0, \qquad t = h_1 \tag{14-30}$$

At the outlet end of the aquifer, line $\overline{cd}$, for which $x = L$,

$$h_{cd} = h_2 = rL + h_1 \tag{14-31}$$

From which,

$$r = -\frac{h_1 - h_2}{L} \tag{14-32}$$

Substituting these values for the constants, the solution becomes

$$h(x, z) = h_1 - \frac{h_1 - h_2}{L} x \tag{14-33}$$

which is illustrated in Figure 14.6b.

We must now inquire whether the solution also satisfies the boundary conditions at the upper and lower impervious boundaries. We recall that an impervious boundary is a limiting streamline along which $\Psi = $ constant. Because they are horizontal in this case, along the impervious boundaries

$$\frac{\partial \Psi}{\partial x} = 0 \tag{14-34}$$

Thus by inspection, or more formally from Equation 14-14, the direction of the tangent to the streamline is

$$\frac{dz}{dx} = 0 \tag{14-35}$$

In order for our solution, Equation 14-33, to be valid, the equipotential lines must be orthogonal to the streamlines at the impervious boundaries, as well as within the medium. From Equation 14-33 we see that

$$\frac{\partial h}{\partial z} = 0 \tag{14-36}$$

When substituted into Equation 14-17, or again by inspection, the tangent to the equipotential line has a slope everywhere equal to

$$\frac{dz}{dx} = -\infty \tag{14-37}$$

Thus the equipotential lines and limiting streamlines are indeed orthogonal, and Equation 14-33 is *the* solution.

There are some interesting additional features of the solution to this simple problem. The first concerns the hydraulic gradient within the aquifer. For the case of flow only in the x direction, the hydraulic gradient obtained by the differentiating Equation 14-33 is

$$i = -\frac{\partial h}{\partial x} = \frac{h_1 - h_2}{L} = i_{avg} \tag{14-38}$$

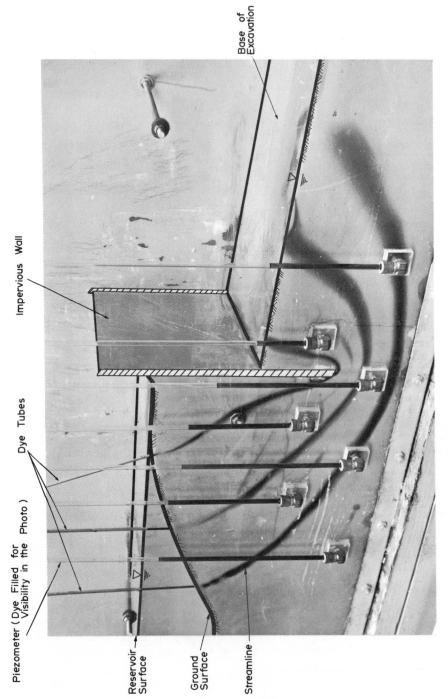

Fig. 14.7—Scale model of flow into excavation.

That is, throughout the flow domain, the hydraulic gradient is *constant*, equal to the *average* hydraulic gradient. This of course, was the assumption we made in analyzing one-dimensional flow problems in Chapter 6. Typical equipotential lines and flow lines are shown in Figure 14.6c. Such a figure, which illustrates the flow pattern, is called a *flow net*.

A second feature of interest is to calculate the quantity of flow from one reservoir into another, per unit distance into the sketch. We do this by considering the change in the value of the stream function between the two boundaries. For this case Equation 14–13 reduces to

$$d\Psi = \frac{\partial \Psi}{\partial z} dz \qquad (14\text{–}39)$$

Substituting Equations 14–10b and 14–38 into this expression gives

$$d\Psi = -k \frac{\partial h}{\partial x} dz \qquad (14\text{–}40)$$

Which, from Equation 14–39, is

$$d\Psi = ki_{avg} dz \qquad (14\text{–}41)$$

To determine the quantity of flow between the boundaries we integrate this result to obtain

$$q = \int d\Psi = \Psi_2 - \Psi_1 = ki_{avg} H \qquad (14\text{–}42)$$

in which H is the thickness of the aquifer. Again, this is the result which we assumed in Chapter 6 (compare, for example, Equation 6–72).

A variety of approaches to obtain the analytical solution for more complicated problems exist. We might, for instance, make other "educated guesses." Alternatively, formal procedures are available for generating closed-form solutions. Such methods, which are beyond the scope of this text, are considered in detail by Harr (1962).

Models and Analogues

For many seepage problems of practical interest, exact solutions are unavailable. In such cases it may be feasible to obtain a solution experimentally.

One approach to this problem is to use a small-scale model. Because the head is proportional to scale for gravity-induced flow, the position of streamlines and equipotential lines in the model are homologous to those in the prototype (provided anisotropic permeability effects in the prototype are also in the model). An example of such a flow model is illustrated in Figure 14.7. This photograph shows flow into an excavation bounded by an impervious sheet-pile wall. The soil in the model is Ottawa sand. Streamlines are determined by injecting dye at the reservoir boundary at convenient points. The head at various points within the flow region is determined by *piezometers* installed in the side of the flow tank, as shown in the figure. These are simply tubes with a screen over the end to prevent soil from washing into them. The water level rises in the piezometers to a point corresponding to the head at locations in the medium where the piezometers are inserted. The piezometer tubes have been filled with dye for visibility in the photograph, although this is not generally necessary. Specific details concerning the conduct of model tests are discussed by Dixon (1967).

While such small-scale models are of pedagogical value, they are often too expensive for all but large and elaborate projects.

An alternative approach to experimental modeling of flow problem is to take advantage of one of the physical analogues for which the governing relationships also reduce to Laplace's equation. One such analogue is the conduction of heat in solids. Another, which is experimentally more useful, is the flow of electrical current in a conducting medium. The correspondence between the flow of electricity and seepage through a porous medium is indicated in Table 14.1. The conducting material for an electrical analogue may vary according to convenience. Salt solution in a shallow tray, coated conducting paper, and absorbent paper soaked with ionic solutions have all been employed. Correspondence between parts of the electrical analogue and the real flow problem is shown schematically in Figure 14.8. Reservoir boundaries are electrical terminals at constant voltage. On conducting paper, these can be produced easily using conductive paint. Impervious boundaries are produced by cutting the paper to the shape desired. In the case of a liquid conducting medium, the shape must be cut from an insulating material.

TABLE 14.1

Correspondence Between Seepage in a Saturated Porous Solid and Flow of Electric Current in a Conducting Medium

Seepage	Flow of Electric Current
Head h	Voltage V
Flow rate q	Current i
Discharge velocity v	Current density j
Coefficient of permeability k	Specific conductivity σ
Darcy's law	Ohm's law
$$\mathbf{v} = -k\frac{\partial h}{\partial \mathbf{s}}$$	$$\mathbf{j} = -\sigma\frac{\partial V}{\partial \mathbf{s}}$$
Laplace's equation	Laplace's equation
$$\frac{\partial^2 h}{\partial x^2} + \frac{\partial^2 h}{\partial z^2} = 0$$	$$\frac{\partial^2 V}{\partial x^2} + \frac{\partial^2 V}{\partial z^2} = 0$$
Impervious boundary-limiting streamline	Insulated boundary-limiting line of current flow

SOURCE: Modified from Muskat (1937).

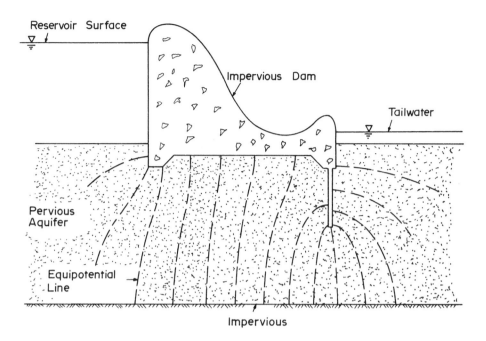

a)- Sketch of Seepage Problem

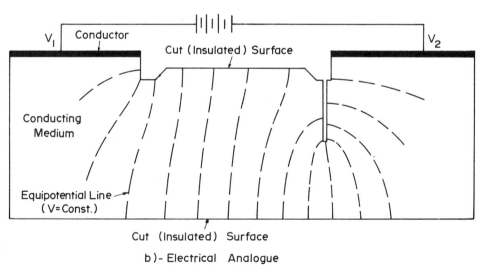

b)- Electrical Analogue

Fig. 14.8—Correspondence between
seepage and flow of electric current.

Equipotential lines are relatively simple to determine in the electrical analogue using a volt meter and small probe. Streamlines can be determined either by *sketching* as described below, or by reversing the problem so that reservoir boundaries become impervious boundaries, and conversely. Then, for the reverse problem the equipotential lines will be the streamlines for the original case.

Other analogies are available, such as the viscous-flow Hele-Shaw model (Brahma and Harr, 1962). However, these lack the general usefulness of the approaches we have described.

Approximate Solutions

Because of the complexity of most real seepage problems of interest, *approximate* solutions are commonly employed. The most useful fall into two categories:

1. Numerical methods such as the finite-difference or finite-element methods, and
2. Graphical solutions, specifically the *flow net*.

In the finite-difference procedure, we replace the governing differential equation (Laplace's equation) by an *algebraic difference* equation in a manner analogous to that introduced in Chapter 7 for the solution of the transient one-dimensional consolidation problem. Details of the procedure are discussed by Harr (1962), Scott, (1963), and Wu (1966). As in the case of the finite-difference solution to one-dimensional consolidation, the finite-difference method is relatively adaptable to use of the digital computer (Tomlin, 1966; Jeppson, 1968).

In the finite-element method, the flow domain is subdivided into small elements within which an approximate form of the solution $h(x, z)$ is assumed. Thus a solution is obtained for each individual element, in which the constants of the solution are obtained by minimizing the energy-dissipation rate, subject to the boundary conditions and continuity requirements between the elements (Zienkiewicz and Cheung, 1965). Although the finite-element method requires the use of a digital computer, it is exceedingly versatile, and is applicable to inhomogeneous and/or anisotropic materials (Zienkiewicz et al., 1966), as well as unusual geometric boundary conditions. Furthermore the method is equally applicable to confined and unconfined flow (Finn, 1967). Thus as the general use of computers increases, and finite-element programs for the solution of flow problems become increasingly available, this versatile approach will probably supersede all others, except for the simplest problems.

A *graphical* representation of the solution of the Laplace equation for a flow problem is called a *flow net* (see Example 14.1) which is a plot of equipotential lines and streamlines on a scale drawing of the given flow region. A simple flow net is illustrated in Figure 14.6c. The flow net may be constructed as an *illustration* of the solution obtained by one of the methods discussed above, as for example, in Figure 14.6c. Alternatively, the requirements which the solution imposes on the flow net can be used to *construct* the solution *graphically*. This is done by *sketching* a trial flow net and observing whether the results satisfy the solution requirements and the boundary conditions. By trial and error, the quality of the flow net can be improved until it does indeed satisfy these requirements, that is the orthogonality of equipotential lines and streamlines, as well as the boundary conditions. When the flow net does this, *it is a plot of the unique solution to the problem.* From the equipotential lines and streamlines so determined, other parameters of interest such as magnitude of the pressure, hydraulic gradient, velocity, and quantity of discharge at any point within the flow domain can be determined.

Because an experienced practitioner can construct a flow net for even complicated boundary conditions, this graphical trial and error procedure is currently the most widely applied method of solving flow problems. Specifics of the procedure and examples are discussed below.

14.6   CONSTRUCTION OF THE FLOW NET

Confined Flow

The solution to Laplace's equation consists of the two orthogonal families of equipotential lines and streamlines. Each "family" contains an infinite number of such curves. In the flow net, we depict only a few of these. For example, a flow net for the simple problem of flow under a sheet-pile cutoff wall embedded in a homogeneous aquifer is illustrated in Figure 14.9. The flow lines* are shown as solid lines, the equipotential lines are dashed. The selection of which specific lines to show was more or less random, but of course, satisfies Laplace's equation and the boundary conditions. By proper selection of which flow lines and equipotential lines to show, we can improve the trial and error process of graphical construction of the flow net. Clearly, there is no fundamental difference between one set of flow lines (or equipotential lines) and another, since a correctly drawn flow net represents a unique solution. However, because construction of the flow net involves visual perception of the extent to which the orthogonality requirements and boundary conditions have been met, it is convenient to select certain specific flow lines and equipotential lines. We note from Figure 14.9b, that the intersection of two adjacent flow lines with two adjacent equipotential lines produces a figure with curved sides which meet at right angles at the corners. Such figures are called *curvilinear rectangles*. Two such curvilinear rectangles with average dimensions a and b, normal and parallel, respectively,

---

* In connection with the construction of flow nets, the term *flow line* is used more commonly than *streamline*.

to the flow lines are shown in Figure 14.9b. It is evident in the figure that the shape of these curvilinear rectangles, as defined by the ratio a/b, varies throughout the diagram.

Let us now assume that rather than selecting equipotential lines at random, we chose to draw only those equipotential lines which represent *equal drops in head* between the inlet and outlet. That is, between any two equipotential lines, the head loss, Δh, is

$$\Delta h = \frac{h_2 - h_1}{n_d} \qquad (14\text{--}43)$$

in which n_d is the desired number of head drops to be sketched within the flow region. Recalling that the quantity of flow across a differential area along an equipotential line is,

$$dq = -k \frac{\partial h}{\partial s} dA \qquad (14\text{--}44)$$

we can estimate the flow between two streamlines by substituting Equation 14–43 for Δh as well as $\Delta s = b$ and $dA = a$. This leads to an approximate expression for the flow between two equipotential lines,

$$\Delta q = \frac{k(h_1 - h_2)}{n_d} \frac{a}{b} \qquad (14\text{--}45)$$

If, for convenience, we then assume that the flow between adjacent equipotential lines is the *same*, so that

$$\Delta q = \Delta \Psi = \frac{q}{n_f} \qquad (14\text{--}46)$$

in which q is the total quantity flow, and n_f is the number of desired flow channels, then the shape of the curvilinear rectangles generated will be

$$\frac{a}{b} = \frac{q}{k(h_1 - h_2)} \frac{n_d}{n_f} \qquad (14\text{--}47)$$

That is, the ratio of the average dimensions of the curvilinear rectangle is the same for all such rectangles: they are the same shape. The total quantity of flow, given any arbitrary selected a/b ratio, is then obtained by rearranging Equation 14–47 to give

$$q = k(h_1 - h_2) \frac{n_f}{n_d} \cdot \frac{a}{b} \qquad (14\text{--}48)$$

It is, of course, simplest to assume $a/b = 1$. Thus the intersecting flow lines and the equipotential lines generate *curvilinear squares*.

In constructing the flow net we take advantage of the converse of the conclusions arrived at above. That is, if the flow net satisfies the requirements of orthogonality and the boundary conditions, *and* consists of curvilinear squares, then—

1. Head drop between adjacent equipotential lines will be the same, and
2. Flow between adjacent flow lines will be the same; and the total flow may be determined from Equation 14–48.

Thus we see that once we have selected the number of flow channels to be drawn, there is one and the only one flow net which will satisfy all of these requirements. The procedure to follow in determining this flow net is as follows:

1. Try a configuration of flow lines and equipotential lines,
2. Identify how this trial flow net violates the requirements,
3. Adjust the trial net to correct these violations. It is likely that the changes will introduce other violations, and further adjustments will be necessary. The process of adjusting and correcting is continued until a satisfactory flow net is obtained.

Details of the approach are described below in connection with an example.

Construction of a Flow Net for Confined Flow

In order to illustrate the method used in constructing a flow net we shall consider the sheet-pile wall partially penetrating an aquifer as shown in Figure 14.9a. Steps in determining the flow net for this problem are:

1. Prepare a scale drawing of the flow region showing all boundary equipotential lines and streamlines.* In the example shown, the upper surface of the soil to the left of the sheet-pile wall is an equipotential line with head H_1. The upper boundary of the soil to the right of the sheet-pile wall is also an equipotential line with head H_2. The surface of the impervious wall and the lower impervious boundary of the aquifer are both limiting flow lines.

2. Sketch a few flow lines (two or three are usually sufficient), such that they are approximately consistent with the boundary conditions and represent reasonable estimates of the flow pattern. Recall that the flow lines are smooth and should intersect equipotential lines, including the boundary equipotential lines, at right angles. Therefore the flow lines cannot intersect each other or the impervious boundary at the bottom of the aquifer, which is itself a flow line. The initial attempt at the flow lines is shown in Figure 14.10a. Note that only half of the figure is completed because the problem is symmetrical. Thus the vertical line from the bottom of the wall to the bottom of the aquifer is an equipotential line, and the trial flow line must have a horizontal tangent at this point.

---

* In constructing the flow net it is usually convenient to draw the boundaries on the *back* side of a sheet of tracing paper. The flow net itself can be constructed on the front side of the page, and erasures and corrections made without disturbing the boundaries.

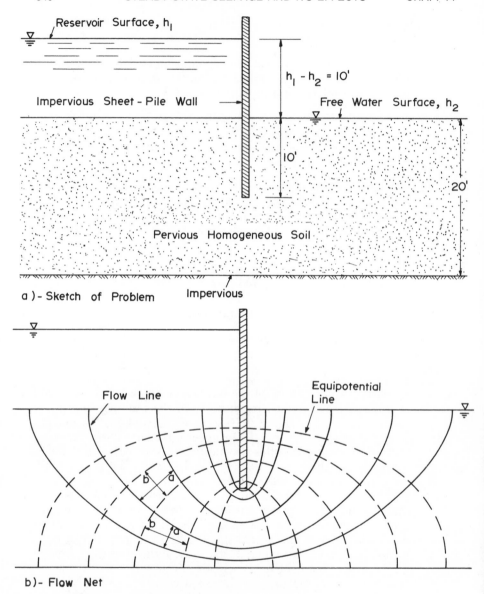

Reservoir Surface, h_1

$h_1 - h_2 = 10'$

Impervious Sheet - Pile Wall

Free Water Surface, h_2

10'

20'

Pervious Homogeneous Soil

a) - Sketch of Problem Impervious

Flow Line

Equipotential Line

b)- Flow Net

Fig. 14.9—Flow net beneath sheet-pile wall.

3. Draw several equipotential lines. These lines must also be smooth and intersect the flow lines at right angles. Equipotential lines are drawn such that adjacent pairs of flow lines and equipotential lines form curvilinear squares insofar as possible. The first trial lines are shown in Figure 14.10a. Not all the shapes produced by intersecting

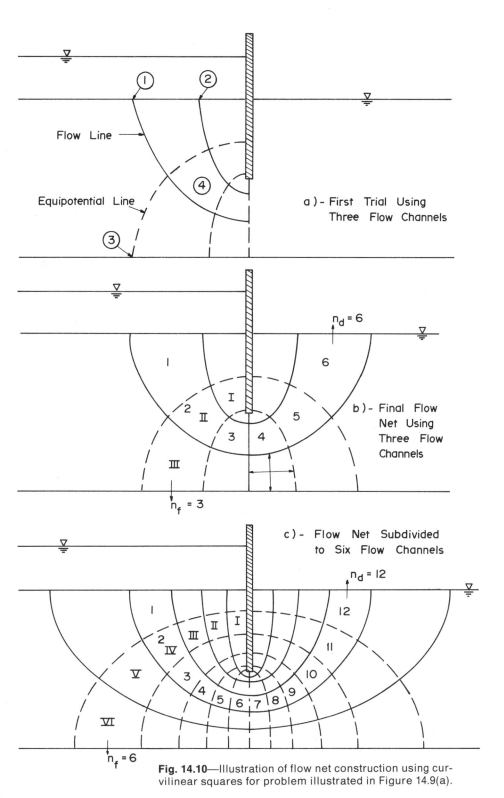

Flow Line

Equipotential Line

a) - First Trial Using
Three Flow Channels

$n_d = 6$

b) - Final Flow
Net Using
Three Flow
Channels

$n_f = 3$

c) - Flow Net Subdivided
to Six Flow Channels

$n_d = 12$

$n_f = 6$

Fig. 14.10—Illustration of flow net construction using curvilinear squares for problem illustrated in Figure 14.9(a).

equipotential and flow lines are four-sided. However if they can be subdivided into three curvilinear squares plus one figure which is the same shape as the original, but smaller, then they can be considered curvilinear squares.

4. The violations of the requirements for the flow net are then identified, and corrections made. In Figure 14.10a at least four such violations are noted by circled numbers. Violations ①, ②, and ③ result from the fact that the orthogonality requirements at the boundary are not satisfied. The shape denoted by ④ is clearly not a curvilinear square. Thus the flow net must be modified. It is best to try to correct the flow net by modifying one line at a time and then changing the entire rest of the flow net in accordance with this modification, before beginning the next series of corrections. That is, it is not desirable to refine one section before carrying through the consequences of the changes to the other sections of the net.

5. After a sufficient number of corrections, the flow net will appear satisfactory, as for example in Figure 14.10b. In the case illustrated, three flow channels were used, and six equal head drops were required in order to maintain curvilinear squares. It may be desirable at this point to further subdivide the flow net into smaller squares which are useful for numerical results, as shown in Figure 14.10c. As the flow net is refined, the need for further modification may become apparent.

In the case of the example shown, there is an integral number of flow lines and equipotential drops. In general this may not be the case. That is, if an integral number of flow lines is enforced then there may be a partial equipotential drop somewhere in the flow net in order to insure that the curvilinear squares exist everywhere else. If this is the case then the a/b ratio for all of the curvilinear rectangles within that partial equipotential drop must be the same. As long as that the a/b ratio is the same, they need not be squares. Alternatively, there may be a whole number of equipotential drops but a partial flow channel. Again in this case, the a/b ratio will be different from one, but it must be constant throughout the length of the partial flow channel.

6. Once the flow net is complete, the magnitude of $h(x, z)$ for each equipotential line can be determined from the known head on the boundaries and the number of head drops, n_d (which may not be an integer). We shall see below how to use the flow net, after we have considered a number of other features involved in its construction.

In the event that there is more than one material in the pervious aquifer, flow occurs across a boundary between two soils of differing permeability. The direction of the flow line is then changed as shown in Figure 14.11. This phenomenon is analogous to the refraction of light rays as they cross the boundary between two materials with different refractive indices. Because the flow lines are bent at the material boundary, the distance between the equipotential lines is changed. Thus, the curvilinear squares in material 1 are changed to curvilinear rectangles in material 2. The widening of the flow channel and shortening of the distance between equipotential drops depicted in Figure 14.11 corresponds to the case in which material 2 is of lower permeability than material 1. These changes are determined by the angles α and β shown in the figure.

Recognizing that the flow rate within the flow channel remains unchanged as the fluid crosses the boundary, we can use Equation 14–45 to describe the flow in both materials,

$$\Delta q = k_1 \, \Delta h \, \frac{a}{a} = k_2 \, \Delta h \, \frac{c}{b} \qquad (14\text{–}49)$$

in which Δh is the head drop across each equipotential drop, and the other parameters are as shown in Figure 14.11. From this expression,

$$\frac{c}{b} = \frac{k_1}{k_2} \qquad (14\text{–}50)$$

We see from the geometry of the sketch in Figure 14.11 that

$$\frac{\tan \beta}{\tan \alpha} = \frac{c}{b} \qquad (14\text{–}51)$$

Substituting this into Equation 14–50, we have

$$\frac{\tan \beta}{\tan \alpha} = \frac{k_1}{k_2} \qquad (14\text{–}52)$$

Thus the deflection of the flow line is determined by the relative permeability of the two strata. We can visualize what happens to a flow line as it crosses the boundary by imagining we are standing on the flow line at a point such that we face the boundary in the flow direction. If the material on the other

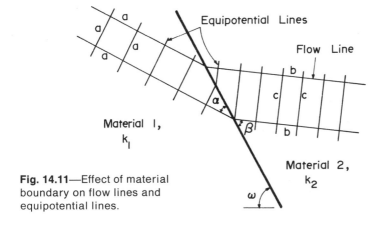

Fig. 14.11—Effect of material boundary on flow lines and equipotential lines.

side of the boundary is of *higher* permeability than the material on which we are standing, the flow line will be bent *away* from the normal to the boundary as it crosses the boundary. If the material on the other side of the boundary is of *lower* permeability than the material on which we are standing, the flow will be bent *toward* the normal to the boundary as it crosses the boundary.

Unconfined Flow

Unconfined flow is illustrated in Figures 14.1b and c, and 14.5b. It is characterized by the fact that the uppermost flow line is a free-water surface whose location is not known *a priori*. In contrast to the confined flow problem, determination of the position of this bounding streamline is one of the major objectives of analysis of unconfined seepage. Because of its importance, this bounding streamline is given a special name, the *line of seepage*. Since the line of seepage is a free-water surface, the head at any point on this line is

$$h = \frac{u}{\gamma_w} + z = z \qquad (14\text{--}53)$$

But for a flow net composed of curvilinear squares, the head loss between equipotential lines, Δh, is a constant. Thus, along the line of seepage,

$$\Delta h = \Delta z = \text{const} \qquad (14\text{--}54)$$

which is illustrated in Figure 14.12.

The effects of entrance, discharge, and material interface conditions on the line of seepage are illustrated in Figure 14.13. This figure shows the effect of the aquifer boundary orientation ω, on the orientation of the line of seepage at the boundary. The derivations of these results are given by Casagrande (1937). Also shown in Figure 14.13b is the point at which the line of seepage exits from the discharge face, denoted as point A. The location

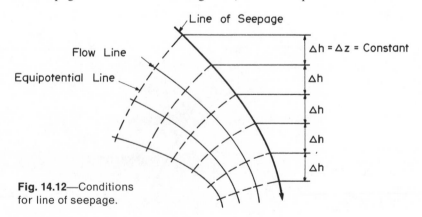

Fig. 14.12—Conditions for line of seepage.

Line of Seepage

Flow Line

Equipotential Line

$\Delta h = \Delta z$ = Constant

Δh

Δh

Δh

Δh

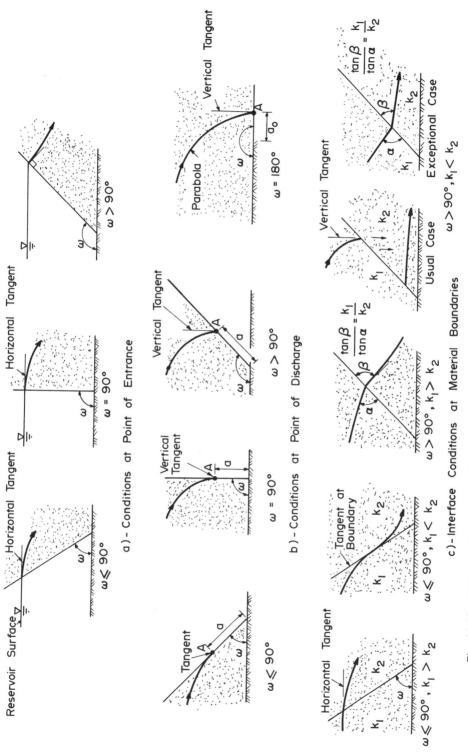

Reservoir Surface

Horizontal Tangent

Horizontal Tangent

Horizontal Tangent

$\omega \leqslant 90°$

$\omega = 90°$

$\omega > 90°$

a) - Conditions at Point of Entrance

Vertical Tangent

Vertical Tangent

Vertical Tangent

Parabola

Tangent

$\omega \leqslant 90°$

$\omega = 90°$

$\omega > 90°$

$\omega = 180°$

b) - Conditions at Point of Discharge

Horizontal Tangent

Tangent at Boundary

$\omega \leqslant 90°, k_1 > k_2$

$\omega \leqslant 90°, k_1 < k_2$

$\dfrac{\tan \beta}{\tan \alpha} = \dfrac{k_1}{k_2}$

$\omega > 90°, k_1 > k_2$

Vertical Tangent

Usual Case

Vertical Tangent

$\dfrac{\tan \beta}{\tan \alpha} = \dfrac{k_1}{k_2}$

Exceptional Case

$\omega > 90°, k_1 < k_2$

c) - Interface Conditions at Material Boundaries

Fig. 14.13—Entrance, discharge and interface conditions for line of seepage. (After Casagrande, 1937.)

645

of this point can be found either by constructing a flow net, or by approximate means described below.

The surface along which discharge takes place is neither a flow line nor an equipotential line, and squares along such a boundary are incomplete. This surface, called the *surface of seepage*, is illustrated in Figure 14.5b as well as in Figure 14.13b (in which the length of the surface of seepage is the distance *a*). Because the surface of seepage is at atmospheric pressure, Equation 14–54 applies there as well as along the line of seepage, even though curvilinear squares are not maintained.

We shall now describe the steps involved in constructing a flow net for an unconfined flow problem, using as our example the homogeneous, isotropic earth dam of typical shape, illustrated in Figure 14.14a. Because the line of seepage is simply the uppermost flow line, when the flow net which satisfies the various requirements of the solution has been determined, the line of seepage will be fixed. These requirements are summarized for reference:

1. At the surface where the water enters the dam, all of the flow lines must be normal to the entrance face.
2. Curvilinear squares must be maintained throughout the flow net (except, of course, along the *surface of seepage*).
3. The head loss between two adjacent equipotential lines along the line of seepage must be a constant, and equal to the elevation head loss (Equation 14–54).
4. At the point of discharge, the line of seepage is parallel to the discharge surface of the dam in the case of a shallow discharge face (Figure 14.13b).

It has been found that the line of seepage is very nearly parabolic in shape except near the points of discharge and entrance. At the point of discharge, the curve deviates to become tangent to the discharge surface. At the point of entrance, the line of seepage deviates from the parabola to become normal to the entrance face. If the parabola were continued back through the entrance face, it would intersect the free-water surface at point P_1, shown in Figure 14.14a, which is approximately one-third of the distance from point P to the vertical projection of the heel of the dam. This distance is denoted in Figure 14.14a by m, and the distance from point P to point P_1 is shown as $m/3$. The h/d ratio, which is the total head loss through the dam divided by the distance from the toe of the dam to point P_1, as shown in Figure 14.14a, may be used as a scale factor to assist in drawing the flow net. The steps are:

1. Construct a base line intersected by the discharge slope as indicated in Figure 14.14b. Also construct a line with slope equal to h/d passing through the toe of the discharge surface.
2. Start by sketching the flow net at the toe working back towards the entrance. When the first two equipotential lines are drawn, as in Figure 14.14c, the constant $\Delta h = \Delta z$ is fixed. Then additional Δz

can be scaled and drawn as horizontal lines on the figure to assist in establishing the location of the line of seepage. Three flow channels are usually sufficient to develop the complete flow net. They may be further subdivided when the flow net is completed.

3. Continue sketching the flow net back toward the entrance, developing the line of seepage using the requirement that along this line $\Delta h = \Delta z = $ constant. Continue the flow net until the line of seepage intersects the h/d line as shown in Figure 14.14d. This intersection is point P_1, and determines the scale of the flow net drawing. The water surface and dam boundaries can then be drawn, as shown by the dashed lines in Figure 14.14d.

4. Correct the flow net for entrance conditions, so that the flow lines are normal to the entrance face as shown in Figure 14.14e. Note from this figure that the required corrections changed the shape of the flow net sufficiently to introduce a partial equipotential drop. As in the case of the confined flow net, this simply means that for the number of flow channels selected, there is not an *integral* number equipotential drops. The final flow net shown is the solution to the equation, and is equivalent to an exact solution within the desired accuracy.

Determination of Water Pressure, Seepage Forces, and Quantity of Seepage

Once we have determined the flow net, we can use it to investigate parameters of interest, especially the distribution of water pressure, seepage forces and the total quantity of seepage. We shall demonstrate this by considering the flow net for the concrete dam with partial cutoff wall shown schematically in Figure 14.8a. The complete flow net for this structure is shown in Figure 14.15a. As sketched, there are 15 equipotential drops and 5.65 flow channels. Had we elected to impose an integral number of flow channels, then there would have been a partial equipotential drop, as for example in Figure 14.14e. Thus from the flow net, we see that the head loss between adjacent equipotential lines, Δh, is

$$\Delta h = \frac{h}{n_d} = \frac{150}{15} = 10 \text{ ft} \qquad (14\text{--}55)$$

Knowing the head at any point we can determine the water pressure from Equation 14–4. For convenience we can select the elevation datum such that the constant is zero. Then

$$\frac{u}{\gamma_w} = h - z \qquad (14\text{--}56)$$

As an example we consider point A in Figure 14.15a. This point is 80 ft below the elevation datum, and on the equipotential line corresponding to

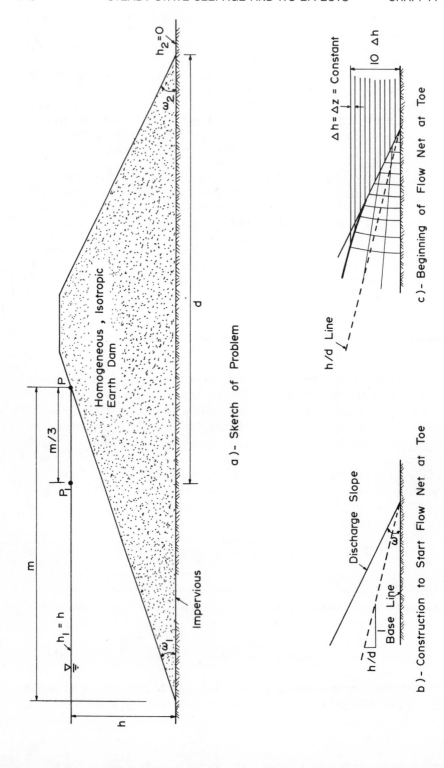

a)- Sketch of Problem

b)- Construction to Start Flow Net at Toe

c)- Beginning of Flow Net at Toe

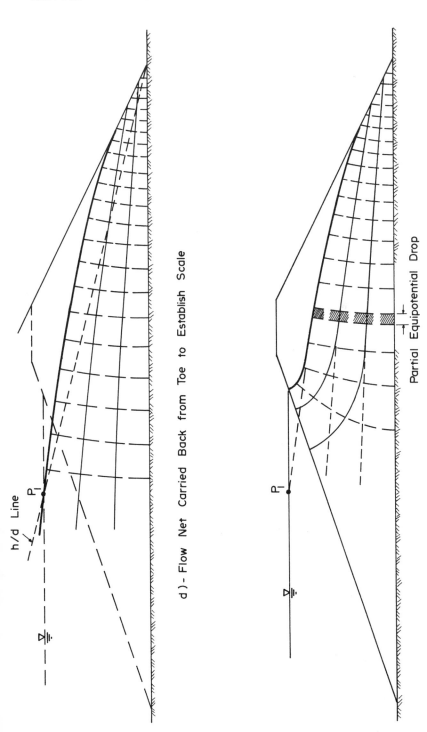

h/d Line

d) - Flow Net Carried Back from Toe to Establish Scale

Partial Equipotential Drop

e) - Completed Flow Net Corrected for Entrance Conditions

Fig. 14.14—Method of constructing flow net for unconfined flow through earth dam.

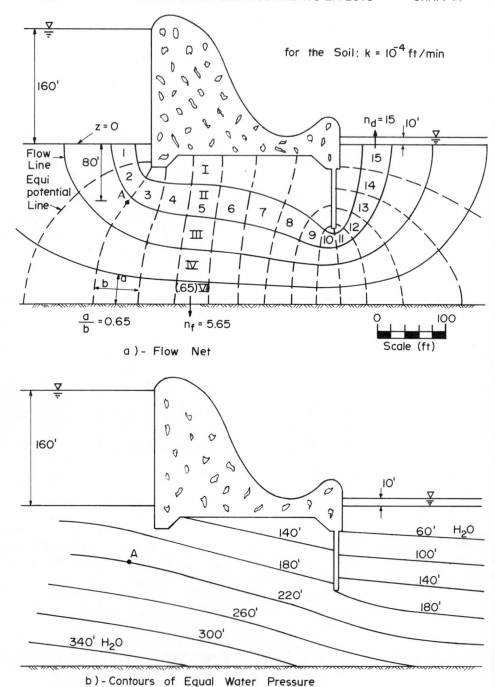

a) - Flow Net

b) - Contours of Equal Water Pressure

Fig. 14.15—Determination of water pressure and hydraulic gradient from flow net. (continued)

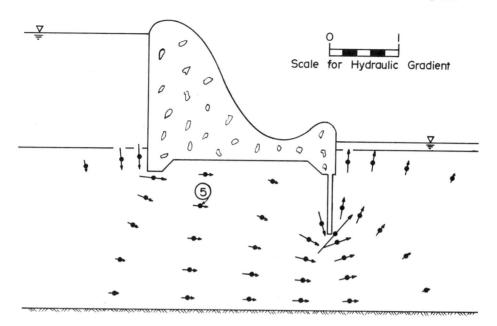

c) - Hydraulic Gradients at Selected Points within Aquifer

Fig. 14.15 (concluded)

$2 \Delta h$. Thus the head at point A is

$$h_A = 160 - 2 \times 10 = 140 \text{ ft}$$

from which the water pressure at A, expressed as a height of water, is

$$\frac{u}{\gamma_w} = 140 - (-80) = 220 \text{ ft}$$

This is illustrated in Figure 14.15b, which shows contours of equal water pressure within the aquifer. Point A is indicated in the diagram. The contours of equal water pressure were determined similarly. We note that the water pressure at a given elevation diminishes in the direction of flow. From such a diagram, the distribution of water pressure on the base of the dam, the side of the cutoff wall, or at other locations of interest can be determined. Of course, the pressure could also be determined directly from the flow net.

Seepage forces due to the flowing water are also important, both in connection with the overall stability of a slope through which the water is flowing, as well as the possibility for a quick condition at the discharge surface. The seepage force per unit of volume, $i\gamma_w$ can be depicted conveniently by the hydraulic gradient, **i**. This is determined approximately from the flow net by considering the hydraulic gradient at the center of a curvilinear square to be equal to the head loss across that square divided by the average distance between the two equipotential lines forming sides of the square. For example, in square number 5 the head loss is 10 ft and the average distance between the appropriate equipotential lines is 40 ft. Thus

$$i_5 \simeq \frac{\Delta h}{\Delta s} = \frac{10}{40} = 0.25$$

This is depicted schematically in Figure 14.15c at the point labeled ⑤. The point is located at the center of square number 5. The length of the vector drawn through that point is equal to the magnitude of the hydraulic gradient there, 0.25; the direction of the vector is the average direction normal to the equipotential lines forming the sides of that square. The hydraulic gradients at a number of other points within the aquifer are shown in a similar manner in the diagram.

It is apparent from Figure 14.15c that the hydraulic gradient is largest in the vicinity of the base of the cutoff wall. However, the magnitude of the gradient becomes most *significant* at the discharge surface, because there the gradient is nearly vertical *and* upward. Consequently, it is in this area where a quick condition, *sand boils*, and *piping* are likely to occur first. We recall that for upward vertical flow through a cohesionless medium, the hydraulic gradient required to produce a condition of zero effective vertical stress (a *quick condition*) is,

$$i_{\text{crit}} = \frac{\gamma'}{\gamma_w} \tag{14--57}$$

in which γ' is the submerged unit weight of the soil, γ_w, is the unit weight of water. Under ordinary conditions, this is near unity. From examination of Figure 14.15c, we see that the vertical upward hydraulic gradient at the toe of the dam is approximately $\frac{1}{3}$. Thus, for a cohesionless aquifer the factor of safety against a quick condition is approximately 3.

The quantity of seepage is also determined from the flow net. Using Equation 14.48, in which $a/b = 1$, the flow rate for the example in Figure 14.15a is

$$q = k(h_1 - h_2)\frac{n_f}{n_d} = 10^{-4}(160 - 10)\frac{5.65}{15} = 5.65 \times 10^{-3} \text{ cu ft/min/ft}$$

in which the flow rate is expressed per foot of length of the dam into the page. This expression is valid, of course, for either confined or unconfined flow.

Approximate Determination of Line of Seepage and Flow Rate

In many cases of unconfined flow we may be interested exclusively in determining the location of the line of seepage, and the total rate of seepage. This may be so, for example, when we establish internal drainage within an earth structure to insure that the line of seepage does not exit from the downstream face of the structure. In such cases, the entire flow net is not required. Approximate methods for finding the line of seepage and flow rate without drawing the complete flow net are described below.

Shallow Slopes ($\omega \leqslant 60°$). An approximate approach, useful for shallow discharge slopes for which $\omega \leqslant 60°$, was developed by Casagrande (1932) based upon earlier work of Schaffernak and Iterson. As indicated in Figure 14.16, the hydraulic gradient at any point on the line of seepage is

$$i = \frac{dz_1}{ds} \tag{14-58}$$

in which s is measured from the toe of the discharge face to the line of seepage, and then back along the line of seepage, and z_1 is measured *upward* from the base of the dam. Casagrande (1932) then assumed that this hydraulic gradient was approximately constant along any vertical section. Thus, the flow rate is

$$q = kz_1 \frac{dz_1}{ds} \tag{14-59}$$

in which z_1 is the height from the base of the dam to the line of seepage. Integrating this gives the equation of a parabola,

$$qs = \frac{1}{2}kz_1^2 + \text{const} \tag{14-60}$$

Boundary conditions for this expression are

$$\begin{aligned} s &= a, & z_1 &= a\sin\omega, & q &= ka\sin^2\omega \\ s &= s_0, & z_1 &= h \end{aligned} \tag{14-61}$$

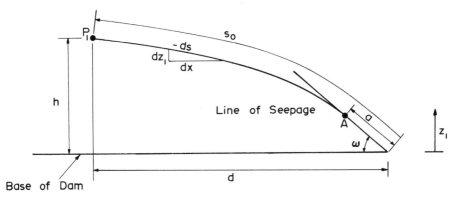

Fig. 14.16—Parameters for approximate determination of line of seepage.

in which a is the distance from the toe of the discharge face to the point where the line of seepage exists, ω is the slope of the discharge face, and s_0 is the distance from the toe of the discharge face along the line of seepage to the point P_1. The location of point P_1 is defined in Figure 14.14a.

Subsituting the boundary conditions into Equation 14–60 leads to an expression for a,

$$a = s_0 - \sqrt{s_0^2 - \frac{h^2}{\sin^2 \omega}} \qquad (14\text{–}62)$$

and the flow rate

$$q = ka \sin^2 \omega \qquad (14\text{–}63)$$

It is convenient, however, to approximate s_0, with only minor error for shallow discharge slopes, as

$$s_0 \simeq \sqrt{h^2 + d^2} \qquad (14\text{–}64)$$

Substituting this assumption in Equation 14–62 leads to an approximate, but useful expression for the distance a:

$$a = \sqrt{h^2 + d^2} - \sqrt{d^2 - h^2 \cot^2 \omega} \qquad (14\text{–}65)$$

Having determined the distance a, we now know two points on the parabola approximating the line of seepage, the tangent to the parabola at one of those points (at the discharge face), and the direction of the parabola axis. Thus the parabola can be constructed by a simple graphical procedure described by Casagrande (1937) and illustrated in Figure 14.17. Two lines are drawn, one through point A parallel to the discharge face, the other through point P_1 parallel to the base of the dam (which is the axis of the parabola). The lines intersect at the point T. The distances $\overline{P_1 T}$ and $\overline{aT}$ are then divided into an arbitrary but equal number of equal parts. Four such parts are shown in Figure 14.17. Through points 1, 2, and 3 are drawn lines parallel to the axis of the parabola; through points I, II, III, and P_1 are drawn lines from point a. The intersection of the lines drawn through points 1 and I, 2 and II, 3 and III, respectively, are then points on the parabola.

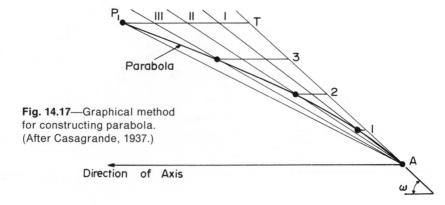

Fig. 14.17—Graphical method for constructing parabola. (After Casagrande, 1937.)

So, for a shallow discharge slope ($\omega \leqslant 60°$) we can locate the line of seepage approximately by the following steps:

1. Draw the dam including point P_1, as for example in Figure 14.14a. Locate point A by determining the distance a by Equation 14–65.
2. Construct the parabolic line of seepage between point P_1 and A as described above.
3. Correct the line of seepage at the upstream face to account for the entrance requirements shown in Figure 14.13a. The correction will be of the same sort we made to the line of seepage when the entire flow net was known, as illustrated in Figure 14.14e. The flow rate is estimated by Equation 14–63.

Steep and Overhanging Discharge Slopes ($60° < \omega \leqslant 180°$). Approximate determination of the line of seepage when the discharge face is steep or even overhanging, is best considered in relation to a problem which has been solved exactly: the case of groundwater flow over an impervious boundary into a horizontal discharge surface. This problem, illustrated in Figure 14.18, was solved by Kozeny (1931). The solution indicates that the flow lines and equipotential lines are two families of confocal parabolas with the focus at the point where the impervious boundary meets the drainage face. The equation for the line of seepage is

$$x = \frac{z^2 - z_0^2}{2z_0} \qquad (14-66)$$

in which x is the horizontal distance from the intersection of the impervious boundary and drainage face as indicated in Figure 14.18, z is measured vertically from the horizontal impervious boundary, and z_0 is the ordinate of the line of seepage at the point $x = 0$. Assuming that the point P_1, shown

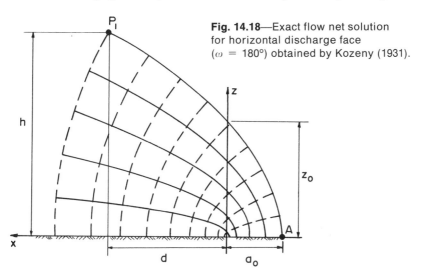

Fig. 14.18—Exact flow net solution for horizontal discharge face ($\omega = 180°$) obtained by Kozeny (1931).

in Figure 14.14a, with coordinates d and h is on the line of seepage, the focal length a_0 and the ordinate z_0 are

$$a_0 = \frac{z_0}{2} = \frac{1}{2}(\sqrt{h^2 + d^2} - d) \qquad (14\text{–}67)$$

Kozeny's solution is of considerable practical importance: first, because horizontal drainage blankets are often used in earth dams, and second, because the line of seepage for a steep or overhanging discharge face is close to Kozeny's parabola over much of its length. This is indicated schematically in Figure 14.19, in which the dashed line, referred to as the *basic parabola*,

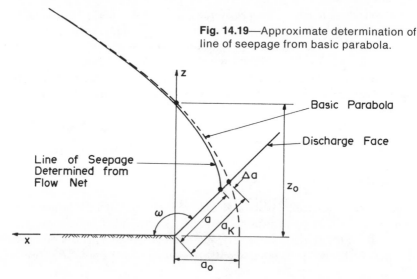

Fig. 14.19—Approximate determination of line of seepage from basic parabola.

a) - Relation between Line of Seepage and Basic Parabola

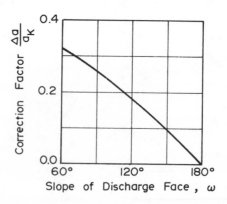

b) - Correction Factor for Line of Seepage Determined by Flow Net
(from Casagrande, 1937)

is the line of seepage determined using the Kozeny (1931) solution, and in which the solid line is the line of seepage determined from a flow net. The distance from the toe of the discharge face to the point at which the basic parabola intersects this face is denoted a_K. As before, a is the distance to the discharge point of the line of seepage. Casagrande (1937) determined, for a number of specific cases, the difference between these two distances, Δa. His results are depicted in Figure 14.19b in which a correction factor $\Delta a/a_K$ is shown as a function of the slope of the discharge face, for steep and overhanging discharge surfaces.

So, for a steep or overhanging discharge face, the procedure for obtaining the line of seepage approximately is:

1. Determine the point P_1, as illustrated in Figure 14.20. Plot the basic parabola using Equations 14–66 and 14–67.
2. Correct the line of seepage at the *entrance* face by making the line satisfy the entrance requirements given in Figure 14.13a.
3. Correct the line of seepage at the *discharge* face so that the line exits at the point A, a distance a from the toe of the discharge face,

$$a = a_K - \Delta a \qquad (14\text{--}68)$$

where Δa is determined from Figure 14.19b.

The rate of seepage can be estimated using Equation 14–63, in which the distance a is determined from Equation 14–68.

Effect of Anisotropic Permeability on Flow

Our discussion so far has assumed that within any homogeneous section of the flow region, the soil is also *isotropic* with respect to its flow properties. But in most natural deposits, as well as compacted earth fills, the horizontal permeability k_x, is greater than the vertical permeability k_z (see Section 6.6).

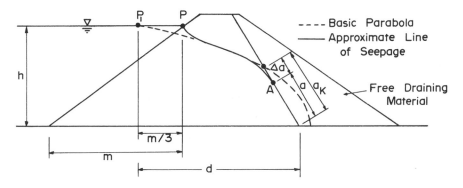

Fig. 14.20—Construction of approximate line of seepage for steep discharge slope.

Consequently, the governing differential equation, Equation 14–6, does not reduce to Laplace's equation. It has been shown however (Samsioe, 1931) that the geometric dimensions can be transformed so that the governing equation reduces to Laplace's equation in terms of the transformed variable. This is most commonly done by transforming the horizontal dimensions x to a new variable $\bar{x}$, defined as

$$\bar{x} = x\sqrt{\frac{k_z}{k_x}} \tag{14–69}$$

so that the governing equation becomes

$$\frac{\partial^2 h}{\partial \bar{x}^2} + \frac{\partial^2 h}{\partial z^2} = 0 \tag{14–70}$$

The way in which we obtain the flow net for this situation is illustrated in Figure 14.21. The steps involved are:

1. The problem is sketched to scale as shown in Figure 14.21a. A shape factor by which the horizontal dimensions are to be modified is calculated as

$$\sqrt{\frac{k_z}{k_x}} = \sqrt{\frac{1}{9}} = \frac{1}{3}$$

2. The problem is redrawn, using the original vertical scale, but with a transformed horizontal scale, in which all of the horizontal dimensions are plotted as the original horizontal dimensions multiplied by the shape factor obtained from Equation 14–69. In the case illustrated, the horizontal dimensions are one-third of the original. Because this transformed situation satisfies Laplace's equation, the flow net may be constructed as if the transformed section were simply a flow problem in an isotropic medium. The flow net for this case is shown in Figure 14.21b.

3. Once the flow net has been generated, the sketch, including the dimensions of the points on the flow net is transformed back to the original natural scale. In this case that means an increasing of the horizontal dimensions by a factor of three. The flow lines and equipotential lines in the flow net so obtained, Figure 14.21c, are no longer orthogonal. Nonetheless, they still possess the various characteristics of flow lines and equipotential lines, and can be used to determine points of discharge, the magnitudes of water pressure, etc.

4. The rate of seepage may be determined from the transformed section, Figure 14.21c, using an equivalent coefficient of permeability, $\bar{k}$ (Casagrande, 1937):

$$\bar{k} = \sqrt{k_x k_z} \tag{14–71}$$

Although the procedure has been illustrated with a confined flow problem, it is equally applicable to a situation in which unconfined flow occurs.

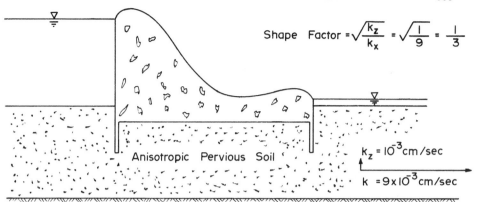

Shape Factor $= \sqrt{\dfrac{k_z}{k_x}} = \sqrt{\dfrac{1}{9}} = \dfrac{1}{3}$

Anisotropic Pervious Soil

$k_z = 10^{-3}$ cm/sec

$k = 9 \times 10^{-3}$ cm/sec

Impervious

a) - Sketch of Problem

b) - Transformed Flow
Reduced to 1/3 of Actual
Net - Horizontal Dimensions

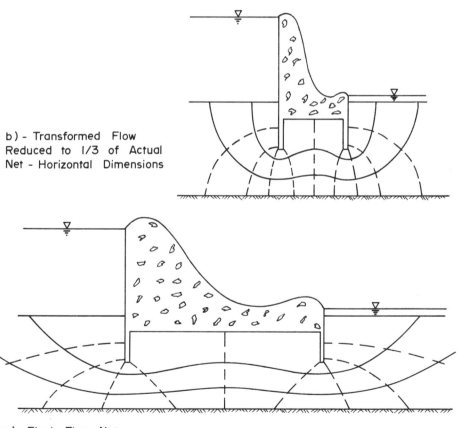

c) - Final Flow Net

Fig. 14.21—Use of transformed flow net
to account for anisotropic permeability.

Transient Seepage

During the transition from one steady-state situation to another, the seepage pattern is transient in nature. This occurs as a result of raising or lowering reservoir levels, seasonal changes in groundwater elevations, extended rainfall or drought conditions, etc. The two-dimensional transient problem is much more difficult to solve than the steady-state case. Analytical solutions are available, however, at least for simple boundary conditions (Brahma and Harr, 1962). In addition, a number of approximate solutions for determining the line of seepage at various times have also been obtained (Browzin, 1961; Newlin and Rossier, 1967). Cedergren (1967) describes an approach to construction of sequential flow nets which has proven satisfactory for predicting changes in the zone of saturation in small-scale models. Thus, while more difficult than the steady-state problem, transient seepage can also be considered.

Capillary Effects

In all the discussion above, we have assumed that the line of seepage constitutes a line of sharp demarcation between a saturated unconfined flow region, and an unsaturated region in which no flow is occurring. This is actually a simplified view of conditions at the interface between these two zones. As discussed in Section 6.4, all soils exhibit capillarity to some degree. Thus, there is a *capillary zone* near the line of seepage in which the soil is fully saturated due to capillarity. Hall (1955) investigated the effect of this capillary zone on the line of seepage in unconfined flow through sand toward a well. He found that the capillary zone raised the line of seepage by an amount which varied along the line of seepage, and was a maximum at the discharge face. In practice, however, the effect of capillarity is neglected. This is done for two reasons: first, for those soils in which the quantity of flow is a significant problem, the effect of capillarity is likely to be small; second, for those soils which exhibit significant capillarity, its effect on flow is not well established.

14.7 APPLICATION OF RESULTS

Effect of Seepage on Stability

The *stability* of an earth mass depends upon the interplay between the forces which tend to produce failure of the mass by inducing large movements along some set of critical surfaces, and the strength of the soil which resists such movements. The shearing resistance along potential failure surfaces is, in part, determined by the effective stresses on these surfaces. Thus, seepage affects the stability of earth masses in *two* ways:

1. The pore water pressure in the flow region changes the effective stresses on the potential failure surfaces from those for the static groundwater condition. Consequently, the shearing resistance is influenced. As indicated in the article in Chapter 1 ("Vaiont Reservoir Disaster," pages 18–24) and in Section 14.1, the changes in effective stress due to changing groundwater conditions contributed significantly to the cause of the Vaiont slide.

2. The seepage forces are *body forces*. Hence, when they act in a direction which tends to produce failure, the stability of an earth mass is adversely affected. Situations which exemplify this are upward flow into the base of an excavation (Figure 14.1a), flow through the downstream slope of an earth dam or a cut in a natural aquifer, and flow out of the upstream slope of an earth dam subject to rapid drawdown of the reservoir.

We illustrate the effect of seepage forces on stability by considering the stability analysis of the downstream slope of the earth dam shown in Figure 14.14. In Section 12.8, we obtained an implicit expression for the factor of safety against sliding along a circular surface using the Bishop (1955) method of slices and effective stress strength parameters (Equation 12–35). The expression was derived on the basis of moment equilibrium of the potential sliding mass. The numerator represents the effect of the shearing resistance along the assumed failure surface; the denominator includes the factors which contribute to the *driving moment*. Thus we can examine the effect of seepage forces on the factor of safety by determining the change in driving moment resulting from the seepage.

Shown in Figure 14.22 is the downstream portion of the dam including the line of seepage determined in Figure 14.14e, and a trial slip circle. For illustrative purposes the sliding mass has been divided into five slices. Those forces on the slices which contribute to the driving moment effects in Equation 12–35 are shown in the figure. Above the line of seepage, the moist weight of the soil, W_i, acts for the ith slice. Below the line of seepage, the saturated weight W_{sat_i}, or the submerged weight W_i' acts. In combination with the submerged weight, the seepage force $\mathbf{i}_i \gamma_w V_i$ acts at a point approximated by the centroid of the portion of the wedge below the line of seepage with volume V_i.

In calculating the driving moment, we can account for seepage in either of two equivalent ways, just as we did in evaluating the quick condition in Section 6.3:

1. By considering the effective weight of the soil, that is, the submerged unit weight below the free-water surface, in combination with the seepage forces. The driving moment about the center of the trial circle, M_D, is then

$$M_D = \sum(W_i \bar{x}_i + W_i' \bar{x}_i' + \mathbf{i}_i \gamma_w V_i \bar{r}_i) \qquad (14\text{–}72)$$

in which $\bar{x}_i'$ is the moment arm of W_i about the center of the trial

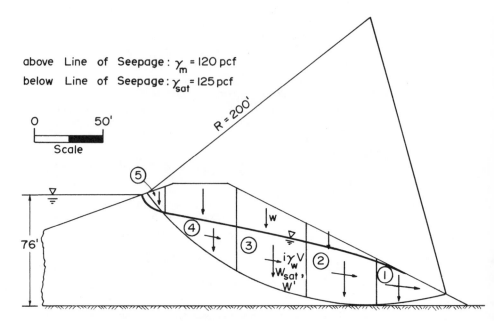

above Line of Seepage: γ_m = 120 pcf
below Line of Seepage: γ_{sat} = 125 pcf

0   50'
Scale

R = 200'

76'

Fig. 14.22—Illustration of effect of seepage on stability of downstream slope of earth dam.

circle, $\bar{x}'_i$ is the moment arm of G'_i, and $\bar{r}_i$ is the moment arm of the seepage force acting on the ith slice.

2. Alternatively, by considering the total weight of each slice and the pore water pressure force acting on the lower boundary. We note, however, that the water pressure on the failure surface produces no moment about the center of the circle. Thus, M_D is

$$M_D = \sum (W_i \bar{x}_i + W_{sat_i} \bar{x}'_i) \qquad (14\text{--}73)$$

The equivalence of these two calculation methods is illustrated in Table 14.2. Parameters used in the table are taken from the flow net for the dam (Figure 14.14e) and the sketch of the trial failure circle (Figure 14.22). The discrepancy between the two calculated versions of M_D is less than 3 per cent. In view of this, and the relative simplicity of the total weight calculation, it is clearly the method of choice in this type of problem.

We can compare the driving moment when seepage is occurring with that for static water conditions. For illustrative purposes, we consider the dam to be completely submerged. In this case, there are no seepage forces and the submerged unit weight of the soil should be used in the calculation.

Thus M_D' is

$$M_D = \sum(W_i \bar{x}_i) \frac{\gamma'}{\gamma_m} + \sum(W_i' \bar{x}_i') \qquad (14\text{-}74)$$

From which we find

$$M_D + (28{,}992) \frac{62.6}{120} + 14{,}476 = 29{,}600 \text{ kip-feet}$$

That is, the driving moment when seepage is present is approximately *twice* that under static submerged conditions. This, of course, leads to a corresponding decrease in the factor of safety.

A second major stability problem arising from seepage forces is the case of *rapid drawdown.* This condition arises when the reservoir level behind an

TABLE 14.2

Effect of Seepage on Calculation of "Driving Moment" M_D

(a) Using Total Weight Below Line of Seepage

Slice	W Above GWT (kips)	$\bar{x}$ (ft)	$W\bar{x}$ (kip-ft)	W_{sat} Below GWT (kips)	$\bar{x}$	$W_{sat}\,\bar{x}$ (kip-ft)
1	—	—	—	98.8	− 21.0	− 2,075
2	39.2	29.5	1,157	230	18.0	4,142
3	116.2	74.0	8,602	230	69.0	15,877
4	145.0	119	17,260	98.8	111.5	11,016
5	13.1	150	1,972	—	—	—
			$\sum = 28{,}992$			$\sum = 28{,}960$

$$M_D = 28{,}992 + 28{,}960 = 57{,}952 \text{ kip-ft}$$

(b) Using Submerged Weight and Seepage Forces Below Line of Seepage

Slice	W' Below GWT (kips)	$\bar{x}$ (ft)	$W'\bar{x}$ (kip-ft)	i_w	$i\gamma_w V$ (kips)	$\bar{r}$ (ft)	$i\gamma_w V\bar{r}$ (kip-ft)
1	49.4	− 21.0	− 1,037	.47	20.8	183	3,799
2	115.0	18.0	2,070	.33	34.5	179	6,187
3	115.0	69.0	7,935	.22	25.1	177	4,449
4	49.4	111.5	5,508	.21	10.1	170.5	1,725
5	—	—	—		—	—	—
			$\sum = 14{,}476$				$\sum = 16{,}160$

$$M_D = 28{,}992 + 14{,}476 + 16{,}160 = 59{,}628 \text{ kip-ft}$$

NOTE: V = volume of portion of slice below line of seepage.

earth dam is lowered suddenly relative to the rate at which the pore water pressure distribution in the dam can adjust to the changed conditions. In such a case, the *upstream* face becomes a surface of seepage, that is, water flows out of the upstream face, leading to a potentially dangerous stability problem. Morgenstern (1963) lists 16 drawdown induced failures of earth dams. Similar conditions prevail in the side slopes of a reservoir subject to rapid drawdown. A complete analysis of this problem requires considering transient flow, as discussed above, and the construction of a series of intermediate flow nets. However, the most critical situation occurs immediately following the drawdown. For an impervious lower boundary the initial flow net in the vicinity of the upstream face can be considered to consist of vertical equipotential lines and horizontal flow lines, as depicted in Figure 14.23. This assumption, while simplified, is conservative. Morgenstern (1963)

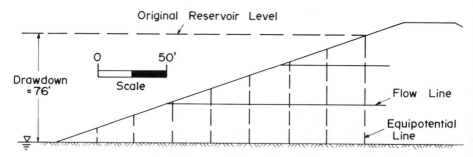

Fig. 14.23—Approximate flow net for instantaneous complete drawdown at upstream face of earth dam.

has tabulated stability charts which indicate the effect on the safety factor of various magnitudes of drawdown for a homogeneous earth dam on an impervious base. For more complex cases, the flow net for the instantaneous drawdown can be constructed, and the factor of safety of the upstream face determined by the method of slices.

The design of earth dams, and other large earth structures, requires many other considerations beyond the scope of this text. Useful references in this connection are Bertram (1966), and Sherard et al., (1963).

Filters To Prevent Piping

Another of the important consequences of seepage is the potential creation of quick conditions followed by *piping*, especially for cohesionless soil. Proper use of free-draining materials may be helpful in reducing the danger of a quick condition in at least two ways:

1. The hydraulic gradient within the free-draining material is likely to be low, so that the danger of a quick condition within the drain itself is small.
2. If the drain overlies a soil in which upward flow is occurring, the weight of the drain produces an increased effective stress which leads to an increase in the vertical hydraulic gradient required to induce a quick condition.

If, however, a coarse free-draining material is in contact with one containing significant quantities of finer soils then the fine-grained material may flow into the interstices of the coarse-grained soil. To prevent this, a *graded filter* is provided to introduce a transition in grain sizes between the two materials. Cedergren (1967) summarizes the criteria to be applied to the grain-size distribution of graded filter between two soils of very different grain size, or between soil and a drainpipe with open joints or slits.

Although the danger of piping is much less in the case of cohesive soil, consideration must still be given to the possibility. Kassiff et al., (1965) discuss this problem.

14.8 SUMMARY OF KEY POINTS

Summarized below are key points discussed in this chapter. Reference is made to the place in the text where the original more detailed discussion can be found:

1. Many problems of practical importance which involve the effects of water flowing through an earth mass can be analyzed assuming that seepage is occurring in two dimensions only, and is steady-state. Features of the problem which are most often of interest are (Section 14.1):
    a. Definition of the flow region;
    b. Determination of the pore water pressure within the flow region;
    c. Evaluation of seepage forces;
    d. Determination of the quantity of seepage.
2. For appropriate restrictions, the governing differential equation of flow in two dimensions is Laplace's equation (Section 14.2).
3. The solution to Laplace's equation and the given boundary conditions consists of two families of orthogonal curves: equipotential lines and streamlines (Section 14.3). The quantity of flow between two streamlines is a constant (Equation 14–27).
4. There are five types of boundary conditions of general importance:
    a. Reservoir boundaries, which are equipotential lines (Section 14.4);
    b. Impervious boundaries, which are limiting flow lines (Section 14.4);

  c. Boundaries between materials of differing permeabilities (Section 14.6);

  d. Line of seepage (Section 14.6);

  e. Surface of seepage (Section 14.6).

The first three pertain to confined flow, in which the flow region is specified *a priori*. All five may pertain to unconfined flow, in which one or more boundaries of the flow region is a free-water surface whose location can be determined only as a part of the solution.

5. Solution of a flow problem may be obtained by a variety of means, including (Section 14.5):

  a. Exact analytical methods;

  b. Small-scale models or analogues;

  c. Approximate methods:

   (1) Numerical solutions such as the finite difference or finite element approaches.

   (2) Sketching of flow nets.

6. Flow net construction consists of sketching a limited number of flow lines and equipotential lines on a scale drawing of the problem such that the resulting curves produce "curvilinear squares," and satisfy the boundary conditions. This is done by trial and error (Section 14.6).

  a. For unconfined flow problems, $\Delta h = \Delta z =$ constant along the line of seepage and the surface of seepage.

  b. Entrance, discharge, and material interface effects on the line of seepage may not be the same as for other flow lines (Figure 14.13). The location of the line of seepage can be determined either from a complete flow net, or by approximate means.

  c. Water pressure and the hydraulic gradient at any point in the flow region can be determined from the flow net (Figure 14.15). The total quantity of seepage can also be calculated (Equation 14–48).

  d. Anisotropic permeability of the soil in the flow region can be considered by constructing a transformed flow net in which one dimension (usually the horizontal) is multiplied by a shape factor given in Equation 14–69.

  e. When transient flow occurs in response to reservoir drawdown or other rapidly occurring effects, flow nets (or just the line of seepage) can be determined for intermediate stages.

  f. Capillary effects on the flow region, especially the line of seepage, are likely to be minor in soils of high permeability, and of unknown significance in fine-grained soils. Such effects are usually ignored.

7. Seepage through slopes in earth masses generally has an adverse effect on stability of the slope. The effect arises from two sources (Section 14.7):

  a. Changes in effective stress, which lead to a reduced shearing resistance along potential failure surfaces;

  b. Seepage forces, which increase the "driving moment" that tends to produce a failure.

The latter effect can be considered simply in the stability analysis by using the saturated unit weight to calculate the driving moment of soils below the free-water surface.

8. Graded filters may be required when materials of widely different grain sizes are in contact (Section 14.7).

PROBLEMS

14.1. A sheet-pile cofferdam was constructed as shown; the existing bottom was removed to a depth of 10 ft by excavating under water, using clamshell buckets. When the excavation was complete, water within the cofferdam area was pumped out.
(a) Sketch the flow net.
(b) To what depth z could the water be removed before sand boils would develop in the excavation?

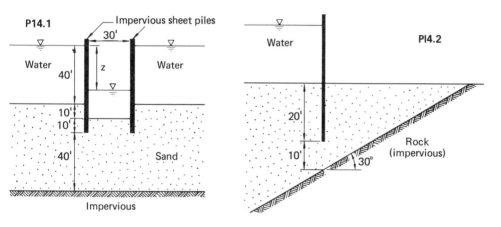

14.2. Sketch the flow net for the case shown.
14.3. For the weir shown, determine—
(a) The quantity of seepage (per foot into the page).
(b) the distribution of uplift pressure on the base of the weir.
(c) the factor of safety against a quick condition.

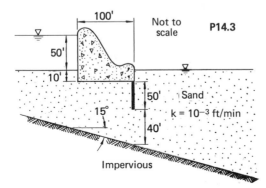

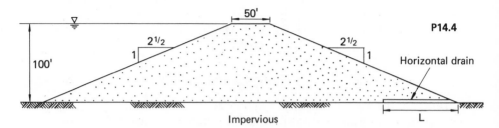

14.4. For the homogeneous earth dam shown, determine the length L of the horizontal drain such that the line of seepage comes no closer than 10 ft to the discharge face of the dam.

APPENDIX

Influence Diagrams for Stresses in Elastic Media

A.1 STRESSES DUE TO SURFACE LOADS

Influence diagrams for stresses within semi-infinite elastic media due to surface loadings of various types are given in Figures A.1–A.7. These diagrams constitute a convenient compilation of frequently used solutions. They do not, however, represent an exhaustive catalog; Scott (1963) and Harr (1966) have assembled a great many additional solutions to which reference can be made.

In the case of all the influence diagrams presented, the magnitude of stress which is determined is that due *only* to the applied loading considered. Preexisting *in situ* stresses must be added to those determined by the use of these diagrams. The diagrams are to some degree self-explanatory. However their use will be illustrated by examples:

Example A.1 ——————————————————————————————————————

Shown in Figure A.8 is a rigid rectangular 5 ft by 5 ft area on the surface of a linear-elastic half-space. The area is subjected to a vertical load of 50 tons, at its centroid. We wish to find the vertical stress, σ_z, at a depth of 10 ft below a point on the half-space located as shown in the figure. We shall consider each of those influence diagrams which may be applicable to this problem, and from which we can obtain at least an approximation to the correct answer.

Solution. The simplest approximation is to assume that the total load is applied to the half-space at the centroid of the loaded area. If the point at which the stress is to be calculated is sufficiently far from the loaded area, St. Venant's principle assures the validity of such an approximation. In this case, in order to use Figure A.1 we note that:

$$P = 50\text{T}, \qquad r = \sqrt{5^2 + 5^2} = 7.07 \text{ ft}, \qquad z = 10 \text{ ft},$$

Hence, $r/z = 0.707$, and we determine from Figure A.1 that the influence value $I = 0.175$.

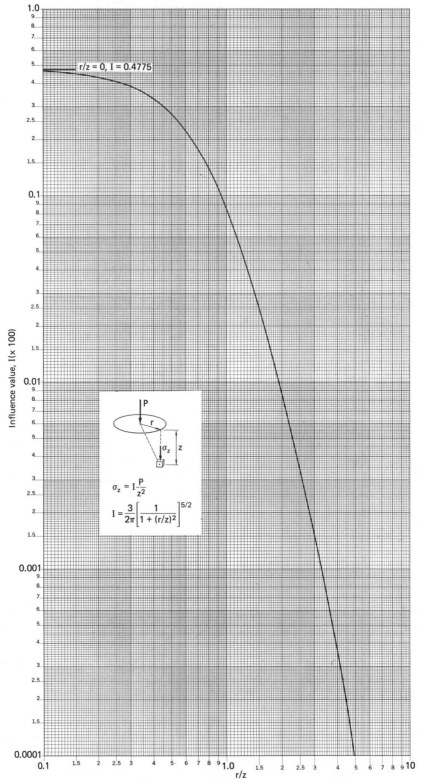

Fig. A.1—Influence diagram for vertical normal stress due to point load on surface of elastic half-space.

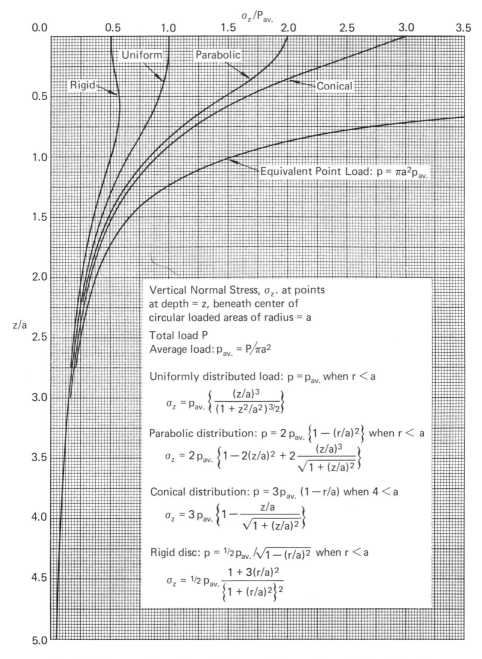

Fig. A.2—Influence diagram for vertical normal stress under center line of circular loaded area on elastic half-space. (After Gray and Hooks, 1948.)

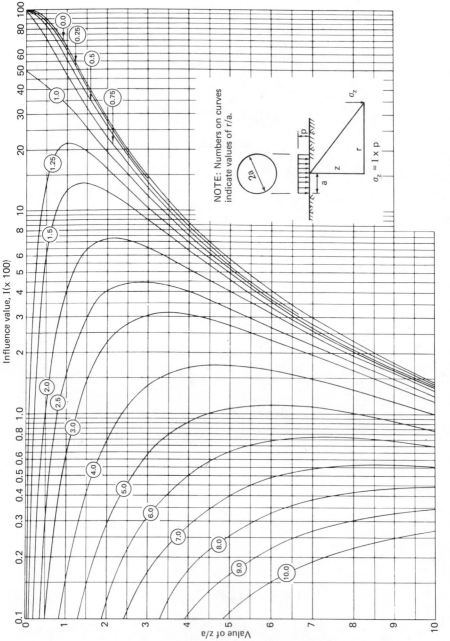

Fig. A.3—Influence diagram for vertical normal stress at various points within an elastic half-space due to a uniformly loaded circular area. (After Foster and Ahlvin. 1954.)

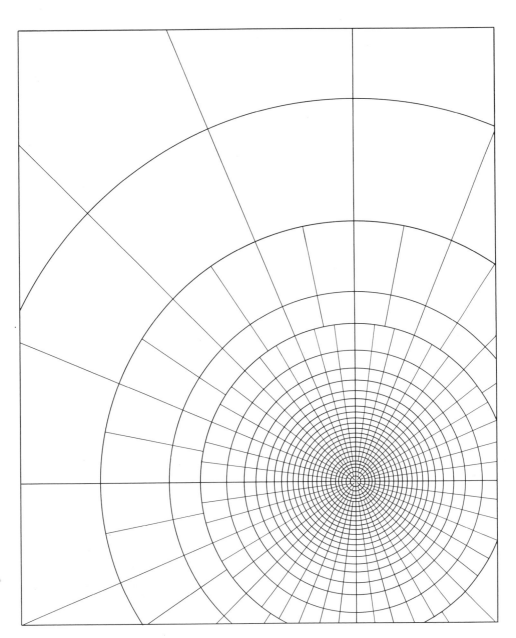

Fig. A.4—Influence diagram for vertical normal stress at a point within elastic half-space due to a uniformly loaded area. (After Newmark, 1942.)

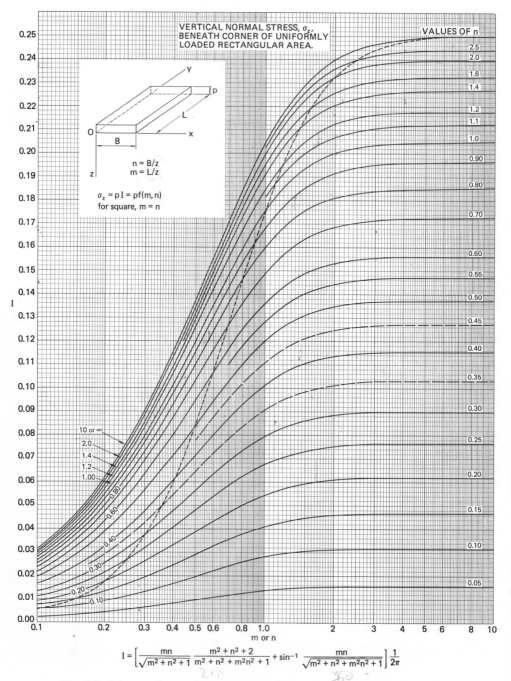

$$I = \left[\frac{mn}{\sqrt{m^2 + n^2 + 1}} \frac{m^2 + n^2 + 2}{m^2 + n^2 + m^2n^2 + 1} + \sin^{-1} \frac{mn}{\sqrt{m^2 + n^2 + m^2n^2 + 1}} \right] \frac{1}{2\pi}$$

Fig. A.5—Influence diagram for vertical normal stress at a point with elastic half-space beneath corner of uniformly loaded rectangular area. (After Fadum, 1948.)

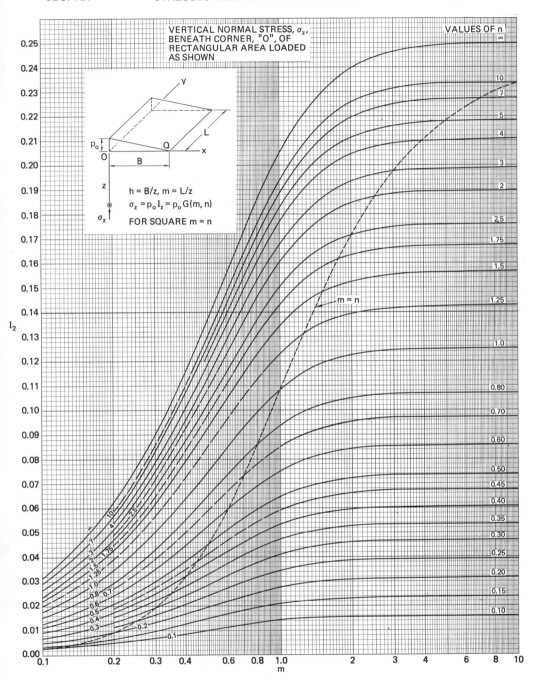

Fig. A.6—Influence diagram for vertical normal stress within elastic half-space beneath corner of rectangular loaded area.

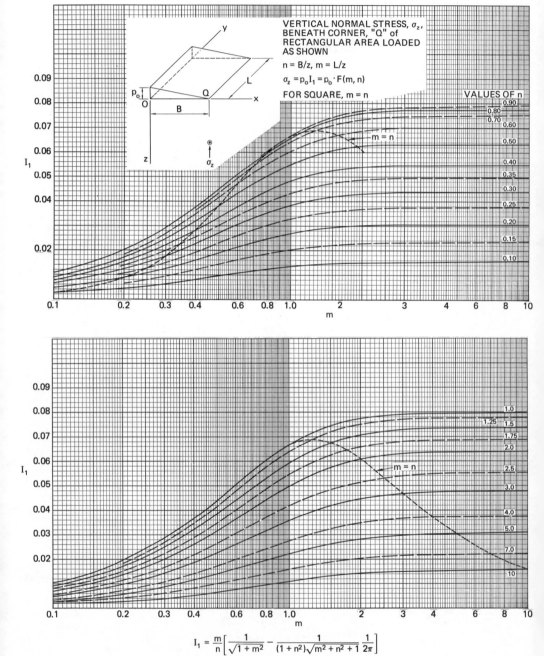

$$I_1 = \frac{m}{n}\left[\frac{1}{\sqrt{1+m^2}} - \frac{1}{(1+n^2)\sqrt{m^2+n^2+1}}\frac{1}{2\pi}\right]$$

Fig. A.7—Influence diagram for vertical normal stress within elastic half-space beneath corner of rectangular loaded area.

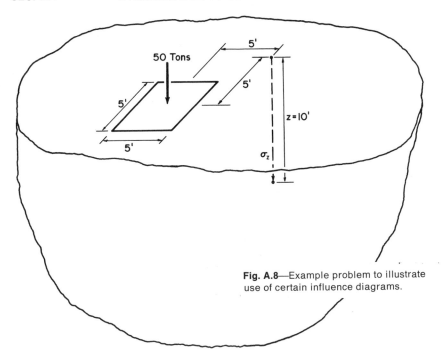

Fig. A.8—Example problem to illustrate use of certain influence diagrams.

This leads to a computed vertical stress of

$$\sigma_z = (0.175)\frac{(50)(2000)}{10^2} = \underline{175} \text{ psf}$$

Figure A.2 cannot be applied directly to this problem because the influence values given by this diagram pertain only to vertical stress under the *centerline* of a circular area, for a variety of distributions of stress on that circular area. However the curve for the uniformly distributed stress in Figure A.2 is a special case of those given by Figure A.3. An approximate answer to the example problem can be obtained from Figure A.3 by replacing the square footing by a circular uniformly distributed load of the same area as illustrated in Figure A.9. In this case the radius of the equivalent area, a, is

$$a = \sqrt{\frac{\text{area}}{\pi}} = \sqrt{\frac{25}{\pi}} = 2.82 \text{ ft}$$

so that $r/a = 2.5$ and $z/a = 3.55$. From Figure A.3 we obtain an influence value, $I = 0.042$. The average vertical normal stress exerted on the surface by the footing is

$$p_{\text{avg}} = \frac{(50)(2000)}{25} = 4000 \text{ psf}$$

so that the vertical stress at the point in question is determined to be

$$\sigma_z = (0.042)(4000) = 168 \text{ psf}$$

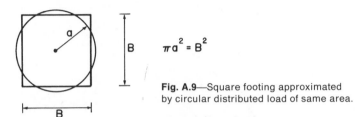

$$\pi a^2 = B^2$$

Fig. A.9—Square footing approximated by circular distributed load of same area.

This use of an equivalent circular area is an excellent approximation if the point at which the stress is sought is not too close to the loaded area.

The Newmark influence chart, Figure A.4, requires significantly more effort in the case of this example problem because the footing must be drawn to the proper scale. This is done in Figure A.10 in which the depth to the point desired, 10 ft is scaled equal to the length OQ in Figure A.4. The footing has been drawn in Figure A.10 to this scale and located the appropriate horizontal distance from the center of the Newmark chart, which represents the horizontal coordinates of the point at which the stress is desired. The relevant part of Figure A.4 is reproduced on Figure A.10. We see that the loaded area occupies 42 subdivisions, each of which has an influence value of 0.001. Thus the desired vertical stress is determined to be

$$\sigma_z = (42)(0.001)(4000) = 168 \text{ psf}$$

In this case, Figure A.3 is evidently simpler to use. However, for areas of relatively complex shape, Figure A.4 is very valuable.

Figure A.5 can be used to determine the vertical normal stress at points beneath the corner of a uniformly loaded rectangle. When the stress is desired at some point other than the corner of the loaded area, the principal of superposition can be used to advantage. This is illustrated in Figure A.11a for the example problem. In this figure the footing is shown as the rectangle *abiha*. The procedure we will use is to imagine that the load is distributed uniformly over a rectangle *abcdefgha* with one corner above the point in question. We then successively subtract and add the effects of rectangles with one corner above the point until the resulting effect is that due to the uniformly loaded footing. The results of this calculation are shown in Figure A.11b. Thus the rectangle *abcdefgha*, for which B/z and L/z are both 0.75 produces an influence value of $+0.137$. We then assume a compensating negative stress over the rectangles *bcdefib* and *defghid*. However we have subtracted the area *defid* twice by doing this, so we must add it back again. This leaves a resulting influence value of 0.044, which leads to the desired stress

$$\sigma_z = (0.044)(4000) = \underline{176} \text{ psf}$$

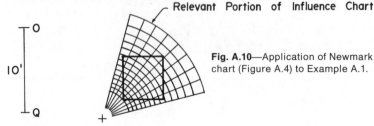

Relevant Portion of Influence Chart

Fig. A.10—Application of Newmark influence chart (Figure A.4) to Example A.1.

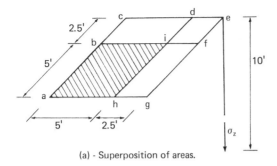

Fig. A.11—Application of Fadum (1948) influence diagram to Example A.1.

(a) - Superposition of areas.

Rectangle	B/Z	L/Z	I
abcdefgha	0.75	0.75	+0.137
bcdefib	0.75	0.25	−0.060
defghid	0.25	0.75	−0.060
defid	0.25	0.25	+0.027
		Sum =	+0.044

(b) - Calculated influence values

In principle this method would be the most correct, if the pressure imposed by the footing on the elastic medium were truly uniform, because the loaded area has not been approximated by some other shape. However the addition and subtraction of numbers leading to a result much smaller than the numbers themselves usually results in inaccuracies. Nonetheless, Figure A.5 is frequently useful, particularly if the stress is desired at a point relatively near the loaded area and within its horizontal boundaries. For such a case, the loaded area would be simply divided into four loaded areas with a common corner at the point in question.

Figures A.6 and A.7 are used in a similar fashion, except that, because the load is not uniform across the area, the sides B and L cannot be interchanged in obtaining the influence value.

A.2 STRESSES DUE TO SELF-WEIGHT

The use of the influence diagrams pertaining to the elastic embankment, Figures A.12 through A.26 are illustrated in Example A.2 which follows.

Example A.2 ———————————————————————

It is desired to determine the distribution of stresses on the base of an embankment which can be approximated as a linear-elastic material. The embankment is shown in Figure A.27 with a height of 30 ft, a half-width at the top of 30 ft, a unit weight of 120 pounds/cubic foot and Poisson's ratio = 0.3. The embankment is assumed to have side slopes of 30°.

Solution. The distribution of vertical normal stress at various vertical sections for this embankment is given in Figure A.13. For each of the four sections shown, the curves

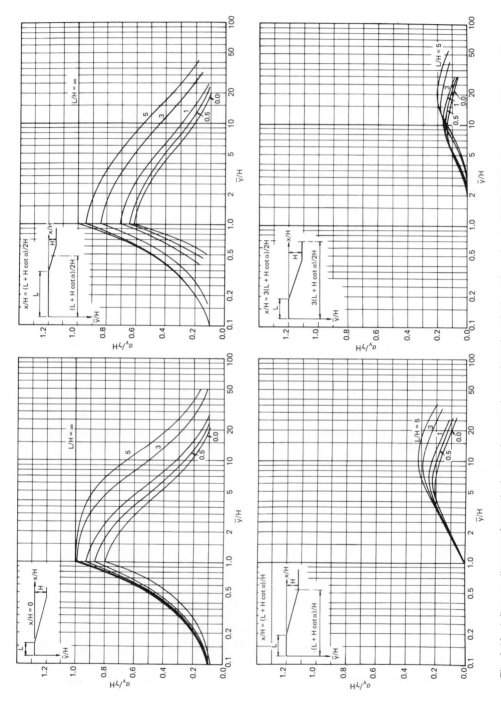

Fig. A.12—Influence diagrams for vertical normal stress along selected vertical sections due to elastic embankment, $\mu = 0.3$, $\alpha = 15°$. (After Perloff, et al., 1967.)

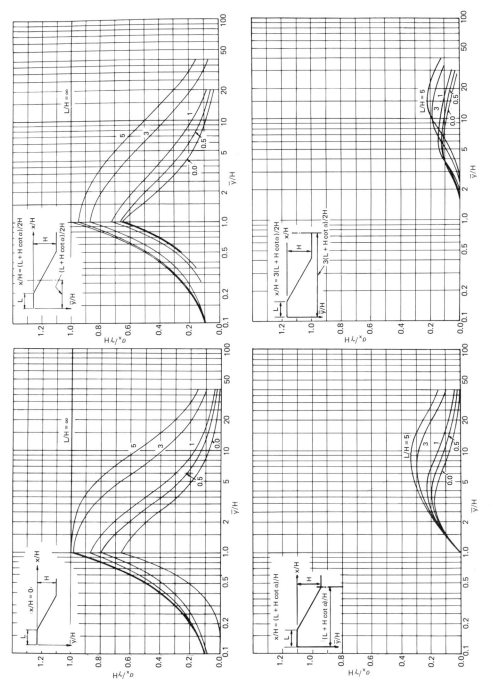

Fig. A.13—Influence diagrams for vertical normal stress along selected vertical sections due to elastic embankment, $\mu = 0.3$, $\alpha = 30°$. (After Perloff, et al., 1967.)

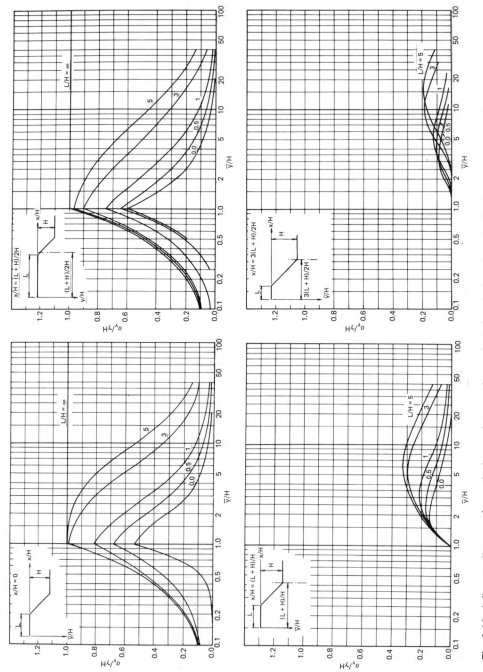

Fig. A.14—Influence diagrams for vertical normal stress along selected vertical sections due to elastic embankment, $\mu = 0.3$, $\alpha = 45°$. (After Perloff, et al., 1967.)

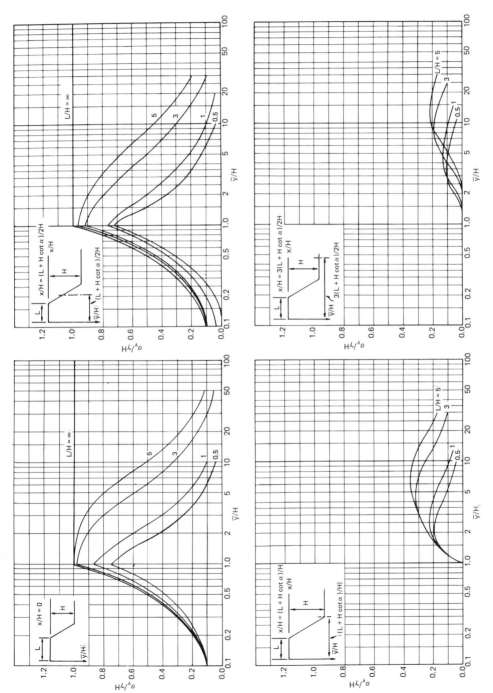

Fig. A.15—Influence diagrams for vertical normal stress along selected vertical sections due to elastic embankment, $\mu = 0.3$, $\alpha = 60°$. (After Perloff, et al., 1967.)

APPENDIX

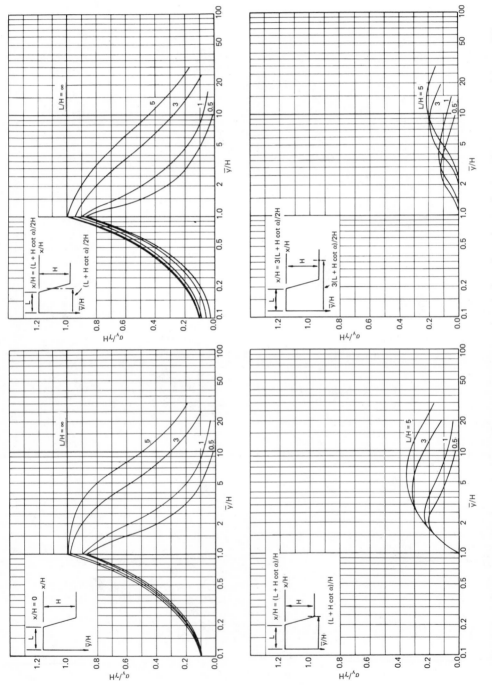

Fig. A.16—Influence diagrams for vertical normal stress along selected vertical sections due to elastic embankment, $\mu = 0.3$, $\alpha = 75°$. (After Perloff, et al., 1967.)

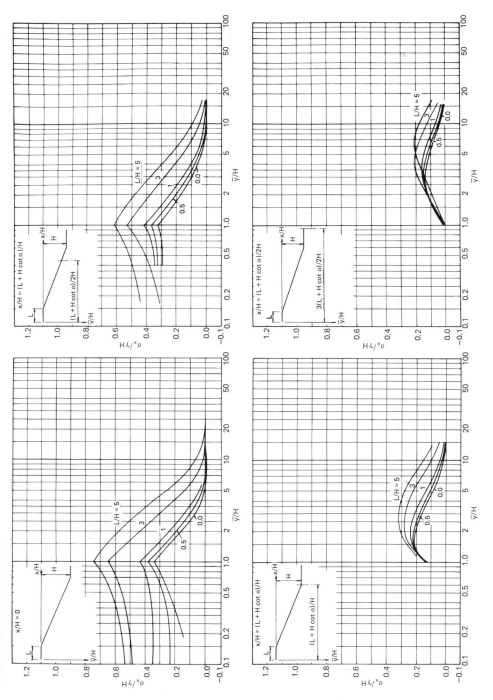

Fig. A.17—Influence diagrams for horizontal normal stress along selected vertical sections due to elastic embankment, $\mu = 0.3$, $\alpha = 15°$ (After Perloff, et al., 1967.)

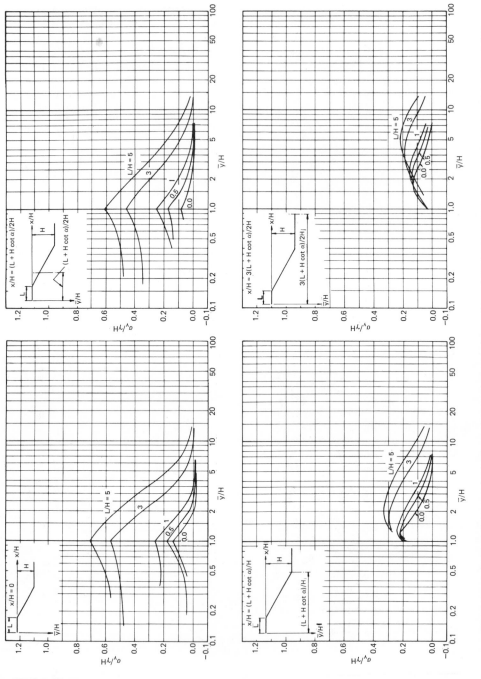

Fig. A.18—Influence diagrams for horizontal normal stress along selected vertical sections due to elastic embankment, $\mu = 0.3$, $\alpha = 30°$. (After Perloff, et al., 1967.)

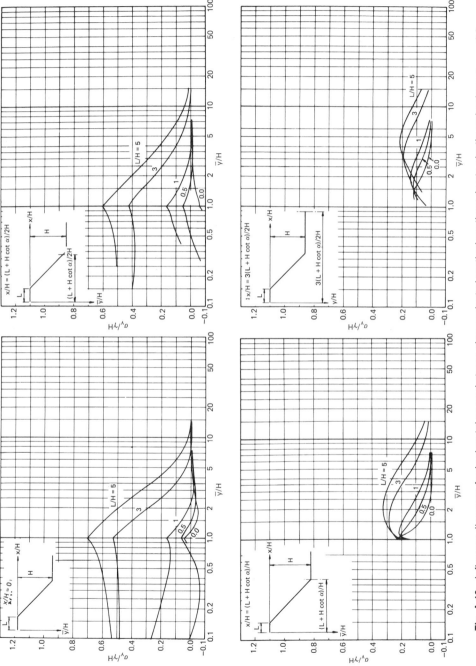

Fig. A.19—Influence diagrams for horizontal normal stress along selected vertical sections due to elastic embankment, $\mu = 0.3$, $\alpha = 45°$. (After Perloff, et al., 1967.)

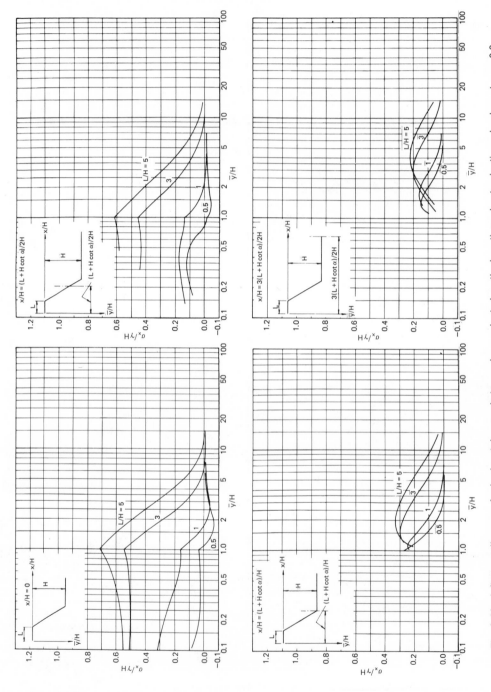

Fig. A.20—Influence diagrams for horizontal normal stress along selected vertical sections due to elastic embankment, $\mu = 0.3$, $\alpha = 60°$. (After Perloff, et al., 1967.)

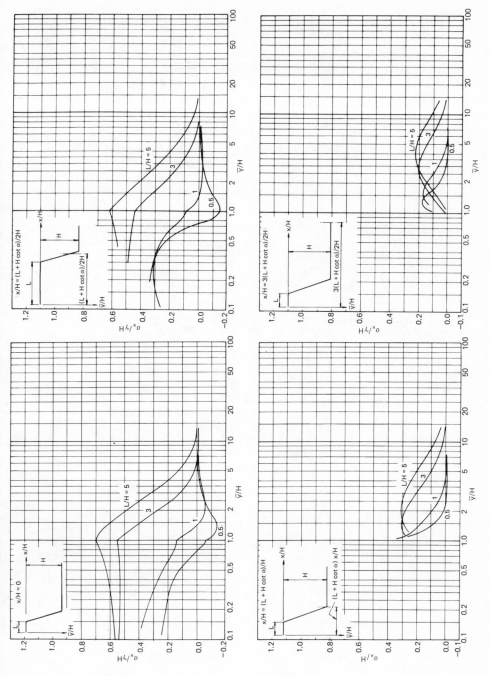

Fig. A.21—Influence diagrams for horizontal normal stress along selected vertical sections due to elastic embankment, $\mu = 0.3$, $\alpha = 75°$. (After Perloff, et al., 1967.)

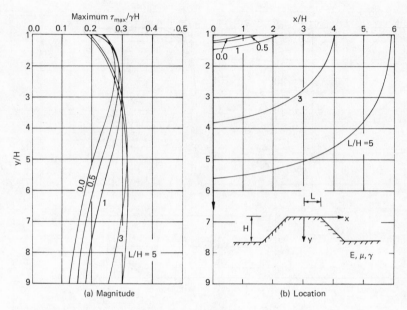

Fig. A.22—Magnitude and location of maximum ($\tau_{max}/\gamma H$) for elastic embankment, as a function of depth, $\mu = 0.3$, $\alpha = 15°$. (After Perloff, *et al.*, 1967.)

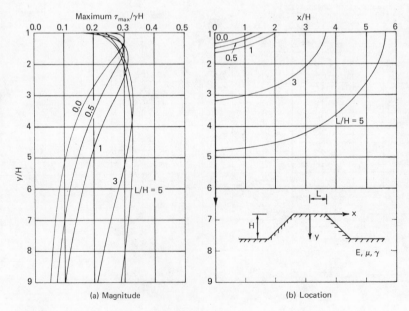

Fig. A.23—Magnitude and location of maximum ($\tau_{max}/\gamma H$) for elastic embankment, as a function of depth, $\mu = 0.3$, $\alpha = 30°$. (After Perloff, *et al.*, 1967.)

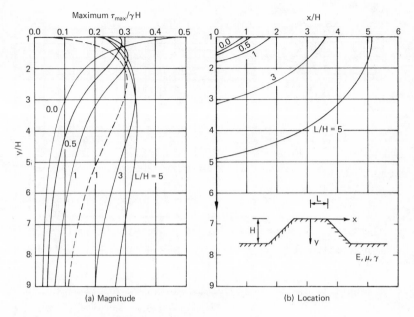

Fig. A.24—Magnitude and location of maximum $(\tau_{max}/\gamma H)$ for elastic embankment, as a function of depth, $\mu = 0.3$, $\alpha = 45°$. (After Perloff, *et al.*, 1967.)

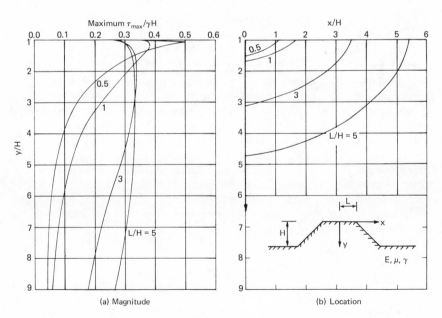

Fig. A.25—Magnitude and location of maximum $(\tau_{max}/\gamma H)$ for elastic embankment, as a function of depth, $\mu = 0.3$, $\alpha = 60°$. (After Perloff, *et. al.*, 1967.)

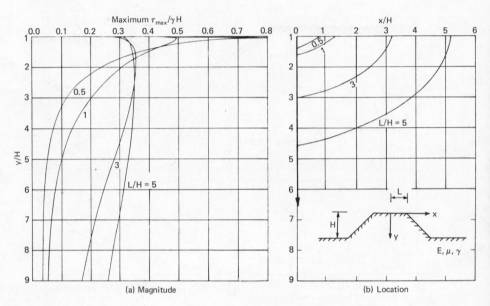

Fig. A.26—Magnitude and location of maximum $(\tau_{max}/\gamma H)$ for elastic embankment, as a function of depth, $\mu = 0.3$, $\alpha = 75°$. (After Perloff, et al., 1967.)

for $L/H = 1$ pertain to this example. These curves show the stresses due *only* to the weight of the embankment material itself. Thus the magnitude of the stress is a maximum at $y/H = 1.0$, which is the base of the embankment. For the section at the centerline, $y/H = 1$, and $L/H = 1$, $\sigma_y/\gamma H = 0.86$. For a 30-ft-high embankment with a unit weight of 120 pcf, this gives a vertical stress of 309 pounds/square foot. This value, as well as that for other sections is illustrated in Figure A.28a. In the upper part of this figure the embankment is shown schematically. In the lower part the distribution of vertical normal stress on the base of the embankment is shown. The stress due to the normal loading approximation is given for comparison purposes. The horizontal stress distribution along the base is determined in a similar fashion from Figure A.18. From the upper left-hand diagram, the horizontal stress factor at the base of the embankment along the centerline, for $L/H = 1$ is determined to be 0.255. This yields a horizontal normal stress of 92 psf. Similar values are obtained for other sections. For example, at the section midway between the centerline and the toe of the slope, illustrated in the

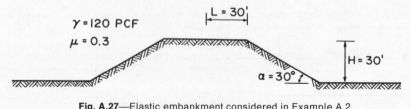

Fig. A.27—Elastic embankment considered in Example A.2.

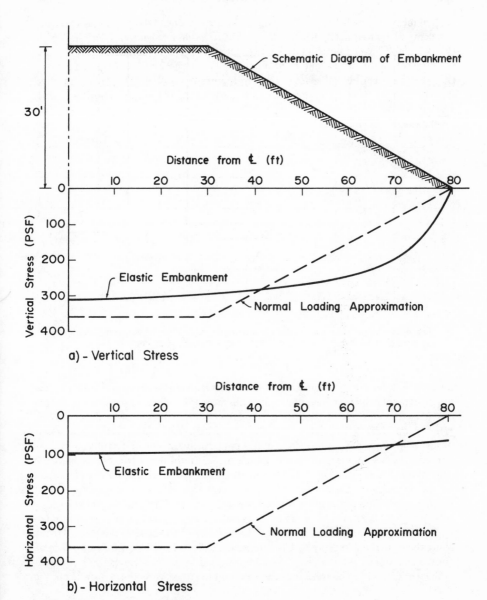

a) - Vertical Stress

b) - Horizontal Stress

Fig. A.28—Distribution of vertical and horizontal stresses on base of elastic embankment in Example A.2.

upper right-hand diagram, the $\sigma_x/\gamma H = 0.25$, which is 90 psf. The distribution of horizontal stress on the embankment is illustrated in a similar fashion to that for the vertical stress in Figure A.28b.

In Figure A.23, the magnitude and location of the largest shear stress are given as a function of depth below the base of the embankment. For example in the case of the embankment for which $L/H = 1$ the largest shear stress due to the embankment occurs at a depth of approximately $1.75H$, which is 52.5 ft. It has a magnitude of $0.3\,\gamma H$, which is 54 psf and is located horizontally on the centerline of the embankment.

The effect on the vertical and horizontal normal and shear stresses of a long excavation in a linear-elastic half-space is given in Figures A.29–A.49. These diagrams give the *reduction* in vertical and horizontal normal stress, and the change in horizontal and vertical shear stress due to the excavation. The figures are used in the same way as those for the elastic embankment.

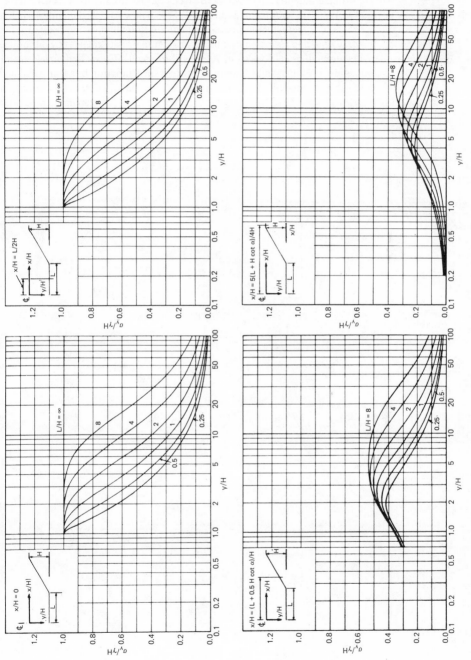

Fig. A.29—Influence diagrams for reduction in vertical normal stress along vertical sections due to excavation in elastic half-space, $\mu = 0.3$, $\alpha = 30°$.

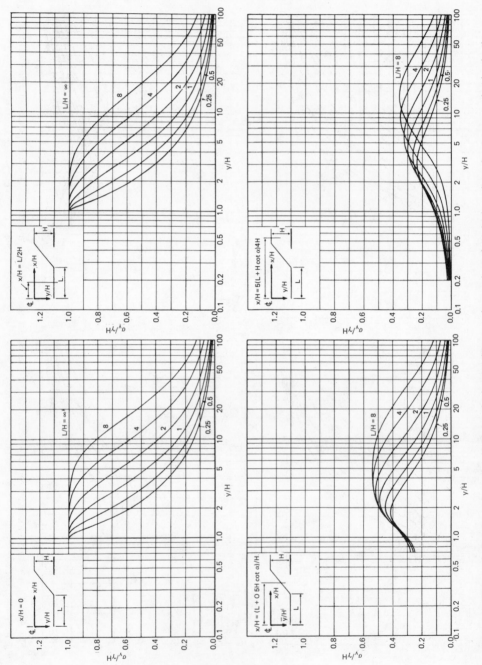

Fig. A.30—Influence diagrams for reduction in vertical normal stress along selected vertical sections due to excavation in elastic half-space, $\mu = 0.3$, $\alpha = 40°$.

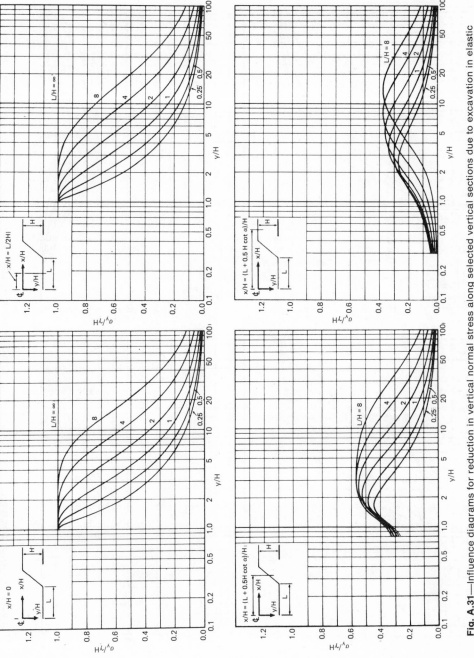

Fig. A.31—Influence diagrams for reduction in vertical normal stress along selected vertical sections due to excavation in elastic half-space, $\mu = 0.3$, $\alpha = 50°$.

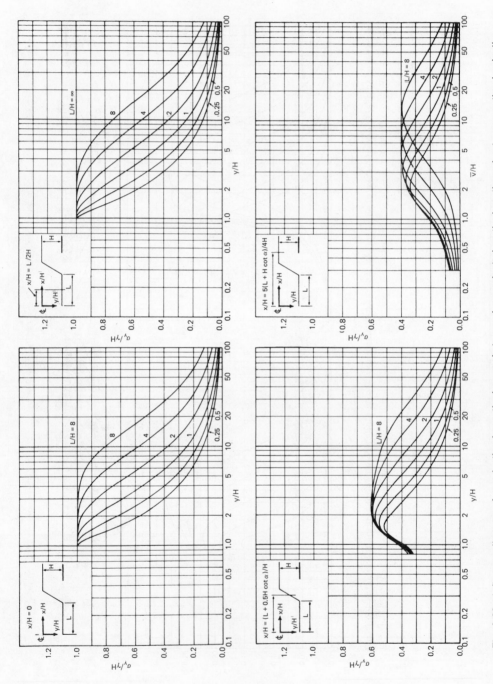

Fig. A.32—Influence diagrams for reduction in vertical normal stress along selected vertical sections due to excavation in elastic half-space, $\mu = 0.3$, $\alpha = 60°$.

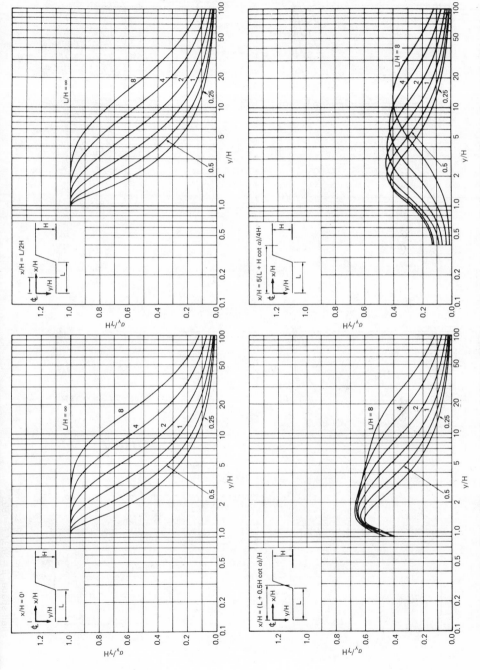

Fig. A.33—Influence diagrams for reduction in vertical normal stress along selected vertical sections due to excavation in elastic half-space, $\mu = 0.3$, $\alpha = 70°$.

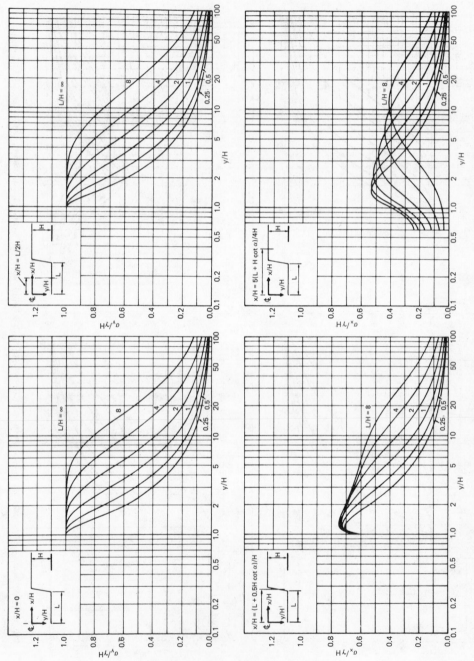

Fig. A.34—Influence diagrams for reduction in vertical normal stress along selected vertical sections due to excavation in elastic half-space, $\mu = 0.3$, $\alpha = 80°$.

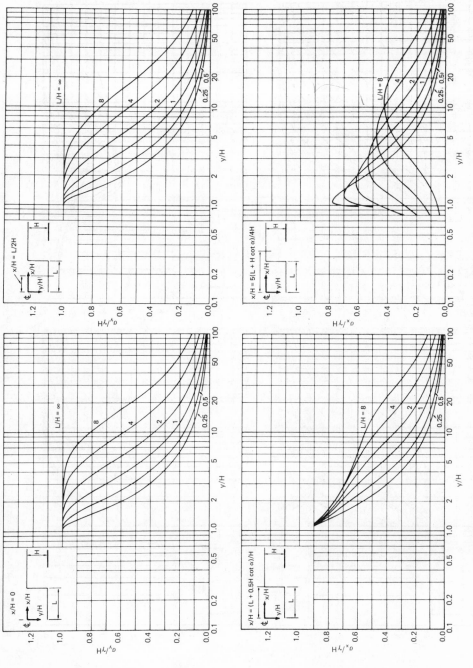

Fig. A.35——Influence diagrams for reduction in vertical normal stress along selected vertical sections due to excavation in elastic half-space, $\mu = 0.3$, $\alpha = 90°$.

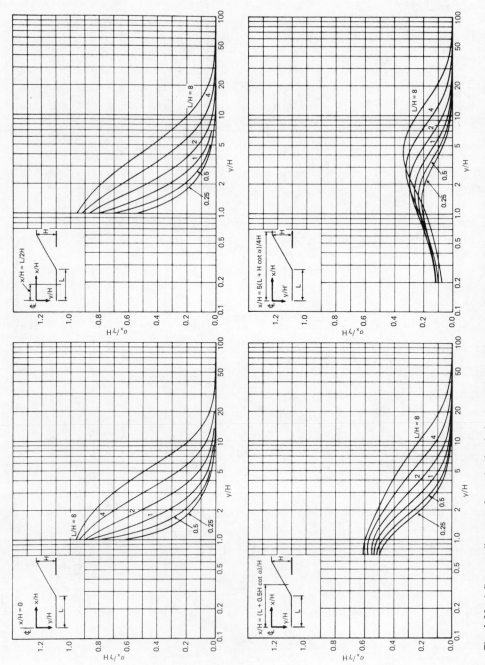

Fig. A.36—Influence diagrams for reduction in horizontal normal stress along selected vertical sections due to excavation in elastic half-space, $\mu = 0.3$, $\alpha = 30°$.

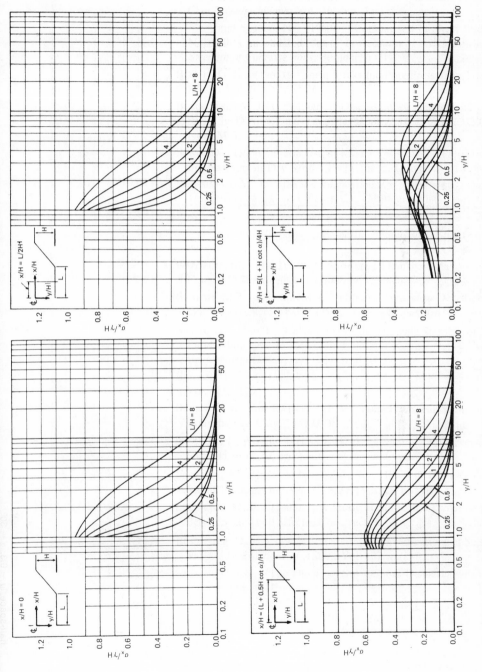

Fig. A.37—Influence diagrams for reduction in horizontal normal stress along selected vertical sections due to excavation in elastic half-space, $\mu = 0.3$, $\alpha = 40°$.

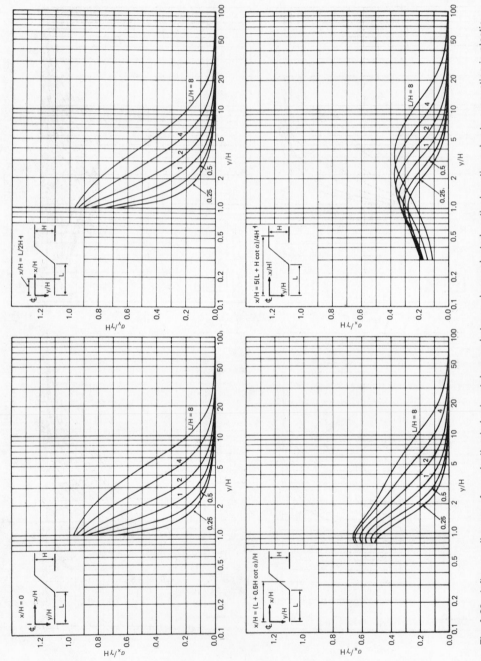

Fig. A.38—Influence diagrams for reduction in horizontal normal stress along selected vertical sections due to excavation in elastic half-space, $\mu = 0.3$, $\alpha = 50°$.

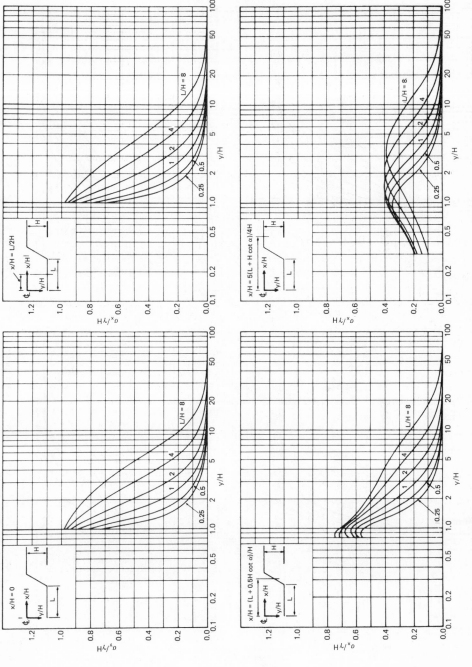

Fig. A.39—Influence diagrams for reduction in horizontal normal stress along selected vertical sections due to excavation in elastic half-space, $\mu = 0.3$, $\alpha = 60°$.

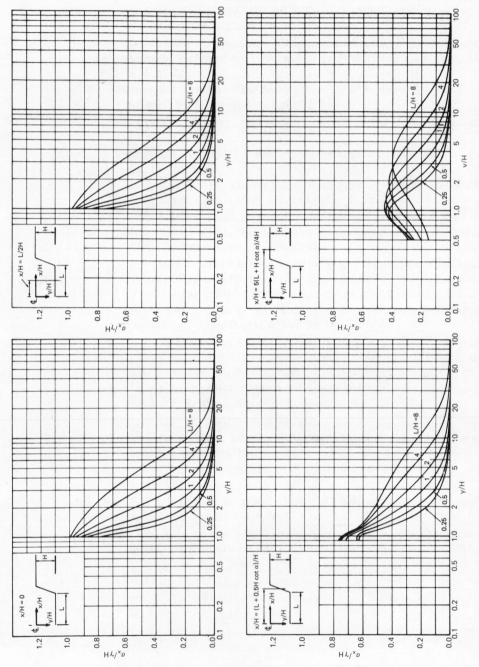

Fig. A.40—Influence diagrams for reduction in horizontal normal stress along selected vertical sections due to excavation in elastic half-space, $\mu = 0.3$, $\alpha = 70°$.

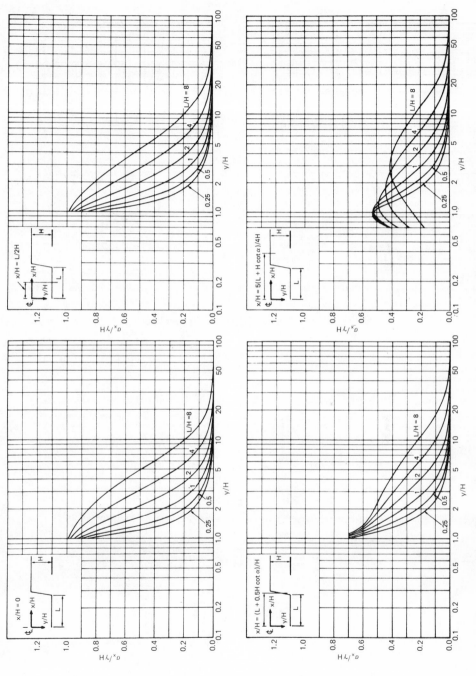

Fig. A.41—Influence diagrams for reduction in horizontal normal stress along selected vertical sections due to excavation in elastic half-space, $\mu = 0.3$, $\alpha = 80°$.

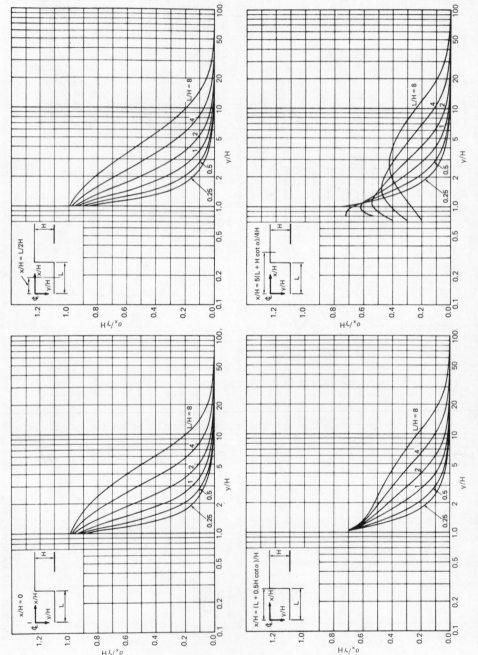

Fig. A.42—Influence diagrams for reduction in horizontal normal stress along selected vertical sections due to excavation in elastic half-space, $\mu = 0.3$, $\alpha = 90°$.

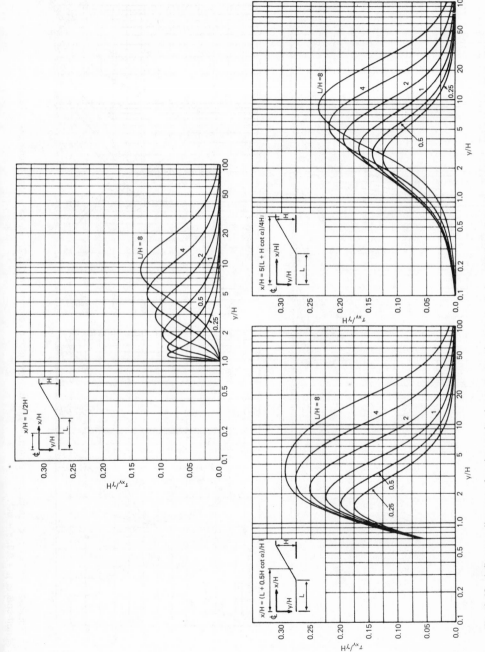

Fig. A.43—Influence diagrams for change in shear stress along selected vertical sections due to excavation in elastic half-space, $\mu = 0.3$, $\alpha = 30°$.

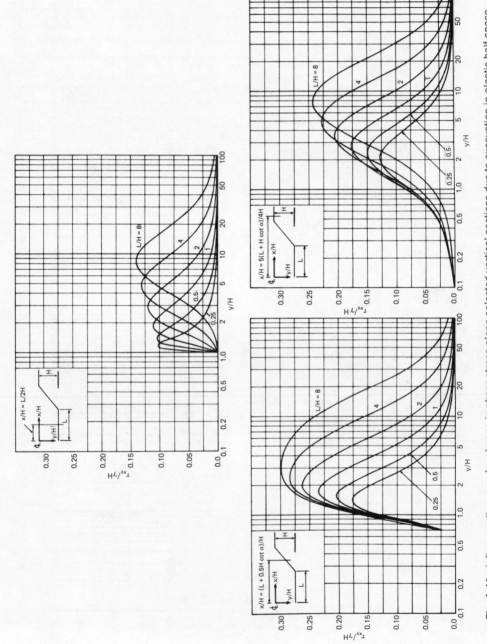

Fig. A.44—Influence diagrams for change in shear stress along selected vertical sections due to excavation in elastic half-space, $\mu = 0.3$, $\alpha = 40°$.

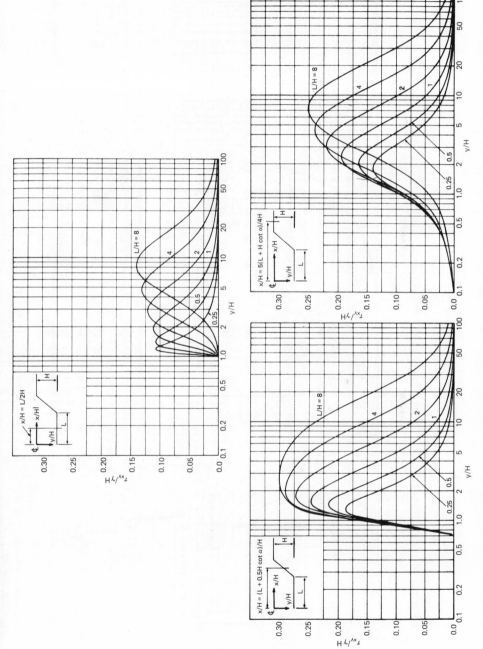

Fig. A.45—Influence diagrams for change in shear stress along selected vertical sections due to excavation in elastic half-space, $\mu = 0.3$, $\alpha = 50°$

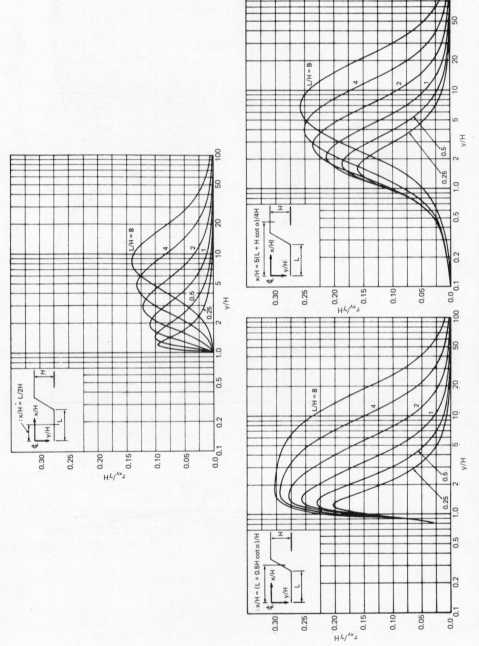

Fig. A.46—Influence diagrams for change in shear stress along selected vertical sections due to excavation in elastic half-space, $\mu = 0.3$, $\alpha = 60°$.

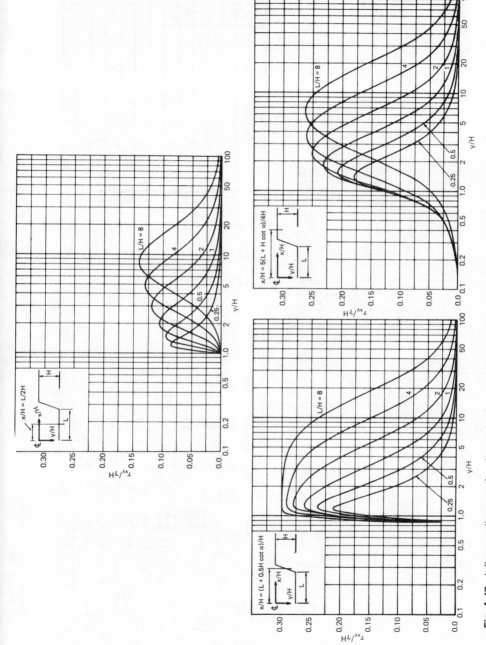

Fig. A.47—Influence diagrams for change in shear stress along selected vertical sections due to excavation in elastic half-space, $\mu = 0.3$, $\alpha = 70°$.

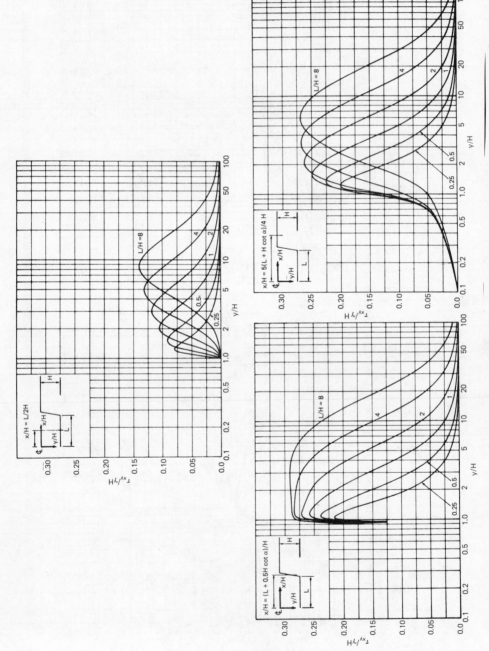

Fig. A.48—Influence diagrams for change in shear stress along selected vertical sections due to excavation in elastic half-space, $\mu = 0.3$, $\alpha = 80°$.

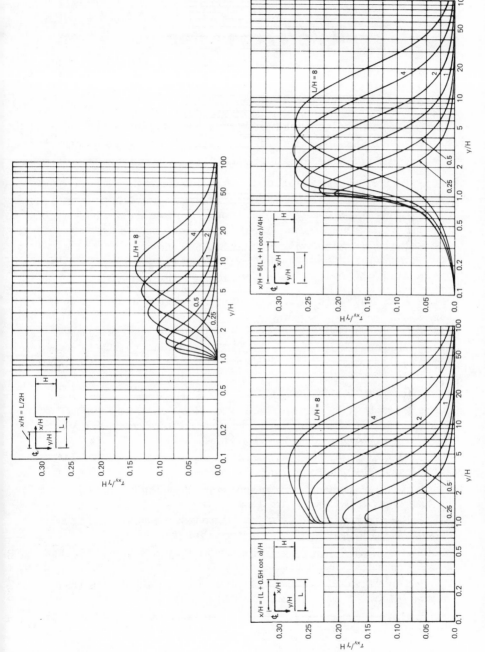

Fig. A.49—Influence diagrams for change in shear stress along selected vertical sections due to excavation in elastic half-space, $\mu = 0.3$, $\alpha = 90°$.

References Cited

The various references cited in the text are listed alphabetically by author below. Certain frequently referenced journals and proceedings are denoted by the following abbreviations:

> ASTM, STP—American Society for Testing and
> Materials, Special Technical Publication.
> ICSMFE—*Proceedings*, International Conference on
> Soil Mechanics and Foundation Engineering.
> JSMFD, ASCE—*Journal* of the Soil Mechanics and
> Foundations Division, *Proceedings*,
> American Society of Civil Engineers.

ACUM, W. E. A., and FOX, L. (1951) "Computation of Load Stresses in a Three-Layer Elastic System," *Geotechnique*, Vol. 2, No. 4.

AHLVIN, R. G., and ULERY, H. H. (1962) "Tabulated Values for Determining the Complete Pattern of Stresses, Strains, and Deflections Beneath A Uniform Circular Load on a Homogeneous Half Space," *Highway Research Board Bulletin 342*, p. 1.

AIRY, G. B. (1862) "On the Strains in the Interior of Beams" (abstract given in *Proceedings of the Royal Society of London*, Vol. 12, 1863, p. 305).

ALDRICH, H. P., Jr. (1952) "Importance of the Net Load to the Settlement of Buildings in Boston," *Journal of the Boston Society of Civil Engineers*, Vol. XXIX, No. 2 (also in *Contributions to Soil Mechanics 1940–1953*).

―――― (1965) "Precompression for Support of Shallow Foundations," *JSMFD, ASCE*, Vol. 91, No. SM 2.

ALPAN, I. (1967) "The Empirical Evaluation of the Coefficient K_o and K_{oR}," *Soil and Foundation (Japan)*, Vol. VII, No. 1.

ANANDAKRISHNAN, M., and VARADARAJULU, G. H. (1963) "Laminar and Turbulent Flow of Water Through Sand," *JSMFD, ASCE*, Vol. 89, No. SM 5.

ASCE Report (1967) "Problems in Design and Construction of Earth and Rockfill Dams," Committee on Earth and Rockfill Dams.

ASTM (1970) "Standard Method of Test for One-Dimensional Consolidation Properties of Soils," *ASTM Standards*, Part II.

BAKER, W. H., and KRIZEK, R. I. (1970) "Mohr Coulomb Strength Theory for Anisotropic Soils," *JSMFD, ASCE*, Vol. 96, No. SM 1.

BALLA, A. (1964) "Stress Conditions in Triaxial Compression," *JSMFD, ASCE*, Vol. 86, No. SM 6.

BARDEN, L., and BERRY, P. L. (1965) "Consolidation of Normally Consolidated Clay," *JSMFD, ASCE,* Vol. 91, No. SM 5.

———, MADEDOR, A. O., and SIDES, G. R. (1969) "Volume Change Characteristics of Unsaturated Clay," *JSMFD, ASCE,* Vol. 95, No. SM 1.

BARRON, R. A. (1948) "Consolidation of Fine-Grained Soils by Drain Wells," *Transactions, ASCE,* Vol. 113.

BATES, T. F., and COMER, J. J. (1955) "Electron Microscopy of Clay Surfaces," *Clays and Clay Minerals,* National Academy of Sciences, Publication 395, p. 1.

BECKETT, R., and HURT, J. (1967) *Numerical Calculations and Algorithms,* McGraw-Hill, New York.

BERTRAM, G. E. (1966) "Design Requirements and Site Selection for Embankment Dams," Soil Mechanics Lecture Series: *Design and Construction of Earth Structures,* Chicago, p. 1.

BISHOP, A. W. (1955) "The Use of the Slip Circle in the Stability Analysis of Slopes," *Geotechnique,* Vol. 5, No. 1.

——— (1961) "The Measurement of Pore Pressure in the Triaxial Test," *Proceedings of Conference on Pore Pressure and Suction in Soils,* London, p. 38.

———, ALPAN, I., BLIGHT, G. E., and DONALD, I. B. (1960) "Factors Controlling the Strength of Partly Saturated Cohesive Soils," *Proceedings of ASCE Research Conference on Strength of Cohesive Soils,* Boulder, p. 503.

———, and BJERRUM, L. (1960) "The Relevance of the Triaxial Test to The Solution of Stability Problems," *Proceedings of ASCE Research Conference on Shear Strength of Cohesive Soils,* Boulder, p. 437.

———, and ELDIN, G. (1950) "Undrained Triaxial Tests on Saturated Sands and Their Significance in the General Theory of Shear Strength," *Geotechnique,* Vol. 2, No. 1.

———, and GREEN, G. E. (1965) "The Influence of End Restraint on the Compression Strength of a Cohesionless Soil," *Geotechnique,* Vol. 15, No. 3.

———, and HENKEL, D. J. (1962) *The Measurement of Soil Properties in The Triaxial Test,* 2nd ed., Edward Arnold, Ltd., London.

———, HUTCHINSON, J. N., PENMAN, A. D. M., and EVANS, H. E. (1969) "Geotechnical Investigation into the Causes and Circumstances of the Disaster of 21st October, 1966," *A Selection of Reports Submitted to the Aberfan Tribunal,* Item 1, H. M. Stationery Office, p. 1.

BJERRUM, L. (1955) "Stability of Natural Slopes on Quick Clay," *Norwegian Geotechnical Institute Proceedings,* No. 10.

——— (1967a) "Engineering Geology of Norwegian Normally Consolidated Marine Clays as Related to Settlements of Buildings," *Geotechnique,* Vol. 17, No. 2.

——— (1967b) "Progressive Failure in Slopes of Overconsolidated Plastic Clays and Clay Shales," *JSMFD, ASCE,* Vol. 93, No. SM 5.

———, and EGGESTAD, A. (1963) "Interpretation of Loading Test on Sand," *Proceedings of the European Conference on Soil Mechanics and Foundation Engineering,* Wiesbaden, W. Germany (also in Norwegian Geotechnical Institute Publication 58, Oslo, 1964).

————, and LANDVA, A.   (1966)   "Direct Simple-Shear Tests on a Norwegian Quick Clay," *Geotechnique*, Vol. 16, No. 1.

————, and SIMONS, N.   (1960)   "Comparison of Shear Strength Characteristics of Normally Consolidated Clays," *Proceedings of ASCE Research Conference on Shear Strength of Cohesive Soils*, Boulder, p. 711.

BLIGHT, G. E.   (1965)   "Shear Stress and Pore Pressure in Triaxial Testing," *JSMFD, ASCE*, Vol. 91, No. SM 6.

BOLOGNESI, A. J. L.   (1969)   "Discussion of 'Settlement of Spread Footings on Sand,' by D'Appolonia *et al.* (May, 1968 *JSMFD, ASCE*)," *JSMFD, ASCE*, Vol. 95, No. SM 3.

BOND, D.   (1956)   "The Use of Model Tests for the Prediction of Settlement Under Foundations in Dry Sand," Ph.D. thesis, University of London.

BOUSSINESQ, M. J.   (1885)   *Application des Potentiels a L'étude de L'équilibre et du Mouvement des Solides Élastiques*, Gauthier Villars, Paris.

BOWLES, J. E.   (1968)   *Foundation Analysis and Design*, McGraw-Hill, New York.

————   (1970)   *Engineering Properties of Soils and Their Measurement*, McGraw-Hill, New York.

BOZOZUK, M.   (1962)   "Soil Shrinkage Damages Shallow Foundations at Ottawa, Canada," *The Engineering Journal*, Canada.

BRAHMA, S. P. and HARR, M. E.   (1962)   "Transient Development of the Free Surface in a Homogeneous Earth Dam," *Geotechnique*, Vol. 12, No. 4.

BRETH, H.   (1967)   "Calculation of the Shearing Strength of a Moraine Subjected to Landsliding Due to Reservoir Filling in Kauner Valley, Austria," *Proceedings of Geotechnical Conference*, Oslo, p. 171.

BRINCH HANSEN, J.   (1953)   "A General Earth Pressure Theory," *3rd ICSMFE*, Vol. II, p. 170.

BROMS, B. B., and RATNAM, M. V.   (1963)   "Shear Strength of an Anisotropically Consolidated Clay," *JSMFD, ASCE*, Vol. 91, No. SM 5.

BROOKER, E. W., and IRELAND, H. O.   (1965)   "Earth Pressures at Rest Related to Stress History," *Canadian Geotechnical Journal*, Vol. 2, No. 1.

BROWZIN, B.   (1961)   "Nonsteady-State Flow in Homogeneous Earth Dams After Rapid Drawdown," *5th ICSMFE*, Vol. II, p. 551.

BRUMUND, W. F.   (1965)   "A Study of Static and Dynamic Coefficients of Friction Between Sand and Selected Construction Materials," M.S. thesis, Purdue University.

BURMISTER, D. M.   (1942)   "Laboratory Investigations of Soils at Flushing Meadows Park," *Transactions of ASCE*, Vol. 107, p. 187.

————   (1943)   "The Theory of Stresses and Displacements in Layered Systems and Applications to Design of Airport Runways," *Proceedings of Highway Research Board*, Vol. 23, p. 127.

————   (1956)   "Stress and Displacement Characteristics of a Two-Layer Rigid Base Soil System: Influence Diagrams and Practical Applications," *Proceedings of Highway Research Board*, Vol. 35, p. 773.

————   (1958)   "Evaluation of Pavement Systems of the WASHO Road Test Layered System Methods," *Highway Research Board Bulletin No. 177*.

——— (1965) "Influence Diagrams for Stresses and Displacements in a Two-Layer Pavement System for Airfields," Contract NBy 13009, Dept. of the Navy, Washington, D.C.

——— (1967) "Applications of Dimensional Analyses in the Evaluation of Asphalt Pavement Performances," *Proceedings of Fifth Paving Conference*, Albuquerque, N.M., p. 22.

BUTTON, S. J. (1953) "The Bearing Capacity of Footings on a Two-Layer Cohesive Subsoil," *3rd ICSMFE*, Vol. I, p. 332.

CARMAN, P. C. (1956) *Flow of Gases Through Porous Media*, Academic Press, New York.

CASAGRANDE, A. (1932) "Research on the Atterberg Limits of Soils," *Public Roads*, Vol. 13, No. 8.

——— (1936) "The Determination of the Preconsolidation Load and Its Practical Significance," *1st ICSMFE*, Vol. 3, p. 60.

——— (1937) "Seepage Through Dams," *Journal of the New England Water Works Association*, Vol. LI, No. 2, (also in *Contributions to Soil Mechanics, 1925–40*, Boston Society of Civil Engineers, 1940).

——— (1947) "The Pile Foundation of the New John Hancock Building in Boston," *Journal of the Boston Society of Civil Engineers*, Vol. 34, No. 4 (also in *Contributions to Soil Mechanics 1941–53*, Boston Society of Civil Engineers).

——— (1965) "Role of the 'Calculated Risk' in Earthwork and Foundation Engineering," *JSMFD, ASCE*, Vol. 91, No. SM 4.

———, and CARILLO, N. (1944) "Shear Failure of Anisotropic Materials," *Journal of the Boston Society of Civil Engineers*, Vol. 31, No. 4, (also in *Contributions to Soil Mechanics, 1941–53*, Boston Society of Civil Engineers).

———, and FADUM, R. E. (1940) "Notes on Soil Testing for Engineering Purposes," Harvard Univ. Grad. School of Engineering Publication 268.

CASAGRANDE, L. (1932) "Näherungsmethoden zur Bestimmung von Art und Menge der Sicherung durch geschüttete Dämme," Thesis, Technische Hochschule, Vienna (cited by Casagrande, 1937).

——— (1952) "Electro-Osmotic Stabilization of Soils," *Journal of the Boston Society of Civil Engineers*, Vol. 39, No. 1.

———, and POULOS, S. (1969) "On The Effectiveness of Sand Drains," *Canadian Geotechnical Journal*, Vol. 6, No. 3.

CAUCHY, A. L. (1828) "Sur les Équations qui Expriment les Conditions d'Équilibre ou les Lois de Mouvement Intérieur d'un Corps Solide," *Exercices de Mathematique*, p. 160.

CEDERGREN, H. R. (1967) *Seepage, Drainage and Flow Nets*, Wiley, New York.

CEPWR (1965) "Jetting Procedure for Vertical Sand Drains," *Civil Engineering and Public Works Review*, p. 217.

CHAPMAN, D. L. (1913) "A Contribution to the Theory of Electrocapillarity," *Philosophical Magazine and Journal of Science*, Series 6, Vol. 25, No. 6, p. 475.

CHURCHILL, R. V. (1960) *Complex Variables and Applications*, 2nd Ed., McGraw-Hill, New York.

CLOUGH, R. W., and WOODWARD, R. J., III (1967) "Analysis of Embankment Stresses and Deformations," *JSMFD, ASCE*, Vol. 93, No. SM 4.

COLLIN, A. (1846) *Recherches Expérimentales sur les Glissements Spontanés des Terrains Argileux Accompagnées de Considerations sur Quelques Principes de la Mechanique Terrestre*, Carilian-Goeury, Paris (translated as *Landslides in Clays*, by W. R. Schriever, University of Toronto Press, Toronto, 1956).

COULOMB, C. A. (1776) "Essai sur une Application des Règles des Maximis et Minimis à Quelques Problèmes de Statique Relatifs à L'Architecture," *Mem. Acad. Roy. Pres. Div. Sav.*, Vol. 7, Paris, France.

CRAWFORD, C. B. (1959) "The Influence of Rate of Strain of Effective Stresses in Sensitive Clay," ASTM, STP No. 254.

CRAWFORD, C. B. (1963) "Cohesion in an Undisturbed Sensitive Clay," *Geotechnique*, Vol. 13, No. 2.

CULMANN, K. (1866) *Die graphische Statik*, Zurich.

D'APPOLONIA, D. J., D'APPOLONIA, E. E., and BRISSETTE, R. F. (1968) "Settlement of Spread Footings on Sand," *JSMFD, ASCE*, Vol. 94, No. SM 3.

———, WHITMAN, R. V., and D'APPOLONIA, E. E. (1969) "Sand Compaction With Vibratory Rollers," *JSMFD, ASCE*, Vol. 95, No. SM 1.

D'APPOLONIA, E., and NEWMARK, N. M. (1951) "A Method for the Solution of Restrained Cylinders Under Compression," *Proceedings of First U. S. Nat'l Conference of Applied Mechanics, ASME*.

DARCY, H. (1856) "Les fontaines publiques de la ville de Dijon," Dalmont, Paris.

DAVIS, E. H., and RAYMOND, G. P. (1965) "A Non-Linear Theory of Consolidation," *Geotechnique*, Vol. 15, No. 2.

DEBEER, E. E. (1948) "Settlement Records on Bridges Founded on Sand," *2nd ICSMFE*, Rotterdam, Vol. 2, p. 111.

——— (1965) "Bearing Capacity and Settlement of Shallow Foundations on Sand," *Symposium on Bearing Capacity and Settlement of Foundations*, Duke University, p. 15.

DIXON, R. K. (1967) "New Techniques for Studying Seepage Problems Using Models," *Geotechnique*, Vol. 17, No. 3.

DOMASCHUK, L., and WADE, N. H. (1969) "A Study of Bulk and Shear Moduli of a Sand," *JSMFD, ASCE*, Vol. 95, No. SM 2.

DUFFY, J., and MINDLIN, R. D. (1957) "Stress-Strain Relations and Vibrations of a Granular Medium," *Journal of Applied Mechanics*.

DUNCAN, J. M., and DUNLOP, P. (1969) "Slopes in Stiff-Fissured Clays and Shales," *JSMFD, ASCE*, Vol. 95, No. SM 2.

DUNCAN, J. M., and SEED, H. B. (1966a) "Anisotropy and Stress Reorientation in Clay," *JSMFD, ASCE*, Vol. 92, No. SM 5.

———, and ———. (1966b) "Strength Variation Along Failure Surfaces in Clay," *JSMFD, ASCE*, Vol. 92, No. SM 6.

Earth Manual (1960) U. S. Bureau of Reclamation.

EGGESTAD, A. (1963) "Deformation Measurements Below a Model Footing on the Surface of a Dry Sand," *Proceedings European Conference on Soil Mechanics and Foundation Engineering*, Vol. 1, p. 233.

EGOROV, K. E. (1958) "Concerning the Question of the Deformation of Bases of Finite Thickness," *Mekhanika Gruntov*, Sb. Tr., No. 34, Gosstroiizdat, Moscow.

FADUM, R. E. (1948) "Influence Values for Estimating Stresses in Elastic Foundations," *2nd ICSMFE*, Vol. 2.

FELLENIUS, W. (1927) *Erdstatische Berechnungen mit Reibung und Kohäision und unter Annahme Kreiszylindrischer Gleitflachen*, Ernst and Son, Berlin.

FILON, N. G. (1902) "The Elastic Equilibrium of Circular Cylinders Under Certain Practical Systems of Load," *Philosophical Transactions of the Royal Society*, A Vol. 198.

FINN, W. D. L. (1967) "Finite Element Analysis of Seepage Through Dams," *JSMFD, ASCE*, Vol. 93, No. SM 6.

FORSYTHE, G. E., and WASOW, W. R. (1960) *Finite Difference Methods for Partial Differential Equations*, Wiley, New York.

FOSTER, C. R., and AHLVIN, R. G. (1954) "Stresses and Deflections Induced by a Uniform Circular Load," *Proceedings of Highway Research Board*, Vol. 34, p. 467.

————, and FERGUS, S. M. (1951) "Stress Distribution in a Homogeneous Soil," Highway Research Board, Research Report No. 12-F.

FUNG, Y. C. (1965) *Foundations of Solid Mechanics*, Prentice-Hall, Englewood Cliffs, N.J.

GENINI, H. E. (1968) "Portsmouth Sand Drains Project," *New Hampshire Highways*.

GIBBS, H. J., and HOLTZ, W. G. (1957) "Research on Determining the Density of Sands by Spoon Penetration Testing," *4th ICSMFE*, Vol. 1, p. 35.

GILBOY, G. (1936) "Improved Soil Testing Methods," *Engineering News-Record*, May 21.

GIRAULT, P. (1960) "A Study on the Consolidation of Mexico City Clay," Ph.D. thesis, Purdue University.

GIRIJAVALLABHAN, C. V., and REESE, L. C. (1968) "Finite-Element Method for Problems in Soil Mechanics," *JSMFD, ASCE*, Vol. 94, No. SM 2.

GOLDER, H. Q., and GASS, A. A. (1963) "Field Tests for Determining Permeability of Soil Strata," ASTM STP 322.

GOULD, J. P. (1968) "Report on Study of Movements of Articulated Conduits Under Earth Dams on Compressible Foundations," Report to U. S. Department of Agriculture Soil Conservation Service, by Mueser, Rutledge, Wentworth and Johnston, Consulting Engineers.

GOUY, G. (1910) "Sur la Constitution de la Charge Électrique À La Surface d'un Électrolyte," *Journal de Physique, Theorique et Appliquée*, Series 4, Vol. 9.

GRAY, D. H., and MITCHELL, J. K. (1967) "Fundamental Aspects of Electro-Osmosis in Soils," *JSMFD, ASCE*, Vol. 93, No. SM 6.

GRAY, H. (1938) "Research on the Consolidation of Fine-Grained Soils," Sc.D. thesis, Harvard University.

———— (1945) "Simultaneous Consolidation of Contiguous Layers of Unlike Compressible Soils," *Transactions, ASCE*, Vol. 110, Paper No. 2258.

————, and HOOKS, I. J. (1948) "Charts Facilitate Determination of Stresses Under Loaded Areas," *Civil Engineering*, Vol. 18, No. 6.

GRIM, R. E. (1962) *Applied Clay Mineralogy*, McGraw-Hill, New York.

—— (1968) *Clay Mineralogy*, 2nd ed., McGraw-Hill, New York.

HALL, H. P. (1955) "An Investigation of Steady Flow Toward a Gravity Well," *La Houille Blanche*.

HAMMER, M. J., and THOMPSON, O. B. (1966) "Foundation Clay Shrinkage Caused by Large Trees," *JSMFD, ASCE*, Vol. 92, No. SM 6.

HAMPTON, D., SCHIMMING, B. B., SKOK, E. L., Jr., and KRIZEK, R. J. (1969) "Solutions to Boundary Value Problems of Stresses and Displacements in Earth Masses and Layered Systems," Highway Research Board Bibliography No. 48.

HANNA, T. H., and ADAMS, J. I. (1968) "Comparison of Field and Laboratory Measurements of Modulus of Deformation of Clay," Highway Research Record No. 243.

HANSBO, S. (1960) "Consolidation of Clay With Special Reference to the Influence of Vertical Sand Drains," *Proceedings of 18th Swedish Geotechnical Institute*, Stockholm.

HANSEN, B. and NIELSEN, H. K. (1965) "Settlement Calculations for a Tunnel Construction in Gothenburg Clay," *6th ICSMFE*, Vol. 2, p. 377.

HARR, M. E. (1962) *Groundwater and Seepage*, McGraw-Hill, New York.

—— (1966) *Foundations of Theoretical Soil Mechanics*, McGraw-Hill, New York.

HAZEN, A. (1911) "Discussion on 'Dams on Sand Foundations,'" *Transactions of American Society of Civil Engineers*, Vol. 72.

HELMHOLTZ, H. (1879) *Wiedemanns Annalen der Physik* (cited by L. Casagrande, 1952).

HENDRON, A. J. (1963) "The Behavior of Sand in One-Dimensional Compression," Ph.D. thesis, University of Illinois.

HENKEL, D. J. (1956) "Effect of Overconsolidation on Behaviour of Clays During Shear," *Geotechnique*, Vol. 6, No. 4.

HOLDEN, J. C. (1967) "Stresses and Strains in a Sand Mass Subjected to a Uniform Circular Load," Departmental Report No. 13, Dept. of Civil Engineering, University of Melbourne.

HOLTZ, W. G., and GIBBS, H. J. (1969) "Discussion of 'Settlement of Spread Footings on Sand' by D'Appolonia et al. (May 1968 *JSMFD, ASCE*)," *JSMFD, ASCE*, Vol. 95, No. SM 3.

HOOKE, R. (1678) *De Potentia Restitutiva*, London.

HORN, H. M., and DEERE, D. U. (1962) "Frictional Characteristics of Minerals," *Geotechnique*, Vol. XII, No. 4.

HORNE, M. R. (1965) "The Behavior of an Assembly of Rotund, Rigid, Cohesionless Particles," *Proceedings of Royal Society*, Vol. 286, Part A, p. 62.

HUNTINGTON, W. C. (1957) *Earth Pressures and Retaining Walls*, Wiley, New York.

HUTCHINSON, J. N. (1965) "The Landslide of February, 1959, at Vibstad in Namdalen," Norwegian Geotechnical Institute Publication No. 61, p. 1.

HVORSLEV, M. J. (1936) "Conditions of Failure for Remolded Cohesive Soils," *1st ICSMFE*, Vol. 3, p. 51.

HVORSLEV, M. J. (1960) "Physical Components of the Shear Strength of Saturated Clays," *Proceedings of ASCE Research Conference on Shear Strength of Cohesive Soils*, Boulder, p. 169.

JACOBSEN, M. (1967) "The Undrained Shear Strength of Preconsolidated Boulder Clay," *Proceedings of Geotechnical Conference*, Oslo, Vol. 1, p. 119.

JAKOBSON, B. (1955) "Isotropy of Clays," *Geotechnique*, Vol. 5, No. 1.

JAKY, J. (1948) "Pressure in Silos," *2nd ICSMFE*, Vol. 1, p. 103.

JANBU, N., BJERRUM, L., and KJAERNSLI, B. (1956) "Veiledning ved Løsning av, Fundamenteringsoppgaver," Norwegian Geotechnical Institute Publication No. 16.

JEFFRIES, C. D., and BATES, T. F. (1960) "Electron Micrographs of Clay Minerals Common in Soils," Circular 51, Mineral Industries Experiment Station, Pennsylvania State University.

JENNINGS, J. E., and ROBERTSON, A. M. (1969) "The Stability of Slopes Cut into Natural Rock," *7th ICSMFE*, Vol. 2, p. 585.

JEPPSON, R. W. (1968) "Seepage Through Dams in the Complex Potential Plane," *Journal of Irrigation and Drainage Division*, ASCE, Vol. 94, No. IR1.

JOHNSON, A. I., and RICHTER, R. C. (1967) "Selected Bibliography on Permeability and Capillarity Testing of Rock and Soil Materials," ASTM, STP No. 417.

JOHNSON, S. J. (1970) "Precompression for Improving Foundation Soils," *JSMFD*, ASCE, Vol. 96, No. SM 1.

JONES, A. (1962) "Tables of Stresses in Three-Layer Elastic Systems," *Highway Research Board Bulletin No. 342*.

JORDAN, J. H. (1949) "A Florida Landslide," *Journal of Geology*, Vol. 57, No. 4, p. 418.

JUMIKIS, A. R. (1956) "Rupture Surfaces in Sand Under Oblique Loads," *JSMFD*, ASCE, Vol. 82, No. SM 1.

——— (1962) *Soil Mechanics*, Reinhold-Van Nostrand, Princeton.

KANE, H., HENDRON, A. J., BROOKER, E. W., and MOHRAZ, B. (1965) "A Study of the Behavior of Soil and Rock Subjected to High Stresses," U. S. Dept. of Commerce, Technical Report No. WL–TR–64–157, Contract AF 29(601)–5817.

KAPP, M. S. (1968) private communication.

———, YORK, D. L., ARONOWITZ, A. and SITOMER, H. (1966) "Construction on Marshland Deposits: Treatments and Results," *Highway Research Record No. 133*.

KENNEY, T. C. (1968) "Stability of the Vaiont Valley Slope," Norwegian Geotechnical Institute Publication No. 74.

KHERA, R. P., and KRIZEK, R. J. (1968) "Effect of Principal Consolidation Stress Difference on Undrained Shear Strength," *Soils and Foundations*, Vol. 8, No. 1, Mar.

KIRKPATRICK, W. M., and BELSHAW, D. J. (1968) "On the Interpretation of the Triaxial Test," *Geotechnique*, Vol. 18, No. 3.

KO, H. Y., and SCOTT, R. F. (1967) "A New Soil Testing Apparatus," *Geotechnique*, Vol. 17, No. 1.

KOZENY, J. (1931) "Grundwasserbewegung bei freiem Spiegel, Fluss- und Kanal-versicherung," *Wasserkraft und Wasserwirtschaft*, No. 3 (cited by Casagrande, 1937).

KRAFT, L. M. (1965) "The Effect of Cyclic Loading on the Stress-Strain Properties of a Cohesive Soil," M.S. Thesis, Ohio State University.

LADD, C. C. (1964) "Stress-Strain Modulus of Clay in Undrained Shear," *JSMFD, ASCE*, Vol. 90, No. SM 5.

——— (1965) "Stress-Strain Behavior of Anisotropically Consolidated Clays During Undrained Shear," *6th ICSMFE*, Vol. 1, p. 282.

———, and LAMBE, T. W. (1963) "The Strength of 'Undisturbed' Clay Determined from Undrained Tests," *Laboratory Shear Testing of Soils*, ASTM STP No. 361.

———, and VARALLYAY, J. (1965) "The Influence of Stress System on the Behavior of Saturated Clays During Undrained Shear," Massachussetts Institute of Technology, Department of Civil Engineering Research Report R65–11, Soils Publication No. 177, Jul.

LAMBE, T. W. (1951) *Soil Testing for Engineers*, Wiley, New York.

——— (1958) "The Structure of Compacted Clay," *JSMFD, ASCE*, Vol. 84, No. SM 2.

——— (1964) "Methods of Estimating Settlement," *JSMFD, ASCE*, Vol. 90, No. SM 5.

——— (1967) "Stress Path Method," *JSMFD, ASCE*, Vol. 93, No. SM 6.

———, and WHITMAN, R. V. (1959) "The Role of Effective Stress in the Behavior of Expansive Soils," Colorado School of Mines, Vol. 54, No. 4.

LANDAU, R. E. (1966) "Method of Installation as a Factor in Sand Drain Stabilization Design," *Highway Research Record No. 133*, p. 75.

LEONARDS, G. A. (1962) "Engineering Properties of Soils," Chapter 2 in *Foundation Engineering*, edited by G. A. Leonards, McGraw-Hill, New York.

——— (1966) "Discussion of 'Settlement of a Mat Foundation on a Thick Stratum of Sensitive Clay,' by L. Casagrande," *Canadian Geotechnical Journal*, Vol. 3, No. 2.

——— (1968) "Predicting Settlements of Buildings on Clay Soils," *Lecture Series on Foundation Engineering*, Soil Mechanics and Foundations Division, Illinois Section ASCE.

———, and ALTSCHAEFFL, A. G. (1964) "Compressibility of Clay," *JSMFD, ASCE*, Vol. 90, No. SM 5.

———, and GIRAULT, P. (1961) "A Study of the One-Dimensional Consolidation Test," *6th ICSMFE*, Vol. 1, p. 213.

LEONOFF, C. E., and RIPLEY, C. F. (1961) "Case History of a Preloaded Foundation," *Proceedings of 15th Canadian Soil Mechanics Conference*, National Research Council of Canada.

LITTLE, A. L., and PRICE, V. E. (1956) "The Use of an Electric Computer for Slope Stability Analysis," *Geotechnique*, Vol. 8, No. 3.

LIVNEH, M., and KOMORNIK, A. (1967) "Anisotropic Strength of Compacted Clays," *Proceedings of Third Asian Regional Conference on Soil Mechanics and Foundation Engineering*, Vol. 1, p. 298.

———, and SHKLARSKY, E. (1965) "Equations of Failure Stresses in Materials with Anisotropic Strength Parameters," *Highway Research Record No. 74*, p. 44.

LO, K. Y. (1965) "Stability of Slopes in Anisotropic Soils," *JSMFD, ASCE*, Vol. 91, No. SM 4.

LOVE, A. E. H. (1944) *A Treatise on the Mathematical Theory of Elasticity*, 4th Ed., Dover, New York.

MANSUR, C. I., and DIETRICH, R. J. (1965) "Pumping Test to Determine Permeability Ratio," *JSMFD, ASCE*, Vol. 91, No. SM 4.

MARGUERRE, K. (1933) "Spannungsverteilung und Wellenausbreitung in der kontinuierlich gestutzten Platte," *Ing. Arch.*, Vol. 4, No. 4.

MEINZER, O. E. (1942) *Physics of the Earth—IX: Hydrology*, McGraw-Hill, New York.

MEYERHOF, G. G. (1951) "The Ultimate Bearing Capacity of Foundations," *Geotechnique*, Vol. 2, No. 4.

——— (1953) "The Bearing Capacity of Foundations Under Eccentric and Inclined Loads," *3rd ICSMFE*, Vol. 1, p. 440.

——— (1955) "Influence of Roughness of Base and Groundwater On the Ultimate Bearing Capacity of Foundations," *Geotechnique*, Vol. 5, No. 3.

——— (1956) "Penetration Tests and Bearing Capacity of Cohesionless Soils," *JSMFD, ASCE*, Vol. 82, No. SM 1.

——— (1957) "Closure to discussion of Meyerhof (1956)," *JSMFD, ASCE*, Vol. 83, No. SM 1.

——— (1961a) "Discussion on 'Foundations Other Than Piled Foundations,'" *5th ICSMFE*, Vol. 3, p. 193.

——— (1961b) "Some Problems in the Design of Rigid Retaining Walls," *Proceedings of 15th Canadian Soil Mechanics Conference*, p. 59.

——— (1963) "Some Recent Research on The Bearing Capacity of Foundations," *Canadian Geotechnical Journal*, Vol. 1, No. 1.

——— (1965) "Shallow Foundations," *JSMFD, ASCE*, Vol. 91, No. SM 2.

MIDDLEBROOKS, T. A. (1942) "Fort Peck Slide," *Trans. ASCE*, Paper No. 2144.

MILLER, R. J., and Low, P. F. (1963) "Threshold Gradient for Water Flow in Clay Systems," *Proceedings of Soil Science Society of America*, Vol. 27, No. 6.

MITCHELL, J. K. (1960) "Fundamental Aspects of Thixotropy in Soils," *JSMFD, ASCE*, Vol. 86, No. SM 3.

———, SINGH, A., and CAMPANELLA, R. G. (1969) "Bonding, Effective Stresses, and Strength of Soils," *JSMFD, ASCE*, Vol. 95, No. SM 5.

———, and YOUNGER, J. S. (1966) "Abnormalities in Hydraulic Flow Through Fine-Grained Soils," *Permeability and Capillarity of Soils*, ASTM STP 417.

OTTO MOHR (1882) *Zinilingenieur* 1882, p. 113.

——— (1900) *Zeitschift der deutschen Ingenieur*, Vol. 44.

MORETTO, O. (1948) "Effect of Natural Hardening on The Unconfined Compression Strength of Remolded Clays," *2nd ICSMFE*, Vol. 1, p. 137.

MORGENSTERN, N. R. (1963) "Stability Charts for Earth Slopes During Rapid Drawdown," *Geotechnique*, Vol. 13, No. 2.

———, and PRICE, V. E. (1965) "The Analysis of the Stability of General Slip Surfaces," *Geotechnique*, Vol. 15, No. 1.

———, and TCHALENKO, J. S. (1967) "Microscopic Structures in Kaolin Subjected to Shear Stress," *Geotechnique*, Vol. 18, No. 4.

MUSKAT, M. (1937) *The Flow of Homogeneous Fluids Through Porous Media*, McGraw-Hill, New York.

NADAI, A. (1963) *Theory of Flow and Fracture of Solids*, Vol. 1, McGraw-Hill, New York.

NARAIN, J., SINGH, B., IYER, N. V., and DEOSKAR, S. R. (1969) "Quasi-Preconsolidation Effects and Pore Pressure Dissipation During Consolidation," *7th ICSMFE*, Vol. 1, p. 311.

NAS (1969) "Seismology, Responsibilities and Requirements of a Growing Science: Part I: Summary and Recommendations," A Report of the Committee on Seismology, Div. of Earth Sciences, National Research Council, National Academy of Sciences, Washington, D.C.

NEWLIN, C. W., and ROSSIER, S. C. (1967) "Embankment Drainage After Instantaneous Drawdown," *JSMFD, ASCE*, Vol. 93, No. SM 6.

NEWMARK, N. M. (1942) "Influence Charts for Computation of Stresses In Elastic Foundations," University of Illinois Eng. Exp. Sta. Bulletin 338.

NOORANY, I., and SEED, H. B. (1965) "In-situ Strength Characteristics of Soft Clays," *JSMFD, ASCE*, Vol. 91, No. SM 2.

OAKES, D. T. (1960) "Solids Concentration On Effects in Bentonite Drilling Fluids," *Clays and Clay Minerals*, Pergamon Press, New York, Vol. 8.

OLSEN, H. W. (1960) "Hydraulic Flow Through Saturated Clays," *Proceedings of Ninth National Conference on Clay and Clay Minerals*.

――― (1965) "Deviations from Darcy's Law in Saturated Clays," *Proceedings*, Soil Science Soc. of America, Vol. 29, No. 2, Mar–Apr.

OVERBEEK, J. Th. G. (1952) "Electrokinetic Phenomena," Chapter V in *Colloid Science*, H. R. Kruyt, Ed., Elsevier, Amsterdam.

PARRY, R. H. G. (1960) "Triaxial Compression and Extension Tests On Remolded Saturated Clay," *Geotechnique*, Vol. 10, No. 4.

PEATTIE, K. R. (1962) "Stress and Strain Factors for Three-Layer Elastic Systems," *Highway Research Board Bulletin No. 342*.

PECK, R. B. (1969) "Advantages and Limitations of the Observational Method in Applied Soil Mechanics," *Geotechnique*, Vol. 19, No. 2.

――――, and BAZARAA, A. R. S. (1969) "Discussion of D'Appolonia et al. (1968)," *JSMFD, ASCE*, Vol. 95, No. SM 3.

――――, and BRYANT, F. G. (1953) "The Bearing Capacity Failure of the Transcona Elevator," *Geotechnique*, Vol. 3, No. 5.

――――, IRELAND, H. O. and TENG, C. Y. (1948) "A Study of Retaining Wall Failures," *2nd ICSMFE*, Vol. III, p. 296.

PERLOFF, W. H. (1962) "The Effect of Stress History and Strain Rate on The Undrained Shear Strength of Cohesive Soils," Ph.D. thesis, Northwestern University.

――――, BALADI, G. Y., and HARR, M. E. (1967) "Stress Distribution Within and Under Long Elastic Embankments," *Highway Research Record No. 181*.

――――, and OSTERBERG, J. O. (1964) "Stress History Effects on Strength of Cohesive Soils," *Highway Research Record No. 48*, p. 49.

――――, and POMBO, L. E. (1969) "End Restraint Effects in the Triaxial Test," *7th ICSMFE*.

PICKETT, G. (1944) "Application of the Fourier Method to The Solution of Certain Boundary Problems in The Theory of Elasticity," *Journal of Applied Mechanics*, ASCE, Vol. 2.

PRANDTL, L. (1920) "Über die Harte plasticher Körper," *Nachr. Ges. Wiss. Göttingen, Math-physik. Kl.*

———— (1921) "Uber die Eindringrings—festigkeit (Harte) plasticher Baustoffe und die Festigkeit von Schneiden," *Zeit. f. Angew. Math. u. Mech*, Vol. 1, No. 15.

POOROOSHASB, H. B., and ROSCOE, K. H. (1961) "The Correlation of the Results of Shear Tests with Varying Degrees of Dilatation," *5th ICSMFE*, Vol. 1, p. 297.

RANGANATHAM, B. V., SANI, A. C., and SREENIVASULU, V. (1969) "Strength Anisotropy on Slope Stability and Bearing Capacity of Clays," *7th ICSMFE*, Vol. 2, p. 659.

RANKINE, W. J. M. (1856) "On The Mathematical Theory of The Stability of Earthwork and Masonry," *Proceedings of Royal Society*, Vol. 8.

RAYMOND, G. P. (1966) "Consolidation of Slightly Overconsolidated Soils," *JSMFD, ASCE*, Vol. 92, No. SM 5.

REUSS, F. F. (1809) in *Mémoires de la Société Impériale des Naturalistes de Moscou*, Vol. 2, p. 327 (cited by Overbeek, 1952).

REYNOLDS, O. (1885) "On the Dilatancy of Media Composed of Rigid Particles in Contact, With Experimental Illustrations," *Phil. Mag.*, Series 5, Vol. 20.

RICHART, F. E., Jr. (1957) "Review of the Theories for Sand Drains," *Transactions ASCE*, Vol. 124, 1959, p.709.

ROBERTS, D. V., and DARRAGH, R. D. (1963) "Area Fill Settlements and Building Foundation Behavior at the San Francisco Airport," *Field Testing of Soils*, ASTM, STP.

ROSCOE, K. H. (1953) "An Apparatus for the Application of Simple Shear to Soil Samples," *3rd ICSMFE*, Vol. I, p. 186.

———— (1970) "The Influence of Strains in Soil Mechanics," *Geotechnique*, Vol. 20, No. 2.

————, and BURLAND, J. B. (1968) "On the Generalized Stress-Strain Behavior of 'Wet' Clay," *Engineering Plasticity*, ed. Heyman and Leckie, Cambridge, p. 535.

————, SCHOFIELD, A. N., and THURAIRAJAH, A. (1963) "An Evaluation of Test Data for Selecting a Yield Criterion for Soils," ASTM, STP No. 361.

————, ————, and ———— (1965) "Discussion of Rowe *et al.* (1964)," *Geotechnique*, Vol. 15, No. 1.

————, ————, and WROTH, C. P. (1958) "On the Yielding of Soils," *Geotechnique*, Vol. 8, No. 1.

ROSENQVIST, I. Th. (1946) "Om Leires Kirkkagtighet," Statens veguesen, Veglaboratoriet, Meddelelse, 4:5 (summarized in "Considerations on the Sensitivity of Norwegian Quick-Clays," *Geotechnique*, Vol. 3, No. 5, p. 195, 1953).

ROWE, P. W. (1962) "The Stress-Dilatancy Relation for Static Equilibrium of an Assembly of Particles in Contact," *Proceedings of Royal Society of London*, Series A, Vol. 269.

———— (1968) "The Influence of Geological Features of Clay Deposits On the Design and Performance of Sand Drains," *Proceedings of Institution of Civil Engineers of Great Britain*, Vol. 39, p. 465.

——— (1969) "The Relation Between the Shear Strength of Sands in Triaxial Compression, Plane Strain and Direct Shear," *Geotechnique*, Vol. 19, No. 1.

———, and BARDEN, L. (1964) "Importance of Free Ends in Triaxial Testing," *JSMFD, ASCE*, Vol. 90, No. SM 1.

———, BARDEN, L., and LEE, I. K. (1964) "Energy Components During the Triaxial Cell and Direct Shear Tests," *Geotechnique*, Vol. 14, No. 3.

RUTLEDGE, P. C. (1944) "Review of the Cooperative Triaxial Research Program of the War Department, Corps of Engineers," Northwestern University Technological Institute.

SAMSIOE, A. F. (1931) "Einfluss von Rohrbrunnen auf die Bewegung des Grundwassers," *Zeit. für angewandte Mathematik und Mechanik*, No. 2, cited by Casagrande (1937).

SCHEIDEGGER, A. E. (1957) *The Physics of Flow Through Porous Media*, Macmillan, New York.

SCHIFFMAN, R. L. (1958) "Consolidation of Soil Under Time-Dependent Loading and Varying Permeability," *Proceedings of Highway Research Board*, p. 584.

———, CHEN, A. T. F., and JORDAN, J. C. (1969) "An Analysis of Consolidation Theories," *JSMFD, ASCE*, Vol. 95, No. SM 1.

SCHLEICHER, F. (1926) "Zur Theorie des Baugrundes," *Der Bauingenieur*, No. 48, 49.

SCHMERTMANN, J. H. (1955) "The Undisturbed Consolidation Behavior of Clay," *Trans. ASCE*, Vol. 120, p. 1201.

——— (1962) "Comparisons of One and Two-Specimen CFS Tests," *JSMFD, ASCE*, Vol. 88, No. SM 6.

——— (1970) "Static Cone to Compute Static Settlement Over Sand," *JSMFD, ASCE*, Vol. 96, No. SM 3.

———, and OSTERBERG, J. O. (1960) "An Experimental Study of the Development of Cohesion and Friction With Axial Strain in Saturated Cohesive Soils," *Proceedings of ASCE Research Conference on Shear Strength of Cohesive Soils*, p. 643.

SCHOFIELD A. and WROTH, P. (1968) *Critical State Soil Mechanics*, McGraw-Hill, New York.

SCOTT, R. F. (1963) *Principles of Soil Mechanics*, Addison-Wesley, Reading, Mass.

SEED, H. B. (1968) "Landslides During Earthquakes Due to Soil Liquefaction," *JSMFD, ASCE*, Vol. 94, No. SM 5.

———, and LEE, K. L. (1966) "Liquefaction of Saturated Sands During Cyclic Loading," *JSMFD, ASCE*, Vol. 92, No. SM 6.

———, and WILSON, S. D. (1967) "The Turnagain Heights Landslide, Anchorage, Alaska," *JSMFD, ASCE*, Vol. 93, No. SM 4.

———, WOODWARD, R. J., and LUNDGREN, R. (1964) "Fundamental Aspects of the Atterberg Limits," *JSMFD, ASCE*, Vol. 90, No. SM 6.

SELIG, E. T., and MCKEE, K. E. (1961) "Static Behavior of Small Footings," *JSMFD, ASCE*, Vol. 87, No. SM 6.

SHERARD, J. L., WOODWARD, R. J., GIZENSKI, S. F. and CLEVENGER, W. A. (1963) *Earth and Earth-Rock Dams; Engineering Problems of Design and Construction*, Wiley, New York.

SHOCKLEY, W. G., and AHLVIN, R. G. (1960) "Nonuniform Conditions in Triaxial Test Specimens," *ASCE Research Conference on Shear Strength of Cohesive Soils.*

SILIN-BEKCHURIN, A. I. (1958) "Dynamics of Underground Water," p. 23, Moscow University (cited by Harr, 1962).

SKEMPTON, A. W. (1953) "The Colloidal 'Activity' of Clays," *3rd ICSMFE,* Vol. 1, p. 57.

——— (1954) "The Pore Pressure Coefficients *A* and *B*," *Geotechnique,* Vol. 4, No. 4.

——— (1957) Discussion of "The Planning and Design of the New Hong Kong Airport," *Proceedings of Institution of Civil Engineers,* Vol. 7, p. 305.

——— (1961a) "Effective Stress in Soils, Concrete, and Rock," *Proceedings of Conference on Pore Pressure and Suction in Soils,* London, p. 4.

——— (1961b) "Horizontal Stresses in an Over-consolidated Eocene Clay," *5th ICSMFE,* Vol. 1, p. 351.

——— (1964) "Long-Term Stability of Clay Slopes," *Geotechnique,* Vol. 14, No. 2.

———, and BISHOP, A. W. (1954) "Soils," Chapter 10 of *Building Materials, Their Elasticity and Plasticity,* ed. M. Reiner, North-Holland Publishers, Amsterdam.

———, and BJERRUM, L. (1957) "A Contribution to the Settlement Analysis of Foundations on Clay," *Geotechnique,* Vol. 7, No. 3.

———, and NORTHEY, R. D. (1952) "The Sensitivity of Clays," *Geotechnique,* Vol. 3, No. 1, Mar.

———, PECK, R. B., and MACDONALD, D. H. (1955) "Settlement Analyses of Six Structures in Chicago and London," *Proceedings of Institution of Civil Engineers,* Pt. 1, Vol. 4.

———, and V. A. SOWA (1963) "The Behaviour of Saturated Clays During Sampling and Testing," *Geotechnique,* Vol. 13, No. 4.

SOKOLOVSKI, V. V. (1954) *Statics of Soil Media* (translated from Russian by D. H. Jones and A. N. Schofield), Butterworths, London, 1960.

SODERMAN, L. G., KIM, Y. D., and MILLIGAN, V. (1968) "Field and Laboratory Studies of the Modulus of Elasticity of a Clay Till," *Highway Research Record No. 243.*

SOWERS, G. B., and SOWERS, G. F. (1967) *Introductory Soil Mechanics and Foundations,* Macmillan, New York.

SOWERS, G. F. (1962) "Shallow Foundations," Chapter Six from *Foundation Engineering,* ed. G. A. Leonards, McGraw-Hill, New York.

———, and VESIC, A. B. (1962) "Vertical Stresses in Subgrades Beneath Statically Loaded Flexible Pavements," *Highway Research Board Bulletin No. 342.*

SPANGLER, M. G. and MICKLE, J. L. (1956) "Lateral Pressures on Retaining Walls Due to Backfill Surface Loads," *Pressure-Deformation Measurements in Earth, Highway Research Board Bulletin 141.*

SWARTZENDRUBER, D. (1962a) "Modification of Darcy's Law for the Flow of Water in Soils," *Soil Science,* Vol. 93.

——— (1962b) "Non-Darcy Behavior and Flow Behavior in Liquid-Saturated Porous Media," *Journal of Geophysical Research,* Vol. 67, No. 13.

——— (1963) "Non-Darcy Behavior and Flow of Water in Unsaturated Soils," *Proceedings of Soil Science Society of America.*

TANIGUCHI, T. (1967) "Landslides in Reservoirs," *Proceedings of Asian Regional Conference on Soil Mechanics and Foundations Engineering*, p. 358.

TAYLOR, D. W. (1937) "Stability of Earth Slopes," *Journal of Boston Society of Civil Engineers*, Vol. 24, No. 3, (also in *Contributions to Soil Mechanics, 1925–40*, Boston Society of Civil Engineers).

——— (1940) "Research on Consolidation of Clays," Massachussetts Institute of Technology, Department of Civil and Sanitary Engineering, Serial 82, Aug.

——— (1948) *Fundamentals of Soils Mechanics*, Wiley, London.

TENG, W. C. (1962) *Foundation Design*, Prentice-Hall, Englewood Cliffs, N.J.

TERZAGHI, K. (1923) "Die Berechnung der Durchlässigkeitsziffer des Tones aus dem Verlauf der Hydrodynamischen Spannungserscheinungen," *Sitzungsberichte, der Akademie der Wissenschaften in Wien*, Mathematischnaturwissenschafliche Klasse, Part II a, Vol. 132, No. 3/4.

——— (1924) "Die Theorie der Hydrodynamischen Spannungserscheinungen und inr Erdbautechnisches Anwendungsgebiet," *Proceedings of First International Congress for Applied Mechanics*, p. 288.

——— (1925) "Principles of Soil Mechanics," *Engineering News-Record*, Vol. 95, Nos. 19–23, 25–27.

——— (1934) "Die Ursachen der Schiefstellung des Turmes von Pisa," *Der Bauingenieur*, Vol. 15, Nos. 1/2, 7/8, 11/12.

——— (1936) "A Fundamental Fallacy in Earth Pressure Computation," *Journal of Boston Society of Civil Engineers* (also in *Contributions to Soil Mechanics, 1925–1940*, Boston Society of Civil Engineers).

——— (1943) *Theoretical Soil Mechanics*, Wiley, New York.

——— (1950) "Mechanics of Landslides," in *Application of Geology to Engineering Practice*, Berkey Volume, Geological Society of America, p. 83.

——— (1951) "The Influence of Modern Soil Studies on the Design and Construction of Foundations," *Proceedings of Building Research Congress*, London, Div. 1, Part III, p. 139.

———, and FROLICH, O. K. (1936) *Theorie der Setzung von Tonschichten: eine Einführung in die Analytische Tonmechanik*, Deuticke, Leipzig.

———, and PECK, R. B. (1948) *Soil Mechanics in Engineering Practice*, Wiley, New York.

———, and PECK, R. B. (1967) *Soil Mechanics in Engineering Practice*, 2nd ed., Wiley, New York.

THURSTON, C. W., and DERESIEWICZ, H. (1959) "Analysis of a Compression Test of a Model of a Granular Medium," *Journal of Applied Mechanics*, Jun.

TOMLIN, G. R. (1966) "Seepage Analysis Through Anisotropic Soils By Computer," *Geotechnique*, Vol. 16, No. 3.

Tribunal Report (1967) "Report of the Tribunal Appointed to Inquire into the Disaster at Aberfan on October 21st, 1966," H. M. Stationery Office, London.

USCE Report (1951) "Investigations of Pressure and Deflections for Flexible Pavements," U.S. Army Corps of Engineers, Waterways Experiment Station, Technical Memorandum No. 3–323.

USCE Report (1953) "The Unified Soil Classification System," U.S. Army, Corps of Engineers, Waterways Experiment Station, Technical Memorandum No. 3–357.

VARNES, D. J. (1958) "Landslide Types and Processes," Chapter 3 in *Landslides and Engineering Practice*, ed. E. E. Eckel, Highway Research Board Special Report 29, p. 20.

WACK, B. (1961) "Distribution of Pressure Along Wall, Supporting a Mass of Soil," *Inzh. Sb.*, Vol. 31.

WAGNER, A. A. (1957) "The Use of the Unified Soil Classification System by the Bureau of Reclamation," *4th ICSMFE*, Vol. 1, p. 125.

WEBB, D. L. (1970) "Settlement of Structures on Deep Alluvial Sandy Sediments in Durban, South Africa," *Conference on In-Situ Investigations in Soils and Rocks*, London, p. 189.

—— (1971) "Discussion of 'Static Cone to Compute Static Settlement Over Sand,' by J. H. Schmertmann," *JSMFD, ASCE*, Vol. 97, No. SM 3.

WHITMAN, R. V., and BAILEY, W. A. (1967) "Use of Computers for Slope Stability Analysis," *SMFD, ASCE*, Vol. 93, No. SM 4.

WOODLAND, A. W. (1969) "Geologic Report on the Aberfan Tip Disaster of October 21st, 1966," *A Selection of Technical Reports Submitted to the Aberfan Tribunal*, Item 4, H. M. Stationery Office, p. 119.

WOODWARD, L. A. (1955) "Variations in Viscosity of Clay-Water Suspensions of Georgia Kaolins," *Clays and Clay Minerals*, ed. W. O. Mulligan, National Academy of Sciences, Publication 395, p. 246.

WU, T. H. (1966) *Soil Mechanics*, Allyn & Bacon, Boston.

WYLLIE, M. R. J., and SPANGLER, M. B. (1952) "Applications of Electrical Resistivity Measurements to Problems of Fluid Flow in Porous Media," *Bulletin*, American Association of Petroleum Geologists, Vol. 36.

ZAPOROZHCHENKO, E. V., and VOLODIN, Ya. F. (1964) "One Case of the Deformation of Buildings Constructed on Quaternary Clays," *Soil Mechanics and Foundation Engineering*, No. 4.

ZIENKIEWICZ, O. C. (1971) *The Finite Element Method in Engineering Science*, McGraw-Hill, London.

——, and CHEUNG, Y. K. (1965) "Finite Elements in the Solution of Field Problems," *The Engineer*, London.

——, MAYER, P., and CHEUNG, Y. K. (1966) "Solution of Anisotropic Seepage by Finite Elements," *JSMFD, ASCE*, Vol. 92, No. EM 1.

Index

Activity, 423–26, 434
ALASKAN EARTHQUAKE, 529
Anisotropic soil properties, 542
Anisotropic permeability and flow, 657–59
Atterberg limits, 431–34
consistency and, 416–23, 431

Backfill, 93
inclination, 149
lateral pressure with surface load, 600
strains in, 570–91
Bearing capacity, 436–76
analysis, 450–56
factors of influence, 459–61, 470
corrections for—
eccentricity, 462–67, 471
geometry, 461–76
inclination, 462, 465–66, 476
factor of safety, 442, 473, 476
groundwater, 470
retaining walls, 612, 617
stratified deposits, 472
Bernoulli's equation, 262–68
Bishop simplified method of slices, 559–61, 562, 581–83, 661–64
Block slide, 613
Body forces, 101, 130, 661
seepage, 269, 300
Boundary value problems, 177
Boussinesq analysis, 178–79, 180, 184, 187

CANADA (EASTERN) FLOWS, 531
Capillarity, 270–77, 300, 666
capillary pressure, 271–72, 279, 280
capillary rise, 270, 272–77
capillary stresses, 237, 280
capillary tension, 218, 496
effect of, 277–82, 660, 666
effective stress, 277–78
free energy, 262
prestress, induced, 496
shrinkage, 282
shrinkage stresses, 281
surface tension, 270–71
volumetric strain, 279

Casagrande—
determination of preconsolidation pressure, 220–22
log-time fitting method, 340–42, 351
liquid limit, test for, 416

Case studies, 1–17
ABERFAN, *coal spoil bank failure, 528*
Inquiry Tribunal Report, *531*
stability analysis, 579–82
BUILDING 301 (PORT NEWARK) *surcharging, 160–67*
loading, duration, 305
vs. magnitude of loading, 519, 521–22
settlement
calculated, 245–52
predicted, 196
and fluid flow, 258
time rate of, 347–50
shearing resistance, 354
CRIB WALL, *failure, 93–100*
analysis of, 149–54
FORT PECK DAM, *slide, 26–34, 49–51*
geologic conditions, 529
liquefaction, 538
mechanical soil behavior, 80–84
pore water pressure, 76
shearing resistance, 354
GOLDEN GATEWAY DEVELOPMENT, *foundation design, 1–2, 4–9*
and deep foundations, 166
LEANING TOWER OF PISA, *settlement, 479–81*
TRANSCONA GRAIN ELEVATOR, *foundation failure, 444–49*
failure prediction, 470–74
VAIONT RESERVOIR, *slope failure, 3, 17–24, 528, 529*
groundwater conditions, 661
slope stability, 564
WEST BRANCH DAM, *electroosmotic stabilization, 2–3, 10–16*
cohesive vs. cohesionless soils, 392
pore water pressure, 415–16
shearing resistance, 354

Clay, 160; *see also* Cohesive soils
activity, 423–26
behavior, 411–16
beneath fill, factor of safety, 568
effect of water on, 416–26
Leda clay, 392
natural, pure, 406–11
normally consolidated, 416–17
    settlement, 222
over-consolidation, 369–73
particles, 396–403
    as colloids, 398–403
    specific surface, 398–401
    structural arrangement; *see* Structural . . .
quick (liquefied), 529, 531
slope failure, 572–73
strength, 3.
Coefficient of—
active earth presure, 112, 113, 148, 592
compressibility, 215, 509
consolidation, 309, 339–46, 509
earth pressure at rest, 124, 127, 218, 590–91
friction, 36, 75, 79, 468–69, 476
permeability; *see* Permeability . . .
secondary compression, 573
uniformity coefficient, 48
wall friction, 142–44, 151, 604
Cohesion, 160
Cohesionless soils
Coulomb analysis, 591–92, 594–95, 604, 617
equilibrium equation, 154
failure of, 27–38
    criteria, 27, 36–38
infinite slope stability, 542–43
mechanical behavior of, 26–86
settlement and design procedure, 489–90, 497–500
shearing resistance; see Shearing . . .
stability analysis, 92–156
    infinite slope, 542–43
strength and total stress, 78
wall friction, coefficient of, 142–44
Cohesive resistance
active earth pressure, 592
factor of safety, 553–55
pore water pressure, 561
slope stability, 543–48
stability number, 544, 546–48, 553–54
Cohesive soils, 160
active earth pressure, 590–600
Coulomb analysis, 591–92, 594–95, 604, 617
earth pressure at rest, 590–91, 600

over-consolidation, 228–36, 361–63
sample disturbance, 278–80
saturated, 480
settlement and compressibility, 196–254
shearing resistance; *see* Shearing . . .
shrinkage; 280–82, 600
slope stability, 543–48
    curved surface, 545–84
structure; *see* Structure . . .
swelling, 600
time-dependent behavior, 599
Compaction, 51, 278
Compatibility, 172, 191
equation of, 173
Compensated foundation, 518, 525
Compressibility, 196-254
coefficient of, 215, 509
effect of sample disturbance, 217–19
Compression, 160
axisymmetric flow, 507–12
coefficient of compressibility, 215, 509
compressible stratum, 302
secondary; *see* Secondary . . .
triaxial test; *see* Triaxial . . .
virgin, 218, 357
Compression index, 218
consolidation settlement, 216, 222–24, 253, 516
liquid limit, 416–17
Compression test, triaxial; see Triaxial . . .
Conservation of—
energy, 262–68
linear momentum, 283–300
mass, 259–62
Consistency, 416
and Atterberg limits, 416–23, 431
Consolidation, 124–25, 160; *see also* Over-, Pre-, Under-consolidation
analytical solution, 328–29
anisotropic, 367–69
axisymmetric fluid flow, 507–12
boundary, 312–13, 318–19
Casagrande log-time fitting method, 340–42, 351
coefficient of, 309
    determination of, 339–46
    for radial flow, 509
compressibility, coefficient of, 215, 509
degree of, 319–28
    average degree of, 321–23
        axisymmetric flow, 509, 511
        calculation of, 335
        in Casagrande method, 340
        at critical plane, 521
        drained stratum, 327

doubly drained stratum, 328–31
  internal drainage layers, 337
  progress of consolidation, 351
  surcharging, 522
  at critical plane, 521
  duration of surcharge, 350
  time-rate of settlement, 334
dimensionless factors, 510, 511, 522
dimensionless parameters, 325–26, 351
equivalent pressure, 356–58, 366
fluid flow, 258
initial conditions, 312–13
isochrones of excess pore pressure, 329
loaded area, size of, 483–89
normally consolidated clays, 363–69,
  416–17, 421
numerical solution, 309–19, 338–39
one-dimensional, 212–45, 306–19, 440,
  522–24
  consolidation test, 213–14, 339, 347–
    49
  pore pressure, 306–8
over-consolidation ratio, 367
pore pressure, 319–20, 328, 339
pressure increment ratio, 241
  effect of, 344
saturated cohesive soils, 480
vs. strength, 375
Taylor's square-root-of-time method,
  341–46, 351
time-rate of settlement, 305–51
volume change, 215
Consolidation settlement, 212–46, 252
for BUILDING 301, 246–52
  Casagrande procedure, 229, 230
  compression index, 216, 516
  field compression curve, 222–28
  one-dimensional consolidation, 212
  preconsolidation pressure; see Precon-
    solidation . . .
  pressure increment ratio, 241
  "quasi" preconsolidation, 514–16, 524
  recompression index, 217, 516
  three-dimensional analysis, 512
Consolidation test, 213–14, 339, 347–49
Constitutive relations, 173–78
  creep compliance, 176
  homogeneity, 175
  isotropy, 175
  linear-elastic material, 174, 175, 184,
    193
  linear-viscoelastic material, 176
  Poisson's ratio, 175, 178
  shear modulus, 175
  Young's modulus, 175, 178
Coulomb analysis; see Earth pressure . . .
Creep, 160, 176, 599

Crib wall, 588–89
  failure (case study), *93–100,* 149–54
Critical void ratio, 76–78, 86
  confining pressure, 76
Culmann construction
  backfill, 151
  crib wall, 152
  retaining wall, 143
  wedge analysis, 136–44

Darcy's law, linear momentum, 283–300
Deep foundations, 2, 443, 475, 517
Degree of saturation, 40
Density
  compaction, 51
  dilatancy, 70
  grain packing, 37–38, 49, 85–86
  maximum, 43
  relative, 42–44, 469, 495
Dessication, 237
Diffusion, general equation, 300–1
Dilatancy, 68
  density, 70
  pore water pressure, 74, 75
Distortion, 160, 197, 493
Distribution of normal stresses
  circular failure surface, 548
  circular arc in a linear-elastic material,
    552
  limiting equilibrium, 105, 107, 109–18
Drainage conditions
  effect of, 567–72
  three-dimensional, 507–13
Draw-down, rapid, 664
Dry unit weight, 40
Dutch cone penetration test, 500–1

Earth pressure
  active, 106, 155
    base projection, 603–4
    bearing capacity, 455
    coefficient of active earth pressure,
      112, 113, 148, 592
    cohesive soil, 590–600
    Coulomb method, 604
    inclined plane, 119
    limiting equilibrium, 106–11
    Mohr's circle, 592
    retaining walls; see Retaining . . .
    submergence, 122–24
    wedge analysis, 130–33, 147, 564–
      65, 617
  at-rest, 106, 124–27, 155
    coefficient of earth pressure at rest,
      124, 127
    cohesive soils, 590–91, 600

Earth pressure—*Continued*
  at-rest—*Continued*
    distortion settlement, 493
    effective stress, 377–78
    limiting equilibrium, 106–9
    normally consolidated soil, 125, 369
    over-consolidated soil, 125, 127, 590–91
    retaining walls, 127, 604–10
    sample disturbance, 218, 279
  Coulomb analysis
    backfill, 151
    bearing capacity, 604–5
    cohesionless soils, 591, 604
    cohesive soils, 591–92, 594–95, 604, 617
    pressure distribution, 601
    stability analysis, 128–49
  Culmann construction; *see* Culmann . . .
  deformation, 590
  failure surfaces
    cohesive soil, 591
    force acting, 595–96
    passive pressure, 604–7
    potential, 601
    trial, 596
  horizontal pressure
    distribution of, 116
    effective, 279
  limiting equilibrium, 106–11, 118–20, 591
  passive, 106, 155
    bearing capacity, 455
    calculation, 617
    coefficient of passive earth pressure, 112, 113, 604–5
    cohesive soils, 590–91
    limiting equilibrium, 108–12
    retaining walls, 604–9, 613
    submergence, 122–24
  Rankine analysis
    assumptions, 591
    backfill-wall stress, 155
    bearing capacity, 452–55
    earth pressure coefficients, 148
    Mohr-Coulomb failure theory, 100, 591–94
    retaining walls, 603–4, 617
    shear stress, 127
    stability analysis, 110–24
    vertical plane, 603
    zone, 607
  relative movement, wall-soil, 146–49
    factors, 151–53
  resultant force, 593
  strain, effect of, 590

  submergence, effect of, 120–24, 144–46, 156
  tensile stresses, 592
    formation of cracks, 596, 617
    zone of, 594
  wall friction, coefficient of, 142, 151
  yielding, 106–9, 120
Earthquakes, 438–39
Effective grain size, 290
Effective normal load, effect on strength, 71
Effective stress, 72–74, 76, 160
  capillarity, 277–78
  compressibility, 215, 218
  consolidation, 305, 308
  earth pressures, 122, 377–78, 592
  equivalent to total stress, 78
  preconsolidation, 237
  quick condition, 267
  seepage, 258, 266, 661
  settlement, 491, 492
  soil sampling, 244, 280
  slope stability, 561, 572
  strength of soil, 51, 73
    Hvorslev parameters, 387
    over-consolidated clays, 369–73, 388
    pore water pressure, 78, 86
    remolded, 393
    vs. total stress, 79
    unconsolidated-undrained test, 373–80
Effective unit weight, 144–46
  elastic constants, 199–212, 252, 253, 359, 499, 500
  evaluation of, 208–11
Elastic media, stresses in, influence diagrams, 671–717
Elastic settlement; *see* Immediate . . .
Elasticity analysis, 176, 198–208, 359
  elastic constants, 209–10
  volumetric strain, 359, 360
Electric analogue, 634–35
Electrochemical forces, 3, 396–98
Electron micrograph, 399–402, 408–10
Electroosmosis, 2–3, 354, 392, 413–16, 431
Embankments, stresses due to, 186–90, 193
Equilibrium
  analysis, 127–28
  equations, 100–2, 127
    cohesionless soils, 154
    solutions of, 103
    three dimensions, 168, 191
  limiting; *see* Limiting . . .
Excavations, 437
  of a cut in clay, safety factor, 570

seepage into, 632
stress changes, 190, 192, 193
Exploration, 447
  field sampling, 209
  subsurface, importance of, 473–74

Factor of safety
  bearing capacity, 442, 473, 476
  clay beneath fill, 568
  cohesion, 553–55
  excavation of a cut in clay, 570
  friction, 553–55
  homogeneous soils, 553–55
  infinite slope stability, 542–43
  method of slices, 557–60
  sliding, 150, 562
  trial failure circle, 583
  varying materials, 573
Failure, pore pressure, 78
Failure, point of, 160
Failure, progressive, 354, 581, 584
  effect of, 573–78
  over-consolidation, 576–77
Failure of a retaining wall, 93–100, 149–54
Failure surface, 136, 155
  curved, 545–48
  plane, 160, 543–45
  stresses, 131
Failure theory; see Mohr-Coulomb . . .
Falls, 528–30, 581
Faults, underground, 438
Fellenius method of slices, 562–63, 584
Field testing
  elastic constants, 210, 211
  plate load test, 253, 473
  sampling, 209
Filters, 664–67
Finite difference analysis, 309–19, 636
Flow
  and anisotropic permeability, 657–59
  confined, 627, 637–44, 666
  porous media, 258–301
  slide, 394–95
  slope failure, 392–95, 528, 580–81
  unconfined, 627–28, 644–47, 666
Flow line and permeability, 644
Flow nets, 636
  construction of, 637–60
  hydraulic gradient, 650
  seepage, 636–60, 663, 664, 666
    quantity of, 647
    seepage forces, 647, 651
  water pressure, 647, 650, 651
Fluid flow and porosity, 258–301
Footings, 443–44
  on sand, 489–504

size and depth, 494
Foundations, 436–44, 474, 475; see also
    Shallow . . .
  beneath a fill, 568
  compensated, 518, 525
  deep, 2, 443, 475, 517
  design, 438–42, 475
  design model, 440–42
  types, 442–44
Free energy, 262–69
Friction
  base friction, 613
  coefficient of, 36, 75, 79, 468–69, 476
  factor of safety, 553–55
  homogeneous soils, 548–53
Frost action, 436–37
  expansive pressures, 600
  retaining walls, 615, 617

Geologic conditions, Fort Peck Dam, 529
Geometry and loading effects, 461–67
Grain shape, 37, 51, 85
Grain size, 44–45
  distribution, 44–51, 86
    cumulative   frequency–distribution
        curve, 45
    distribution curves, 47–49
    effective grain size, 49
    frequency–distribution curve, 45
    gap graded, 45
    histogram, 45, 46
    poorly graded, 45
    sieve analysis, 44, 45
    uniformity coefficient, 49
    well-graded soil, 45
  equivalent diameters, 45, 48
Grain-to-grain interaction, 37–38
Granular soils, properties, 142
GREENSBORO (FLORIDA), SLIDE, 531, 534–
    35, 538
Groundwater, 437–38
  table, 120, 121, 470, 597

Half-space, surface displacements, 199–
    202
Head, 620–21
Hele-Shaw model, 636
Hooke's law, 174
Hvorslev envelope, 372
Hvorslev hypothesis, 355–59
  parameters, 358–59, 380, 384, 387, 388
  shearing resistance, 355–59, 363–64,
    369, 372–73, 387
Hydraulic fill, 26, 529
Hydraulic gradient, 263–64
  compressible stratum, 307
  Darcy's law, 284, 288

Hydraulic gradient—Continued
  flow nets, 650
  laminar flow, 413
  seepage, 620, 650–53, 665
  seepage force, 269

Immediate (elastic) settlement, 160
  cohesive soils, 198–212, 522
  components of, 197
  footings on sand, 489–504
  foundation, 252, 480
  laboratory testing, 252
  size of loaded area, 482
  surcharged, 245–46
Influence diagrams for stresses in elastic
  media
  influence diagrams—elastic half-space
    circular loaded area—vertical nor-
      mal stress under centerline, 673,
      679
    elastic embankment—horizontal nor-
      mal stresses along selected vertical
      sections, 687–91, 694
    elastic embankment—magnitude and
      location of maximum $(\tau_{max}/\gamma H)$,
      692–94
    elastic embankment—vertical nor-
      mal stresses along selected vertical
      sections, 681–86
    long excavation—change in shear
      stress along selected vertical sec-
      tions, 711–17
    long excavation—reduction in hori-
      zontal normal stress along selected
      vertical sections, 704–10
    long excavation—reduction in verti-
      cal normal stress along vertical
      sections, 697–703
    point load on surface—vertical nor-
      mal stresses, 671–72
    uniformly loaded area—vertical nor-
      mal stresses at a point within the
      half space (Newmark Chart), 675,
      679–80
    uniformly loaded circular area—ver-
      tical normal stress at a point
      within the half space, 674, 679
    uniformly loaded rectangular area—
      vertical normal stresses beneath
      corner, 676–78, 680–81
  stress
    changes due to excavation, 696–717
    self-weight, 681–717
    surface-load, 671–81
Influence values, 179, 181
  Newmark Chart, 181
  stresses in an earth mass, 179, 193

stresses in elastic media, 671–717
Intergranular stress, 72
Ion exchange, 403–5

JAPAN, FLOWS, 531

Kozeny-Carman equation, 285–92

Laboratory testing, 209
  immediate settlement, 252
  triaxial tests, 380
Laplace's equation, seepage, 300, 622–27,
  637, 658, 665
Layered systems
  effects of, 184–86, 193
  surface displacement, 202–8
Leda clay, 392
Limit design, 450
Limiting equilibrium, 104–18, 343
  active earth pressure, 106–11
  in analysis of earth pressure, 118–20,
    591
  development of, 106–9
  displacements, 107
  earth pressure at rest, 106–9
  generalized analysis, 127–28
  Mohr's circle, 104–5, 109, 118
  Mohr-Coulomb failure theory, 104–5,
    127, 155, 591
  stress distribution, 105, 107, 109–18
Linear-elastic problems, solutions to,
  178–90
Liquefaction, 392–93, 531, 538
Liquid limit
  Atterberg limits, 431–34
  Casagrande test for, 416
  compression index, 416–17
  equivalent consolidation pressure, 357
  plasticity, 421
  shear strength, 423
Liquidity index, 422–23
Loading effects, geometry and, 461–67;
  see also Case studies—BUILDING 301

Mineral types, 37, 85
Mohr's circle, 54–59
  active earth pressure, 592
  limiting equilibrium, 104–5, 109, 118
  pole method, 56–58, 118
  saturated undrained test, 74
  shearing resistance, 358, 363–66, 369–
    73, 380–83, 385
  triaxial compression test, 67
  two-dimensional stress, 62
Mohr-Coulomb failure theory, 58–60,
  64n, 86
  Coulomb analysis, 594

failure surfaces, 60, 64
limit design, 450
limiting equilibrium, 104–5, 127, 155, 591
φ-circle, 548
Rankine analysis, 100, 591–94
shearing resistance, angle of, 58, 59
slope stability, 548
total stresses, 78–79
two-dimensional stability, 168
wedge analysis, 591

Negative pore pressure, 74, 79, 86, 218, 373
Nonlinear soil behavior, 177
Nonlinear stress-strain behavior, 566
Normal loading approximation, 187
Normally consolidated soils, 218–21
clay, 363–69, 416–17, 421
coefficient of earth pressure at rest, 124, 127, 218, 590–91
earth pressure at rest, 125, 369
horizontal effective earth pressure, 279
relative settlement, 496
shearing resisstance, 355–56, 361, 363–69, 372, 377

Octahedral stresses, 359, 491–92
One-dimensional compression, 213, 253, 524; see also Compression
One-dimensional consolidation; see Consolidation, Consolidation settlement
Over-consolidation, 124–25
assumption of, 524
capillarity-induced prestress, 496
clays, 369–73
cohesive soils, 228–36
earth pressure at rest, 125, 127, 590–91
coefficient of, 218
factors, 237
horizontal stresses, 577
Hvorslev envelope, 372
preconsolidation pressure, 219
progressive failure, 576–77
settlement prediction, 524
shearing resistance, 355-57, 361, 369–73, 388
Over-consolidation ratio, 124–25, 490–91
cohesive soils, 361–63
effect of, 374
footing size and depth, 494
normally consolidated soil, 367
Overturning, 613–14, 617

Permeability, 283–99
coefficient of, 283–84

anisotropic materials, 296–300, 657–59, 666
closed form solution, 308
Darcy's law, 620
electroosmotic, 415
horizontal, 509
isotropic, 657, 658
measurement of, 292–96
rate of seepage, 658
typical and common values of, 290–92
Darcy's law, 283–300
equivalent coefficent of, 658
flow line, 644
inhomogeneous, 296–300
Kozeny-Carman equation, 285–92
tests, 293–95
Phase relations, 39–44
phase diagrams, 39–42, 86–87
relative density, 42–44
Piezometers, 265, 512, 634
Plane; see Principal . . .
Plastic limit, 416, 417, 430
Plasticity chart, 419, 421
Plasticity index, 418, 421, 424, 431, 434
Pole method, 56–58, 118
Pore pressure, 70–76, 359–63
cohesive resistance, 561
consolidation, 319–20, 328, 339
dilatancy, 74, 75
effect of volume change, 71–76, 78–79, 86
electroosmotic stabilization, 415–16
excavation of a cut in clay, 570
excess, 313–14, 320, 505, 508, 524
at failure, 78
foundation beneath a fill, 568
negative pressures, 74, 79, 86, 218, 373
one-dimensional consolidation, 306–8
seepage, 619, 647-52, 661
shearing resistance, 359–63
slope stability, 584
analysis, 561–62, 580, 584
Porosity, 41, 43
fluid flow, 258–301
Precompression, with sand drains, 507–12
Preconsolidation, apparent or "quasi," 237–40, 253
effect of, 514–16, 524
Preconsolidation pressure
apparent or "quasi," 237–41, 253
determination of, 218–22, 253
over-consolidation, 219
profiles, 237
Preloading, 245–46, 305, 519; see also Case studies—BUILDING 301

Principal planes, 56, 62, 103, 592
Principal stresses, 56, 62, 64, 377, 388, 491, 568
Progressive failure, 354, 573–78
    over-consolidation, 576–77

Quick (liquefied) clays, 529, 531
Quick condition, 267–68, 652, 664–65

Rapid draw-down, 664
Rankine analysis; see Earth pressures . . .
Recompression index, 217, 218, 229, 236, 253, 516
Recurrence formula, 309–12
Relative density, 42–44, 469, 495
Remolding, 218, 392, 411, 419, 422
Residual strength, 581
Retaining walls, 82, 587–617
    adhesive force, 595, 601, 608
    backfill
        inclination, 149
        lateral pressure with surface load, 600
        strains in, 590–91
    base friction, 613
    bearing capacity, 612, 617
    cantilever wall, 588, 589, 603
    coefficient of wall friction, 142–44, 151, 604
    crib wall, 588–89
        failure (case study), 93–100, 149–54
    design of, 587–88
    earth pressure
        active, 118–24, 127, 129–33, 148, 591–600
        and base projection, 603–4
        passive, 133–36, 604–10
        at rest, 127, 604–10
        resultant force, 600–2
        earth pressure analysis, 118–20
    earth pressure distribution, 600–3
        and magnitude, 590–610
    factor of safety, sliding, 613
    failure of, 93–100, 149–54
    flexible structures, 588–90, 617
    frost action, 615, 617
    gravity wall, 588–89
    inclined wall, 149, 150
    key, 613–14
    movement, relative wall-to-soil, 146–49
    overturning, 613–14, 617
    rigid structures, 588–89, 617
    settlement, 615–17
    sliding along base, 612–13, 617
    slope stability, 614, 617

soil deformation, 599–600, 604
stability analysis, 610–14
submergence, effect of, 120–24, 144–46
tension cracks, 592, 594, 596, 617
types of, 588–590
water pressure force, 597

Safety factors; see Factor . . .
Sample disturbance, 278–80
    effect on compressibility, 217–19
    settlement, 210
    settlement analysis, 242–45
Sand
    boils, 619, 652
    dense, 68–70, 74–76
    distortion, 489–504
    loose, 68–70, 74–76
    precompression, 507–12
Sand drains, 519, 524
    axisymmetric flow, 507–12
Saturation, degree of, 40
Scour, 438
Secondary compression, 197, 239, 252, 480, 513–14
    coefficient of, 513
Seepage, 258–301, 619–667
    Bernouli's equation, 259, 263, 300, 620
    body force, 269, 300
    boundary conditions, 627–29, 637, 639, 665
        confined flow, 627, 637–44, 665
        impervious boundaries, 628–29, 636, 655, 664, 665
        reservoir boundaries, 628, 636, 665
        unconfined flow, 627–28, 644–47, 666
    capillary effects, 660, 666; see also Capillarity . . .
    continuity, 259, 300, 620
    Darcy's law, 283–300, 313, 620, 622
        anisotropic material, 299
    diffusion equation, 300–1
    discharge velocity, 261, 620
    equipotential lines, 623–25, 665
        effect of boundaries, 642–43
        flow net, 636–37, 658
        head loss, 638–39, 646
        intersecting, 640–42
        reservoir boundaries, 628
        seepage force, 652
    filters, 664–67
    flow into excavation, 632
    flow nets, 636–60, 663, 664, 666
    flow rate, 652
    flow velocity, 624, 652
    fluid flow through porous media, 258–301

force, seepage, 101, 130, 268–300, 620, 647, 651, 661
free surface, 619, 644
hydraulic gradient, 620, 650–53, 665
Laplace's equation, 300, 622–27, 637, 658, 665
line of, 644–47, 652–57, 666
piping, 652, 664–65
pore water pressure, 619, 647–52, 661
quantity of, 620, 625–26, 647–52, 658
quick condition, 652, 664–65
solution procedures, 629–37
  approximate, 636–37
  closed-form, 629–33
  models and analogues, 633–36
stability, effect on, 660–64, 666–67
steady-state, 619–67
streamlines, 623–26, 665
  effect of boundaries, 642–43
  flow net, 636–37, 658
  reservoir boundaries, 628
surface of seepage, 664, 666
three-dimensional flow, 260
transformed section, 658
transient seepage, 660, 664, 666
velocity, 261
Seismic probability, 439
Sensitivity, 392, 411–12, 422, 430
Settlement; see also Case studies—BUILD-ING 301
cohesionless soils, 489–90, 497–500
cohesive soils, 196–254
components of, 197–98
consolidation; see Consolidation . . .
and design procedure, 497–500
distortion, sands, 489–504
effective stress, 491–92
elastic distortion, 482–83
limiting settlements, 516–17, 525
loaded area
  shape of, 504
  size of, 482–97, 522
normally consolidated clays, 222
prediction of, 222–28
over-consolidation, 524
reduction of, 517–21
relative, 496
retaining walls, 615–17
sample disturbance, 210, 242–45
secondary compression; see Secondary . . .
shallow foundations; see Shallow . . .
structural distortion, 196, 516–17, 525
three-dimensional, 504–7, 524
  general effects, 512–13
time rate of settlement, 197–98, 252, 258, 305–51

Shallow foundations
  bearing capacity, 436–76, 475–76
  and settlements, 497–525
Shear failure, 35–36, 42, 529
Shear strength, 58, 72–76, 421–23, 441
  liquid limit, 423
  undrained, 374, 387, 594
Shear stresses, 52, 58, 127
Shearing resistance
  angle of, 36, 38
    density of packing, 51
    factors of influence
      grain-size distribution, 49, 51
      grain-to-grain interaction, 37–38
      granular soil, 142
      relative density, 42–44
    granular soil, 142
    Mohr-Coulomb theory, 58, 59
    slope stability, 548, 576, 583
  cohesionless soils, 27, 58, 74, 76
    angle of, 36
    bearing capacity, 468–70
    effect of water, 67–70, 79
    factors of influence, 36–38
    horizontal stress, 108
    total stresses, 78–80
  cohesive soils, 354–88
    equivalent consolidation pressure, 356–58, 366
    Hvorslev hypothesis, 355–59, 363–64, 369, 372–73, 387
    Hvorslev parameters, 358–59, 380, 384, 387, 388
    Mohr's circle, 358, 363–66, 369–73, 380–83, 385
    normally consolidated, 355–56, 361, 363–69, 372, 377
    over-consolidated, 355–57, 361, 369–73, 388
    pore water pressure, 359–63
    Skempton "A-factor," 361–63, 366–67, 369–71, 385
    testing
      consolidated drained test, 365, 367, 383–84
      consolidated undrained test, 365–67, 384–87
      unconsolidated undrained, 373–80
  water content, effect of, 67–70
Shrinkage, 272, 280–82, 437, 600
Shrinkage limit, 280–82
Sieve analysis, 44
Silt, 160
Site conditioning, 438–39, 447
Skempton "A factor," 505–7
  shearing resistance, 361–63, 366–67, 369–71, 385

Skempton-Bjerrum correction, 361–63, 366–67, 369–71, 385, 504–7
Slides, 528, 529, 533, 581
Slope failure
  causes of, 538, 540–41, 583
  clay soils, 572–73
  creep, 160, 176, 528, 599
  falls, 528–30, 581
  flows, 392–95, 528, 529–38, 580–81
  progressive failure, 354, 581, 584
    effect of, 573–78
  slides, 528, 529, 533, 581
Slope stability, 528–84
  cohesive resistance, 543–48
  effect of different materials, 573
  effective stress, 561, 572
  infinite slope, 542–43
  pore pressure, 584
  and retaining walls, 614, 617
  shearing resistance, 548, 576, 583
  stress-strain behavior, 573
Slope stability analysis, 538–84
  assumptions of, 538–42
  cohesionless soil, infinite slope, 542–43
  cohesive soil failure, 543–48
    curved surface, 545–84
    plane surface, 543–45, 547
  effects of—
    drainage conditions, 567–72
    pore water pressure, 561–62, 580, 584
    progressive failure, 573–78
    rapid draw-down, 664
    seepage, 660–64, 666–67
  factor of safety, infinite slope, 542–43
  failure surfaces, 573
    composite, 563–65, 583
  flow slide, 394–95
  homogeneous soils possessing friction, 548–53
  maximum shear stress method, 565–66, 576, 584
  Mohr-Coulomb failure theory, 548
  ϕ-circle method, 548–53, 555
  prediction and observation, 572
  slices method, 556–63
    Bishop simplified, 559–61, 562, 581–83, 661–64
    Fellenius method, 562–63, 584
    generalized, 556–59
  slip circle analysis, 581
  slope with a load, 571
  stability number, 544, 546–48, 553–54
  strength, residual, 575–78
  strength parameters, selection of, 567–78, 584
  Taylor's stability chart, 549, 553, 554
  tension cracks, 565, 592, 594, 596, 617

Soil classification, 426–29
  systems of, 426–27
    Unified Soil Classification System, 427–29, 432–33
Soil exploration, 447
  field sampling, 209
  subsurface, importance of, 473–74
Soil mechanics, 17
Specific gravity, 40
Stability analysis
  cohesionless soils, 92–156
  effect of seepage, 660–64, 666–67
  retaining walls, 610–14
  slopes, 538–42; see also Slope . . .
Stability number, 544, 546–48, 553–54
Standard penetration test, 468–69, 497n, 501
Strain, 51–80, 168–73
  compatibility with displacement, 172–73
  distribution, 499, 500
  normal, 169, 176, 191
  plain, 450
  shear, 171–72, 191
  small, 169–72, 178
  volumetric, 214–15, 262, 279, 359, 360
Strength, 51–80
  vs. consolidation, 375
  clay soils, 3
  parameters, selection of, 567–78
  residual, 581
  shear, 58, 72–76, 421–23, 441
  strain, 553, 575–76
  total stresses, 78–80
  ultimate, 69
  volumetric, 214–15, 262, 279, 359, 360
Stress, 51–80, 167–68, 191, 380–87
  capillary, 237, 280
  definition, 52
  distribution, limiting equilibrium, 105, 107, 109–18
  excavations, 190, 192
  failure surface, 131
  Mohr circle of; see Mohr's circle . . .
  normal, distribution, 548, 552
  octahedral, 359, 491–92
  at a point, 51–53, 62
  principal directions, 56
  principal stresses, 56, 62, 64, 377, 388, 491, 568
  shear stresses, 52, 58, 127
  sign convention, 55
  state of, 62, 64
  total, 73–74, 78–80, 371
Stress path, 380–87, 388
  method of analysis, 507
Stress-strain relationship, 68, 573
  nonlinear behavior, 566

Stresses in elastic media, *influence diagrams,* 671–717
Stresses within an earth mass, 160–93
  approximate solution, 186–88, 193
  deformation, 168, 191
  due to—
    embankments, 186–90, 193
    excavations, 190, 192, 193
  influence values, 179, 193
  layered systems, effects of, **184–86, 193**
  Newmark influence chart, 181
  pressure bulb, 181
  solutions to linear elastic problems, 178–90
  solutions by superposition, 180
  stress contours (isobars), 183
  stress distribution, 167, 181–86, 193, 487, 548
    approximate, 186
Structural arrangement of clay particles
  disperse structure, 405–6, 409, 412, 430
  fabric, 406–11
  flocculant structure, 405–6, 412, 430–31
Structure of cohesive soils, 392–434
  double-layer theory, 402–3, 411, 413, 430
  electrochemical forces, 396–98, 427, 430
  ion exchange, 403–5, 430
Submerged unit weight, 597
Submergence, effect of, retaining walls, 120–24, 144–46
Subsurface exploration, 473–74
Superposition (linearity), 174
  solutions by, 180–81
Surcharging, 245–46, 305, 519; *see also* Case studies, BUILDING 301
Swelling of soils, 437, 600

Taylor's square-root-of-time method, 341–46, 351
Tensile stresses, 592
Tension cracks, 565
  retaining walls, 592, 594, 596, 617
Three-dimensional drainage, 507–13
Three-dimensional settlement analysis, 504–7, 512, 524
Thixotropy, 412–13, 430
Time-dependent response, 177, 599
Time rate of settlement, 197–98, 252, 258, 305–51
Total stress, 73–74, 371
  Mohr-Coulomb failure theory, 78–79
  strength of cohesionless soils, 78
Triaxial compression test, 60–64, 86
  deformations, 62

  drained test, 68, 74, 76
  elastic constants, 209–10
  linear elastic solid cylinder, 64–65
  Mohr's circle, 67
  nonlinear elastic cylinder, 64
  on sand, 67, 73
  state of stress, 62, 64
  stress paths, 383–87
  undrained test, 68, 74
  volume change, 77

ULLENSAKER (NORWAY) FLOW, 531
Unconfined compression test, elastic constants, 209
Unconsolidated-undrained test, 373–80
Under-consolidation, 241–42, 253
Undrained shear strength, 374, 387, 594
Unified Soil Classification System, 427–29, 432–33
Unit weight, 38, 40–42, 85, 101
  submerged, 597
  of water, 40
Ultimate strength, 69

van der Waals forces, 396
VIBSTAD (NORWAY), SLIDE, 529
Virgin compression, 218, 357
Viscoelastic analysis, 176
Void ratio, 40
  critical, 76–78, 86
  maximum and minimum, 43, 49, 50
Volume change, 218, 258
  consolidation, 215
  dense sand, 68–70, 74–76
  effect of initial void ratio, 77
  effect on pore pressure, 71–76, 78–79, 86
  loose sand, 68–70, 74–76
  triaxial compression test, 77
  zero volume change, 76
Volume compressibility, 211
Volumetric strain, 214–15, 262, 279, 359, 360

Wall friction, coefficient of, 142–44, 151, 604
Water content, 40
  effect of on shearing resistance, 67–70
Water pressure, 647, 650, 651
  force, retaining walls, 597
Wave action, 438
Wedge analysis, 128–49, 155, 600
  active earth pressure, 130–33, 147, 564–65, 617
  Mohr-Coulomb failure theory, 591

Young's modulus, 175, 178
  equivalent, 500–3